Geometrical Investigations

Illustrating the Art of Discovery in the Mathematical Field

The royalties from this book have been donated to

COMMUNITY AID ABROAD,

An Australian-based agency working through
practical-aid local community programs
in developing countries
and with Australian aborigines.

CAA is the Australian Associate of OXFAM.

Headquarters:
75 Brunswick Street, Fitzroy, Australia, 3065.

Geometrical Investigations

Illustrating the Art of Discovery in the Mathematical Field

John Pottage

Department of History and
Philosophy of Science
University of Melbourne
Victoria, Australia

Foreword by
Stillman Drake

1983

ADDISON-WESLEY PUBLISHING COMPANY, INC.
Advanced Book Program/World Science Division
Reading, Massachusetts

London • Amsterdam • Don Mills, Ontario • Sydney • Tokyo

Library of Congress Cataloging in Publication Data
Pottage, John, 1924-
Geometrical investigations.

Bibliography: p.
Includes index.
1. Mathematics—Philosophy. 2. Mathematics—History. I. Title.
QA8.4.P67 1983 510'.1 82-11357
ISBN 0-201-05733-6

American Mathematical Society (MOS) Subject Classification Scheme: 00-00, 00-01, 00-A25, 01-00, 01-A15, 01-A20, 01-A45.

Manufactured in the United States of America

ABCDEFGHIJ-HA-898765432

INDAGATIONES GEOMETRICAE
QUIBUS
ARS INVENIENDI IN RE MATHEMATICA
ILLUSTRATUR

GEOMETRICAL INVESTIGATIONS

ILLUSTRATING
THE ART OF DISCOVERY
IN THE MATHEMATICAL FIELD

A DIALOGUE IN THE GALILEAN STYLE

in which *Salviati, Sagredo,* and *Simplicio,*
Being Returned from the Shades to Discuss a Problem
Concerning PERIMETER, AREA, and VOLUME RATIOS
are Thereby Led to Examine in an Elementary Fashion
The RECTIFICATION of the ELLIPSE
and the Nature of OVALS OF UNIFORM BREADTH
and to Touch upon the PROBLEM OF ISOPERIMETRY:
The Whole Providing the Modern Reader
with an Introduction to Mathematical Heuristic
in general,
and especially to
Certain Classical Modes of Geometric Thought
(Supplemented by Occasional Items of
Elementary Algebra & Infinitesimal Calculus,
and Utilizing a Transparent & Convenient Symbolism)
With the addition of Numerous Annotations and Problems
by the Author

John Pottage

School of History and Philosophy of Science
University of Melbourne

MCMLXXXII

To DOROTHY
and to all who are endeavoring
to promote forms of education
that will sustain
and not subjugate
the human spirit

CONTENTS

FOREWORD

by Stillman Drake

Heuristic is the name of an art—the art of discovery—which we all greatly admire and would like if possible to master. Yet it is an art whose pursuit, in contrast with its products, is extraordinarily difficult to pin down. If there is a logic of discovery in the physical sciences, it is probably that of mathematics rather than that of the syllogism, as Galileo had long remarked before he had Simplicio say in *Two New Sciences:* "Truly I begin to understand that although logic is a very excellent instrument to govern our reasoning, it does not compare with the sharpness of geometry in awakening the mind to discovery."

Books have been written recounting the history of some actual discoveries, which on the whole have been more illuminating than attempted logical analyses of the heuristic art. But histories necessarily suffer from a kind of particularity that shows us only how past discoveries were reached and not how we are to go on to new ones. There is also a tendency, in writing up past discoveries, to select a few of great consequence, with the result that what is reported seems far beyond our powers, and thus we are confronted with an inspirational literature and not with instruction we can put to use.

Dr. Pottage offers us an introduction to mathematical heuristic in general, based on classical modes of geometric thought. The approach is novel and, to me at least, it is a welcome relief from the customary abstract analyses now usually employed in the attempt to formulate some theory of procedure. The problem used as an example is one easily understood by nonmathematicians; its complete solution appears very obscure at the outset; the lines of attack taken up are generally simple and plausible, and as in real life some are successful and others illusory. In presenting them Dr. Pottage adopts the dialogue form utilized by Galileo, who was a master of the art of discovery. There is also another good reason for this choice, since Galileo likewise wrote at a time when abstract analyses had replaced, among scholars, simple and direct methods of approaching problems in the fields of particular interest to him, namely, physics and astronomy. A few words about his situation are appropriate here.

Aristotle's magnificent scheme to weld all knowledge into a single integral system had so captivated medieval thinkers that attention to particular problems of science had been replaced by ingenious reconciliations between classes of phenomena and interpretations of basic Aristotelian principles. Galileo's piecemeal approach—his concentration on separate and well-defined problems for particular scrutiny—resulted in discoveries ranging from the law of free fall to the existence of previously unknown heavenly bodies, discoveries that could neither be fitted into the overall scheme nor yet made the basis of some new and equally comprehensive closed system of all knowledge. Accordingly they were rejected by defenders of the unity of philosophy and science, who likewise rejected his conception that knowledge supported only by "sensate experiences and necessary demonstrations" was to be cherished in science whether or not it entailed any general system of all human knowledge.

It seems to me that there is a striking parallel between such a state of affairs and our present situation with respect to heuristic. Discovery in mathematics has its elements of difference from discovery in science, and again from discovery in the arts; nevertheless, the fact that we are still far from having any general system of heuristic procedures presently enhances the value of piecemeal studies of processes of discovery. This being so, the consideration of the special case, not requiring acquaintance with modern mathematics or symbolic logic, is not necessarily rendered trivial. On the contrary, let us recall Galileo's insistence that science as he conceived it be tied again and again to elementary direct observations of the most commonplace kinds, so that however limited in application that science might be, there could be no doubt of its correctness within well-defined boundaries. John Pottage's insistence that each step in his example be subjected to elementary geometrical conceptions gives us a similar guarantee that, however limited in scope the discoveries reached, the procedures used in reaching them have at all times been clear and in the end have been fruitful.

Some of the greatest discoverers in the mathematical field, including C. F. Gauss and Albert Einstein, were convinced that anyone could have done what they did by simply concentrating as hard and as long on the same problems they solved. That view seems so bizarre to the rest of us, who know very well how absurdly flattering it is, that we might dismiss it as the naïveté of genius or as false modesty. Yet we hesitate to dismiss anything said by such thinkers as being entirely without foundation. Hence we are entitled to assume that the process of discovery, at least as seen from the inside, contains no distinguishable elements that mark it off sharply from the everyday processes of problem-solving by means of good sense. Now, we should be hard put to explain how we go about that activity, though in fact we are always at it to some extent; hence it should not surprise us that even the great discoverers usually have found nothing worth saying to tell us about their "method" or the "logic" of their procedures. Such words may indeed name no more than modern counterparts of the ancient occult qualities habitually invoked by systematizers of all knowledge to explain things not really understood, but merely needed for some system, as were Nature's horror of the void and a tendency of every body to seek its natural place, in the old physics. What was apparent to Gauss and Einstein was the necessity of long and hard work in order to reach any discovery, something that is no less apparent in this book. The patient work required for any discovery may be quite disproportionate to the importance of the result, but it is always considerable. Indeed, discoveries that may seem small to outsiders may have cost their discoverers extraordinary amounts of time and effort, and they may be genuinely astonished to see them disparaged in comparison with some others that had in fact occasioned much less trouble.

Even this one rule on which some great discoverers have agreed—that nothing new is found without patience and hard work—seems to be controverted by instances in which important discoveries appear to have been stumbled upon accidentally. Examples sometimes cited in the history of science are Roentgen's discovery of X rays and Galileo's discovery of the satellites of Jupiter. Good luck has certainly played a role in most scientific discoveries; in Galileo's case it did not only in the matter of Jupiter's satellites but also in his discovery of the law of free fall. He cheerfully admitted the fact to G. B. Baliani not long after he had presented the law of free fall to the public as a rigorous deduction from a certain definition of "uniformly accelerated motion." It is only because we look at discoveries from the outside and not from the inside that some appear merely lucky and others devoid of any element of luck, or some seem simple and others very difficult. This is hardly the place to discuss the enormous difference between luckily noticing Jupiter's satellites and rationally discovering the fact that they must orbit around that planet. Perhaps it is only the latter that ranks as a proper scientific discovery, though the former (which would normally be called "discovery of the satellites") had to precede it and cannot have been planned or foreseen. Again, that a body falls from rest in times proportional to the square roots of the distances traversed was known to Galileo in 1604. That counts as a scientific discovery, though Galileo had no rigorous proof for the proportionality concerned until 1630 or thereabouts. This latter constituted a separate discovery, much more difficult for him but in a sense much less important, since the proof was of no further use once it had been found, while the law of falling bodies became a cornerstone of modern physics.

It would be pleasant to have space to compare the concept of discovery with that of invention in mathematics, but I can only invite the reader's special attention to such things. Thus one may ask whether Simplicio, at a certain point in this book, discovered the curious case of the regular hexagon "inscribed" in an equilateral triangle, rather in the way that Galileo first happened to notice certain points of light near Jupiter and neatly aligned, or whether he invented a new meaning for the word "inscribed" to bring this case within the problem under discussion. In physical science it is usually easy to distinguish discovery from invention, but it is worth noticing that the more mathematical science becomes, the less sharp the distinction—and the more important for us to have the general introduction to mathematical heuristic that follows.

ACKNOWLEDGMENTS

I wish to express my appreciation for assistance received from colleagues and others who have read all or part of the typescript. In particular, I sincerely thank Wesley Salmon, Norman Gulley, Keith Hutchison, Ken Pledger, Gordon Smith, St. John Kettle, John Barton, and Harry Kannegiesser for their suggestions and encouragement, and especially Hazel Maxian for her expert assistance in matters of translation.

My deepest gratitude is due to Professor Stillman Drake, not only for contributing the Foreword, but for reading and carefully commenting upon the penultimate draft. Substantial improvement has undoubtedly resulted wherever I have been able to act on his suggestions; naturally I am entirely responsible for the shortcomings that remain.

The project has received much appreciated support from the Research Development Fund of the Faculty of Arts, University of Melbourne. The diagrams were lovingly and meticulously prepared by my niece, Anne Pottage of the Centre for the Study of Higher Education at this university. Nonie Holman and Barry Cheney kindly assisted with the correction of the proofs.

I am grateful also to the publishers of the *Archive for History of Exact Sciences,* Springer Verlag, for permission to use again (in notes 17, 177, 178 and in Problem 29) material that first appeared in my article in vol. 12, no. 4 (1974) of that journal. Finally, I should like to add what too often goes without saying, namely how appreciative authors of Addison-Wesley Advanced Book Program books are of this company's scrupulous production of their works.

John Pottage

INTRODUCTION

The aim of this book is to present mathematics not only as a cumulative, self-correcting discipline, but also as a truly liberal art, as it was once seen to be. Although mathematics can hardly be counted among the humanities, since it lacks the element of moral concern, at least it is possible to humanize mathematics education. I have not presumed to write for professional mathematicians, though, if they are interested in the history of their subject or in the history and philosophy of science generally, they may find in these pages an easy entrance to studies adjacent to their own. Many mathematical items, with which they will undoubtedly be familiar, are provided with historical contexts.

Those who are accustomed to the austere, disciplined style of orthodox mathematical writing are warned that both in the dialogue and the notes they will find wide-ranging excursions going far beyond the immediate needs of the argument at hand. For such digressions I make no apology. They serve to connect particular facts with truths of wider significance and sometimes to suggest analogies between mathematical and other kinds of discovery or invention. In any case, I have aimed, like Fontenelle nearly three centuries ago in his famous *Conversations on the Plurality of Worlds,* to write in a manner "that would not be found too dry and insipid for the common man, nor too mean and trifling for the scholar."

Present-day students of mathematics, the sciences, engineering (and, in some places, even architecture) may find that their mathematical studies are so restricted to abstract "modern" topics that they are almost strangers to the beauties of classical geometry—to the aesthetic and practical virtues of a subject that, until a few decades ago, was regarded as central to education for their professions. This book provides an implicit demonstration, as well as an affirmation, of the continued value of much of the older material. As Thom has remarked, "the spirit of geometry circulates almost everywhere in the immense body of mathematics, and it is a major pedagogical error to seek to eliminate it."[1]

A study of the history of mathematics suggests that an adequate appreciation of the generality of theorems and the attainment of an acceptable rigor of demonstration have typically been approached only slowly via the insights of numerous mathematicians over an extended period. By now it should be abun-

[1]René Thom, "Modern Mathematics: Does It Exist," in A. G. Howson (ed.), *Developments in Mathematical Education: Proceedings of the Second International Congress on Mathematical Education,* Cambridge, England: Cambridge University Press, 1973, p. 208.

dantly clear, even to philosophers preoccupied with logical analysis and systematization, that mathematical discovery does not, except very rarely, amount merely to the working out of the consequences of explicitly stated postulates. This must always have been clear to productive mathematicians themselves, but that it continues largely to be hidden from students points to a grave deficiency in customary pedagogical methods. This deficiency along with others related to it is understandable, though not excusable, in view of the desire for cut and dried answers and a general unwillingness to tolerate the discomforts of uncertainty.

Philosophers, psychologists, historians of ideas, and generalists interested in learning about mathematics in a reflective way will find in this work fuller explanations than they are likely to encounter elsewhere. Although a minimum of prior mathematical knowledge is assumed, it would be misleading to suggest that this will be an easy book for those who have only sketchily-remembered school mathematics to build upon. Serious nonspecialist students will find plenty to challenge them, yet even casual readers should gain some appreciation of the nature of mathematics in the making. No productive mathematician expects to be able to bypass the untidy stages of conjectural inductivism that are the fundamental means of obtaining and checking provisional insights leading in the end (it is to be hoped) to elegant, valid proofs. It is this process, involving analogical surmise, testing, proof construction, and analysis, that the examples discussed are intended to illustrate.

As a contribution to mathematics education rather than to mathematics itself, this work is above all intended for teachers. It is all too easy to pay lip service to the heuristic art but far from easy to put it into practice in actual teaching situations. Of course, teaching would hardly be required if students could be imbued with the enthusiasm of the participants of our dialogue. But resources are needed, and mathematics is a written craft. Orthodox textbooks present mathematics as a fait accompli. In the interests of a very dubious efficiency, they suppress the natural dialogue between cooperating searchers after truth and dogmatically "substitute the communiqué."[2]

Although as yet there appears to be no widespread "radical mathematics" movement, vast numbers of harassed students (taught by harassed instructors) overreach themselves trying to learn the right recipe to follow, too often adopting a rule-of-thumb approach, in their desperate efforts "to get through the

[2]In a not entirely unrelated context, Camus observed: "We no longer say as in simple times: 'this is my opinion. What are your objections?' We have become lucid. For the dialogue we have substituted the communiqué. 'This is the truth,' we say. 'You can discuss it as much as you want; we aren't interested. But in a few years there'll be the police to show you I'm right'." (Albert Camus, *The Fall,* translated by Justin O'Brien, Harmondsworth, England: Penguin Books, 1963, p. 35.) Already in the 1920s Ortega asserted that "the 'new thing' in Europe is 'to have done with discussions' . . . . All the normal processes are suppressed in order to arrive directly at the imposition of what is desired." (Ortega y Gasset, *The Revolt of the Masses,* anon. trans. Norton, New York, 1957, p. 74.) Have not these observations a disturbing applicability to some mathematics classrooms? Mathematics itself may be ethically neutral, mathematics education is not.

course."[3] As well as exhibiting mathematics as less isolated from the mainstream of our cultural heritage than is usually implied, this book may contribute to the finding of alternative educational practices of more permanent value to the student. The reader will not be presented with theorems to be laboriously mastered and applied. Instead, the aim has been to show how mathematical theorems are arrived at, and to show at the same time that satisfying and effective independent mathematical education is not nearly as difficult or mysterious as it is usually assumed to be.

The approach to mathematical education attempted here is not original; Plato represents Socrates as having foreshadowed it (*Meno,* 82b–85e). And not every mathematician has striven to erase every error and suppress every side issue or been content to write more like a computer than a human being. A few, such as Kepler and Galileo, have shared with their readers something of their own journeying through error or confusion to understanding. Of recent writers, George Polya has been influential among the more receptive teachers, and Imre Lakatos has stirred the philosophers. I have been sustained too by the perceptive views of numerous writers appreciative of the value of the historical approach. For R.G. Collingwood, "history did not mean knowing what events followed what. It meant getting inside other people's heads, looking at their situation through their eyes, and thinking for yourself whether the way they tackled it was the right way."[4] This and Freud's remark that "every discovery is made more than once and none is made all at once,"[5] are well exemplified in the best writings on the history of mathematics. Lancelot Hogben asserted that the historical approach, by emphasizing the gradual growth of mathematics, "assuages and restores the self-confidence shattered by repeated assertions that a statement is *obvious.* For ordinary mortals it is reassuring to find that a supposedly obvious statement had defied the collective effort of all the best intelligences for several centuries."[6]

When, in the 1930s, Hogben produced his *Mathematics for the Million* and *Science for the Citizen,* he called these "Primers for the Age of Plenty," suggesting that he was writing for an age less straitened and more just than his own and also, perhaps, in the hope that his readers might be encouraged to prepare for

[3]"In practice, most university mathematics courses are designed by lecturers who want to get to the final oasis as fast as possible (and preferably faster than the people in the next university). In driving their buses over the desert they forget that most of their passengers have been thrown off or numbed to sleep and the consequences for mathematics education are disastrous." (Brian Griffiths, "The Language of Mathematical Communication," *Mathematical Gazette, 60* (1976): 298–301, on p. 299.) Reasons for these disastrous consequences are examined at length in Morris Kline, *Why the Professor Can't Teach: Mathematics and the Dilemma of University Education*, New York: St. Martin's Press, 1977.

[4]*An Autobiography,* London: Oxford University Press, 1939, p. 58.

[5]In Lecture 17 of *Introductory Lectures on Psychoanalysis,* Harmondsworth, England: Penguin Books, 1974.

[6]Lancelot Hogben, "Clarity Is Not Enough," *Mathematical Gazette, 22* (1938): 105–122, on p. 115. Cf. J. E. Littlewood, *A Mathematician's Miscellany,* London: Methuen, 1953, p. 131.

such a future. In the same decade Hermann Hesse was writing his *Glasperlenspiel,* in which he has a historian of the future look back on our era of "almost untrammelled individualism," to find "truly astonishing examples of the intellect's debasement, venality, and self-betrayal." Reflecting upon this "Age of the Feuilleton,"[7] from which now, forty years later, we can hardly be judged to have begun to extricate ourselves, Hesse found his fellow citizens busying themselves with various superficial pastimes: card games, crossword puzzles, motoring—activities that "sprang from their deep need to close their eyes and flee from unsolved problems and anxious forebodings of doom into an imaginary world as innocuous as possible."[8]

I trust that I will not be judged presumptuous if I describe the present work as a "Primer for the Post-Feuilleton Age," as a book written in the spirit of the educational province of Castalia, as Hesse described it, several centuries into the future. While avoiding antiquarianism, Castalian scholars were represented as striving to extract the best from the past and to relate in a creative synthesis insights previously understood only imperfectly or in an isolated way. Of the traditional disciplines and arts, mathematics and music were the most highly regarded by Hesse's Castalians. But from their point of view the various academic specialities of the preceding centuries were to be recognized as disparate antecedents of the subtlest and most noble of their creations: the mysterious, magnificent glass bead game itself.

Despite reservations we may have concerning the elitism and the puritanism of the Castalian scholars, we can only admire their dedication to the idea of bringing together the separate strands of our best intellectual traditions. To promote the unity of knowledge, to help reconcile "estranged" science and ethics, logic and imagination, reason and emotion, is, one hopes, to make some contribution toward the unification of mankind. And perhaps there is even yet some justification in seeing, as Bertrand Russell tells us the nineteenth-century mathematician W. K. Clifford saw, "all knowledge, even the most abstract, as part of the general life of mankind, and as concerned in the endeavor to make human existence less petty, less superstitious, and less miserable."[9]

---

[7] A *feuilleton* is a short journalistic item, specifically a ruled-off article or episode of a serial in a French newspaper, here it symbolizes the showy false coin of pseudoscholarship.

[8] *The Glass Bead Game,* translated by R. and C. Winston, Harmondsworth, England: Penguin Books, 1972, p. 24.

[9] Russell's preface to the 1946 edition of William Kingdon Clifford's *Common Sense of the Exact Sciences,* 1885, reprinted by Dover, 1955, p. ix.

THE CHARACTERS

Included in this work is a more detailed investigation of matters admirably introduced by Galileo in his *Two New Sciences* (First Day, 97–104). For this reason and in order the better to preserve the classical dialectic flavor of mathematics in the making, it seemed appropriate to resurrect the characters of Galileo's own dialogues. With the progress of the work, however, a strange feeling was induced in the writer, which will perhaps be shared by the reader, namely, that these characters did not so much need to be brought to us as we to them—that they were alive in another abode, more real, more lasting, more serene than our own, and that they had been able to keep in touch with much going on in our world, especially with matters connected with their own interests three and a half centuries ago.

Two of the three characters (in the *Two Chief World Systems,* 1632, as well as the *Two New Sciences,* 1638), SALVIATI and SAGREDO, bear the names of revered friends of Galileo; their death "deprived Venice and Florence of . . . two great luminaries in the very meridian of their years."[1] The third labored under the pseudonym "Simplicio," after Simplicius, toward whose *Commentaries* he showed "excessive affection." SALVIATI was the mouthpiece for Galileo's best thoughts. "His was a sublime intellect which fed no more hungrily upon any pleasure than it did upon fine meditations." SAGREDO, "of noble extraction and trenchant wit," had the role of a receptive layman eager to understand the new ideas. Poor, unsubtle SIMPLICIO presented the Peripatetic views; his "greatest obstacle in apprehending the truth seemed to be the reputation he had acquired by his interpretations of Aristotle."

In the present dialogue SIMPLICIO shows a not unintelligent enthusiasm for elementary mathematics, whereas SAGREDO remains the apt pupil who, in minor matters, is sometimes able to proceed competently on his own. SALVIATI is the patient teacher. He often allows the other two to take the initiative, to discover much for themselves and to learn from their mistakes.

[1] The quotations in this paragraph are all from Galileo's own introduction to the *Dialogue Concerning the Two Chief World Systems—Ptolemaic and Copernican*, trans. Stillman Drake, foreword by Albert Einstein (Berkeley: University of California Press, 1953), p. 7. Cf. Drake's introduction to his edition of the *Two New Sciences Including Centres of Gravity and Force of Percussion* (Madison: University of Wisconsin Press, 1974), pp. xiii–xiv, for a comparison of the roles of the interlocutors in the two main works of Galileo.

Geometrical Investigations

Illustrating
the Art of Discovery
in the Mathematical Field

I

Heuristic, or heuretic, or "ars inveniendi" was the name of a certain branch of study, not very clearly circumscribed, belonging to logic, or to philosophy, or to psychology, often outlined, seldom presented in detail, and as good as forgotten today. The aim of heuristic is to study the methods and rules of discovery and invention.

—George Polya

We know from experience that, when it is proving very difficult to make any headway in some investigation, the first efforts usually cast very little light on our problem; and it is only by trying again and again, and by considering the same thing from several points of view, that we arrive at complete understanding.

—Leonhard Euler

Proficiency in enquiry, if it can be acquired at all, will be furthered much more through specific living examples, rather than through pallid abstract formulae that in any case need concrete examples to become intelligible.

—Ernst Mach

The polished presentations in the [usual mathematics] courses fail to show the struggles of the creative process, the frustrations, and the long arduous road mathematicians must travel to attain a sizable structure. Once aware of this, the student will not only gain insight but derive courage to pursue tenaciously his own problems and not be dismayed by the incompleteness or deficiencies in his own work.

—Morris Kline

For me at least, the monologue, spoken or printed, is of small value as a vehicle for communication of ideas. Dialogue is essential and the simplest form of dialogue is a conversation between two people, or perhaps three or four, certainly not a crowd.

—J. L. Synge

HORA PRIMA

Simplicio: How good it is to see you Salviati! Sagredo has just been telling me that our new author has brought us together again at this time in order to reconsider and perhaps to develop in a more general way one or two of the results established by Archimedes. We learned from you in previous discussions that our original author had a respect for Archimedes probably exceeding that which he accorded any other of his predecessors. In those days he showed through you a lack of respect for Aristotle, and this caused some friction between us. Yet I have to admit that his judgment in many matters has been largely vindicated by the developments of the last three and one-half centuries.

Salviati: I hope, Simplicio, that you have kept abreast of these developments sufficiently to appreciate how remarkable that vindication has been, so far at least as Galileo's general approach to physics is concerned. No one is infallible, as that difficult matter of the tides illustrates.[1]* And, of course, there were other less defensible hypotheses. We can now see that we were led to assume an over-simple stress distribution within the variously loaded beams that we discussed —fancy overlooking those forces of compression![2] The more important fact, however, is that in this, as in so many other of the investigations that Galileo shared with us, his crystal clear quantitative analysis was, within reasonable limits, empirically testable without too much difficulty and readily modifiable. In surveying the historical development of other sciences, our successors have not unnaturally been led to inquire when and in what circumstances these found their Galileos.

Sagredo: I have myself been following certain of these developments with interest. It does appear that some of the social sciences are in need of their Galileos at this very time. May we not say that a Galileo is one able to deliver a science from the machinations of precursors who, being unable to tolerate the incompleteness of knowledge, attempted the impossible task of laying out a Great System of Final Truth?

Salviati: Well yes, Sagredo. Of course, too much should not be claimed for any one thinker. A common mistake made by youthful students is to expect too much from their heroes. As for the great pioneering system-builders, the wisest doubtless appreciated the incompleteness and probable inaccuracy of their construc-

---

*Bracketed superior numbers refer to the notes, which begin on p. 261.

tions far more clearly than the many who came to embrace their explanations so uncritically. Let us remember, Simplicio, that it was for the uncritical followers of Aristotle, rather than for the philosopher himself, that our first author showed his scorn.[3]

Simplicio: I have not been idle in my efforts to understand something of the development of the sciences in recent centuries. And I confess that much should have caused our former enthusiasms to be mellowed. In mathematics I have not been very successful in following recent trends, and I welcome the opportunity of returning to consider anew the discoveries of the ancient genius of Syracuse. And strange as it might seem to you Sagredo, I do have a suggestion concerning the very matter about which you were speaking as Salviati arrived.

Sagredo: Do you mean the discovery of Archimedes for which he himself wished to be especially remembered?

Simplicio: The same. In asking that a representation of a sphere and its circumscribing cylinder be placed upon his tomb,[4] Archimedes surely missed the opportunity of including along with those two solids another, namely, the right cone inscribed in the cylinder so as to share its base. For the volumes of the cone, the sphere, and the cylinder are as 1 : 2 : 3—a fundamental relation of such purity as to warm the heart of any with Pythagorean inclinations.

Sagredo: But, while the relation of the sphere to the cylinder was Archimedes' own discovery, that of the cone to the cylinder (or of a pyramid to its circumscribing prism) was very well known before Archimedes' time, having been treated in detail by Euclid in the twelfth book of his *Elements* and before him by Eudoxus, who is credited with the original formal demonstration. Earlier still Democritus had given the results as conclusions of some nonrigorous argument or other, the details of which are lost. Archimedes would not have believed it fitting, therefore, to include the cone on his own memorial, though admittedly the more complete diagram or model that you suggest would have greater mathematical significance. Do you not agree, Salviati?

Salviati: Not entirely. Neither of you appears to appreciate that not only did Archimedes establish that the *content* of the cylinder is $1\frac{1}{2}$ times that of the sphere, he established also that the *surface* of the cylinder (including the bases) is $1\frac{1}{2}$ times the surface of the sphere.[5]

Sagredo: Ah yes, I was forgetting—a quite unique relationship. Apparently, then, the introduction of the cone would have detracted from the perfect simplicity of the relation between the surfaces.

Simplicio: I admit that I had not been thinking about surface areas. But do I understand you to be claiming that the surfaces of the cone, the sphere, and their common circumscribed cylinder are not also, like their volumes, as 1 : 2 : 3? I think that my intuition would have led me to assume that the areas would have conformed to these same ratios.

Sagredo: The matter is easily settled by calculation. Let us not rely upon intuition where certainty can be achieved so readily. I hope we can convince you that the sphere and the cylinder are unique solids in that the ratio of the volumes is equal to the ratio of their surfaces. I am grateful to you, Salviati, for reminding me that this discovery so rightly captivated Archimedes.

Salviati: Not so fast Sagredo! You are the one who has asserted the uniqueness. I suggest that this question should now occupy our attention, if that meets with your approval.

Sagredo: It does indeed.

Salviati: I must warn you though, Sagredo, that you will discover that your own belief concerning the uniqueness is as ill-founded as Simplicio's regarding the conformity of the cone. But now, Simplicio, kindly let us have the calculation of the ratio of the surface of the cone to that of the cylinder.

Simplicio: The base of the cone is congruent with that of the cylinder, and the area of each is πr^2, to use the modern form of expression.[6] The slant surface of the cone has as its development a sector of a circle of radius equal to the slant height H, say, of the cone and having its arc equal to the circumference of the base. The area of a sector is half the product of the arc and the radius, just as the area of the whole circle is half the product of the circumference and the radius, as Archimedes established.[7]

Salviati: So far so good Simplicio; but why do you not proceed?

Sagredo: Since Simplicio hesitates, I may be allowed to point out that it can be seen already that not only is the total surface of the cone more than one-third that of the enclosing cylinder but that it is greater even than one-half of it. For the cone has one base whereas the cylinder has two and the curved surface of the cone is to *half* the curved surface of the cylinder as H is to h in the diagram (Fig. 1).

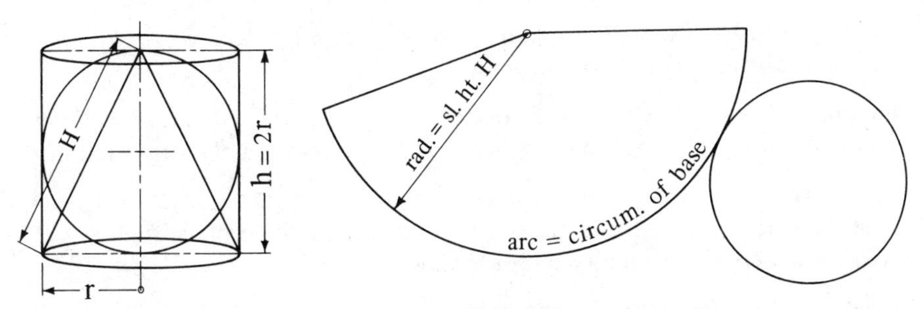

Development of conical surface with base

Fig. 1

Simplicio: I suppose you are right, but I do not understand the last part of your claim quite clearly enough. Permit me to set down the reasoning step by step. The curved surface of the cylinder is given by the product of its circumference and its height,[8] to which must be added the area of the two ends or bases: in all, $2\pi rh + 2\pi r^2$, where $h = 2r$, so that the total surface of the cylinder is measured by $(2\pi r)(2r) + 2\pi r^2$; that is, $6\pi r^2$.

Sagredo: Agreed; this is truly $1\frac{1}{2}$ times the $4\pi r^2$ for the surface of the sphere of equal radius,[9] but what of the cone, Simplicio?

Simplicio: The cone has a total surface of

$$\frac{1}{2}\text{(base circum.)(slant height)} + \text{area of base, viz., } \pi rH + \pi r^2,$$

that is $\pi r(H + r)$, where H can be calculated by the Pythagorean relation.

Sagredo: It is exactly as you say; and because H exceeds h, that is, $H > 2r$, the measure of the total surface of the cone exceeds $3\pi r^2$. Thus the area is more than half that of the cylinder.

Simplicio: I have to admit it. $H = \sqrt{(r^2 + h^2)}$, which comes to $r\sqrt{5}$ when $h = 2r$. Hence, for the cone, the total area is given by $\pi r^2(\sqrt{5} + 1)$.

Sagredo: The surfaces of the cone and cylinder are therefore as $(\sqrt{5} + 1) : 6$, certainly not a ratio of whole numbers, let alone small whole numbers such as 1 and 3. This does not surprise me, but I must confess that I am disconcerted by Salviati's claim that the sphere and its circumscribing cylinder are not unique in having their area and their volume ratios equal to each other. I should indeed be most interested to hear of any other pair of solids that share this remarkable property.

Salviati: There are examples without end as I am sure you yourself will see in due course. But tell me first, Sagredo, which pair of plane figures is analogous to the sphere circumscribed by the cylinder?

Sagredo: The circle and its circumscribing square, surely.

Salviati: And has this pair an analogous property? I mean, is the ratio of the areas equal to the ratio of the perimeters?

Sagredo: This is indeed so; the former ratio is $(\pi/4)d^2 : d^2$ and the latter is $\pi d : 4d$, each simplifying to $\pi : 4$, or $11 : 14$ approximately, as Archimedes showed.[10]

Salviati: And are you of the opinion that no other pair of plane figures will be found to possess this property of equal ratios?

Sagredo: That was my belief, but now it occurs to me that perhaps an ellipse and its circumscribing rectangle might also be as intimately related.

Simplicio: I think that this would very probably be found to be the case.

Salviati: This is indeed an interesting problem. Neither of you claims that the solution is provided intuitively, I notice. Perhaps this is because you realize that many different rectangles may be circumscribed about a given ellipse! (Fig. 2)

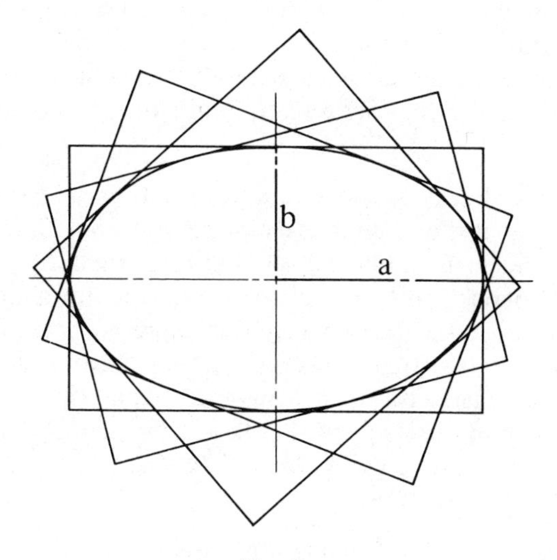

Fig. 2

Sagredo: I had in mind only the one having its sides parallel to the axes. I do not think that we have to be concerned with these other rectangles, all of which are larger and more nearly square, are they not?

Salviati: They are as you say. That the rectangle with sides parallel to the axes is the smallest both in area and perimeter is a result that can be established on the basis of the following theorem: *The locus of the intersection of perpendicular tangents to an ellipse is a circle.* This circle, called the *director circle* of the ellipse, thus circumscribes all the rectangles that circumscribe the ellipse.

Sagredo: Good. And when Archimedes established that the area of an ellipse is given by πab, as it would now be expressed, he made use of the proposition that all chords of an ellipse that are parallel to the minor axis are the same fixed fraction of the corresponding (collinear) chords of the circle having the major axis as diameter.[11]

Salviati: And what about the *perimeter* of the ellipse?

Simplicio: Since Sagredo does not answer, I would suggest that, on analogy with the formula for finding the circumference of a circle, which is in these times symbolized as $C = \pi d$, we will have to write $p_e = \pi d'$, where d' is the average length of the axes. Thus, if a, b denote the semiaxes, the perimeter would be given by $p_e = \pi(a + b)$.

Salviati: You do not appear altogether satisfied, Sagredo, but Simplicio is eager to continue.

Simplicio: The ratio of the areas of the ellipse, A_e, and of the rectangle, A_r, is equal to $\pi ab:(2a)(2b)$; that is, $\pi:4$. The ratio of the perimeters is $\pi(a+b)$ to $2(2a+2b)$; that is, $p_e:p_r = \pi:4$. See, it *is* correct! The ellipse and the rectangle of length and width equal to the axes have their area ratio and their perimeter ratio equal to each other and indeed equal to the very same ratio obtaining in the case of the circle and its circumscribing square. Does this not meet with your approval, Sagredo?

Sagredo: I wonder that Archimedes did not determine the perimeter as well as the area of the ellipse if the result is as simple as you would have it. You have not justified your perimeter formula. To say that $p_e = \pi(a+b)$ holds for an ellipse because it is "analogous" to $C = \pi d$, or $C = \pi(r+r)$, for a circle is certainly no demonstration. In the special case where the ellipse becomes a circle, $b = a$ and both formulae may be written $C = 2\pi a$, but this agreement does not validate $p_e = \pi(a+b)$ as a general formula. I might ask you: Why not $p_e = \pi'(a+b)$, where π' depends upon the shape of the ellipse? Or again, why do you not claim that the *area* of the ellipse is given by $A_e = (\pi/4)(a+b)^2$, on analogy with the circle formula $A = (\pi/4)d^2$?[12]

Salviati: Sagredo, you are to be congratulated on your wholly justifiable skepticism! If we were to assume $p_e:p_r = A_e:A_r$—that is, $p_e = (A_e/A_r)p_r$— and then substitute the known expressions for the terms on the right-hand side, we *would* obtain $p_e = \pi(a+b)$. This shows only that the assumption of (1) the equality of the perimeter and area ratios and of (2) the putative formula $p_e = \pi(a+b)$ are equivalent. They are assumptions none the less. That they are consistent with each other as well as with results known to be correct for the special case of the circle and its circumscribing square understandably gave you prima facie encouragement, Simplicio. Nevertheless, I am afraid that confidence in your too simple perimeter formula cannot be justified, as we shall have to show.

Sagredo: In such matters partial success is still to be counted as failure, as I have found not a few times to my own chagrin.

Salviati: Well, from a narrow, abstract point of view, an inexact answer *is* an erroneous one, but just to assert this really does leave out too much to be satisfactory.

Simplicio: Not long ago I overheard a modern philosopher say something to the effect that interesting error is preferred to dull truth, but I suppose he was speaking facetiously.

Sagredo: Or ironically! In this age of mass communication, the written and uttered word is being produced and reproduced for a predominantly undiscriminating market. Too often are distortions, embroideries, and half-truths the stock in trade of those whose job it is to see to it that ephemeral publications and programs are filled as attractively as possible while conforming to the prevailing myths and to the policies of those whose interests they serve. But all this is poles apart from the austere and incorruptible world of mathematics, where beauty and truth dwell in eternal harmony.

Salviati: I am inclined rather to think that Simplicio's philosopher was making an oblique reference to those interesting errors in connection with which philosophical analysis might be expected to be applied profitably. These philosophers, you know, welcome opportunities to exercise their subtle intellects by clearing up confusions arising from unacceptable or debatable assumptions and value judgments or from invalid deductions and unjustifiable associations.

Sagredo: You are probably right. But I am still anxious to know what you find unsatisfactory about my view that in mathematics exactness is not to be compromised and that inaccurate surmises have to be exposed for what they are, namely errors. Surely, while Truth in mathematics is one and eternal, the errors that might be made in respect of any one truth are diverse and innumerable.[13] For that reason I should have thought that mathematicians could hardly be expected to concern themselves with mistakes and erroneous hypotheses, except, of course, to point them out. Psychologists, on the other hand, might find an investigation of false mathematical conjectures (as well as true ones) illuminating in their study of analogy, imagination, and creativity.[14]

Salviati: Not only psychologists but mathematicians too! In many situations conjectures are not so easily come by that they ought to be rejected out of hand merely because we discover them to be inaccurate or mistaken.[15] Erroneous conjectures ought, if possible, to be used in a positive way—that is, as stepping stones to demonstrably correct hypotheses, or theorems. The impressiveness of the elegant, austere, "finished" productions of the great mathematicians may lead those without first-hand experience to overlook the fact that the process of mathematical discovery is a creative human activity, for the most part a tortuous one, with many setbacks and apparent blind alleys.[16] Often conjectures arise that are both interesting and false; for a time, they may have a quite persuasive plausibility. They involve the implied, albeit temporarily unrecognized, challenge to show not only *that* they are erroneous but *why* or in which circumstances. For they might be false only in the sense that their range of applicability was misjudged, and the recognition of this can lead to a more precise understanding, a sharpening of concepts. I know, Sagredo, that you accept that the sciences generally grow by way of conjectures and attempted refutations,[17] but you were led momentarily, and I believe against your better judgment, to make an exception in the case of mathematics. The transition from one conjecture to a more promising one is not by way of an arbitrary jump; rather, it is something arising out of a pattern of experience in the face of loss of confidence in a previous standpoint. Essential to this process is the bringing of the mind to bear upon assumptions or intuitions that have been taken uncritically to be "given" somehow, "in the nature of the case." I do not wish to suggest that it is simply by some conscious effort of will that the mathematical innovator is able to bring crucial assumptions into focus, or even that he is thus able to examine these appropriately when they are brought to his attention. At least, such a description fails to do justice to the more interesting and original innovations. Rather, as numerous writers of this Freudian age have depicted it, illumination comes to consciousness, if at all, after the unconscious has done its work, often after a very considerable period of incubation.[18]

Simplicio: I think I can see that Sagredo agrees with your emphasis upon the importance of conjectures in mathematics, Salviati, yet you both were dissatisfied with my conjecture that the perimeter of an ellipse is given by $p_e = \pi(a + b)$. Although I admit that I do not have a proof of this formula, you have not shown me a refutation of it either. I should like to return to this matter, for all that your discussion has raised other questions in my mind.

Salviati: Well then, let us put your formula to the test without further delay. It is clear that $p_e = \pi(a + b)$ gives a correct result for the special case of a circle, since, when $a = b$, $a + b = d$, and the formula simply reduces to $C = \pi d$. But moving toward the other extreme, what happens in the case of ellipses with greater and greater eccentricities?

Sagredo: Do you mean where b becomes smaller and smaller in relation to a?

Salviati: Yes, or as a becomes smaller and smaller in relation to b. Now clearly, the perimeter of the ellipse is greater than twice its major axis, and less than the perimeter of the enclosing rectangle. Do you agree that this is so, Simplicio?

Simplicio: Yes; it is evident from a diagram (Fig. 3) that the arc XY is longer than OX and yet shorter than $XA + AY$.[19] Multiplying by 4, we may write $p_e > 4a$ and $p_e < 4(a + b)$, which is as you have said.[20]

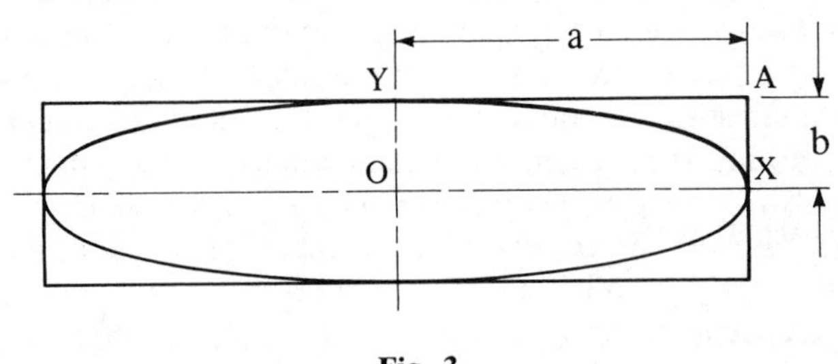

Fig. 3

Salviati: Very well. Now, Simplicio, what happens if we take the first of these inequalities, the undoubtedly true $p_e > 4a$, and divide the left side by p_r and the right side by the value of p_r, viz. $4(a + b)$?

Simplicio: Surely we obtain the undoubtedly true relation

$$p_e : p_r > a : (a + b).$$

Salviati: Precisely. Now suppose that b is rather small compared with a; suppose, for example, that b is one-seventh of a.

Simplicio: Then the value of the ratio $p_e : p_r$ is greater than $7:8$.

Salviati: But $7/8 = 3\frac{1}{2}/4$, and this is clearly greater than $\pi/4$.

Sagredo: Ah! So we have a clear demonstration that $p_e/p_r > \pi/4$, and, by substituting $4(a + b)$ for p_r, we obtain

$$p_e > \pi(a + b).$$

Salviati: So you see, Simplicio, to calculate p_e according to your formula, $p_e = \pi(a + b)$, would be to obtain an underestimate of the true measure of the perimeter of the ellipse. I suggest that you review the reasoning on your own for other small values of the ratio $b:a$. The foregoing simple test is decisive whenever $b:a < 3:11$ approximately. More refined tests could readily be devised to provide falsification of the conjecture that $p_e/p_r = \pi/4$ or its equivalent, $p_e = \pi(a + b)$, for values of $b:a$ closer to $1:1$, and to indicate that your conjectured formula would yield values for p_e underestimating the true measure of the elliptical perimeter in all cases (other than where $b = a$).

Simplicio: I see. And I understand that the ratio of the *areas* of the ellipse and the circumscribing rectangle is not in doubt?

Salviati: It is not. That $A_e : A_r = \pi : 4$ follows from the result, rigorously proved by Archimedes using a double *reductio ad absurdum* demonstration, that "the area of any ellipse is to that of the auxiliary circle as the minor axis to the major." In our notation, $A_e : \pi a^2 = b : a$. Thus $A_e = \pi ab$, and since $A_r = 4ab$, we obtain $A_e : A_r = \pi : 4$.

Sagredo: You have shown us very clearly that neither $p_e = \pi(a + b)$ nor $p_e : p_r = A_e : A_r$ can be true. I must say that I was suspicious of Simplicio's conjecture as soon as he proposed that $p_e = \pi d'$, "where d' is the average length of the axes," and then immediately assumed that the appropriate mean would be $a + b$, that is $\frac{1}{2}(d_{\max} + d_{\min})$. Why the arithmetic mean rather than the geometric mean, or the harmonic mean, or the root mean square, or some other mean?[21] But you have shown us the answer to our main question, Salviati; $p_e : p_r = A_e : A_r$ if and *only if* $p_e = \pi(a + b)$, and yet $p_e \neq \pi(a + b)$, and therefore

$$p_e : p_r \neq A_e : A_r.$$

It appears that this result provides support for my original belief that the circle and its circumscribed square are unique in having equal perimeter and area ratios. For I cannot imagine two figures more allied to the circle and the square than the ellipse and the rectangle.[22] And yet, Salviati, you have indicated otherwise. I eagerly await some illustration of your claim.

Salviati: You shall have them by the barrowful soon enough, Sagredo; but I see that Simplicio has something to say.

Simplicio: I have to admit the inaccuracy arising from the use of the arithmetic mean in the perimeter formula. But, before we leave the ellipse, I should like to make another attempt to find the correct expression for the perimeter. It now seems to me that the exactness of the area formula, $A_e = \pi ab$, rigorously shown by Archimedes' demonstration, provides us with an indication of the appropriate mean to be used in the perimeter formula. The areas of the two circles in the following diagram are given by πa^2 and πb^2, respectively. As has been mentioned, the area of the ellipse is to the area of the outer circle as $b : a$; it is as though the outer circle has been compressed to this fraction, b/a, of its original area. For example, if $b : a = 1 : 2$, then the ellipse has one-half the area of the circle of radius a (Fig. 4).

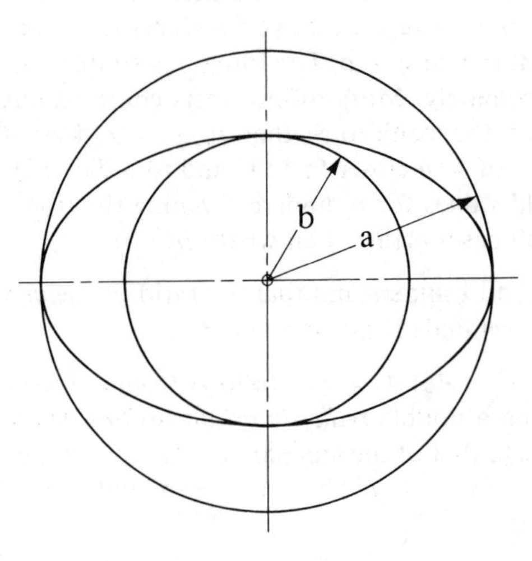

Fig. 4

Sagredo: And twice the area of the circle of radius b.

Simplicio: Precisely. This means that the areas of the inner circle, the ellipse, and the outer circle are as $1 : 2 : 4$, or in general as $b^2 : ba : a^2$. The elliptical *area* is thus the mean proportional between the areas of the two circles. Surely then, in like manner, the *perimeter* of an ellipse is the mean proportional between the circumferences of the two circles, for the small circle can be transformed into the ellipse by stretching in one direction, and the ellipse in turn can be transformed into the larger circle by stretching in a perpendicular direction in exactly the same ratio. Thus

$$\frac{\text{perimeter of ellipse}}{\text{circum. of small circle}} = \frac{\text{circum. of large circle}}{\text{perimeter of ellipse}}.$$

Sagredo: According to this argument, the perimeter of the ellipse is the geometric mean of the circumference of the inscribed and circumscribed circles, or, to express it another way, $p_e = 2\pi r'$, where $r' = \sqrt{(ab)}$.

Simplicio: Yes, exactly as for the area: $A_e = \pi r'^2$, where $r' = \sqrt{(ab)}$. What do you say to this?

Sagredo: I say that you are still arguing by analogy, though perhaps this time the analogy is an appropriate one.[23] What do you think of it Salviati?

Salviati: I cannot help thinking, Simplicio, of how you once avowed that if you were beginning your studies again you would endeavor "to follow the advice of Plato and commence from mathematics, which proceeds so carefully, and does not admit as certain anything except what it has conclusively proved."[24] Do you remember?

Simplicio: I did express myself in some such way when under the spell of a particularly clear and convincing proof you had shown us.

Salviati: Rigorous demonstrations are often very elusive, and apart from this the mathematician needs to have in advance some idea of what it is that he is seeking to prove![25] Hence it comes about that, where proofs are not yet glimpsed, it is appropriate to entertain conjectures of various kinds, preferably comparing one conjecture against another, and to try deliberately to expose their consequences to tests that will enable the false conjectures to be eliminated if not provide some indication as to how these might be suitably modified. To do this is to show the true spirit of science. Now, Simplicio, you have been good enough to offer us a second conjecture. Perhaps you feel that it is more than a conjecture, that a perimeter relation apparently so in harmony with the indubitable result for the area must be true.[26] I am pleased to see that you have paid some attention to the geometry of the situation, but I think that you will agree that a rigorous proof would require a much more thorough examination of the precise nature of the change undergone by the boundary of the figure obtained when a circle is "simply" elongated or compressed. But in this case, as with your first suggestion for a perimeter formula, we may proceed most easily by testing the consistency of the proposed measure with the undoubted fact that the perimeter of any ellipse exceeds twice its major axis. Now $2\pi\sqrt{(ab)} \gtreqless 4a$ according as $4\pi^2 ab \gtreqless 16a^2$, which in turn is true according as $b:a \gtreqless 4:\pi^2$, or $2:5$, approximately. (We are concerned only with positive values.) This means that if b is less than about two-fifths of a, then $2\pi\sqrt{(ab)} < 4a$, which shows that the formula $p_e = 2\pi\sqrt{(ab)}$ cannot possibly provide correct perimeter measures in general, since it so clearly fails when the eccentricity increases beyond a value corresponding to $b/a < 0.4$ approximately.[27] A somewhat less crude test could be provided by noticing that the true measure of the perimeter of an ellipse must exceed $4\sqrt{(a^2 + b^2)}$, as can be seen from Fig. 5. Notice as well, that as b approaches zero, $2\pi\sqrt{(ab)}$ also approaches zero instead of approaching $4a$ as a correct perimeter measure would have to do.

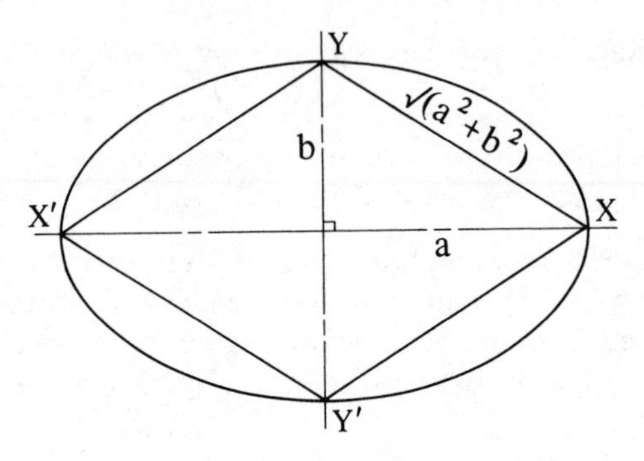

Fig. 5

Sagredo: Oh, what a sudden death to a promising hypothesis! I was hoping that Simplicio was right this time, but alas I fear that the elliptical form is somewhat subtler than either of us has realized.

II

The gods did not reveal all things to mortals in the beginning; but by searching, men find out better in the course of time.

—Xenophanes

In order to learn, we must make mistakes, and the most fruitful mistake which nature could have implanted in us would be the assumption of even greater simplicities than we are likely to meet in this bewildering world of ours. . . . To probe a hole we first use a straight stick to see how far it takes us. To probe the visible world we use the assumption that things are simple until they prove to be otherwise.

—E. H. Gombrich

No sooner did an Ionian philosopher learn half-a-dozen geometrical propositions, and hear that the phenomena of the heavens recur in cycles, than he set to work to look for law everywhere in nature, and, with an audacity almost amounting to ὕβρις, *to construct a system of the universe.*

—John Burnet

There is no empirical method without speculative concepts and systems; and there is no speculative thinking whose concepts do not reveal, on closer investigation, the empirical material from which they stem.

—Albert Einstein

The line of advance in geometry will then surely be from activities involving real things to the abstractions they suggest, coherent physical behaviour contributing to the coherence of ideas, thinking in the early stages being "thought experiment" (Mach's word), "interiorising" (Piaget's word) potential action.

—A. G. Sillitto

HORA SECUNDA

Sagredo: I have just now been considering these tests applied to eccentric ellipses, and especially to the case where b approaches zero, while a remains unchanged—tests *in extremis,* as we might say. They have given me an idea that I hope you will not think is too worthless. It has occurred to me that conjectures about human behavior, about motivation say, if they were to have a claim to generality, would have to apply to "eccentric" as well as to "normal" people or at least to ordinary people in unusual situations.

Simplicio: Well, I know it is sometimes said that if you want to know what people are really like, you should observe them either when they are somewhat intoxicated or when they are subject to unusual stress.

Sagredo: Such as under inquisition with threat of imprisonment or torture? Or, we do have many less unhappy examples in legends and fairy stories, where heroes prove their merit against what appear to be overwhelming odds. Many of these tales, so it seems to me, amount to exercises in understanding human behavior in the face of extreme changes of fortune.

Simplicio: Yes, and such stories often remind us that good fortune, no less than bad, can bring about a person's downfall.

Salviati: In real life, as well as in fiction, individuals often seem driven to test themselves under extreme conditions. The question that comes to my mind when I consider such persons is whether they are effectively reaching for the stars, or whether, on the other hand, they are really overreaching themselves—testing themselves, and perhaps others, to destruction! But now, Sagredo, your reflections remind me how the abnormal sometimes proves valuable in illuminating the normal,[28] famous ethical examples being provided in the discussions of Socrates and of Kant.[29] Of course, in complex cases, it is difficult to be sure that qualitatively different modes of behavior will not be induced at some stage, as conditions are made to deviate more and more from the accustomed ones. Needless to say, authentic interpretations involving extreme cases are liable to be far more readily obtained in mathematics than in fields such as psychology.

Sagredo: We will not easily forget how Galileo was able so effectively to isolate relevant underlying factors from more superficial influences with which they tended to be confounded in ordinary circumstances. By comparing the observable behavior of falling bodies in various liquids and then in air, he was able to extrapolate to the extreme case of truly free fall, where the resistance is reduced to zero.

Simplicio: Yes, I have good reason to remember how he handled my incredulity at the thought that in a vacuum even a lock of wool would fall as fast as a piece of lead.[30]

Sagredo: And that if there were a frictionless level plane, a moving body would move along it forever with unchanging velocity.[31]

Simplicio: These were certainly impressive arguments. But now, if you are both ready, I should like to return to our own humble example. For while I cannot deny the force of your demonstration that the expression $2\pi\surd(ab)$ cannot in general provide correct measures of elliptical perimeters, neither can I understand *why* the perimeter is not multiplied by the same factor for each of the transformations shown in Fig. 6:

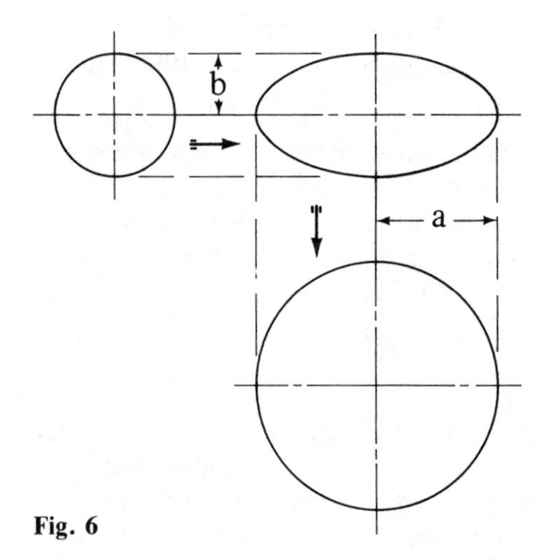

Fig. 6

Sagredo: I know, Simplicio, that you realized all along that while the larger circle has just twice the perimeter of the smaller, it has four times the area. Nevertheless, inspired by the undoubted truth that the area of the ellipse is the geometric mean of the areas of the two circles related to it in the manner indicated in Fig. 4 or Fig. 6 (giving $A_e = \pi ab$), you surmised, erroneously as we have since learned, that the perimeter of the ellipse is the geometric mean of the circumferences of the same two circles (leading you to $p_e = 2\pi\surd(ab)$). When you offered us this conjecture, I think I admitted that it might well be correct. At first I felt that your speculation was rather shrewd.

Simplicio: But you also asserted that I was merely arguing by analogy. I admit this, and, more importantly, I acknowledge that Salviati has shown us a decisive falsification, at least in the case of somewhat elongated ellipses. But what I am now appealing for is some *insight* to counter what still seems intuitively compelling to me—in spite of the unacceptable perimeter estimates that the formula $p_e = 2\pi\sqrt{(ab)}$ has been shown to yield.

Sagredo: I can see that we are going to need Salviati's help in this. But I should like first to clarify something that occurs to me. Did we not earlier come to the conclusion that your first conjectured perimeter formula, $p_e = \pi(a + b)$, provides underestimates for the perimeters of ellipses (other than circles)?

Simplicio: We did. I accepted this and then proposed my improved suggestion in which the "average diameter" estimate, $a + b$, was replaced by $2\sqrt{(ab)}$.

Sagredo: But is it not a fact that the geometric mean of any two different (positive) magnitudes is *less* than the arithmetic mean? So that if your first formula underestimates the true measure of the perimeter, your second one, far from being an improvement, must yield estimates that fall still further short of the true values. Look Simplicio, we can show that $2\pi\sqrt{(ab)} < \pi(a + b)$ by showing that $\{2\sqrt{(ab)}\}^2 < (a + b)^2$, or that $(a + b)^2 - \{2\sqrt{(ab)}\}^2$ is positive. This last expression $\equiv (a^2 + 2ab + b^2) - 4ab \equiv a^2 - 2ab + b^2 \equiv (a - b)^2$; thus the expression is zero if $b = a$ but is positive in all other cases. Hence, for $b \neq a$, the geometric mean $\sqrt{(ab)}$ is less than the arithmetic mean $\frac{1}{2}(a + b)$, and $2\pi\sqrt{(ab)}$ is less than $\pi(a + b)$, which is less than the true measure of the elliptical perimeter. Perhaps too, Simplicio, you remember Fig. 7 from your former studies. The semicircle has radius r equal to $\frac{1}{2}(a + b)$. From the adjacent similar right-angled triangles, we have $a : g = g : b$, giving $g = \sqrt{(ab)}$. The figure clearly shows that $g < r$, provided $b \neq a$.

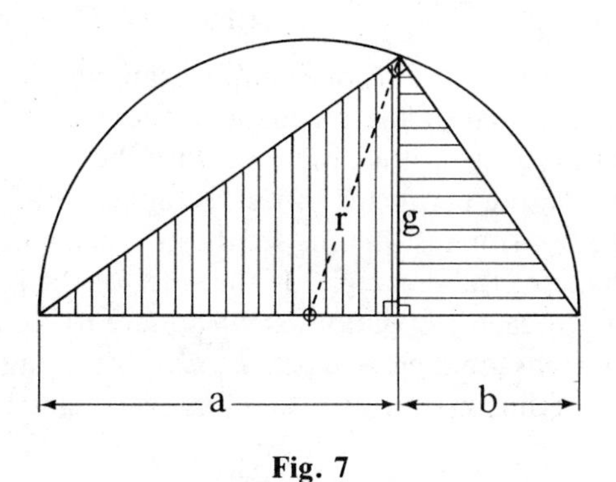

Fig. 7

Simplicio: I follow all that you have said. Yet even now as I look back at Fig. 6, it seems to suggest the $2\pi b$, $2\pi\sqrt{(ab)}$, $2\pi a$ sequence of perimeters as the natural accompaniment of the πb^2, πab, πa^2 area sequence. To be quite free from the seductiveness of this hypothesis, I should still appreciate a somewhat less remote insight than you have just given me.

Salviati: Look Simplicio, if we elongate this square of side s [Fig. 8], so as to transform it into a rectangle of length $2s$ and width s, and then transform the rectangle into a square of side $2s$ as indicated, tell us what happens to the area and to the perimeter.

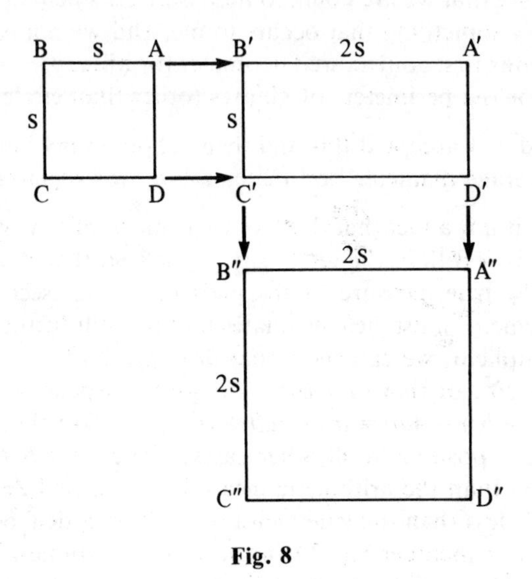

Fig. 8

Simplicio: The area is simply doubled each time. As for the perimeter,

$$\frac{\text{perim } A'B'C'D'}{\text{perim } ABCD} = \frac{6s}{4s} = \frac{3}{2}; \quad \text{and,} \quad \frac{\text{perim } A''B''C''D''}{\text{perim } A'B'C'D'} = \frac{8s}{6s} = \frac{4}{3}.$$

Sagredo: Bravo Salviati! With this one simple example, the mystery has all but vanished. Although the *area* of the rectangle is the geometric mean $2s^2$ of the areas s^2 and $4s^2$ of the squares, it is otherwise with the *perimeters.* Indeed, since these are $4s$, $6s$, and $8s$, the perimeter of the rectangle is the *arithmetic mean* of the perimeters of the two squares. Let me see if this is generally true—I mean true regardless of the shape of the rectangle. If the sides of the rectangle are as $1:k$, this defines the ratio of each elongation, small square to rectangle and rectangle to large square. The areas are then given by s^2, ks^2, k^2s^2, thus forming a geometric progression. The perimeters are $4s$, $(2s + 2ks)$, $4ks$, clearly in arithmetic progression.

Salviati: Yes, the perimeter of any *rectangle* certainly is the arithmetic mean of the perimeters of squares on a pair of adjacent sides. But of course the square-rectangle-square transformation is a very rough model for the circle-ellipse-circle transformation, so far as changes in the boundary lengths are concerned. So it should not be supposed, on the evidence of the perimeter relations you have just established, that the formula $p_e = \pi(a + b)$, based on the arithmetic mean, will be exact for the elliptical perimeter. And of course we have found that it is not. But you were quite correct in your demonstration that the formula $p_e = 2\pi\sqrt{(ab)}$, based on the geometric mean, is even less accurate. I think, Simplicio, that a somewhat better indication of the perimeter changes that you are seeking to understand will be provided by the transformations of Fig. 9. If, for convenience of illustration, a is taken equal to twice b, we obtain for the perimeter measures of these rectilinear figures: $4b\sqrt{2}$, $4b\sqrt{5}$, and $8b\sqrt{2}$. We may now compare the geometric mean and the arithmetic mean of the first and last of these measures with the actual measure of the perimeter of the intermediate rhombus: the geometric mean is $8b$ exactly; the arithmetic mean $6b\sqrt{2} \approx 8.485b$; whereas the true measure of the intermediate perimeter is known to be $4b\sqrt{5}$, which is $8.944b$ correct to the same number of significant figures.

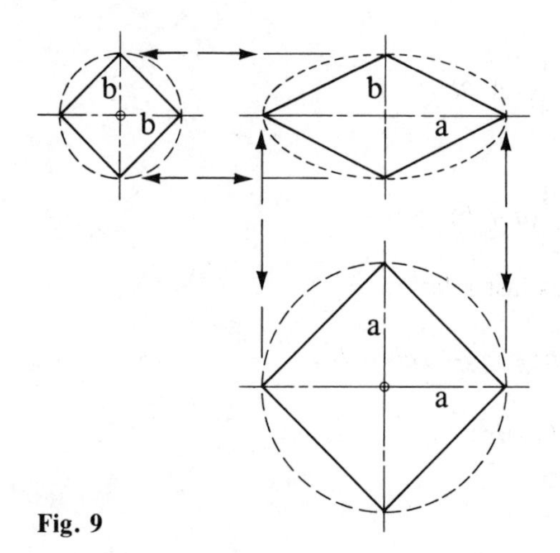

Fig. 9

Simplicio: Thank you, Salviati; I am satisfied. Earlier the thought had occurred to me that perhaps a weighted average of the two estimates, $\pi(a + b)$ and $2\pi\sqrt{(ab)}$, might give the correct measure of the elliptical perimeter, but since both of these expressions provide underestimates, it appears that an average based upon them cannot help us.

Sagredo: Possibly the root mean square of a and b should be employed; at least this mean is, if I remember correctly, always greater than either the geometric or the arithmetic mean.

Salviati: First, let us consider Simplicio's remark. Suppose that A is a slight underestimate of some true value T, whereas G more seriously underestimates T, so that the errors associated with G are, say, three times those associated with the corresponding values of A. Could we not then determine T when A and G are known?

Sagredo: The problem can be solved as soon as it is expressed symbolically: $T - G = 3(T - A)$ implies $3A - G = 2T$, and hence $T = \frac{1}{2}(3A - G)$.

Salviati: That is right. As you see, the problem is simply one of extrapolation, very closely related to the problem of averaging, which is one of interpolation. And now, Sagredo, let us look at your suggestion. Would you show us in general terms why it is that the root mean square exceeds the arithmetic mean? It will be sufficient for our purposes if you consider the simple case where just two positive measures, a and b say, are being combined.

Sagredo: We can show that

$$\sqrt{\left\{\frac{1}{2}(a^2 + b^2)\right\}} > \frac{1}{2}(a + b)$$

by proving that

$$\frac{1}{2}(a^2 + b^2) > \frac{1}{4}(a + b)^2,$$

or that

$$\frac{1}{2}(a^2 + b^2) - \frac{1}{4}(a + b)^2 > 0.$$

The left side of this last relation

$$\equiv \frac{1}{4}(2a^2 + 2b^2 - a^2 - 2ab - b^2)$$

$$\equiv \frac{1}{4}(a^2 - 2ab + b^2)$$

$$\equiv \frac{1}{4}(a - b)^2,$$

which is zero only if $b = a$ and is positive in all other cases. Thus, since all the steps are reversible, the r.m.s. value is shown to be greater than the a.m., except where the quantities being combined are equal to each other (in which case *any* kind of average coincides with the common value of the individual quantities).

Salviati: Very good. Also, that r.m.s. > a.m. > g.m., for $b \neq a$, may be shown by means of a diagram composed of a rectangle and squares. But it is unnecessary to dwell on such details as you will be able to supply for yourselves. (See Problem 5d.)

Sagredo: I notice that if we were to write $p_e = 2\pi R$, where $R = \sqrt{\{\frac{1}{2}(a^2 + b^2)\}}$, then p_e would reduce to $2\pi a$ when $b = a$, as is correct for a circle of radius a, but at the other extreme, when $b = 0$, it comes to $\pi a\sqrt{2}$, approximately equal to $4.443a$, instead of to $4a$, which is the measure of the major axis counted twice.

Salviati: Yes, and for ellipses that are not too elongated, it can be shown that the r.m.s. approximation, $2\pi\sqrt{\{\frac{1}{2}(a^2 + b^2)\}}$, overestimates the perimeter by about as much as the a.m. approximation $2\pi A$, where $A = \frac{1}{2}(a + b)$, underestimates it. And the g.m. approximation $2\pi G$, where $G = \sqrt{(ab)}$, underestimates the perimeter by about three times as much as the a.m. approximation underestimates it. Thus it comes about that one or other of two formulae:

$$p_e = 2\pi\left\{\frac{1}{2}(A + R)\right\}, \quad \text{i.e.,} \quad p_e = \pi\left[\frac{1}{2}(a + b) + \sqrt{\left\{\frac{1}{2}(a^2 + b^2)\right\}}\right],$$

and

$$p_e = 2\pi\left\{\frac{1}{2}(3A - G)\right\}, \quad \text{i.e.} \quad p_e = \pi\left\{\frac{3}{2}(a + b) - \sqrt{(ab)}\right\},$$

are sometimes used where improved approximations are required. Some idea of the way in which these estimates work out can be gained from their counterparts for the elongated hexagon of Fig. 10. Denoting the perimeters by $p_1(= 6s)$, p_2, and p_3 $(= 12s)$; we have as very rough estimates of p_2:

$$R = \sqrt{\left\{\frac{1}{2}(p_1^2 + p_3^2)\right\}}, \qquad A = \frac{1}{2}(p_1 + p_3), \qquad G = \sqrt{(p_1 p_3)},$$

from which $R = \sqrt{(90s^2)} \approx 9.487s$, $A = 9s$, and $G = \sqrt{(72s^2)} \approx 8.485s$. Hence $\frac{1}{2}(A + R) \approx 9.243s$, and $\frac{1}{2}(3A - G) \approx 9.257s$. That the latter two estimates are very much more accurate than the crude R, A, and G values can be seen from the independent evaluation of p_2, which gives $2s(2 + \sqrt{7})$, or $9.292s$ correct to four significant figures.

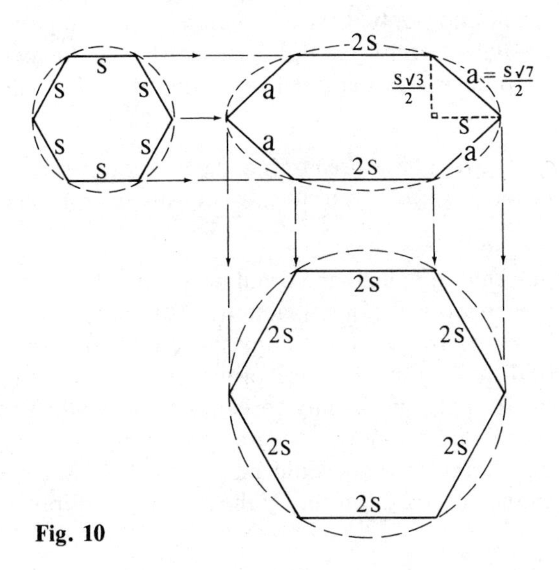

Fig. 10

Simplicio: I find your illustrations involving rectilinear figures helpful since they enable us to check any hypothesis for finding the p_2, given p_1 and p_3, by an independent exact calculation of p_2; but in the case of the ellipse, this independent knowledge of the perimeter is just what we seem to be lacking.

Sagredo: Let us not give up yet.

Simplicio: I have no further hypotheses. But I do notice something that might be worth mentioning concerning these rectilinear cases. When a square is elongated or compressed in a direction parallel to a pair of sides, the figure so obtained (a rectangle) is equiangular but not equilateral (Fig. 8), whereas when a square is elongated or compressed in the direction of a diagonal, the new figure (a rhombus) is equilateral but not equiangular (Fig. 9). Finally, when a regular hexagon is elongated or compressed in a direction parallel to a pair of sides, and hence to a diagonal as well, the new hexagon so obtained is neither equiangular nor equilateral (Fig. 10)!

Salviati: Your observations are not without relevance to an appreciation of what happens to the boundary line when a circle (or a many-sided polygon circumscribable or inscribable with respect to a circle) is elongated to produce an ellipse (or the corresponding polygon that would circumscribe an ellipse or that could be inscribed within one).[32] A comparison of Figs. 8, 9, and 10 suggests that it is only in the most exceptional cases that equality of angles or of sides could be preserved. Among regular polygons, it is only from the square that either equiangular or equilateral figures can be obtained by simple (uniform, unidirectional) elongation or compression. If you will reflect upon the matter for a moment you will see that, in general, squares elongate to *parallelograms,* which are rectangles or rhombuses only if the elongation is directed parallel to a pair of sides or to a diagonal respectively. But I notice that Sagredo is quite lost in thought; he seems to have been occupied with other matters while we have been speaking.

Sagredo: I do apologize. I have been trying to connect together certain thoughts that I dare to hope might yet enable us to solve the abstruse problem of the elliptical perimeter.

Salviati: I am sure that Simplicio and I will be delighted if you will share these ideas with us. Allow me to make one final point, however, before leaving the examples of Figs. 8, 9, and 10. It is this: if any suggestion for a perimeter formula for ellipses is based upon some assumed principle (of stretching, for example) that is not shown to apply specifically to ellipses but would apparently apply equally well to other figures, such as elongated polygons, and if any of the corresponding perimeter formulae that would be obtained for these figures can be shown to fail, then the proposed formula in the case of the ellipse must be deemed highly suspect.

Sagredo: Yes, I accept this. I shall not claim that my ideas are specifically connected with the ellipse, but I do believe that the underlying principle that I shall use will turn out to have such generality as to apply alike to ellipses and to any rectilinear or curvilinear cases that you might care to devise in order to test it or its consequences. I am not yet confident, however, that I can state the general principle and I am not quite sure how best to organize the ideas that have only now come together. So please allow me some time to develop my proposal with you. The basic ideas came to me in connection with circles and parts of circles, not with ellipses at all. This occurred in two recent dreams, which, if it had not been for our present discussion, would probably have remained unrecovered and unavailable for sharing with you. If you will bear with me, I should like to recount this background to you now, and then, in our next hour, examine the relevance of the undoubtedly simple insights to the problem of the elliptical perimeter.

Salviati: I feel sure that Simplicio will be as agreeable as I am to your proceeding as you suggest.

Simplicio: This particular perimetric problem has so caught hold of me that I welcome the opportunity of giving further consideration to it, especially since I do not see any other way in which we ought to be proceeding toward understanding the relations between areas and perimeters except by getting to the bottom of such problems as this.

Sagredo: Neither do I.

Salviati: Well, you have chosen a difficult figure to begin with. But please proceed, Sagredo.

Sagredo: In the first dream I found myself in a modern bookshop. After marveling at the range of stock available, I fell to wondering how to calculate the total length of paper on the large roll from which is drawn off what is required for wrapping the books. The length of paper would depend only upon the inside and outside diameters and upon the thickness of the paper. The latter I judged to be approximately one two-hundredth part of an inch by recollecting that a book made from paper of about the same thickness would contain something like 400 pages, that is, 200 leaves, per inch of its thickness. The inner diameter of the roll appeared to be about 2 inches and the outer diameter about 9 inches. It seemed reasonable to assume that the cylinder of paper when unrolled undergoes a change in shape only, not a change of volume. The length of the paper could, of course, be found empirically after first unwinding it so that the spiral formed by an edge becomes a straight line. But it occurred to me that the calculation that would render such an inconvenient practical measurement unnecessary was indeed a very simple one. As the annulus of area $(\pi/4)(9^2 - 2^2)$, or approximately $(11/14) \times 77$ square inches, could be transformed by unwinding into the very long and very narrow rectangle, of length L inches (which we seek) and width t inches (the thickness of the paper), we may write

$$L \times \frac{1}{200} \approx \frac{11}{14} \times 77.$$

From this we find the length to be 12,100 inches, a little over 1,000 feet.

Simplicio: I must admit that your calculation is most apt. I should have supposed the spiral winding to be replaced by a series of concentric cylindrical surfaces—or, rather, a series of very thin hollow cylinders—in order to get an amenable subject for calculation. But your method is clearly less involved.

Sagredo: The procedure you suggest occurred to me also, and although I did not then attempt to carry it out, I am sure that it will involve no difficulty. Adjacent layers differ in diameter by twice the thickness of the paper. Hence the required length, which is the sum of the circumferences $C_1 + C_2 + C_3 + \ldots + C_{700}$,

$$= \pi(d + t) + \pi(d + 3t) + \pi(d + 5t) + \ldots + \pi(d + 1{,}399t)$$

$$= \pi\left\{700 \times \frac{1}{2}(d + t + d + 1{,}399t)\right\}, \text{ since the terms are in A.P.,}$$

$$\approx \frac{22}{7} \times 700(d + 700t), \text{ where } d = 2 \text{ and } t = \frac{1}{200} \text{ (inches),}$$

$$= \frac{22}{7} \times 700\left(5\frac{1}{2}\right).$$

Thus we obtain 12,100 inches for the length of the paper, as before.

Simplicio: Very good. I dimly perceive from your first calculation that you are on the track of a method connecting the area with the length of the boundary or with lines concentric with the boundary, which you hope will be applicable to the elliptical case.

Sagredo: You perceive correctly. But please bear with me. We ought to proceed very carefully if we are to be properly assured of our conclusions. Also, I wish to indicate the origins of the idea that has so recently come to me concerning the elliptical perimeter. I want nothing to be hidden so that if there is any error it can be exposed without trouble. Further results occurred to me in my first dream, of which I now remember only the following. As you know, for circles having diameters as $1:2:3:4:5: \ldots$, the areas are as $1:4:9:16:25: \ldots$. Hence, if such circles are arranged concentrically, the annular spaces between them will be as $3:5:7:9: \ldots$. That the differences between successive terms of such a sequence of square numbers form an arithmetic progression should be familiar to us from our first author's discussion of uniformly accelerated motion.[33]

In my second dream, I found myself watching some athletes training on a running track. The task they had set themselves was to run a single lap of the

400-meter track in exactly one minute. This they did several times with only a short spell after each attempt, and rarely did they differ from their target time by more than a single second.

Simplicio: Your mention of the running track, Sagredo, and the expectations I have of hearing about your reflections there, remind me of how it was held in Greek philosophy that the contemplative spectator at the Olympian festival was more worthily engaged than even the competitors themselves.

Sagredo: You mean, because the philosophic mind could draw inspiration from such displays of athletic excellence and, by comparison with other kinds of excellence, reach out and, perhaps, lay hold of true excellence, perfection and being, per se? And that such experiences could afford opportunities to ascend from *πρᾶξις* to *θεωρία*, from the particular to the general and from transient substance to eternal form?

Simplicio: Well possibly, but I don't recall that it was quite this that was offered as a reason. Rather, it had a connection with the Aristotelian doctrine of the "three lives,"[34] did it not, Salviati?

Salviati: Yes it did, though the doctrine goes back to Plato, if not to Pythagoras. Plato points out that, while some men seek money before all else (as offering them the prospect of satisfying their manifold desires) and others seek satisfaction from victorious achievement (or from the consequent respect or adulation of others), there is yet a third class, few in number, whose preoccupation it is to seek knowledge and wisdom, the desire for which has no need of ulterior purposes, being complete and good in itself.[35] Later writers attributed this same classification to Pythagoras.[36] In order to explain the superiority of the contemplative vision of the intellect, of philosophy, over gain-seeking and honor-seeking pursuits, Pythagoras—so the tradition or legend has it—told as a parable how, while some are attracted to an athletic festival in order to buy or sell for profit and others in the hope of achieving victory and honor, there remain some few among the spectators who desire only to partake contemplatively of the experience, relishing and understanding it and, perhaps, through reflection upon tacit *μῆτις* to make some approach to definitive *ἐπιστήμη*.

Sagredo: Ah yes, I do recall something of this now; the trichotomy appeared to me rather simplified, and yet I do find it interesting that while all human beings grow up with some capacity to derive satisfaction in a variety of ways, many do come to adopt a style of life that is centered on one focus, and they tend to have little sympathy or understanding for those who are otherwise motivated. And I find it fascinating to observe the ways in which different cultures encourage or discourage particular styles of life. I cannot help being reminded by your last remarks, Salviati, of the lines of an English poet writing at the beginning of the nineteenth century: "Getting and spending, we lay waste our powers:/Little we see in Nature that is ours . . ."[37]

Salviati: It is fitting that we remind ourselves of such matters; they really are of perennial interest. Themistocles may have preferred to have been an Achilles rather than a Homer, but twenty-two centuries later a British general held to an opposite view, declaring on the eve of his death that he would rather have written a certain poem than achieve glory by defeating the enemy on the morrow.[38] And then there is Socrates' remark, on the very last day of his life: "Wars are occasioned by the love of money, and money has to be acquired for the sake and in the service of the body; and by reason of all these impediments we have no time to give to philosophy."[39]

Simplicio: You remind me of how Socrates and Plato recommended rigorous mathematical studies as going some way toward freeing the soul from the distractions occasioned by bodily passions—and of how arithmetic, geometry, astronomy, and harmonics (all understood as abstract, theoretical disciplines) were especially advocated as best leading the soul from the world of appearances to that of ideas.[40]

Sagredo: But do you know, I actually overheard a French scientist of the present century declare that without ambition and without vanity, no one would enter a profession such as his own, so contrary was it to the natural appetites![41]

Simplicio: It is to be hoped that such sentiments have not become too widespread. But let us not be led on now to discuss the troubles of the present age, Sagredo. May we return instead to your proposal to solve the particular problem with which we are confronted?

Sagredo: As you wish. In my second dream I felt something of the stimulation that Galileo experienced on his visits to the Venetian arsenal.[42] It was while watching the four runners training together, one in each of four adjacent lanes of the track, that I realized that some of the results I had arrived at for circles would apply also in the case of certain other figures without the slightest inaccuracy. I began by considering the problem of finding the different distances that would have to be covered by the runners in different lanes if each were required to make one complete circuit. As you know, in conducting races, an appropriate compensation is made by staggering the starts (see Fig. 11). In my dream the groundsman informed me that for a one-lap race of 400 meters he advanced the starting mark by 7.854 meters for each lane of width $1\frac{1}{4}$ meters. Thereupon I asked him the reason why this was the correct distance between the successive starting lines. All that he was able to tell me, however, was that if one were to measure 400 meters back from the finishing line along the inner white line of the inside lane then one would come back exactly to the same point again, so that, in the first lane, the starting line and the finishing line coincide. But in the second lane, the same procedure carried out along the next longer white line would not bring one back to the starting line but to a point 7.854 meters from it. In the third lane, the gap would be *twice* 7.854 meters; in the fourth lane it would be *three* times 7.854 meters; and so on. I could see that he was not able to give a theoretical justification for the distance of 7.854 meters, although he did say that this would have to be "adjusted in proportion" if lanes wider or narrower than $1\frac{1}{4}$ meters were to be used.

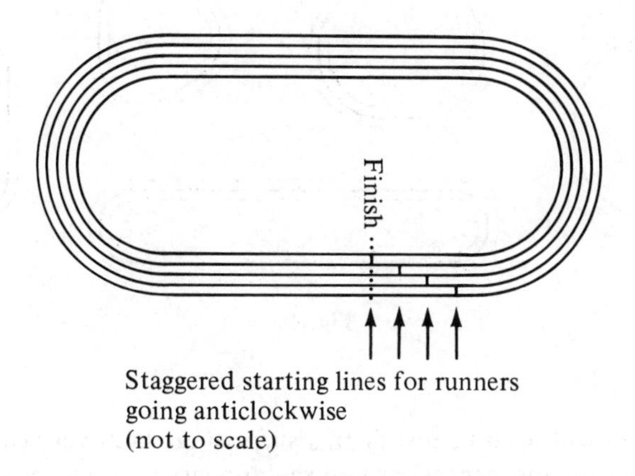

Fig. 11

Simplicio: Is this 7.854 meters simply calculated by taking 2π times the $1\frac{1}{4}$ meters? Let me see: $2\frac{1}{2} \times \pi \approx 3.1416 + 3.1416 + 1.5708 = 7.8540$.

Sagredo: Quite correct.

Simplicio: Well, I suppose that would be the appropriate calculation if the track were a perfect circle. But I must admit that I am not absolutely clear about all this, especially as the running track is *not* simply circular. It is made up of two straights joined by semicircular ends, so that a runner covers about the same distance along the straights as he does around the bends. I cannot see that this has been taken into account.

Sagredo: The shape is as you say, but it makes not the slightest difference whether the track is perfectly circular or whether it has two short straights with large semicircular bends or two long straights with small semicircles at the ends. Or again, three or more circular arcs may join an equal number of straights as indicated in Fig. 12. I say that in each of these cases the lengths of the outer dotted line exceeds that of the full line by exactly the same amount as the full line exceeds the inner dotted line, provided only that these lines are equally spaced. For any two of the lines distance t meters apart, the difference in length is simply $2\pi t$ meters. This turns out to be the case whether the track is large or small; it is true alike for an indoor track of ten or more laps to the kilometer and for a motor racing circuit of several kilometers per lap or, if you like, for an imaginary track across space from here to the moon.

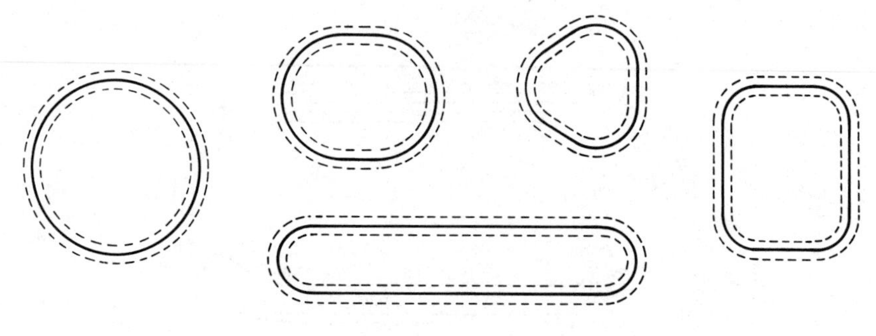

Fig. 12

Simplicio: You will need to justify this surprising claim very carefully before I can accept it. At the moment what you say appears to me to be hardly possible.

Sagredo: In the second calculation for the length of the wrapping paper, you seemed to accept that for equally spaced circles the increments in the circumferences would be uniform, regardless of the actual sizes of the circles. If $C_1 = \pi d_1$ and $C_2 = \pi d_2$, then $C_2 - C_1 = \pi(d_2 - d_1)$; that is, the difference in the circumferences of any two circles is simply equal to π times the difference in their diameters. Circles of diameters 101 and 102 units differ in their circumferences by π units, just as do circles of diameters 1 and 2 or circles of diameters 10,001 and 10,002.

Simplicio: I see; and do the straights in the compound forms of Fig. 12 make no difference?

Sagredo: If there are two parallel straights, they are complemented by two semicircles. If there were two *non*parallel straights, then one bend would be more than a half circle and the other less; if one were two-thirds of a circle, the other would be one-third.

Simplicio: Yes, but the longer bend would be two-thirds of a *large* circle, the shorter bend one-third of a *small* circle.

Sagredo: This does not matter at all; neither does it make any difference how many circular arcs are involved. Let me demonstrate these claims with reference to the diagram in Fig. 13, which I think you will agree is sufficiently representative of the general case.

Length of full line

$$= s_1 + \frac{\alpha}{360} \text{ of } 2\pi r_1 + s_2 + \frac{\beta}{360} \text{ of } 2\pi r_2 + s_3 + \frac{\gamma}{360} \text{ of } 2\pi r_3$$

$$= s_1 + s_2 + s_3 + \pi\left\{\frac{\alpha r_1 + \beta r_2 + \gamma r_3}{180}\right\}.$$

Clearly, the length of the outer dotted line is obtained by writing $r_1 + t$ in place of r_1, $r_2 + t$ in place of r_2, and $r_3 + t$ in place of r_3, while the length of the inner dotted line is given by substituting instead $r_1 - t$, $r_2 - t$, $r_3 - t$, in place of r_1, r_2, r_3, respectively. Thus the lengths of the dotted lines are given by

$$s_1 + s_2 + s_3 + \pi\left\{\frac{\alpha(r_1 \pm t) + \beta(r_2 \pm t) + \gamma(r_3 \pm t)}{180}\right\}.$$

It follows by subtraction that the length of the outer dotted line exceeds that of the full line by precisely the same amount as the full line exceeds the inner dotted line, the common difference being $\pi t\{(\alpha + \beta + \gamma)/180\}$. You will observe from the diagram, Simplicio, that the angles dimensioned $\alpha°$, $\beta°$, $\gamma°$, total 360°.

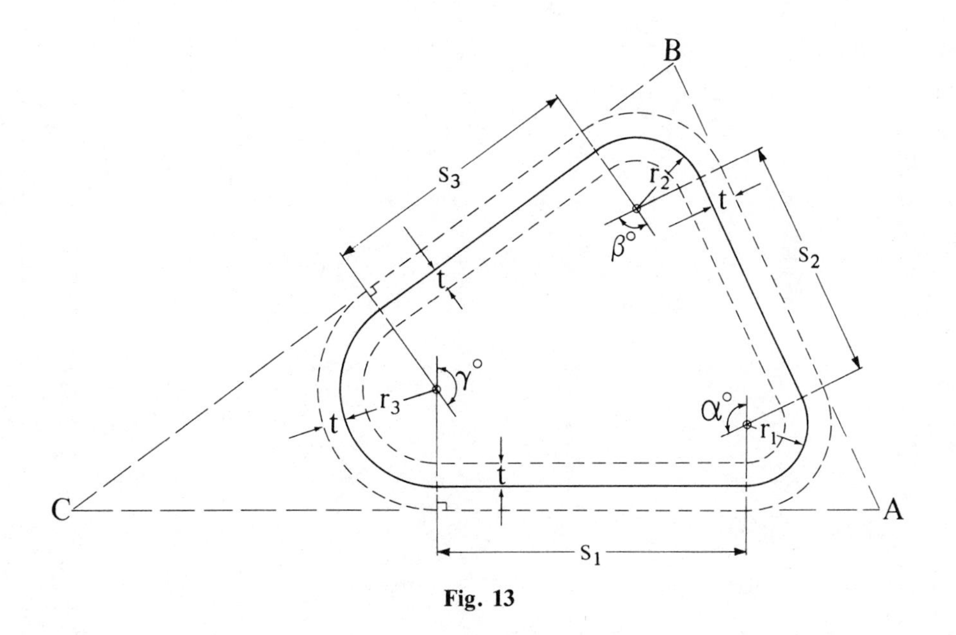

Fig. 13

Simplicio: I see that these three angles are respectively the supplements of the angles of triangle ABC. Hence $\alpha + \beta + \gamma = 180 - A + 180 - B + 180 - C$, and since A, B, and C sum to 180 degrees, it follows that $\alpha° + \beta° + \gamma° = 360°$.

Sagredo: Quite so. You might also notice that the six arms of these angles dimensioned $\alpha°$, $\beta°$, $\gamma°$ can be brought together without change of direction so as to form just three rays from a common vertex, thereby exhibiting that three angles of these magnitudes are together equivalent to a single rotation. Clearly then, our expression, $\pi t\{(\alpha + \beta + \gamma)/180\}$, for the common difference in the lengths of the adjacent equally spaced lines demarking the double track, simplifies to $2\pi t$. Notice how this expression is independent both of the lengths of the straights and the radii of the bends. It is precisely the same as the result that applies for the difference between the circumferences of concentric circles having *any* radii that differ by t.

Simplicio: Yes, I am now convinced of the correctness of all that you have claimed.

Salviati: Sagredo, you are to be congratulated alike on the accuracy and on the clarity of your analysis.

III

The interest of science lies in the art *of making science.*

—Paul Valéry

Would that I could as easily discover the truth as point out error.

—Cicero

The human understanding is like a false mirror, which, receiving rays irregularly, distorts and discolors the nature of things by mingling its own nature with it.

—Francis Bacon

Like those who walk alone in the dark, I resolved to proceed so slowly and with such meticulous circumspection, that, though my advance be but small, I should at least guard myself from falling.

—Descartes

False facts are highly injurious to the progress of science for they often endure long; but false views, if supported by some evidence, do little harm, for everyone takes salutory pleasure in proving their falseness.

—Charles Darwin

Every sentence I utter must be understood not as an affirmation, but as a question.

—Niels Bohr

As science develops, there must always be an area just beyond its borders where imagination rules, where analogy and simplicity and coherence compensate for lack of specific empirical verification [or, in mathematics, lack of deductive proof].

—Ernan McMullin

HORA TERTIA

Sagredo: I am pleased that the first stage of my exposition met with your approval. May we now pass to the next stage, in which area considerations are combined with perimetric ones?

Salviati: Simplicio and I will give you our fullest attention.

Sagredo: Your vigilance is greatly appreciated, I assure you. Let me begin by asserting that in all such cases as we have been considering (Figs. 12 and 13), *the area lying between the dotted lines* (distant t on either side of the full line) *is measured by the product, length of full line times 2t.* This may be established by independently obtaining an expression for the double-track area and showing that it is equivalent to the product of the form $p \times 2t$. The independent determination of the area may be carried out either by finding the difference of the areas bounded by the outer and by the inner dotted lines or else by adding together the rectangles and the sectors of annuli composing the double track.[43] In the case represented in our general diagram (Fig. 13), the area between the two dotted lines is composed of three rectangles of total area $(s_1)(2t) + (s_2)(2t) + (s_3)(2t)$, together with three sectors of annuli having total area

$$\frac{\alpha}{360} \times \pi\{(r_1 + t)^2 - (r_1 - t)^2\} + \frac{\beta}{360} \times \pi\{(r_2 + t)^2 - (r_2 - t)^2\}$$
$$+ \frac{\gamma}{360} \times \pi\{(r_3 + t)^2 - (r_3 - t)^2\}.$$

Altogether, therefore, we have for the entire double-track area:

$$2s_1t + 2s_2t + 2s_3t + \frac{\pi \cdot \alpha(2r_1)(2t)}{360} + \frac{\pi \cdot \beta(2r_2)(2t)}{360} + \frac{\pi \cdot \gamma(2r_3)(2t)}{360},$$

that is,

$$2t\left[s_1 + s_2 + s_3 + \pi\left\{\frac{\alpha r_1 + \beta r_2 + \gamma r_3}{180}\right\}\right],$$

where the expression within the square brackets is precisely that previously obtained for the length of the full line. Hence, the *area of the double track is given by perimeter of middle line times constant total width.*

This insight, though not of course all of the details, occurred to me in the second dream.[44] Let us now proceed to examine the application of the same line of reasoning to the determination of the perimeter of the ellipse. (See Fig. 14.)

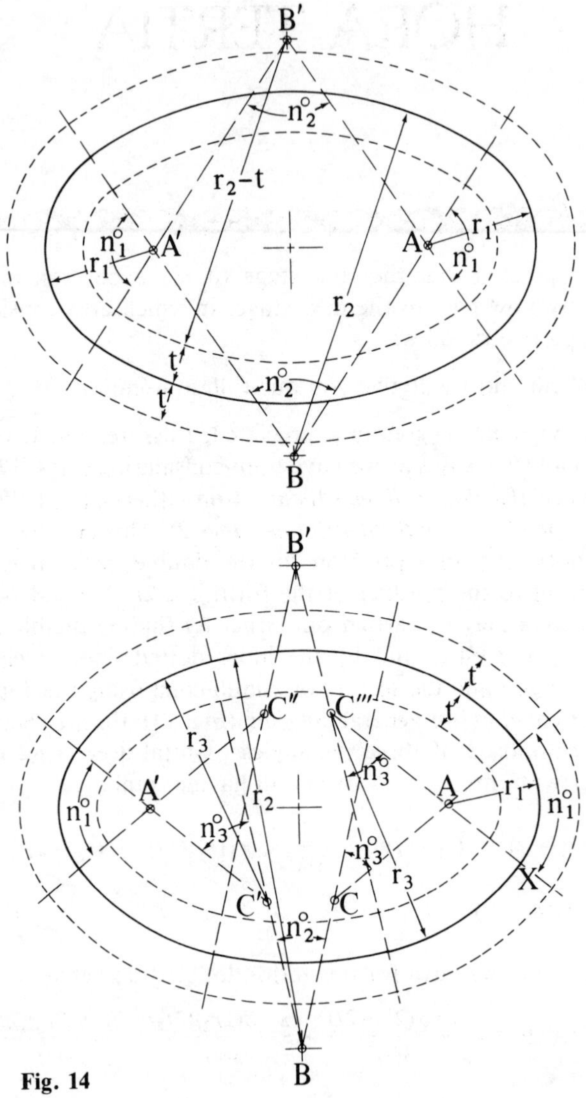

Fig. 14

First of all we notice that the same general result for the double-track area will apply for any of those centrally symmetrical constructions in which four, eight, or more arcs of circles blend with each other so as to form closer and closer approximations to a true ellipse with given axes. Two examples are shown in the diagrams of Fig. 14.[45] In the second diagram, if the arcs are described in an anti-clockwise direction starting from X, the centers are A, C, B, C', A', C'', B', C''', and the angles swept out (n_1, n_3, n_2, n_3, n_1, n_3, n_2, n_3 degrees) together amount to one complete rotation. In the working that follows, for the eight-centered case, the substitution of zero for n_3 gives the corresponding expressions for the four-centered construction of the first diagram. The perimeter of the full line

$$= 2 \times \frac{n_1}{360} \text{ of } 2\pi r_1 + 2 \times \frac{n_2}{360} \text{ of } 2\pi r_2 + 4 \times \frac{n_3}{360} \text{ of } 2\pi r_3$$

$$= \pi(n_1 r_1 + n_2 r_2 + 2n_3 r_3)/90.$$

I say then that the expression for the area of the double track is

$$2t \times \pi(n_1 r_1 + n_2 r_2 + 2n_3 r_3)/90.$$

In order to verify this by a difference-of-areas calculation, we use the expression $n/360$ of $\pi(r_o^2 - r_i^2)$ for the area of the sector of an annulus, where r_o, r_i are the measures of the outer and inner radii, respectively. The double-track area lying between the dotted boundary lines is thus given by

$$\frac{2n_1}{360} \text{ of } \pi\{(r_1 + t)^2 - (r_1 - t)^2\} + \frac{2n_2}{360} \text{ of } \pi\{(r_2 + t)^2 - (r_2 - t)^2\}$$
$$+ \frac{4n_3}{360} \text{ of } \pi\{(r_3 + t)^2 - (r_3 - t)^2\}.$$

This

$$= \frac{2\pi}{360}\{n_1(2r_1)(2t) + n_2(2r_2)(2t) + 2n_3(2r_3)(2t)\}$$

$$= 2\pi t(n_1 r_1 + n_2 r_2 + 2n_3 r_3)/90.$$

Simplicio: The agreement is exact. Bravissimo, Sagredo!

Sagredo: Yes, Simplicio, the agreement is exact and does not merely approach exactness for vanishingly small values of t. And with, say, 24 or 48 suitably chosen short circular arcs properly blended with each other, the deviation from a truly elliptical form could be made quite negligible. In our theoretical analysis, of course, we are not restricted to any particular number of approximating arcs. And is it not obvious from the working just given that the same general relation, area of double track $= 2pt$, must continue to hold no matter how many circular arcs are involved? In every such case, then, there will be two equally valid methods of determining the area of the same double track of width $2t$ going right round the figure. And so there would be in the case of the double track enclosed between concentric similarly placed ellipses with semiaxes $a + t$, $b + t$, and $a - t$, $b - t$, respectively, if we but knew the perimeter of the ellipse with semiaxes a, b. But one method is enough. Let us determine the area of a double elliptical track by the difference method and then, by working backward, find the elliptical perimeter itself.

Simplicio: Oh excellent, excellent! This is surely the path we have been seeking. By means of the indubitably known area formula, we shall obtain the true measure of the double track and then equate this to the alternative expression involving the perimeter. I cannot wait to see how this will work out.

Sagredo: Nor I! As I have explained, the idea of applying this to the ellipse has occurred to me only under the stimulus of our present discussion, and so I have not performed the calculation before. The area of the double track is measured by $\pi(a + t)(b + t) - \pi(a - t)(b - t)$; alternatively, it is given by $2p_e t$. Hence,

$$
\begin{aligned}
p_e &= \frac{\pi}{2t}(ab + at + bt + t^2 - ab + at + bt - t^2) \\
&= \frac{\pi}{2t}\, t(2a + 2b) \\
&= \pi(a + b).
\end{aligned}
$$

Oh, look what has happened! But why, why?

Simplicio: From the indubitably correct has issued the indubitably incorrect, from the incontrovertible has come the impossible!

Sagredo: I am no less disconcerted than you, Simplicio, but we must respond rationally and say either that from the apparently incontrovertible has issued the impossible or that from the truly incontrovertible has issued the apparently impossible.

Salviati: And which alternative do you favor, Sagredo?

Sagredo: We know that the formula $p_e = \pi(a + b)$ is erroneous,[46] so I suppose that I must choose the first alternative, but please do not expect me to find the error. I have tried to proceed with the utmost care. Now, I can only appeal to you for assistance.

Salviati: And what do you think, Simplicio?

Simplicio: You previously mentioned two elliptical perimeter formulae more accurate than $p = \pi(a + b)$, but I think you indicated that these were still inexact (see p. 23). I was then about to ask you to tell us, once and for all, the true formula for the perimeter of the ellipse, if such is discoverable and has been discovered.[47]

Salviati: As I shall be glad to indicate in due course, the true perimeter is measured by the sum of a certain infinite series. This form is not very well adapted to practical use, and the challenge to find convenient close approximations has been met by various mathematicians. The following are some noteworthy examples[48]:

$$p = \pi\left\{a + b + \frac{1}{2}(\sqrt{a} - \sqrt{b})^2\right\} \tag{1}$$

$$p = \pi[3(a + b) - \sqrt{\{(a + 3b)(3a + b)\}}] \tag{2}$$

$$p = \pi\left\{a + b + \frac{3(a-b)^2}{10(a+b) + \sqrt{(a^2 + 14ab + b^2)}}\right\} \tag{3}$$

$$p = \pi\left[\frac{9}{8}(a+b) - \frac{5}{8}\sqrt{(ab)} + \frac{3}{8}\sqrt{\left\{\frac{1}{2}(a^2 + b^2)\right\}}\right] \tag{4}$$

Sagredo: These are interesting formulae, and I suppose that the approximate coincidence of the results they produce is not merely coincidental but will be understandable in terms of some theory of the ellipse. I am preoccupied, however, with two outstanding questions: Where did my reasoning go astray? How is the exact elliptical perimeter to be determined? As you know, I am of a cautious disposition, as a result of my studies if not by nature, and I have tried to follow the advice that Plato has one of his characters give about guarding against being misled by slippery resemblances.[49] You will understand, therefore, how disconcerting it is for me to realize that, in applying the double-track formula in the case of concentric and similarly aligned ellipses, I have somehow strayed beyond its range of applicability. Yet this formula, area = $2pt$, is assuredly true for the blending circular arc constructions that may be brought into as close agreement as we care to arrange with any given ellipses.

Salviati: It is because I know you to be a man remarkably free of conceits that interfere with the understanding that I am confident you will yet discover the error that has led you astray. I should like you first to consider whether your double-track formula applies in the case of such simple figures as rectangles, rhombuses, or other quadrilaterals.

Sagredo: I feel sure that it *is* exactly applicable in all these cases, and I say this because I think that I can give the proof in the general case of any double polygonal track, which covers all the simpler cases you have mentioned.

Salviati: Well, Sagredo, before you show us your general proof, let me ask you or Simplicio to apply your calculation method to the case of three concentric and similarly aligned rhombuses with semidiagonals $a + h$, $b + h$; a, b; $a - h$, $b - h$. What will be the area contained between the inner and the outer boundaries?

Simplicio: Since any rhombus is precisely half of the enclosing rectangle (drawn so that the sides of each figure are parallel to the diagonals of the other), the area of the rhombus is given by half the product of its diagonals—or by twice the product of its semidiagonals. Hence the required double-track area is measured by the difference

$$\begin{aligned} &2(a+h)(b+h) - 2(a-h)(b-h) \\ &\equiv 2(ab + ah + bh + h^2 - ab + ah + bh - h^2) \\ &\equiv 4h(a+b). \end{aligned}$$

Salviati: And now, please Simplicio, calculate this same area by Sagredo's alternative method.

Simplicio: The side length is given by $\sqrt{(a^2 + b^2)}$, and the perimeter by four times this, and so the required area is . . .

Salviati: Go on Simplicio.

Simplicio: Well, of course, I was going to say, area $= p$ times $2h$, but this gives $8h\sqrt{(a^2 + b^2)}$, which certainly does not look as though it could equal the correct answer just found to be $4h(a + b)$.

Salviati: It is indeed a simple formality to show that, for nonzero values of h, a, b, and $a \neq b$, the expressions $8h\sqrt{(a^2 + b^2)}$ and $4h(a + b)$ cannot be equal. But can you tell us, Sagredo, the *source* of the discrepancy between these two answers?

Sagredo: I am convinced that it is simply due to a misapplication of the $2pt$ method and that a diagram (Fig. 15) will surely show us what has gone wrong. The outer area $= 2(a + h)(b + h)$, and the inner area $= 2(a - h)(b - h)$, the difference coming to $4h(a + b)$, as we have already seen. But now—why of course! You must see what has happened, Simplicio. When we come to write down the product *perimeter times width of the double track*, we must notice that the latter dimension, the width, is not $2h$ at all but some smaller value $2t$, where, by similar triangles, $t : h = a : \sqrt{(a^2 + b^2)}$. Hence, the value of $2pt$ comes to $2\{4\sqrt{(a^2 + b^2)}\}\{ha/\sqrt{(a^2 + b^2)}\}$. Accordingly, the area expression simplifies to $8ha$. Oh! I am sorry, I must have made some silly mistake. This is a ridiculous answer; it does not even contain b.

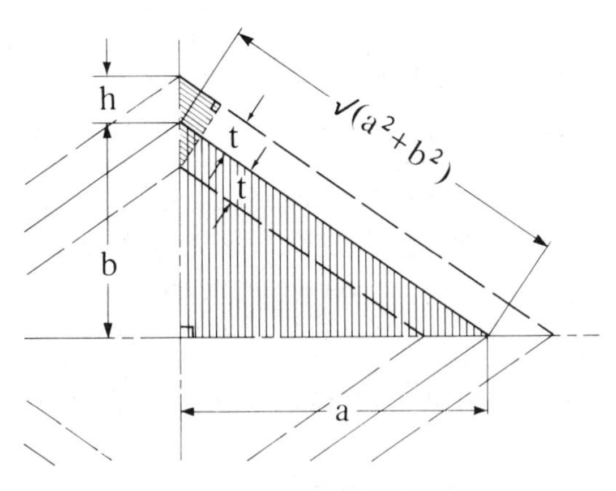

Fig. 15

Salviati: You have just said that a diagram would enable us to detect the previous error. I invite you now to detect your own error by examining the figure you have drawn. But it might take you some little time to find the source of the trouble since I fear that you are staring at your diagram under the spell of a preconceived interpretation.

Simplicio: I must be equally ensnared; I simply cannot see what has gone wrong.

Salviati: You were expecting the $2pt$ calculation to give the same answer as the difference method, were you not?

Simplicio: Yes.

Salviati: But you find that you get $8ha$ instead of $4h(a + b)$. Now observe that if $b = a$—that is, for the special case where the rhombuses are concentric *squares*, with semiaxes as before—the expressions $4h(a + b)$ and $8ha$ *are* identical. There is, then, no discrepancy between the two methods in this particular case. Now I put it to you both that you have imported into your interpretation of the diagram (Fig. 15)—indeed I ought to say, Sagredo, into your very drawing of it—a characteristic that is true for squares but for no other rhombuses. The constructiveness of your visual perception, or, if you like, the way you have attended to one part of the diagram and ignored the indications of another part, might be said to have paralleled expectations formed uncritically in your mind's eye.

Sagredo: I hope you are not playing games with me, Salviati. I can see nothing wrong with the diagram. In fact, I have taken care not to have it misshapen since I know from past experience that one can easily overlook essential properties of a geometric configuration if one sketches it too roughly. I have carefully indicated the double symmetry of the figure. What more is wanting?

Salviati: Please, Sagredo, could we leave this for a couple of minutes? I believe it will help if you tell us about the demonstration you had in mind for showing that the $2pt$ measure is correct for the area of any double polygonal track of uniform width, that is, for the class of figure that I gather you supposed would include the rhombus-shaped tracks we have just been considering.

Sagredo: The general diagram is Fig. 16. The demonstration is, I submit, as follows. $ABCD \ldots$ is any polygon; $A'B'C'D' \ldots$ and $A''B''C''D'' \ldots$ are drawn with their respective sides parallel to those of $ABCD \ldots$ so that the distances between the corresponding sides are constant as shown. The double track is composed of the trapezia $A'A''B''B'$, $B'B''C''C'$, $C'C''D''D'$, etc. Hence the area of the double track is given by

$$\frac{1}{2}(A'B' + A''B'')(2t) + \frac{1}{2}(B'C' + B''C'')(2t) + \frac{1}{2}(C'D' + C''D'')(2t) + \ldots$$
$$= (AB)(2t) + (BC)(2t) + (CD)(2t) + \ldots$$
$$= (AB + BC + CD + \ldots)(2t)$$
$$= 2pt,$$

where p is the total length of the middle line $ABCD$ . . . , and t is the width of each track.

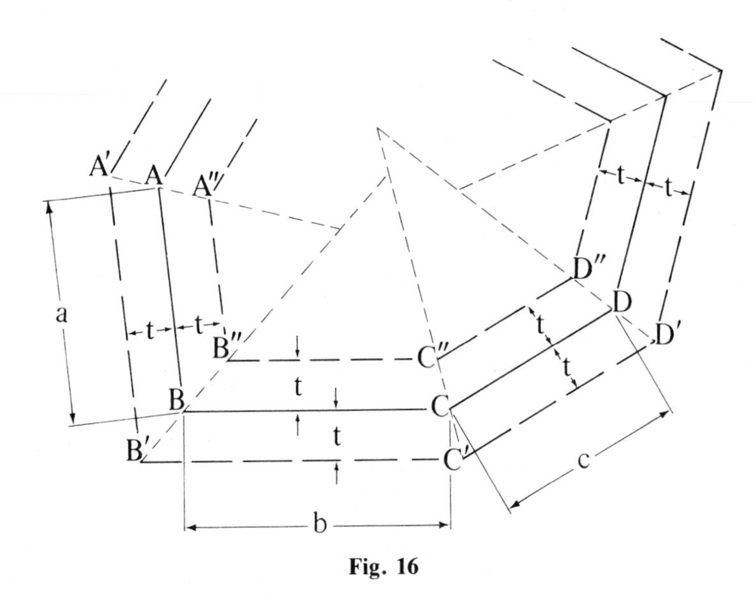

Fig. 16

Simplicio: A very clear demonstration that must surely apply to any set of three similarly aligned and similarly shaped polygons whether or not they are regular. Yet I cannot understand what went wrong with the rhombuses.

Salviati: The demonstration is perfectly in order, but some care is needed if misunderstanding of its range of applicability is to be avoided. Your remark about similar polygons seems to suggest that perhaps you have erroneously assumed that the polygons $ABCD$ . . . , $A'B'C'D'$, . . . , $A''B''C''D''$ . . . (Fig. 16), are geometrically similar. Perhaps the simplest counterexample is provided by a double track formed by a set of three *rectangles*. The rectangles are clearly *not* similar if the track is of constant width (Fig. 17i). If they are similar (Fig. 17ii), then the track changes its width ($t_2 \neq t_1$), except where the rectangles are squares.[50]

Sagredo: And, more generally, it would seem that constancy of track width and similarity of the polygons will coexist only in the case of regular polygons.

Salviati: This conjecture has a certain prima facie plausibility. But tell me, Sagredo, could not similar concentric rhombuses be separated by tracks of uniform width?

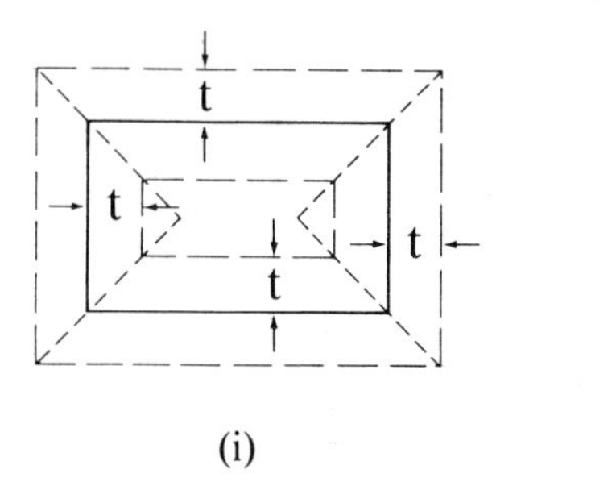

(i)

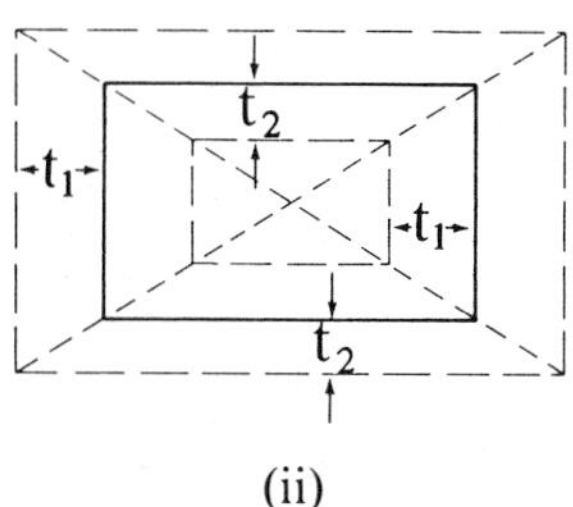

(ii)

Fig. 17

Sagredo: I am not quite sure. But I can see that we might have set up variously related rhombuses; however, since you specified at the outset that the semi-diagonals of the three rhombuses were to be given by $a + h$, $b + h$; a, b; $a - h$, $b - h$, then these quadrilaterals could not be similar, since $(a + h):(b + h) \neq a:b \neq (a - h):(b - h)$.

Salviati: Very good. You doubtless realized that I suggested we consider rhombuses so related in the hope that we might find some pointers to the relation between three *ellipses* having semiaxes corresponding to these semidiagonals. Now what about the other question that must be asked about these rhombuses. Are they or are they not separated by tracks of uniform width?

Sagredo: I did take it that the width of the tracks would not vary, but apparently this is where I blundered. Yes, Salviati, I now appreciate your invitation to us to reconsider this hasty and uncritical assumption. Let me correct my erroneous diagram (Fig. 15). I start with the semidiagonals, a and b, and increase and reduce each of these by h, as in Fig. 18. At last it is clear to me how the boundaries of the tracks must diverge from the parallel as a consequence of the requirement that the semiaxes of the outer and inner figures be $a \pm h$, $b \pm h$. Let me draw in these four small shaded right-angled triangles. Although each has a hypotenuse h units long, it is clear that t_1, t_2, t_3, t_4 cannot be equal so long as a and b are unequal. By similar triangles,

$$t_1 : b = h : s; \qquad t_2 : a = h : s; \qquad t_3 : (b + h) = h : s'; \qquad t_4 : (a + h) = h : s',$$

where

$$s' = \sqrt{\{(a + h)^2 + (b + h)^2\}} \quad \text{and} \quad s = \sqrt{(a^2 + b^2)}.$$

Incidentally, since $s'' = \sqrt{\{(a - h)^2 + (b - h)^2\}}$, it does not appear to me that s'', s, and s' could be in arithmetic progression, as the circumferences of circles with radii $r - h$, r, $r + h$ would be.

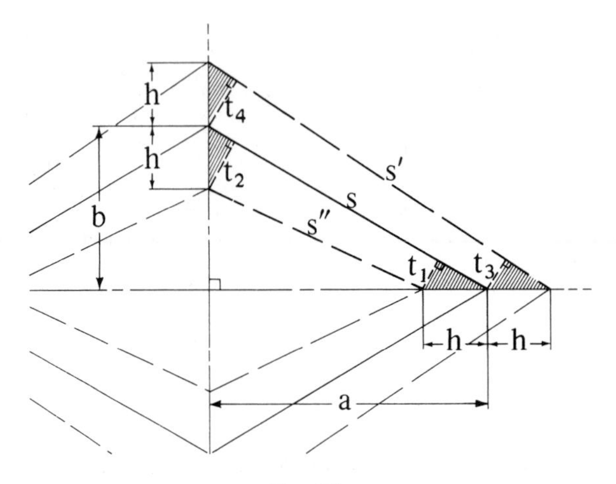

Fig. 18

Salviati: And the perimeters of our concentric ellipses?

Sagredo: I am unable to answer that.

Simplicio: Following on from the distinction you drew between the two different sets of concentric rectangles, Salviati, I should like to spend a minute looking at an alternative set of three rhombuses, related according to a different principle of construction. As in your general diagram (Fig. 16), Sagredo, so in my arrangement of three rhombuses (Fig. 19), the dashed lines are simply made parallel to the full line and equidistant from it. Will not all three rhombuses be exactly similar and will not their perimeters be in arithmetic progression? All four shaded triangles will be similar to triangles with sides a, $b, \sqrt{(a^2 + b^2)}$; hence, assuming a value of t at pleasure, the values of u and v are obtainable from

$$u : \sqrt{(a^2 + b^2)} = t : b, \quad \text{and} \quad v : \sqrt{(a^2 + b^2)} = t : a.$$

Sagredo: Yes, this is all correct, and the product $2pt$ will provide a correct measure of the double-track area, where p is given by $4\sqrt{(a^2 + b^2)}$. The demonstration given in connection with Fig. 16 applies to this example as a simple particular case. I have to admit, however, that I had not realized that similarity of the polygons and uniformity of track width could both be present in a case involving irregular polygons. You remember that under the stimulus of Salviati's rectangle example (Fig. 17), I surmised that the only similar polygons that could be arranged so as to be separated by uniform tracks were regular polygons. I see now that I was mistaken; your diagram (Fig. 19) presents a clear counterexample to this conjecture. But it is all too easy for us to be led away with all these side issues. Our main concern is to try to understand why the $2pt$ procedure fails in the case of the concentric ellipses; it seemed to offer such a promising means of

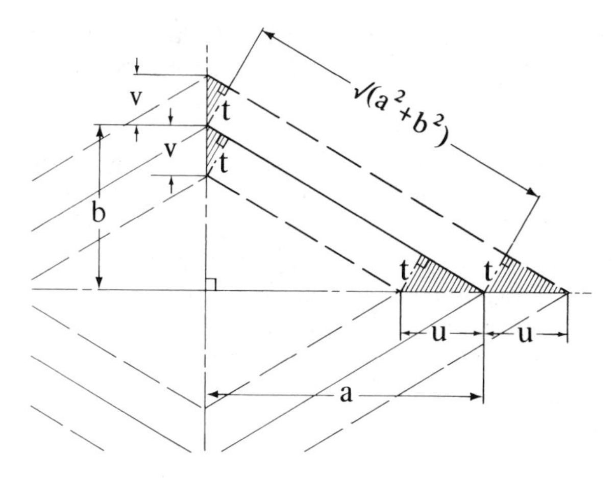

Fig. 19

determining the unknown elliptical perimeter. Having seen that this same procedure is demonstrably applicable in the case of the blending circular arc constructions (Fig. 14), we proceeded to apply it to the case of three concentric and similarly aligned ellipses, with semiaxes a, b and $a \pm t$, $b \pm t$. Clearly, the width of each track where it crosses each axis is exactly t units in this case. I assumed that this width (measured perpendicularly to the ellipses) would be constant all the way around, just as it assuredly is in the case of the approximating circular arc constructions.

Salviati: If you will give somewhat further consideration to the rhombuses of Fig. 18, I believe you will find reason to revise your intuitions in the case of the ellipses.

Sagredo: I assure you, Salviati, that I do desire to be purged of any erroneous judgments and I accept that it is Fig. 18 rather than Fig. 19 that is appropriate to our purpose because it is in thc former diagram that the intercepts on the axes are all equal.

Salviati: Very well then, returning to this figure, let me ask you what kind of average, τ say, of t_1, t_2, t_3, t_4, do you consider ought to be used in order that $2p\tau$ shall measure the double-track area correctly?

Sagredo: I realize that I have no reason simply to suppose that the arithmetic mean would be the appropriate average. For one thing, there was something arbitrary about taking the widths of the tracks in precisely the positions shown in Fig. 18. It suited the convenience of calculation to take t_3 and t_4 as measures of perpendiculars to the *outer* boundary. If instead all four widths had been taken at right angles to the middle line, the measures corresponding to t_3 and t_4 would have been slightly greater than the more easily calculated values that we obtained. It is up to us to decide upon representative measures of the track widths and our decision clearly influences the way in which these ought to be averaged for a given purpose. But wherever it is decided to take these widths, their measures will be expressible in terms of a, b, and h. So that, it seems to me, your question might be reduced to asking what function of a, b, and h will measure the double-track area. This function we have found by the difference method to be $2(a + h)(b + h) - 2(a - h)(b - h)$, simplifying to $4h(a + b)$. Thus, even if we do not know how to average the representative t values so as to obtain the appropriate value of τ, this latter, being defined by the relation $2p\tau = 4h(a + b)$, is easily determined. Since $p = 4\sqrt{(a^2 + b^2)}$, we find

$$\tau = \frac{h(a + b)}{2\sqrt{(a^2 + b^2)}}.$$

Salviati: You have answered well. And I take it that we are now all agreed that, even for such a clearly defined rectilinear configuration as that of Fig. 18, if we did not already know the double-track area by the difference method, it would be no simple matter to extract the proper value of τ from individually calculated width measures such as t_1, t_2, t_3, and t_4. How much more ought we to be prepared for difficulties in the case of such a subtle form as that composed of the elliptical and related curves!

Simplicio: Well, it appears that knowing the area and the uniform (or true average) track width, we could readily obtain the perimeter of the middle line; on the other hand, if we knew both the track area and the length of the middle line then we could find the track width. But if we knew neither the perimeter nor the true average track width, it would not be possible to find these two unknowns by the $2pt$ method, even though we knew the double-track area exactly by the difference method.

Sagredo: Yes Simplicio, apparently the mistake I made in the elliptical case was to assume that I really did know the track widths as well as the track area. You will remember that my approach to the ellipses was not by way of analogies—if I may say so, *crude* analogies—with such rectilinear figures as rhombuses, but by way of a hypothetical sequence of more and more accurately approximating circular arc constructions (Fig. 14). In every one of these constructions the widths of the two tracks were uniform and equal to each other. It never entered my head that this would not also be true for the double elliptical tracks. Indeed, I confess that I still do not really understand how this uniformity of track widths could fail to be preserved in the limiting case of the ellipses.

Simplicio: Nor do I.

Sagredo: And yet, Simplicio, you remember that we have two indubitable facts to reconcile. One is that the double elliptical track area is measured by $\pi(a + t)(b + t) - \pi(a - t)(b - t)$,[51] simplifying to $2\pi t(a + b)$. The other fact is that the perimeter measure p of the middle ellipse is greater than $\pi(a + b)$.[52] Forming now the equation $2p\tau = 2\pi t(a + b)$, it follows that, since $p > \pi(a + b)$, the mean track width τ must be less than t. Although I see that this is necessarily indicated, I am still anxious to have some more direct way of showing that the ellipses are closer together in intermediate positions than they are where they intersect the axes.

Salviati: Although it is true that any given ellipse, $x^2/a^2 + y^2/b^2 = 1$, say, can be approximated as closely as may be required by some theoretically specifiable construction of tangentially blending circular arcs having appropriately chosen centers, it is false to suppose that other concentric ellipses, such as our $x^2/(a \pm t)^2 + y^2/(b \pm t)^2 = 1$, could be found to be similarly approximated by a construction made up of "parallel" circular arcs, that is to say, made up of arcs having the same centers as were appropriate for the first ellipse. I know you will agree that a straight line perpendicular at its point of intersection to any one of a set of concentric circular arcs must likewise be perpendicular to every one of the other arcs.

Sagredo: Yes, this is simply because any normal must pass through the common center.

Salviati: True; but notice now that the ellipses to which we have been referring as concentric, are concentric only in the sense that their axes, being collinear, intersect at a common point. Except at the axes, the ellipses $x^2/a^2 + y^2/b^2 = 1$, $x^2/(a \pm t)^2 + y^2/(b \pm t)^2 = 1$, possess no common normals. The distance between these curves is not constant. These three ellipses no more run parallel to each other than do the three simply elongated octagons shown in Fig. 20.

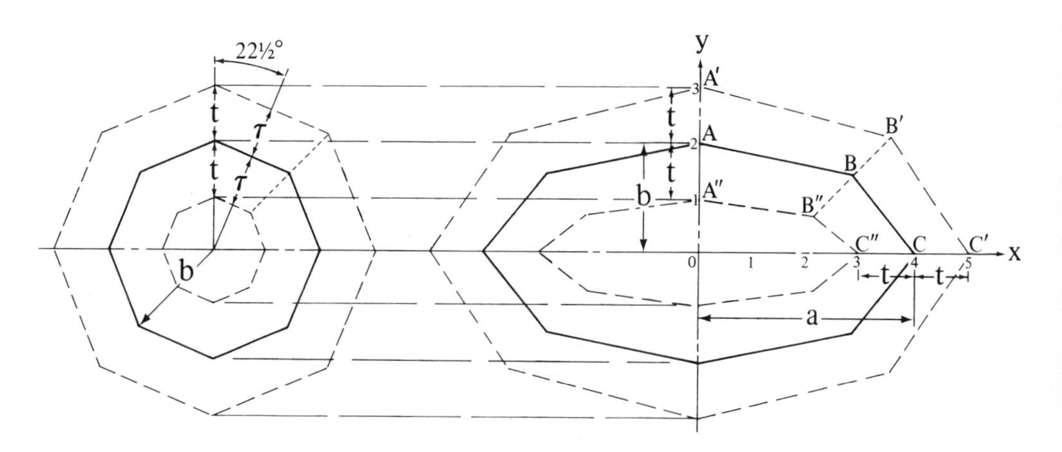

Fig. 20

Simplicio: I notice that, proceeding from A', A, A'' to B', B, B'', and again from B', B, B'' to C', C, C'', the boundaries appear to converge.

Salviati: This nonparallelism could be demonstrated easily by showing the inequality of the calculated gradients of the sides $A'B'$, AB, $A''B''$, and so on. Relative to the axes shown, point B has its y coordinate equal to the y coordinate of the corresponding point on the middle regular octagon, viz. $\sqrt{2}$, and the elongation is such that the x coordinate of B is twice the y coordinate. Thus B is the point $(2\sqrt{2}, \sqrt{2})$. Like reasoning shows B' and B'' to be the points

$$\left(\frac{5}{2}\sqrt{2}, \frac{3}{2}\sqrt{2}\right) \quad \text{and} \quad \left(\frac{3}{2}\sqrt{2}, \frac{1}{2}\sqrt{2}\right),$$

respectively. Hence it turns out that $B'B = BB'' = 1$ $(= t)$, and then it is evident that everywhere else the widths of the tracks must be less than t, provided the measurements are made at right angles to the intermediate octagonal line. I do not think it is necessary to go into further details now. I leave it to you to calculate at your leisure the double-track area by the difference method, the perimeter, and hence the mean width, τ_E say, of the elongated tracks.[53] But perhaps I should mention that here (as indeed in the case of the rhombuses shown in Fig. 18) the two elongated tracks will not usually have exactly the same width when measured along the same normal to the middle polygonal line at a randomly chosen point. And although we should not expect the mean width of the inner track to be the same as the mean width of the outer track, the quotient double-track area divided by twice midperimeter may still be used to define the combined mean track width.

Sagredo: Well, Salviati, you have opened our eyes to the difficulties involved in appreciating the character of these converging tracks. The uniform tracks are far easier to understand. I notice, for example, that in the case of the regular polygonal tracks represented on the left in Fig. 20, the uniform track width τ is given immediately by $\tau/t = \cos 22\frac{1}{2}°$. Hence the value of the expression $2p\tau$ can be calculated and shown to be equal to the independently evaluated difference of the measure of the areas of the outer and inner regular polygons.

Simplicio: As for the coaxial rhombuses and elongated octagons, it dawns on me now, Salviati, that perhaps you introduced these not merely as remote analogues to the three ellipses but as sets of polygons that may be regarded as belonging to a whole sequence of sets of polygons inscribed within the respective ellipses, each polygon having 2^n sides with n taking larger and larger values. But, like Sagredo, I am especially concerned to reach a better understanding of the relation between the true ellipses and the blending circular arc constructions of the kind shown in Fig. 14.

Salviati: Then I recommend the diagram in Fig. 21 to your attention. The radius of curvature of any (noncircular) ellipse varies continuously from point

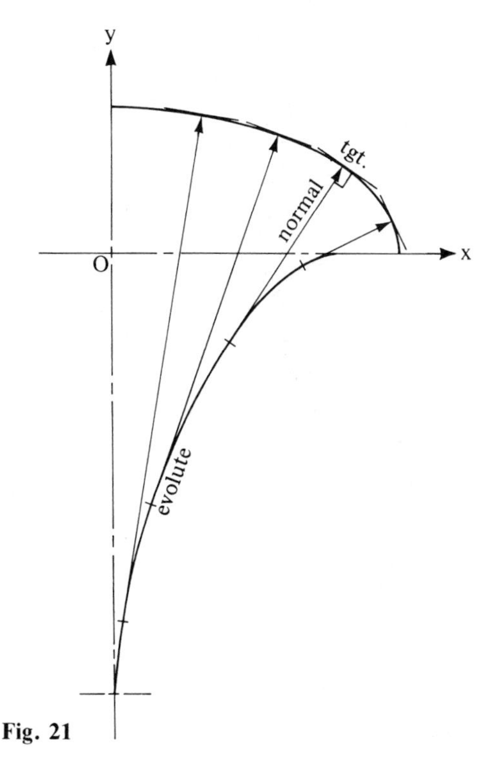

Fig. 21

to point, and the locus of the centers of curvature is called the "evolute." For the ellipse $x^2/a^2 + y^2/b^2 = 1$, the evolute is the curve $(ax)^{2/3} + (by)^{2/3} = (a^2 - b^2)^{2/3}$.[54] This diagram shows one quadrant of an ellipse, having its axes as 2 : 1, together with the corresponding quadrant of its evolute and a few representative radii drawn from it to the ellipse. A given curve is called the "involute" of the curve that is its evolute. Thus the ellipse $x^2/a^2 + y^2/b^2 = 1$, is the involute (or, to be more precise, one of the involutes) of the astroid $(ax)^{2/3} + (by)^{2/3} = (a^2 - b^2)^{2/3}$.[55] In the geometry of plane curves, an involute may be defined as the locus of a point P in an inextensible string kept taut as it is unwound from another curve (its evolute) lying in the same plane. And other points in this string, at fixed distances from P, will generate separate involutes equidistant from each other. In the next diagram, Fig. 22, P' and P'', at fixed distances from P will generate curves parallel to each other and to the curve generated by P, since there is a constant distance between them, measured along their common normal. Figure 22, then, shows the quadrants of three parallel curves. I keep the one generated by P the same as in Fig. 21, the other curves parallel to this ellipse being generated by P' and P'', where $PP'' = P'P = \frac{1}{4}OA = \frac{1}{2}OB$.

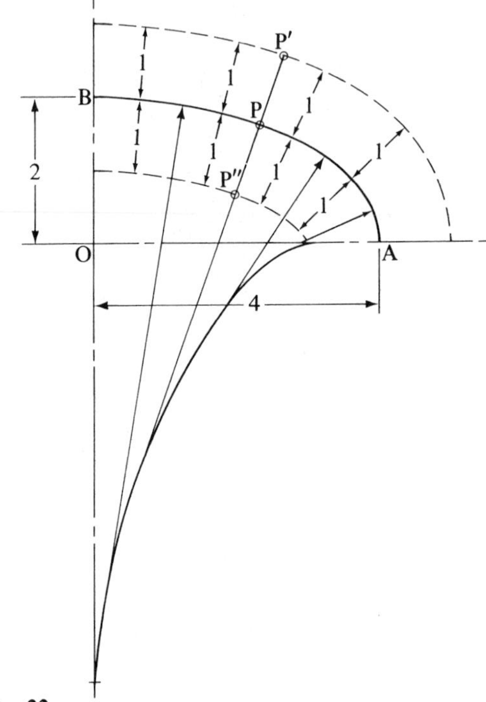

Fig. 22

Simplicio: Now I *am* confused! A little while ago you asserted—did you not?—that the three ellipses, with semiaxes $a + t$, $b + t$; a, b; $a - t$, $b - t$, have no common normals and do not run parallel to each other, being in this regard analogous to the approximating inscribed polygons rather than to the approximating circular arc constructions. But here, in describing Fig. 22, you appear to be contradicting what you have said earlier.

Salviati: If there seems to be a contradiction, it is probably because you are assuming that the dashed lines in Fig. 22 are *elliptical* arcs. *But they are not;* only the middle arc is elliptical. The construction that I have just described, by ensuring that the two track widths are uniform and equal, produces two new arcs that assuredly deviate from true elliptical form. In the cases that Sagredo examined initially, the components were curves of uniform curvature—circular arcs in Fig. 14, straight line segments in Fig. 16, and both of these in Fig. 13. Consequently, in those cases, the constructions that provided for the uniformity of the track widths simultaneously guaranteed that the nature of the curves would not be altered. I do not mean by this that the curves running parallel to each other were geometrically similar; in fact we saw that they were differently proportioned. I mean only that the components of the outer and inner boundary lines were straight line segments or concentric circular arcs according as the corresponding components of the middle line were straight or circular.

Sagredo: I can see that my mistake was to suppose that what is associated in these simple cases would also be associated in the case of curves of varying curvature. And what resistance I offered to dissociating in thought what are dissociated in fact!

Salviati: Do not reproach yourself, Sagredo. It is fitting that we proceeded as we did. We are now well prepared to further our investigations.

IV

They say that habit is second nature. Who knows but nature is only first habit?

—Pascal

Superior intellect is a large development of the faculty of association by similarity.

—Alexander Bain

Genius means little more than the faculty of perceiving in an unhabitual way.

—William James

He who seeks for methods without having a definite problem in mind seeks for the most part in vain.

—David Hilbert

Men who have excessive faith in their theories are not only ill prepared for making discoveries; they also make very poor observations.

—Claude Bernard

Not to be absolutely certain is, I think, one of the essential things of rationality.

—Bertrand Russell

Natural science does not simply describe and explain nature; it is part of the interplay between nature and ourselves; it describes nature as exposed to our method of questioning.

—Werner Heisenberg

What science deals with is an imagined *world.*

—D. O. Hebb

HORA QUARTA

Salviati: Well! Such activity! What are those equations, Simplicio? And what is that diagram, Sagredo?

Sagredo: We have just been consolidating the conclusions reached in the last hour; but I fancy that Simplicio is not completely satisfied.

Simplicio: I think I understand the situation correctly. You told us, Salviati, that the evolute of the general ellipse, $x^2/a^2 + y^2/b^2 = 1$, is the curve

$$(ax)^{2/3} + (by)^{2/3} = (a^2 - b^2)^{2/3}.$$

Therefore, the evolutes of the ellipses $x^2/(a \pm t)^2 + y^2/(b \pm t)^2 = 1$ must be the curves

$$\{(a \pm t)x\}^{2/3} + \{(b \pm t)y\}^{2/3} = \{(a \pm t)^2 - (b \pm t)^2\}^{2/3}.$$

Hence, corresponding to the three coaxial ellipses with semiaxes $a + t$, $b + t$; a, b; $a - t$, $b - t$, there would be three distinct evolutes!

Salviati: That is perfectly correct, Simplicio.

Simplicio: But only *one* evolute was used in the construction of Fig. 22.

Salviati: Does that surprise you?

Simplicio: I must admit I was somewhat puzzled a minute or two ago, but now I take it that the fact that there is just one evolute for all three curves is indicative that just one of the curves is an ellipse—the middle curve. An annulus, bounded by two concentric circles, is of uniform width; conversely, a curve "parallel" to a circular arc is a concentric circular arc. But it is a mistake to try to generalize this to noncircular ellipses.

Sagredo: I judge from your remarks that my diagram (Fig. 23) is superfluous, but now that I have finished it, let me submit it to you anyway. Inspired by Fig. 22, I have simply increased the uniform track width until the nonelliptical nature of the parallel curves is immediately evident, at least in the case of the inner curve with its four cusps and two nodes.

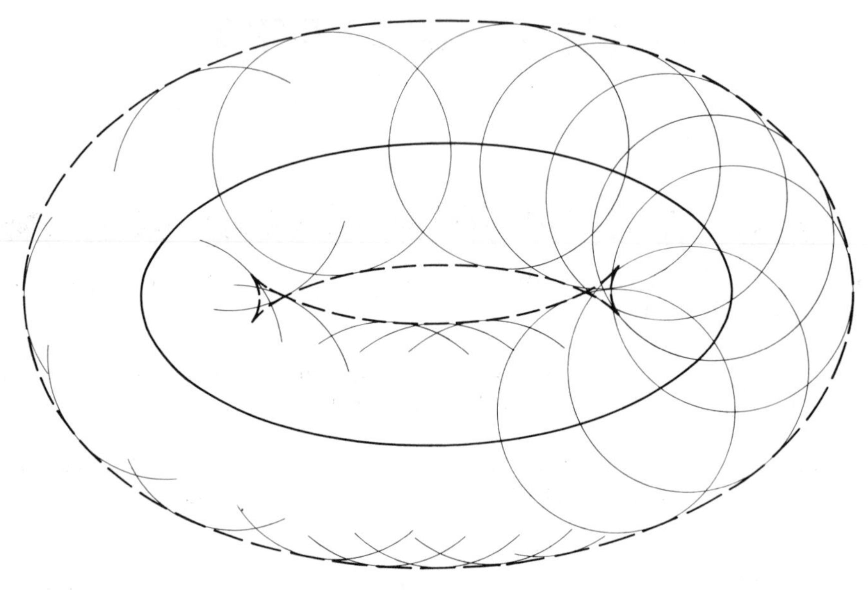

Fig. 23

Simplicio: What an admirable illustration! It appears that by fixing on a different radius for the arbitrary circle the character of the "parallel" curves obtained from a given ellipse will be altered. And to think that from any given curve any number of parallel curves may be generated! I take it that I am correctly following your lead, Salviati, in using the term *parallel* in this extended sense?

Salviati: Yes, the usage is quite customary. I expect that we shall be speaking later of the value of extending definitions. Curves in general are said to be parallel to one another if they are equidistant at all points when measured normally. Sagredo has used a practical method for obtaining curves parallel to a given curve. The dashed lines in Fig. 23 are called the *envelopes* of the family of circles he has indicated.

Simplicio: The outer envelope in this diagram may have the appearance of an ellipse to the untrained eye,[56] but I can now believe that it is not. Just before our discussion for this hour began, I too was starting to make a drawing in response to the diagram that you left with us at the end of the last hour (Fig. 22). Taking the same first-quadrant arc of the ellipse (1) : $x^2/4^2 + y^2/2^2 = 1$ as you had used, I plotted, with respect to the same coordinate axes, the first-quadrant arcs also of the ellipses (2) : $x^2/5^2 + y^2/3^2 = 1$ and (3) : $x^2/3^2 + y^2/1^2 = 1$, in order the better to understand how the nonelliptical dotted arcs in Fig. 22 differed from these outer and inner elliptical arcs. I got the drawing to the stage shown in Fig. 24 when I realized that each of my three ellipses would have to have its own particular evolute, and I suspected that probably it would be tedious to plot these.

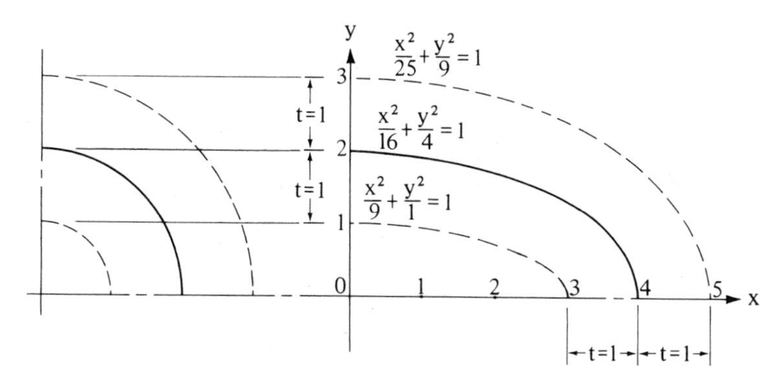

Fig. 24

Salviati: Well, it is easy to determine the cusps of the three evolutes. These, you will have realized, correspond to the points on the ellipses of maximum and of minimum curvature. For $x^2/a^2 + y^2/b^2 = 1$, the evolute, $(ax)^{2/3} + (by)^{2/3} = (a^2 - b^2)^{2/3}$ has cusps given by putting x and y in turn equal to zero (since the cusps are on the axes). For x equal to zero,

$$(by)^{2/3} = (a^2 - b^2)^{2/3}, \quad \text{so} \quad y = \pm(a^2 - b^2)/b.$$

Consequently, Simplicio, for your quadrants of ellipses having (1) $a = 4$, $b = 2$, (2) $a = 5$, $b = 3$, (3) $a = 3$, $b = 1$, the cusps corresponding to the maximum radii of curvature are at $(0, -6)$, $(0, -5\frac{1}{3})$, $(0, -8)$, the respective radii of curvature being 8, $8\frac{1}{3}$, and 9 units. Correspondingly, for the cusps on the x-axis, $y = 0$ gives

$$(ax)^{2/3} = (a^2 - b^2)^{2/3}, \quad \text{so} \quad x = \pm(a^2 - b^2)/a.$$

so the cusps on the positive x-axis are at (3,0), $(3\frac{1}{5}, 0)$, $(2\frac{2}{3}, 0)$, indicating minimum radii of curvature of 1, $1\frac{4}{5}$, and $\frac{1}{3}$ unit, respectively.

Simplicio: This information enables me to distinguish the three evolutes clearly enough and I can now roughly sketch them in on the diagram of Fig. 25.

Sagredo: I suppose the evolute of a curve is the envelope of its radii of curvature or of the normals to the curve.

Salviati: Indeed it is. As a curved or a straight line moves—in this case, as the normal to an ellipse takes up successive positions—the curve it always touches is called the envelope of the moving curve. So the evolute, which we previously defined as the locus of the center of curvature, may also be described as the envelope of the normals.

Sagredo: Thank you, Salviati; that is as I expected. And now, Simplicio, will you permit me to add to your diagram?

Simplicio: Of course.

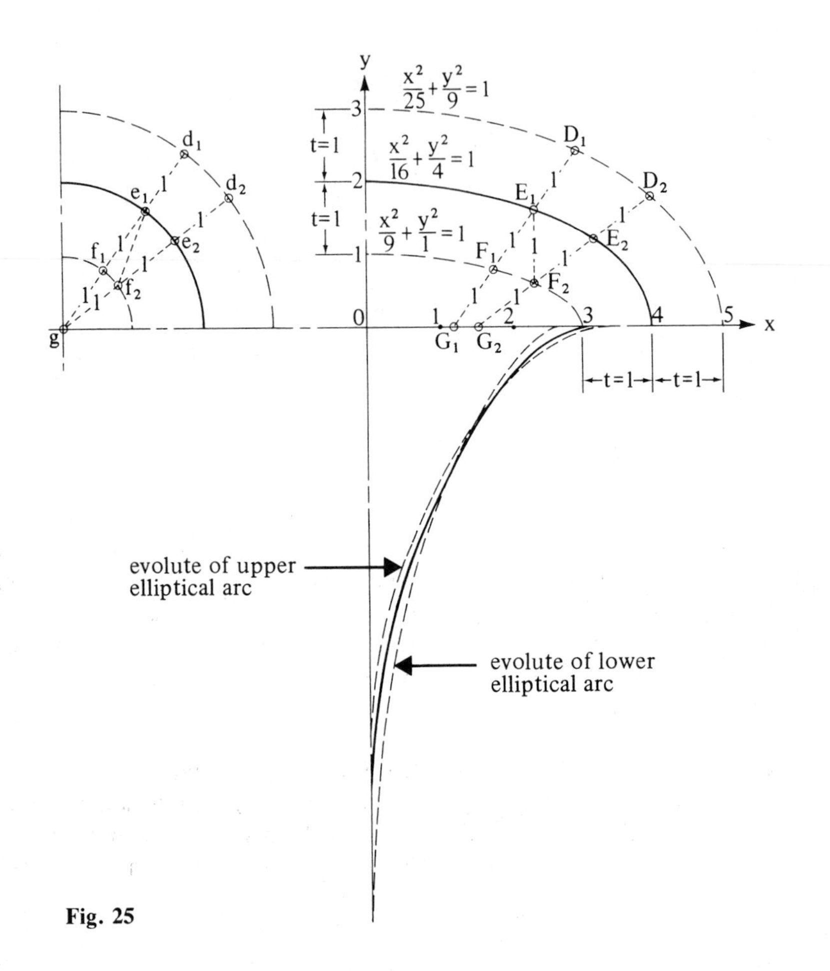

Fig. 25

Sagredo: It occurs to me that we might make use of the property of the 3-4-5 triangle in order to calculate the distances between selected points on the different ellipses, and so, perhaps, in this way show that the curves are not exactly parallel. If, as would seem most natural, the distances between the ellipses were to be measured perpendicularly to the intermediate ellipse, the calculation could become somewhat involved. But if I make use of the following points,

$$D_1:\left(3,\frac{12}{5}\right),\quad E_1:\left(\frac{12}{5},\frac{8}{5}\right),\quad F_1:\left(\frac{9}{5},\frac{4}{5}\right);\quad D_2:\left(4,\frac{9}{5}\right),$$

$$E_2:\left(\frac{16}{5},\frac{6}{5}\right),\quad F_2:\left(\frac{12}{5},\frac{3}{5}\right),$$

then it is evident that $E_1F_2 = 1$ unit, and that all the other intervals between adjacent points that are connected by dotted lines are also unit distances, since these are equal to hypotenuses of right-angled triangles with sides $\frac{3}{5}$, $\frac{4}{5}$ of a unit. But these unit distances are not the shortest distances between the ellipses. It is, for example, immediately evident that the distance from E_1 to points on the inner ellipse between F_1 and F_2 must be less than one unit.

Simplicio: This is very impressive, Sagredo. I must say that you showed keen foresight in selecting the particular points that you did.

Sagredo: Well, I must admit that things worked out even more easily than I anticipated, but the rationale underlying the choice of points is easily explained. To correspond to the points $(\frac{3}{5}a, \frac{4}{5}a)$ and $(\frac{4}{5}a, \frac{3}{5}a)$ on the circles $x^2 + y^2 = a^2$ ($a = 1, 2, 3$ in the case of the quadrants illustrated), I simply took the points $(\frac{3}{5}a, \frac{4}{5}b)$ and $(\frac{4}{5}a, \frac{3}{5}b)$ on the ellipses $x^2/a^2 + y^2/b^2 = 1$. I might also have included the points $G_1 : ((6/5), 0)$ and $G_2 : ((8/5), 0)$, collinear with D_1, E_1, F_1, and with D_2, E_2, F_2, respectively. G_1 and G_2 may be regarded as points on the segment of the x-axis from $(-2, 0)$ to $(+2, 0)$ considered as the initial or as a limiting and degenerate member of the family of ellipses $x^2/(2 + h)^2 + y^2/h^2 = 1$, $h > 0$, having in each case the major axis along the x-coordinate axis and exceeding by 4 units the minor axis along the y-coordinate axis.

Simplicio: I cannot help noticing that the dotted intervals D_1E_1, E_1F_1, F_1G_1 are equal and parallel to d_1e_1, e_1f_1, f_1g_1, and likewise that D_2E_2, E_2F_2, F_2G_2 are equal and parallel to d_2e_2, e_2f_2, f_2g_2.

Sagredo: A pertinent observation, Simplicio.

Simplicio: I am surprised that these equalities should hold. I am not clear whether it is through chance or necessity.

Sagredo: I too am surprised; this is what I meant a minute ago when I referred to things working out more easily than I had expected. I am sure that $D_1D_2 > d_1d_2$, and the fact that $E_1F_2 < e_1f_2$ does not attract our attention. But I agree that there is cause for scrutiny when we find that the lengths and the gradients of the segments of D_1G_1 and of D_2G_2 "happen" to be equal to the lengths and the gradients of d_1g_1 and of d_2g_2.

Salviati: We have here a very humble example of the challenge of a "happening" of the kind to which the mathematical mind is so responsive.

Sagredo: I do not claim to possess a mathematical mind, for all that I find my elementary studies satisfying and our discussion so enjoyable. But I have found myself wondering about unexpected happenings and coincidences on a number of occasions.

Salviati: Such things can well cry out for explanation, for the demonstration that certain situations "cannot be otherwise than they are"—for proof showing that the coincidences are after all necessary and so are not merely coincidences, as I think you expressed it earlier. This is not to say that many coincidences will not be judged to be trivial. But in mathematics the most significant necessary coincidences or the most interesting and surprising ones receive expression as the standard propositions of the subject. Thus, for example, the formula

$$\sum_{1}^{n} r^3 = \left(\sum_{1}^{n} r \right)^2$$

encapsulates an infinite set of necessary coincidences, as does the statement that the medians of any triangle are concurrent, the point of concurrency coinciding with one of the points of trisection of each median.

Simplicio: I have found myself wondering how our understanding of those situations that we do not regard as accidental differs from our understanding of those that we do.

Salviati: The short answer seems simply to be that only in the case of the former are we aware of plausible explanatory devices, theories and the like, for connecting together occurrences, or propositions about occurrences, which more superficially (or less imaginatively) viewed appear essentially independent and only accidentally or gratuitously associated.

Sagredo: That answer surely applies in respect to situations of a very diverse kind. It appears to me to apply no less to our understanding of the phenomena of the natural world than to that of the formal relations of mathematics.

Salviati: I would agree, though of course it can never really be shown that natural phenomena cannot be other than they are. To seem to do so is to leave more fundamental questions undisturbed; it is merely to point out consequences of an already accepted system of beliefs. When not dealing with formal systems, we have learned to be satisfied with lower levels of explanation, such as that "*A* causes *B*," or that "*A* is (seems) invariably a concomitant of *B*," where the ultimate links between *A* and *B* remain unknown and inaccessible to us, for all that we may have evolved an agreed upon manner of speaking about them.

Simplicio: I do not find such views as disquieting as I would have once.

Sagredo: It seems that in substantive science we are still obliged to rely upon occult qualities that we ourselves have invented. But now, may we return to our particular problem?

Salviati: Of course.

Sagredo: The question before us is whether certain equalities of distance and slope are consequences of some general characteristic of circle-to-ellipse transformations or whether they are simply the outcome of the particular choice of the coordinates or of the proportions of the ellipses involved.

Simplicio: I suggest we provide ourselves with a new simplified diagram to help us focus on the essential relations more clearly. These two elliptical arcs with semiaxes a, b and a', b' projected from these concentric circular arcs of radii b, b'

(Fig. 26) will suffice, will they not? Now obviously we cannot take just *any* two points, one on each ellipse (let alone two points on the one ellipse) and expect the distance between them to be equal to that between their counterparts on the circles.

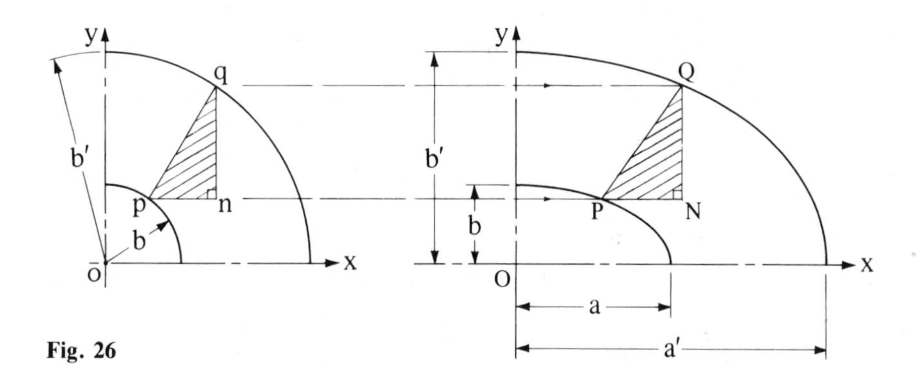

Fig. 26

Sagredo: I think that we have implicitly accepted that we restrict our attention to points p, q, say, on the concentric circles and the *projections* (parallel to the aligned x-axes) of these points onto the corresponding ellipses, points P, Q, say.

Simplicio: Very well. Now I note as you add these points to the diagram that what we really have to find is the condition for which PN will be equal to pn.

Sagredo: Precisely; given that the ordinates to P and Q are respectively equal to the ordinates to p and q, the necessary and sufficient condition for PQ to be both equal and parallel to pq is the same as that which guarantees that $PN = pn$, as you say, for then the triangles PQN, pqn will be congruent.

Simplicio: Good. And now the conjecture that occurs to me is that the required condition is none other than that p, q, be collinear with the common center, o; this, I believe is what is necessary and sufficient for the equality and parallelism of PQ and pq. But I cannot see how to prove it.

Sagredo: Allow me to try. Let p be any point, (h, k) say, on the circle of radius b; then if q is the point on the concentric circle of radius b' collinear with o, p, it is the point $(b'h/b, b'k/b)$, and so n is the point $(b'h/b, k)$. Further, let P be the "horizontal" projection of p onto the ellipse with semiaxes a, b, related to the circle of radius b as indicated in Fig. 26, then P is the point $(ah/b, k)$. And let Q be the point on the ellipse with semiaxes a', b' projected in like manner from q on the circle of radius b'; then the abscissa of Q is a'/b' times the abscissa of q, that is, a'/b' times $b'h/b$; hence Q is the point $(a'h/b, b'k/b)$ and N is the point $(a'h/b, k)$. Consequently

$$PN = a'h/b - ah/b = (a' - a)h/b,$$

whereas,

$$pn = b'h/b - h = (b' - b)h/b$$

Thus we find, Simplicio, that PN is *not* without reservation equal to pn; the further necessary and sufficient condition for this (given our preceding assumptions) is here clearly indicated, $PN = pn$ iff $a' - a = b' - b$. The condition that the distance between the ellipses along the x-axis is to be equal to the distance between them along the y-axis was satisfied by any pair of the ellipses we were considering earlier, in Figs. 24 and 25. It is interesting to notice, is it not, Simplicio, that although in stating your conjecture you overlooked or failed to mention the proviso that $a' - a$ and $b' - b$ must be equal, the need for this was unmistakably exposed as a result of the proof itself, a nice illustration of the self-corrective nature of such mathematical work.[57]

Simplicio: You seem to have clarified the whole situation very satisfactorily.

Salviati: I notice, Sagredo, that you implied that the necessity and/or sufficiency of the final condition was relative to the assumptions already made. Would you like to say something about that?

Sagredo: Well, it occurred to me that there were, perhaps, other ways of obtaining PQ equal and parallel to pq, without having o, p, q collinear at all. Let me try to show this on a new diagram (Fig. 27). First I trace the curves from Simplicio's diagram since he has drawn them so carefully. Now I place p_1q_1 here, say, *not* in line with center o. Then it seems to me that there will a position for P_1Q_1 between the elliptical arcs so that it will be both equal and parallel to p_1q_1. I am not concerned to try to find a straightedge-and-compasses construction. With the aid of a set square and a couple of rulers, I think I have succeeded in locating P_1Q_1 accurately enough for it to be seen that it was not necessary for o, p, q to be collinear. Let us try again. Corresponding to p_2q_2, placed here say, we may draw in a parallel and equal line segment P_2Q_2 in about this position. And not only does this show that the collinearity requirement is unnecessary, it appears to indicate also that the requirement that $a' - a$ and $b' - b$ be equal may likewise be dispensed with.

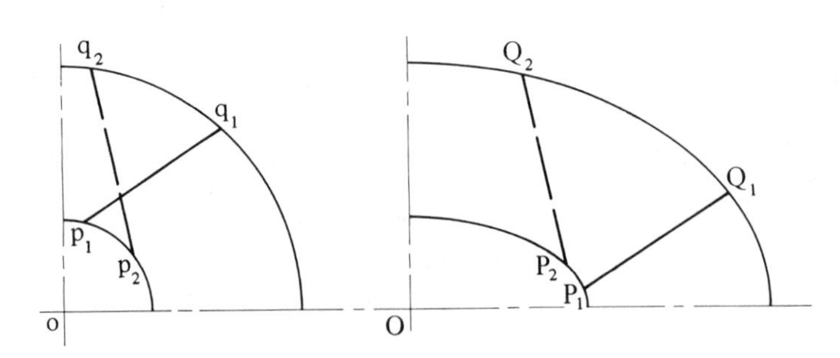

Fig. 27

Salviati: And no doubt the requirement that the curves be circles and ellipses is an unnecessary restriction too?

Sagredo: It certainly seems so.

Simplicio: And somehow the condition that P, Q, be projections of p, q (parallel to the axis common to the circles and the ellipses) fell out of consideration also.

Sagredo: I have to admit it; my effort to gain generality by removing restrictions does appear to have misfired. We seem to have been left with a situation so general as to be devoid of interest.

Salviati: Do not give in too easily, Sagredo.

Sagredo: Well, it now seems to me that we have replaced a mathematically attractive situation consisting of a precisely connected constellation of conditions and conclusions by a collection of isolated possibilities in some undefined situation. For it will doubtless be possible for equal and parallel line segments to be placed so as to have their ends lying on curves such as we might produce by scribbling randomly on a sheet of paper. And what would be the interest in that?

Simplicio: None at all, I should say. Previously, Salviati, you mentioned, as an example of a significant necessary coincidence, the proposition about the concurrency of the medians of a triangle. Now surely there are *insignificant* as well as significant ways of arranging for lines through the vertices of a triangle to be concurrent. One significant way you have indicated: join the vertices to the midpoints of the opposite sides. Another significant way is to draw in the altitudes of the triangle, I believe. But if we were to dispense with such a prior condition sufficient to guarantee concurrency—and were merely to draw nonparallel straight lines through two vertices of a triangle and then add in the line determined by their point of intersection and the third vertex—then surely we would have utterly trivialized the situation!

Sagredo: You mean because there is no other property connected with the concurrency?

Simplicio: Exactly.

Salviati: Simplicio, I must tell you that the "example" you have chosen is rather a counterexample in relation to the general position that you and Sagredo have so reasonably adopted.

Simplicio: I do not see how that can be.

Salviati: Let us look at your suggestion in detail. Take any triangle ABC and draw in lines at pleasure through A, B so as to intersect at P, say, and to meet the opposite sides at X, Y, respectively. Finally, add the line through C, P, meeting AB at Z. Now I admit that the very generality of the specification of this figure does not encourage one to hope that anything of interest could be shown to flow from it. That is doubtless the reason why, so far as is known, none of the great geometers of antiquity anticipated the discovery of our countryman, Giovanni Ceva, who showed in his *De lineis rectis*, published in 1678, that for any figure of the kind just described,

$$\frac{AZ}{ZB} \cdot \frac{BX}{XC} \cdot \frac{CY}{YA} = 1.$$

(See Fig. 28.) This easily proved proposition has come to be known as Ceva's theorem. Its converse, namely, given that

$$\frac{AZ}{ZB} \cdot \frac{BX}{XC} \cdot \frac{CY}{YA} = 1,$$

then AX, BY, CZ are concurrent, is especially important.

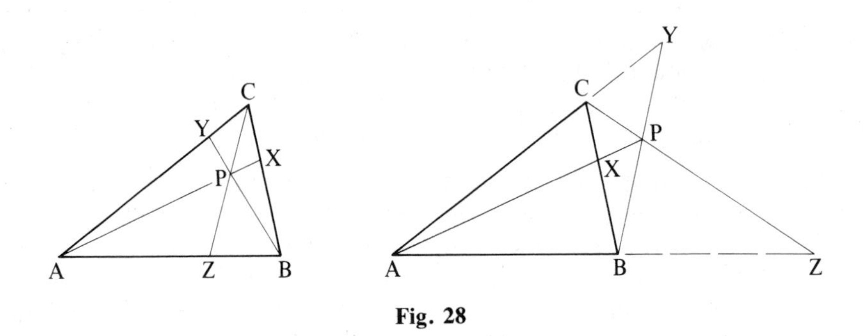

Fig. 28

Sagredo: This seems to be a remarkably general theorem. I notice that the concurrency of the medians of any triangle follows as a special case. For then X, Y, Z are the midpoints of the sides, and so each of the component ratios is equal to unity; hence their product is unity. Consequently given what you have just told us, the medians are proved concurrent.

Salviati: Yes, clearly the medians of any triangle are cevian lines, but so too are the altitudes and the bisectors of the angles and a number of other independently recognized sets of lines through the vertices. Before Ceva's discovery, proofs for some of these special cases had been separately devised, but with his theorem a unification of this work became possible.

Simplicio: So I was wrong in supposing that merely to pass a line from the third vertex of a triangle through the point of intersection of lines from the other two vertices would be to trivialize the notion of concurrency.

Salviati: That would depend upon what significance could be seen to attach to the resulting configuration. It is possible that, far from depriving the concurrency property of any interest, the procedure you suggest would have provided a suitably prepared mind with an opportunity for discovering, or rediscovering, the valuable general relation that now carries Ceva's name.

Sagredo: And this theorem of Ceva's is not hard to prove?

Salviati: Not at all—provided, that is, that one has some idea of what is to be proved! In this, we are at a tremendous advantage when we are the heirs to previous work. At the least we must be led to our proofs by apt conjectures and promising intuitions, but in this case you have the result to be proved before you and even the diagrams drawn (Fig. 28). Given that AX, BY, CZ have a common point of intersection, P say, we will find that

$$\frac{AZ}{ZB}\cdot\frac{BX}{XC}\cdot\frac{CY}{YA}=1.$$

Show us, Simplicio, in either of the representative diagrams of Fig. 28, two triangles having their areas in the ratio $AZ:ZB$.

Simplicio: Triangles AZP and ZBP, of course.

Salviati: And two other triangles in the same ratio?

Simplicio: Oh yes, AZC and ZBC.

Salviati: And yet two more?

Simplicio: I can see no others.

Sagredo: Are not triangles APC, BPC in the same ratio?

Salviati: Tell us why, Sagredo.

Sagredo: Let me see. $\triangle AZC:\triangle ZBC=\triangle AZP:\triangle ZBP=AZ:ZB=k:1$, say. Then

$$\begin{aligned}\triangle APC:\triangle BPC &= (\triangle AZC-\triangle AZP):(\triangle ZBC-\triangle ZBP)\\ &= (k\cdot\triangle ZBC-k\cdot\triangle ZBP):(\triangle ZBC-\triangle ZBP)=k:1.\end{aligned}$$

And looking at the diagrams again, I now notice that triangles APC, BPC on a common base PC have their areas proportional to the altitudes to PC, which will be in the ratio AZ to ZB. Thus, either way, $AZ:ZB=\triangle APC:\triangle BPC$.

Salviati: Very good. And likewise,

$$BX:XC=\triangle BPA:\triangle CPA,\quad\text{and,}\quad CY:YA=\triangle CPB:\triangle APB.$$

Consequently,

$$\frac{AZ}{ZB}\cdot\frac{BX}{XC}\cdot\frac{CY}{YA}=\frac{\triangle APC}{\triangle BPC}\cdot\frac{\triangle BPA}{\triangle CPA}\cdot\frac{\triangle CPB}{\triangle APB}=1.\quad \text{Q.E.D.}$$

For the converse, we start with the following datum:

> X, Y, Z are points such that either all are on the sides of a $\triangle ABC$, or just one of them is on a side and two are on *sides produced*, and in either case such that
>
> $$\frac{AZ}{ZB}\cdot\frac{BX}{XC}\cdot\frac{CY}{YA}=1. \qquad (1)$$

We have to prove that AX, BY, CZ are concurrent. If P is the point of intersection of AX and BY, let CP, or CP produced, meet AB at Z'; we will prove that Z and Z' are the same point, and thus that CZ coincides with CZ'. Since AX, BY, CZ' are concurrent, it follows by Ceva's theorem that

$$\frac{AZ'}{Z'B} \cdot \frac{BX}{XC} \cdot \frac{CY}{YA} = 1. \tag{2}$$

From (1) and (2), Z and Z' necessarily coincide. Thus AX, BY, CZ are concurrent. Q.E.D.

If, for example, X, Y, Z are chosen as the feet of the altitudes from A, B, C, respectively, then it is easy to show with the aid of some obvious pairs of similar right-angled triangles, that

$$\frac{AZ}{ZB} \cdot \frac{BX}{XC} \cdot \frac{CY}{YA} = 1,$$

and hence that the altitudes of any triangle are concurrent. It is better that I leave the details for you to fill in at your leisure. Now I think that we ought to return to the matter of the reservations you both expressed concerning the possibility of gaining greater generality through some relaxation of the assumed conditions.

Sagredo: It seemed to us that chaos would be more likely to result, but doubtless we did not take a broad enough view. The example of the equal and parallel line segments intercepted between the related pairs of curves indicated first of all that it is sometimes easy to overlook a condition that is required to complete a set of necessary and sufficient conditions. Conversely, it suggested that it would sometimes be difficult to notice if a condition that had been accepted as necessary were not really necessary at all.

Salviati: This last fact particularly is of great importance over a wide range of creative thinking. The role of the association of previously unconnected ideas is never omitted from discussions of creativity. Hardly less important is the strangely neglected activity of dissociation.[58] Of course, the ability to consider the possibility of separating ideas that custom has bound together is less rare in this age than it was in *our* times.[59] As for your fear of "chaos," is not successful generalization the way we reduce chaos to order?

Sagredo: I suppose so. Perhaps the concept of the general triangle, with its infinitude of possible shapes, would once have seemed quite chaotic—I mean in the sense of seeming to be unamenable to mathematical treatment.

Salviati: To judge from the clay tablets, it seems likely that the earliest mathematicians of the Old Babylonian period cast only the most tentative glances in the direction of the general triangle. Yet they showed a remarkable knowledge of right-angled triangles, being able to generate "Pythagorean" triads over one thousand years before Pythagoras' lifetime. In this phase of mathematics, right-angled triangles seem first to have been obtained by the diagonalwise division of rectangles into pairs of matchable triangles that were then found to be capable of rearrangement to produce isosceles triangles. Probably, symmetrical figures such as isosceles trapezia or regular polygons attracted the attention of investigators before the idea of juxtaposing a pair of *non*congruent right-angled triangles to form a scalene triangle suggested itself. Indian mathematics, from Brahmagupta to the second Bhāskara, had as one of its main themes the generation of right-angled triangles with integral measures for the side lengths, and the combination of such triangles to form certain scalene triangles, and especially certain particular forms of simple quadrilateral. All other kinds of quadrilateral not so amenable to mensurational treatment were conspicuously neglected.[60] We know from more recent examples how mathematicians can show impressive sophistication in one direction while ignoring indications of potentially exciting materials in another direction that hindsight would suggest were before their very eyes.

Sagredo: Were not the mathematicians of Europe "in the presence of" complex numbers for centuries without "recognizing" them?[61]

Salviati: Perhaps we should rather say, without "legitimizing" them. Mathematicians still to some extent under the spell of Euclid's definition of number[62] could hardly be expected to have readily embraced such radically new extensions to the number system.

Sagredo: I cannot help thinking that in this they were like miners, ingenious in their extraction of a particular metal, yet unaware that other material, perhaps even more valuable, was being rejected in the slag. But excuse my digression. I think you were about to sum up.

Simplicio: Digressions, digressions! And to the *n*th order! They are our chief occupational hazard, I fear. Yes, Salviati, it would help us a great deal if you would be so good as to sum up this present part of our discussion.

Salviati: Then I will do what I can. The fundamental point I take to be the desirability of regarding the "necessity," as well as the "sufficiency," of conditions not as properties of those conditions in isolation but as properties they possess in relation to entire constellations of relevant background conditions. To overlook this notion of "necessity (or of sufficiency) in the circumstances" is to restrict one's chances of glimpsing possible innovations. It is certainly not only in mathematics that attention to the circumstances is less often given than is appropriate. Exceptional imaginative powers are, of course, sometimes required to dissociate characteristics previously assumed to be necessarily connected. In a mathematical proof the introduction of any unnecessary restriction implies an unnecessary reduction in the generality of the resulting conclusion.[63] But there are two desiderata, the pursuit of which sometimes leads to conflict: one is the maximization of generality; the other is the enhancement of amenability to mathematical treatment. By the latter I mean to include both the facilitation of discovery and the simplification of proof structure. This is where your example of the equal and parallel chords between pairs of related curves is illuminating. I suggest that you automatically imposed certain restrictions on the situation not so much because of some insight that these were really necessary (they were not), but rather because you felt that they defined an antecedent demonstrably sufficient to guarantee the consequence of parallelism and equality. I think that the dominant motive was the more or less unconscious wish to avoid intractable complexity, or "chaos" as you were to put it when we began our reflections upon what had been done.

Sagredo: I notice that articles and even whole books are still being written on the right-angled triangle, for all that the geometry of the general triangle is over two thousand years old.[64]

Salviati: True. But the properties that are being written about in these works are, for the most part, specific to right-angled triangles. Perhaps a better example is provided by the geometry of the circle, also still a subject for special treatises,[65] even though it has been recognized since the fourth century B.C. that circles are a species of ellipse. Although some theorems about circles can be extended to apply to ellipses and even to other conics, conic section theory has never absorbed the geometry of the circle. It has not been in the interest of general convenience for this to have occurred.

Sagredo: I think I can see why that might be, given the apparent difficulty of determining the perimeter of the ellipse, which reminds me just how far we have strayed from our initial problem!

Salviati: You are right, Simplicio. Let me simply conclude by remarking that it is possible that some abstract minds will be inclined to dismiss our second desideratum, the enhancement of amenability, as merely of pedagogical interest, and unworthy to be set alongside the first. And some mathematicians do seem prepared to devote themselves to the pursuit of ever greater generality, whatever the cost.

Sagredo: Is there not something paradoxical in this, if, as I suspect, the ascent to greater and greater generality has become the preserve of those who have become more and more specialized in their interests?

Salviati: Perhaps we should say that they are more prepared to concentrate their powers. I have not found such mathematicians to be noticeably less interested in worthwhile matters of a general nature than are other highly intelligent persons. But, be this as it may, what I wish to suggest is that the second desideratum is as important in directing the course of mathematical effort as is the first. I cannot help thinking here of a twentieth-century scientist's remark: "If politics is the art of the possible, research is surely the art of the soluble."[66] In mathematical practice, the art of the soluble calls for a delicate balance between efforts to achieve the two desiderata.

I fear these observations are somewhat premature. Doubtless they would have had much more meaning for you if delayed until later in the day, when, I expect, you will be in a position to reflect upon the reasons for your adoption of such a roundabout approach to the problem taken up at the outset of our present discussion. But now, Simplicio, it is your turn to direct us back to the path from which we have strayed.

Simplicio: We began this morning with the problem of finding pairs of solids, apart from the Archimedean sphere with its circumscribing cylinder for which the volume ratio $V_2 : V_1$ would equal the area ratio $A_2 : A_1$ and, analogously, pairs of plane figures, other than the circle and its circumscribing square, for which the area ratio $A_2 : A_1$ would be equal to the perimeter ratio $p_2 : p_1$. Sagredo was originally of the opinion that there was something unique about these pairs of figures and that no other pairs would be found for which the ratio of the contents would be equal to the ratio of the boundaries. But you asserted that there is no end to further pairs of figures related in this way. Sagredo responded by suggesting that perhaps the equality of the ratios was maintained in the case of any *ellipse* and its circumscribing rectangle with sides parallel to the axes. After some misguided conjecture on my part, we found that this would be the case if and only if the expression $\pi(a + b)$ provided correct perimeter measures for ellipses with semiaxes a, b. But you convinced us, Salviati, of the inaccuracy of this supposed perimeter expression. At this point, we might well have turned away from our ellipses and rectangles in order to search out other pairs of figures that do possess equal area and perimeter ratios; you assured us, Salviati, that they were to be found "by the barrowful." But I think we were both at a loss as to how these could be found. In any case, we went on to make further attempts to determine the correct expression for the elliptical perimeter and Sagredo proposed his ingenious and promising double-track method. But unfortunately this turned out to be inapplicable owing to the subtle and unexpected variation in the distance between the aligned concentric ellipses even though their minor axes were chosen to differ by the same amount as their major axes. I assure you, Salviati, we should be greatly obliged if you would reveal to us the correct formula for the general ellipse in order that we may proceed with free minds to the main part of our investigation.

Salviati: You shall have it then, as the first item of our next hour.

V

Nature is not embarrassed by difficulties of analysis.

—Augustin Fresnel

It behoves us always to remember that in physics it has taken great men to discover simple things. They are very great names indeed which we couple with the explanation of the path of a stone, the droop of a chain, the tints of a bubble, the shadows in a cup.

—D'Arcy Thompson

Discovery consists of seeing what everybody has seen and thinking what nobody has thought.

—Albert Szent-Gyorgyi

One had to be a Newton to notice that the moon is falling, when everyone sees that it doesn't fall.

—Paul Valéry.

The paradox is now fully established that the utmost abstractions are the true weapons with which to control our thoughts of concrete fact.

—A. N. Whitehead

One of the chief duties of the mathematician in acting as an adviser to scientists . . . is to discourage them from expecting too much from mathematics.

—Norbert Wiener

Physics is much too hard for physicists.

—David Hilbert.

HORA QUINTA

Salviati: You have charged me with the task of delivering up to you the correct formula for the perimeter of an ellipse. Bearing in mind your desire to have done with this rectification problem, I will be brief and follow the concise procedure to be found in the standard calculus textbooks of this modern age. I think that to return to the pioneering treatment of James Gregory or of Newton[67] would be to involve you in even more difficulties than will the outlining of the method that I propose to follow.

Sagredo: I have made only a very slight study of the infinitesimal calculus and I have not come across anything that I recognize as related to our particular problem. I fear that our lack of technical preparation might prove to be an embarrassment to us, but you may be assured that we are thirsty for this knowledge.

Salviati: Then let us begin. You know how it is sometimes convenient to express x and y in terms of some third variable. For example, instead of the equation $x^2/a^2 + y^2/b^2 = 1$, we may write $x = a \cos\phi$ and $y = b \sin\phi$. In these equations a and b are the parameters whose values determine the particular ellipse, while ϕ is the parameter whose value determines a particular point on the ellipse. This new specification is justified because, whatever value is assigned to ϕ, the substitution of $a \cos\phi$ for x and of $b \sin\phi$ for y in the equation $x^2/a^2 + y^2/b^2 = 1$ reduces it to the identity $\cos^2\phi + \sin^2\phi \equiv 1$, showing that whatever the value of ϕ, the coordinates of the point $(a \cos\phi, b \sin\phi)$ satisfy the equation of the ellipse. The significance of ϕ is indicated in Fig. 29. The coordinates of point P, sometimes referred to simply as "the point ϕ" are

$$x = MN = MQ \cos\phi = a \cos\phi, \quad \text{and} \quad y = NP = N'P' = b \sin\phi,$$

where ϕ is the measure of what is called "the eccentric angle" corresponding to the point P of the ellipse.[68] Arc lengths s are obtained from:

$$\left(\frac{ds}{d\phi}\right)^2 = \left(\frac{dx}{d\phi}\right)^2 + \left(\frac{dy}{d\phi}\right)^2.$$

Here,

$$\left(\frac{ds}{d\phi}\right)^2 = a^2 \sin^2\phi + b^2 \cos^2\phi,$$

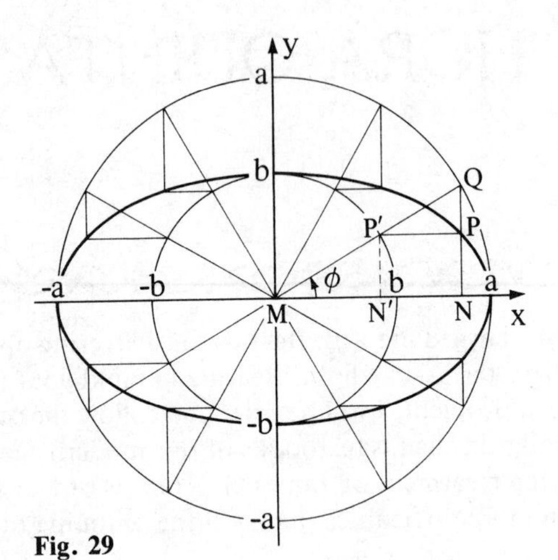

Fig. 29

which comes to

$$\left(\frac{ds}{d\phi}\right)^2 = a^2(1 - \epsilon^2 \cos^2\phi), \quad \text{where[69]} \quad \epsilon^2 = 1 - b^2/a^2.$$

Hence, the whole perimeter is measured by

$$p = 4a\int_0^{\pi/2} (1 - \epsilon^2 \cos^2\phi)^{1/2}\, d\phi.$$

Expanding by the binomial theorem:

$$p = 4a\int_0^{\pi/2}\left(1 - \frac{\epsilon^2}{2}\cos^2\phi - \frac{\epsilon^4}{8}\cos^4\phi - \frac{\epsilon^6}{16}\cos^6\phi - \frac{5\epsilon^8}{128}\cos^8\phi - \ldots\right)d\phi,$$

$$\text{where } \int_0^{\pi/2} \cos^n\phi\, d\phi = \frac{1 \cdot 3 \cdot 5 \ldots (n-1)}{2 \cdot 4 \cdot 6 \ldots n} \times \frac{\pi}{2}, \quad \text{for } n \text{ even.}$$

Hence,

$$p = 2\pi a\left(1 - \frac{1}{4}\epsilon^2 - \frac{3}{64}\epsilon^4 - \frac{5}{256}\epsilon^6 - \frac{175}{16{,}384}\epsilon^8 - \ldots \text{ ad inf.}\right)$$

So there is your formula!

Sagredo: We will have to take your word for the details of the working, Salviati, but at least we can reflect upon the *result* in order to understand it more clearly.

Simplicio: I notice that it may be checked in the special case of a circle, for which $\epsilon = 0$, since then the formula reduces to $p = 2\pi a$, that is, $C = 2\pi r$ in the usual symbols.

Sagredo: And I suppose it is now possible to compare the accuracy of the various approximate formulae for the elliptical perimeter, such as those you mentioned earlier (pp. 23, 38–39).

Salviati: Indeed yes. In serious work, it is expected as a matter of course that an indication of the error be provided in connection with an approximate formula. For example, the expression

$$\pi(a + b) = \pi a(1 + b/a) = \pi a\{1 + (b^2/a^2)^{1/2}\} = \pi a\{1 + (1 - \epsilon^2)^{1/2}\}.$$

$$\therefore \pi(a + b) = \pi a\left\{ 1 + 1 + \frac{1}{2}(-\epsilon^2) + \frac{(1/2)(-1/2)}{1 \cdot 2}(-\epsilon^2)^2 \right.$$

$$+ \frac{(1/2)(-1/2)(-3/2)}{1 \cdot 2 \cdot 3}(-\epsilon^2)^3$$

$$\left. + \frac{(1/2)(-1/2)(-3/2)(-5/2)}{1 \cdot 2 \cdot 3 \cdot 4}(-\epsilon^2)^4 + \ldots \right\}$$

$$= 2\pi a\left\{ 1 - \frac{1}{4}\epsilon^2 - \frac{1}{16}\epsilon^4 - \frac{1}{32}\epsilon^6 - \frac{5}{256}\epsilon^8 - \ldots \right\}.$$

For convenience of comparison with the exact formula given above, this may be written as

$$\pi(a + b) = 2\pi a\left(1 - \frac{1}{4}\epsilon^2 - \frac{4}{64}\epsilon^4 - \frac{8}{256}\epsilon^6 - \frac{320}{16{,}384}\epsilon^8 - \ldots \right).$$

Hence this expression is seen to be too small by

$$2\pi a\left(\frac{1}{64}\epsilon^4 + \frac{3}{256}\epsilon^6 + \frac{145}{16{,}384}\epsilon^8 + \ldots \right);$$

thus it *underestimates* by $(2\pi a/64)\epsilon^4$ approximately for ϵ near zero. As a further example, we may express $2\pi\sqrt{\{\frac{1}{2}(a^2 + b^2)\}}$ as $2\pi a(1 - \frac{1}{2}\epsilon^2)^{1/2}$, which expands to

$$2\pi a\left(1 - \frac{1}{4}\epsilon^2 - \frac{2}{64}\epsilon^4 - \frac{2}{256}\epsilon^6 - \frac{40}{16{,}384}\epsilon^8 - \ldots \right).$$

Comparison with the exact formula, shows that this expression is too large by

$$2\pi a\left(\frac{1}{64}\epsilon^4 + \frac{3}{256}\epsilon^6 + \frac{135}{16{,}384}\epsilon^8 + \ldots \right);$$

thus it *overestimates* by $(2\pi a/64)\epsilon^4$ approximately for ϵ near zero.

Sagredo: It is surely remarkable that these errors will almost annihilate each other if $\pi(a + b)$ and $2\pi\sqrt{\{\frac{1}{2}(a^2 + b^2)\}}$ are averaged, for a value too small by

$$2\pi a\left(\frac{1}{64}\epsilon^4 + \frac{3}{256}\epsilon^6 + \frac{145}{16{,}384}\epsilon^8 + \ldots \right)$$

when averaged with a value too large by

$$2\pi a\left(\frac{1}{64}\epsilon^4 + \frac{3}{256}\epsilon^6 + \frac{135}{16{,}384}\epsilon^8 + \ldots \right)$$

gives a value too small by a very minute quantity, approximately $2\pi a((5/16{,}384)\epsilon^8)$. Thus it appears that the error attending the use of the formula

$$p = \pi\left[\frac{1}{2}(a+b) + \sqrt{\left\{\frac{1}{2}(a^2+b^2)\right\}}\right]$$

would be very small indeed.

Salviati: This is so. In fact, you have now justified a remark I made when briefly introducing this formula earlier (p. 23). Perhaps I should indicate here why it is sometimes necessary to consider more than one term in such error expressions as those just obtained. A numerical illustration using the terms

$$\frac{1}{64}\epsilon^4, \ \frac{3}{256}\epsilon^6, \ \frac{145}{16{,}384}\epsilon^8$$

will suffice. The common factor $2\pi a$ may be omitted because it does not affect the relative sizes of the terms:

Example (i), taking $\epsilon = 0.1$ say,

$$\frac{1}{64}\epsilon^4 = 0.000001562500$$

$$\frac{3}{256}\epsilon^6 = 0.000000011718\ldots$$

$$\frac{145}{16{,}384}\epsilon^8 = 0.000000000088\ldots$$

Example (ii), taking $\epsilon = 0.9$ say,

$$\frac{1}{64}\epsilon^4 = 0.01025\ldots$$

$$\frac{3}{256}\epsilon^6 = 0.00622\ldots$$

$$\frac{145}{16{,}384}\epsilon^8 = 0.00381\ldots$$

(From $b/a = \sqrt{(1-\epsilon^2)}$, we find that for $\epsilon = 0.1$, $b/a \approx 0.995$, whereas for $\epsilon = 0.9$, $b/a \approx 0.436$.) It will be seen that for ϵ close to zero it is appropriate to say that $\pi(a+b)$ measures the perimeter with errors of approximately $(2\pi a/64)\epsilon^4$, owing to the rapidity with which the successive terms diminish, though this is too crude an estimate of the error if ϵ is near to unity.

I leave it as an exercise for you to show that Peano's formula,

$$p = \pi\left\{a + b + \frac{1}{2}(\sqrt{a} - \sqrt{b})^2\right\},$$

which we wrote earlier (see pp. 23, 39) in the form

$$p = \pi\left\{\frac{3}{2}(a+b) - \sqrt{(ab)}\right\},$$

expands to

$$p = 2\pi a\left(1 - \frac{1}{4}\epsilon^2 - \frac{3}{64}\epsilon^4 - \frac{5}{256}\epsilon^6 - \frac{172}{16{,}384}\epsilon^8 - \ldots\right),$$

and hence that the expression for the error begins with the term $2\pi a((3/16{,}384)\epsilon^8)$, the approximation being the excess.

Sagredo: If, as you have indicated, we have the two formulae,

$$p = \pi\left[\frac{1}{2}(a+b) + \sqrt{\left\{\frac{1}{2}(a^2+b^2)\right\}}\right],$$

$$\text{underestimating by } 5\left(\frac{2\pi a}{16{,}384}\epsilon^8\right) + \ldots,$$

$$p = \pi\left\{\frac{3}{2}(a+b) - \sqrt{(ab)}\right\},$$

$$\text{overestimating by } 3\left(\frac{2\pi a}{16{,}384}\epsilon^8\right) + \ldots,$$

then what an extremely accurate formula would be produced by a suitably weighted average of these two estimates! For, by adding $\frac{3}{8}$ of the underestimate to $\frac{5}{8}$ of the overestimate there would be no error associated with terms at least up to and including that in ϵ^8. The formula would be

$$p = \frac{3}{8}\pi\left[\frac{1}{2}(a+b) + \sqrt{\left\{\frac{1}{2}(a^2+b^2)\right\}}\right] + \frac{5}{8}\pi\left[\frac{3}{2}(a+b) - \sqrt{(ab)}\right]$$

$$= \frac{\pi}{8}\left[9(a+b) + 3\sqrt{\left\{\frac{1}{2}(a^2+b^2)\right\}} - 5\sqrt{(ab)}\right]$$

Was this one of those very accurate formulae that you listed previously, Salviati?

Salviati: Yes, it was the one published, apparently for the first time, by R. Goormagtigh in the year 1930,[70] and I think that he would have arrived at it in just the same way as you have. In any case, by comparing his formula in its expanded form with the exact series expression, as has been done above in the case of the simpler approximations, Goormagtigh showed that there was in fact complete agreement up to and including the term in ϵ^{10}. The first term in the error expression for this formula is $2\pi a(7\epsilon^{12}/2^{20})$. Still smaller errors are associated with Ramanujan's approximations given earlier.[71]

Simplicio: Although I cannot claim to have understood every detail of the theory of the rectification of the ellipse that you have given us, now I can at least appreciate the necessity for the approximate agreement between the various formulae that have been suggested. And it is satisfying to have seen not only how a prospective formula can be analyzed, once one has it, but how it becomes possible to take the initiative and deliberately devise better and better formulae by more and more appropriate combinations of simple functions of the data. But interesting as all this is, it does not seem to offer us any help with the problem of trying to find a pair of plane figures other than the circle and its circumscribing square for which the area ratio is equal to the perimeter ratio. Already, in the first hour, we found that *if* for an ellipse and its circumscribing rectangle,

$A_e : A_r = p_e : p_r$—that is, $p_e = p_r(A_e/A_r)$—this would imply that $p_e = \pi(a + b)$, and it was this false consequence that exposed the falsity of the conjecture that $A_e : A_r = p_e : p_r$. So we must surely be ready now to leave the ellipse.

Sagredo: Before finally doing so, perhaps you will bear with me if I briefly mention something that passed through my mind earlier in connection with the roll of wrapping paper used in the bookshop. It concerns the great resistance that would be offered to attempts to flatten even slightly such a tightly wound hollow cylinder when removed from its roller. (I need hardly mention those other smaller and more loosely wound cylindrical rolls of paper that appear to be among the most familiar and convenient of twentieth-century mass-produced items for personal use; they are readily transformed from hollow circular form to hollow, approximately elliptical form without difficulty because the slightly creped paper offers so little resistance to stretching and to changes of distance between the layers.) It seems that it would be worth asking what would be involved in transforming the previously considered roll of wrapping paper, 9 inches outside diameter and 2 inches inside diameter, into a roll with cross-section having an ellipse with axes 10 inches, 8 inches for its peripheral boundary and an ellipse with axes 3 inches, 1 inch for inner boundary. The original cross-sectional area

$$= \pi\left(\frac{9}{2}\right)^2 - \pi(1)^2 = \frac{77}{4}\pi \text{ square inches;}$$

the final cross-section would have area

$$= \pi(5 \times 4) - \pi\left(1\frac{1}{2} \times \frac{1}{2}\right) = \frac{77}{4}\pi \text{ square inches}$$

exactly as before.

Simplicio: Excuse me Sagredo, but this appears to be another of your necessary coincidences! Are you not surprised . . . ?

Sagredo: You are asking me, I take it, whether I am surprised that $\pi ab - \pi a'b' = \pi(R^2 - r^2)$, given that $a = R + k$, $b = R - k$, $a' = r + k$, $b' = r - k$. The elliptical annulus has area given by

$$\pi(R + k)(R - k) - \pi(r + k)(r - k),$$

which comes to $\pi(R^2 - k^2 - r^2 + k^2)$ or $\pi(R^2 - r^2)$—as you see, the measure of the area of the circular annulus.

Simplicio: Oh very well; I will not ask you whether you knew this in advance.

Sagredo: I was not really conscious of it, nor should I claim that the choice of the assumed dimensions was a somnambular act.[72] If the two annular areas had not worked out to the same value, I would have adjusted the data so that they did, for I believe that it is a common assumption that when bodies change their *shape* under stress, their *volume* remains approximately constant. Now I think we can return to the particular calculation. Notice first, though, that *if* we believed that the elliptical perimeters were measured by the expression $\pi(a + b)$, so that changes in shape from the circles to the ellipses in our wound roll could take place, not only without changes in area, but also without changes in perimeter, then it would be difficult to explain the source of the resistance offered by the roll. The circumferences of the inner and outer circles are 2π and 9π inches. I should like to know what stretching of the paper would be involved if these circles—I mean the inner and outer layers—were transformed into ellipses of the dimensions already assumed. For the inner ellipse, with semiaxes $a = 3/2$, $b = 1/2$, we have, using Ramanujan's very accurate formula,

$$p = \pi[3(a + b) - \sqrt{\{(a + 3b)(3a + b)\}}],$$

that the perimeter is $\pi[6 - \sqrt{15}]$ inches. If you would not mind evaluating $\sqrt{15}$ for me, Simplicio, I will proceed with the other calculation.

For the perimeter of the outer layer, the same formula, with $a = 5$, $b = 4$, gives $p = \pi[27 - \sqrt{(17 \times 19)}]$. Since $17 \times 19 = 323$ and $18 \times 18 = 324$, we find by division that $18 \times 17\frac{17}{18} = 323$, the first factor, 18, slightly exceeding $\sqrt{323}$, the second factor, $17\frac{17}{18}$, correspondingly falling short of $\sqrt{323}$. Hence we take the arithmetic mean of 18 and $17\frac{17}{18}$, namely $17\frac{35}{36}$, as an extremely accurate estimate of $\sqrt{323}$.[73] So the outer perimeter must be very close to $9\frac{1}{36}\pi$ inches. While Simplicio is completing the calculation of $\sqrt{15}$, I note that $15 = 5 \times 3 = 4 \times 3\frac{3}{4} \approx 3\frac{7}{8} \times 3\frac{7}{8}$, indicating that $3\frac{7}{8}$ (= 3.875) will only slightly overestimate the true value of $\sqrt{15}$.

Simplicio: $\sqrt{15}$ comes to 3.873 correct to four significant figures by the long method.[74] But how did you know that your short method would be an overestimate?

Sagredo: $\sqrt{15}$ is the geometric mean between 4 and $3\frac{3}{4}$. By taking instead the arithmetic mean, $3\frac{7}{8}$, we have a value slightly in excess. Remember that we proved this in general terms in our second hour (p. 19). And now we have for the perimeter of the ellipse with axes 3 inches and 1 inch, $\pi(6 - \sqrt{15}) \approx 2{\cdot}127\pi$ inches, compared with just 2π inches for the circle of equal area. The foregoing calculations suggest that while all the turns of our flattened roll of paper would have to stretch, the shorter turns would have to stretch the most; from 2π to 2.127π inches is an increase of more than 6 percent. I cannot believe that wrapping paper could be stretched as much as this without being torn. But there would have to be *some* stretching of the paper and, by way of compensation, a reduction in its thickness so that the average distance from the outer to the inner surface of the roll would be somewhat less than $\frac{1}{2}(9 - 2)$ inches. From our earlier investigations (for example, in connection with Fig. 25), we know that there will be a reduction in the average distance between ellipses so arranged.

Salviati: Perhaps the cross-section of the roll would no more have elliptical boundaries than the double track of Fig. 22 had.

Sagredo: Perhaps not, but I believe that the same kind of difficulty would still be present; I mean the apparent need for a greater stretching of the paper than we can believe possible.

Salviati: Then no doubt your conclusion is that the roll could not be deformed as much as you had assumed without the paper being torn asunder.

Sagredo: I suppose so.

Simplicio: I too cannot believe that ordinary paper can be stretched by more than 6 percent of its natural length without its ripping apart. But what would compel this increase if the ends of the paper are free?

Sagredo: The extreme ends of the paper may be assumed to be free, but I think that the friction between adjacent layers is such that it would make little difference to the rigidity of the roll whether the ends were free or not, or even if the layers were glued together here and there throughout the roll.

Salviati: Your hypothetical analysis rests upon two evident assumptions: (1) that the volume of the roll (and consequently the area of its annular cross-section) does not alter, and (2) that the initially circular cylinder becomes elliptical. The precise shape attained will obviously depend upon the distribution of the applied forces, and you have given no specification of these. But then you said that you believed the difficulty of the excessive stretching would still occur whether or not the cross-section of the deformed roll were elliptical.

Sagredo: I had in mind the impossibility of simultaneously preserving both the area and the perimeter unchanged when a circle is transformed into another shape, whether elliptical or not.

Simplicio: Well *there* is an important assumption, and only just now made explicit! I hope it can be proved, for I feel that it must bear upon our original problem of possible pairs of figures having equal area and perimeter ratios—though I do not see how.

Sagredo: I have to admit that I am not at the moment able to give you a satisfactory answer.

Salviati: Let us restrict ourselves for the time being to your two specific simplifying assumptions: that of an *unchanged cross-sectional area* when the roll is deformed and that of a *resulting ellipticity* of the annular form. Although I accept the former assumption as an appropriate first approximation, I am much less satisfied with the latter. But even if we were to accept both of these assumptions, a quite different analysis might be given, such as the following. We could begin by supposing that the inner, unsecured turn has its perimeter equal to the circumference of the original circle, 2π inches. Then, if it were of elliptical form with its axes in the ratio 3 : 1, the lengths of these axes would be about 6 percent less than 3 inches and 1 inch. For, from your previous working, we have for the major axis 2/2.127 of 3 inches $\approx$ 2.82 inches and for the minor axis 2/2.127 of 1 inch $\approx$ 0.94

inch. The area of such an ellipse would be given by $\pi \times 1.41 \times 0.47 \approx 0.66\pi$ square inches. Assuming the cross-section bounded by two ellipses to be equal in area to the annular cross-section of the original cylindrical roll, the size of the peripheral ellipse can be found. The original cross-section you correctly calculated as $77\pi/4$, which comes to 19.25π square inches. Adding to this annular area the area of the inner ellipse, we obtain 19.91π square inches for the area of the outer ellipse. If its semiaxes were as 10 : 8, as you assumed, these would be determined from $b = \frac{4}{5}a$, $\pi ab = 19.91\pi$. Thus $\frac{4}{5}a^2 = 19.91 \Rightarrow 4a^2 = 99.55$, from which $2a = \sqrt{(9.955 \times 10)} \approx 9.9775$. Certainly it appears that a is not more than 4.99, and b is not more than four-fifths of this, so the perimeter of the outer ellipse will not be more than 4.99/5 of the previously calculated value of $9\frac{1}{36}\pi$ inches.

Sagredo: I see that this is only a trifling amount more than the original 9π inches for the undistorted cylinder. Your analysis does seem preferable to mine.

Salviati: And yet I am sure that it is unsatisfactory. Notice, for instance, that whereas the distance between the circumferences of the concentric circles was a uniform 3.5 inches, the semimajor axes of the elliptical boundaries differ by $4.99 - 1.41 = 3.58$ inches approximately, while the semiminor axes differ by about $3.99 - 0.47 = 3.52$ inches. Can you accept that the latter difference could actually be found to hold in the severely flattened roll?

Sagredo: I must admit that it seems very unlikely, bearing in mind that it is greater than the original 3.5 inches and yet is in the direction of the applied compressive forces, being in the very line of action of their resultant.

Salviati: If this elliptical model were true, there would be a slight increase in the thickness of the paper where the turns pass close to the axes (perhaps even a slight separation of the layers near the ends of the major axis) and a marked compression of the layers in the intermediate positions where the distance between the bounding ellipses would have to be considerably less than 3.5 inches. Now, of course, such a distortion of the paper would be a mark of induced resisting forces, but no explanation suggests itself as to why the paper should be so strangely contorted, suffering four compression-expansion cycles in each of its turns! To cause the paper to behave in this way, it would be necessary, I think, to forcibly apply rigid elliptical formers from within and without. In the absence of some such artificial constraint, why should we expect to get closer to an elliptical configuration than to one more like that illustrated in Fig. 22?

Sagredo: You have convinced me that we shall have to give up the elliptical model. I have to admit that my unreflecting supposition that the circles would be transformed into ellipses was not made, as it should have been, as a consequence of some insight as to how the equilibrium conditions would indicate a resulting elliptical form. The supposition arose instead out of an unconscious desire for amenability to mathematical treatment. How well this indicates the truth of your earlier remarks, Salviati! (See p. 68.)

Salviati: Without more theoretical knowledge than we have, we can hardly make useful conjectures as to the precise behavior to be expected; neither have we the opportunity to carry out actual measurements of variously loaded cylinders of paper.

Sagredo: Yes, it is always nice to be able to predict an appearance, but it seems that this is a case where we do first need to know what the appearance is before we attempt to explain it.[75]

Simplicio: The distinction between knowledge of the "that" and knowledge of the "why" reminds me of Aristotle's discussion in the *Analytica Posteriora*[76] where he speaks of optics as being "subordinate to" (that is, explicable by) geometry, of harmonics as subordinate to arithmetic, and so on, and how "it is the business of the empirical observers to know the fact, of the mathematicians to know the reasoned fact" and indeed to know the reason even if they be unacquainted with the fact, for insight into a universal does not depend upon knowledge of all its instances. Then he ends with an example that seems especially relevant to our investigations: "it is the physician's business to know that circular wounds heal more slowly, the geometer's to know the reason why."

Sagredo: I agree that it is the business of the geometer to try to understand why, as well as to know that, a circle has less perimeter than any other figure of equal area.

Simplicio: And consequently that there would be less adjacent undamaged surface tissue in the case of a circular wound.

Sagredo: That is something that a mathematician might point out to a physician, but I cannot believe, Simplicio, that Aristotle would wish to claim that the geometer could provide the physician with a very profound explanation of healing merely by connecting that phenomenon with a geometric theory. The physician or certainly the physiologist would require a much more comprehensive explanation; but that is another matter.[77] Once, I more or less satisfied myself that the circle has less perimeter than any other plane figure of equal area and greater area than any other isoperimetric figure by a number of piecemeal considerations involving comparisons between rectangles and squares, between scalene and isosceles triangles, between squares and regular octagons, between polygons with a greater or a fewer number of sides when inscribed or circumscribed about circles, and so on. It was against this background that I stated my belief a little while ago that it is impossible simultaneously to preserve both the area and the perimeter unchanged when a circle is transformed into another shape, whether elliptical or not. I do not yet see how this fact, for such I believe it to be, can be reconciled with Salviati's assertion that there are many other pairs of figures, besides the circle and its circumscribing square, for which the ratio of the areas is equal to the ratio of the perimeters. I hope that we can concentrate our attention upon the why of these matters in the coming hours. But at the moment a practical illustration occurs to me. You know how the people of the twentieth century use tubes of toothpaste, as they call it, and how in this age a thousand other substances are similarly contained for easy dispensing. Not only can the contents be squeezed out by flattening the tube somewhat—without changing the perimeter of the oval cross-section—but sometimes, if too much has been squeezed out, it is possible to draw a little of the contents back into the tube simply by pressing it in such a way that the oval cross-section becomes more nearly circular again.

Salviati: Everybody learns how certain manipulations bring about particular results, and for ordinary persons that is usually the end of the matter. On the other hand, a few, more theoretically inclined than the rest, are motivated to explain why these actions achieve what they do; indeed it is usually only this reflective minority that perceives that there is something requiring explanation. We hear a lot about the practical applications of mathematics. I am just as interested in the mathematical inspiration that is occasionally generated in a prepared mind contemplating practice—when it finds itself urged, for the sake of the understanding, to attempt an abstract formulation and theoretical treatment of problems suggested by the activities of artisans of various kinds, of mapmakers, engine shunters, itinerants, customers queued at service counters, chessplayers or gamblers.

Simplicio: I am reminded of the famous passage where Aristotle describes how our awareness of universals develops out of relevant sense perceptions, memories, and experiences by a process "like a rout in battle stopped first by one man making a stand and then another, until the original formation has been restored."[78]

Sagredo: I suppose that this means "until the essential pattern, or law, is apprehended" or, more plausibly, until some conjecture suggests itself. It seems reasonable to say that appropriate experience is a necessary if not a sufficient condition for conscious, formulated, specifically human knowledge of the general kind that is the beginning of science.[79] Men probably carried liquids in skins for thousands of years with nothing more than the immediate awareness that care is needed if loss of some of the contents through inadvertent pressure on the skin is to be avoided. But eventually, after a suitable background of relevant ideas had arisen, this experience would engender explicit items of scientific significance, for example, that the capacity of a container can be varied merely by changing its shape and that a given skin has a greater capacity when its form is rounded like a plum than when it is flatter or is crinkled like a prune; then it might come to be conjectured that a sphere has a greater volume for a given surface area than any other geometrical form.

Salviati: Quite so. But, as I was about to add before, for all that my own mathematical interests have been especially connected with natural phenomena and technological practice, I have to admit that the most frequent stimulus to new mathematical work has been provided for a long time past by what has already been incompletely achieved within mathematics itself. Doubtless it was the realization of this that led Poincaré to give it as his opinion that in the genesis of mathematical discovery "the human mind seems to borrow least from the exterior world."[80] It appears that mathematics was already a largely self-sustaining discipline in Babylonia early in the second millennium B.C.; certainly it had this character in Greek times. No doubt you remember the saying of the Pythagoreans to the effect that a theorem ought to be regarded as a platform enabling further progress to be made. Extensions, generalizations, analogies between problems, or between whole systems are not merely logically interesting abstractions made *post factum* by commentators. Mathematicians themselves have long been conscious that these names do signify quite characteristic and fundamental features of the development of mathematics. What I had to say in the last hour concerning the desire for amenability to mathematical treatment (which is always relative to the knowledge and the techniques available) was certainly not meant to deny the attractiveness of generalization within mathematics wherever it can be elegantly achieved.

Extramathematical influences, though relatively rare, are still in evidence and occasionally they are of decisive importance. Apart from scattered individual problems, most of which are illustrations rather than sources of theory,[81] whole new fields of mathematical investigation have been opened up by mathematical minds reflecting upon daily occurrences. Stimulated by a dispute with a wine merchant over the accuracy of the customary methods of gauging casks, Kepler was led to make a comprehensive study of the proper determination of the volumes not merely of wine barrels but of numerous solids of revolution, including ninety-two not treated by Archimedes. It was in this way that his *Nova stereometria doliorum vinariorum* of 1615 came to be written, a work that you know stimulated Kepler's contemporaries to work along related lines for several decades. The early use of sphere-pack models to explain crystalline forms in terms of arrangements of spherical atoms, most notably by Kepler and by Dalton, belong to just that era, from the sixteenth century to the nineteenth, when stacks of cannonballs were a not unfamiliar sight in Europe.[82] Such lines from Virgil as "The anchor drops, the rushing keel is staid," are said to have sent Euler enquiring into the dynamics of the situation, and of course the bridges of Königsberg provided him with a problem pregnant with later significance.[83] The famous four-color problem, unsolved until 1976, is commonly supposed to have first occurred to a young student of mathematics in 1852 while reflecting upon the coloring of an actual map.[84] Other well-known examples are provided by game theory,[85] queueing theory, linear programming, and other techniques of operations research in the mid-twentieth century, theory of statistics from its nineteenth-century beginnings, and, of course, the early investigations into the mathematization of probability by Pascal and Fermat.[86]

Sagredo: I find these very interesting examples. Naturally, it is the mathematicians whose minds are best prepared to respond to potentially mathematical aspects of the physical and social environment, but another characteristic would perhaps significantly detract from this advantage; I mean the tendency of at least some of them to live in a world of mathematics with little thought for the things of the world of everyday concerns.[87]

Salviati: Perhaps you are right, though in my experience not only are mathematicians peculiarly well fitted to respond to the relevant items of the external world as these present themselves, they are also *motivated* to so respond, and to seek out, and abstract from, items of daily experience that will provide apt illustrations of mathematical structures already in their possession or in the process of being understood or created. If we keep this in mind, we will be less likely to accept simplistic accounts of practical needs determining the course of mathematical progress. Where empirical influences are operative we should expect a dynamic interaction between them and the internal momentum of the evolving patterns of mathematical thought.

Sagredo: You had me thinking at first of those old pseudopractical problem types involving such things as cisterns with inlet and outlet pipes simultaneously open, the division of inheritances, the separation of false from true coins by the balance, the ferrying of incompatible items across rivers, the times of coincidence of hands of clocks, the summation of grains of wheat doubling from unity as many times as there are squares on a chessboard, and so on.[88] But I now perceive that you have in mind more revolutionary developments in mathematics where economic or military needs may be claimed as influencing the direction of mathematical growth.

Salviati: It has to be admitted that such needs are sometimes important because they induce a particular concentration of effort toward the solution of technological or scientific problems and this in its turn may engage the attention of mathematicians and encourage them to construct the relevant mathematical models.[89] Kepler's laborious attempts to obtain close approximations for the elliptical perimeter were indisputably part and parcel of his astronomical studies.[90] Indeed, the whole development of trigonometry up to the seventeenth century was intimately connected with astronomy.[91] Huygens's discovery of the tautochronism of the cycloid and his theory of evolutes stemmed directly from his work with the pendulum clock.[92] The mutually beneficial interactions between mathematics and the natural sciences have been more widely recognized—they are now part of the popular imagination—than the internal interactions within the subject matter of mathematics itself, for all that the latter are more commonly influential in motivating mathematicians.

Simplicio: Perhaps the first point to be made in any consideration of the "sociology of mathematics" is the obvious one that in recent centuries, in so many societies, it has been economically possible for certain pure mathematicians to flourish in academia while engaged, so far as can be judged by the usually prevailing criteria, on completely *useless* pursuits. But now another hour has ended. We have wandered far from our initial problem and our investigations continue to proceed rather erratically,[93] yet I am conscious of having learned much and I am grateful to you both.

VI

It looks as if the scientific, like other revolutions, meant to devour its own children; as if the growth of science tended to overwhelm its votaries; as if the man of science of the future were condemned to diminish into a narrow specialist as time goes on.

—T. H. Huxley

Mathematics . . . discovered in new men those very qualities that the perception of God had discovered in their predecessors.

—C. L. Morgan

The fatal tendency of mankind to leave off thinking about a thing when it is no longer doubtful is the cause of half their errors.

—J. S. Mill

The important thing is not to stop questioning.

—Einstein

Civilization advances by extending the number of important operations which we can perform without thinking.

—A. N. Whitehead

The real danger is not that computers will begin to think like men, but that men will begin to think like computers.

—S. J. Harris

I state my case, even though I know it is only part of the truth, and I would state it just the same although I knew it was false, because certain errors are stations on the road to truth.

—Robert Musil

HORA SEXTA

Simplicio: Are we any nearer to finding even one pair of figures, other than the circle and its circumscribing square, for which the ratio of the areas is equal to the ratio of the perimeters? If it were not for Salviati's assurances that lots of such pairs exist, I think that you and I, Sagredo, would still confidently believe that the circle and its enclosing square form a combination unique in this respect.

Sagredo: Our experience with the ellipse and rectangle, obtained by simply elongating the circle and square, does seem to have added to the plausibility of our initial assumption.

Simplicio: If we could explain the loss of equality between the perimeter and area ratio in this case, we might gain a useful insight.

Sagredo: But surely, Simplicio, we have done that in considerable detail already, first by showing that $p_e : p_r = A_e : A_r$ if and only if $p_e = \pi(a + b)$, and then showing that $p_e \neq \pi(a + b)$. What else do you want?

Simplicio: I think that I am looking for something more directly geometrical. The radius of curvature and also the diameter of an ellipse continually vary as we proceed around the figure. In ellipses of sufficiently large eccentricity, in positions not too far from the minor axis, the radius is actually greater than the diameter, elsewhere it is less, and, in the vicinity of the major axis, the radius is less than half the diameter. But for a circle and only for a circle, the radius is everywhere exactly half the diameter, and every diameter is an axis of symmetry. So it appears that we have several unique properties possessed by circles and by no other figures. Consequently we could, I suppose, just as well select some property other than the uniformity of radius as the defining characteristic.

Salviati: Tell us, Simplicio, what would you say about the following propositions?

1. All figures of uniform diameter are circles.
2. All coins are circular.

Simplicio: The first I judge to be true unreservedly, the second true only with respect to some restricted time and place. The first has an absolute, mathematical necessity, a necessity of the kind Plato called divine, since no god could gainsay, much less alter, such a truth.[94] A truth of this kind, as Aristotle would have expressed it, just cannot be otherwise, its denial involves a self-contradiction. As to coins being circular, this is a matter subject to human decision, which I still believe to be free.

Salviati: Then you will not be surprised if Sagredo tells you about a noncircular coin now in use in the United Kingdom, for I can see by the sketch he is starting to prepare that he has knowledge of this particular coin.

Simplicio: My only surprise would be to find that Sagredo had been dreaming about money! You know what our condition is: we now dream realities, from time to time we intimately share the great thoughts of Galileo's successors, we have had access to some of the finest works while they were being created, and, for relaxation, we occasionally mingle with the commonality of mankind. But, as you know only too well, we are restricted in our dreaming to just such things as are directly connected with matters about which we showed genuine concern during our own lifetimes. (And, speaking for myself, I am sure that I promptly forget most of what I have dreamed.) I should hardly have expected to find that Sagredo had been dreaming about money. In our day he was in that comfortable position of having just so much wealth that he was not obliged to concern himself with matters of material necessity, an expression I use in its popular sense, and I well remember how he was wise enough to devote the greater part of his energies to the acquisition of knowledge.

Sagredo: Some would say "selfish enough" where you say "wise enough," Simplicio.

Simplicio: Well, few indeed in your situation have turned away from so much that would distract them in the pursuit of truth.

Salviati: I am sure we will not easily forget the occasion when you told us, Sagredo, that already in your youth you had discovered that the wealthy were usually so possessed by their possessions and so filled were their lives with trivial pastimes that it appeared they had adopted their way of life for the express purpose of avoiding any significant thought or action.[95]

Sagredo: Well gentlemen, you have allowed me to finish my drawing. This coin (Fig. 30), an English fifty-pence piece, constitutes a counterexample to the two propositions Salviati asked you to consider, Simplicio.

Simplicio: It appears to be quite a pretty coin. I did not deny the existence of noncircular coins you know.

Sagredo: No, but by asserting that the first proposition was necessarily true, you were denying the existence of noncircular figures of uniform diameter. Yet the edges of this coin are just such curves, as can be shown in various ways. Using

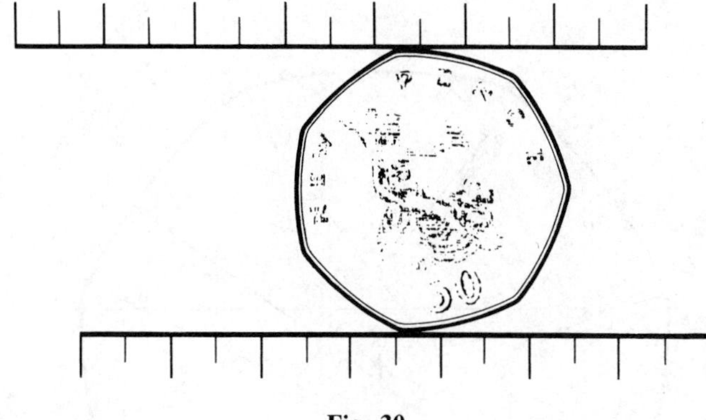

Fig. 30

two or more of these coins as rollers placed between a pair of rulers or other straight-edged objects, there would be no bumpiness in the motion; they would serve quite as well as rollers of circular form!

Simplicio: I would very much like to see a theoretical analysis of this curve. Of immeasurably greater value than the coin would be the understanding of its contour. If it turns out to have the property you claim for it, I shall have to admit that it is a most remarkable figure. Am I right in suspecting that it is not a Pythagorean pentagram but a heptagram that will prove to be the key to this tantalizing curve? For several millennia the circle has been taken as the symbol of the sun, heaven, the sacred, as contrasted with the square, symbolic of the earth, the secular, or man-made. Are we now to believe that it is the number *seven*—divine, complete, perfect, the number of the seven gifts of the spirit[96]—that somehow lies at the heart of the mystery of this circle that is not a circle? But what a profane use this twentieth century has found for it! Had it been known in our age it probably would have been employed very differently: How fitting it would have been to have used it in the designs for the rose windows of the cathedrals of Europe.[97]

Sagredo: Perhaps so, Simplicio. But now, if you will give this curved figure your attention, you will see how readily its construction can be carried out with blending circular arcs, and how the regular heptagram does indeed have a perfectly explicable relevance to this particular construction.[98] Referring to Fig. 31, it is clear that the distance between any pair of parallel tangents is equal to $R + r$ units. Hence the circumscribing rectangle is a *square* of this side length, and perimeter given by $p_s = 4(R + r)$, whereas the perimeter of the uniform diameter figure itself is clearly 7 times the arc JK together with 7 times arc KM. These arcs, of radii R, r, respectively, subtend angles at their centers, A and E, of $\frac{1}{2}$ of $360°/7$, that is, $\frac{1}{7}$ of $180°$. Hence the perimeter is the sum of arcs that could be rearranged to make up a semicircle of radius R together with a semicircle of radius r; so the perimeter is given by $p_u = \pi(R + r)$.

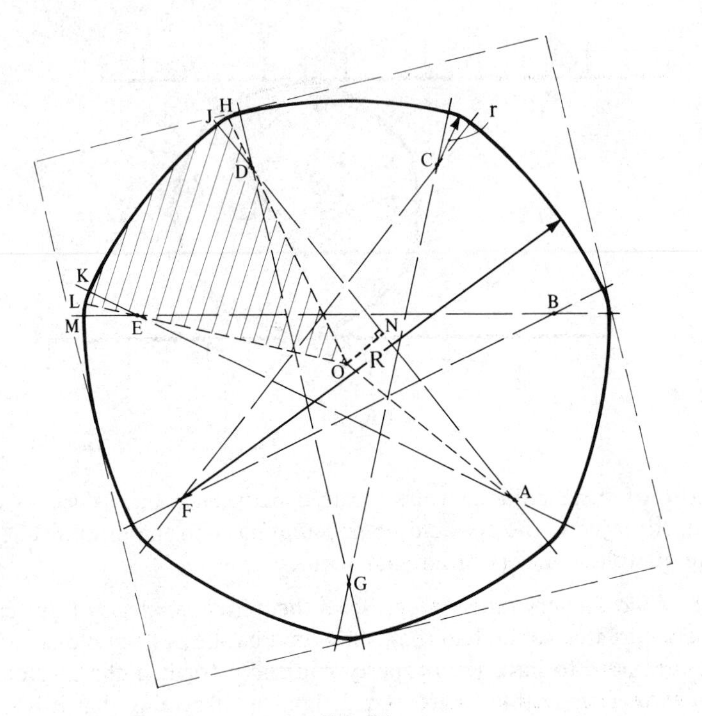

Fig. 31

Simplicio: You mean to say that the perimeter is π times the uniform diameter, exactly as is the case for a circle!

Sagredo: That is how it works out. And notice, Simplicio, there is no special virtue in the number 7 in this context; other figures of the same general character could certainly be constructed using the pentagram as the skeleton. Of course, the shape obtained is dependent upon the value chosen for $R:r$ as well as upon the number of points used for the centers. I think you can see that *any odd* number of equally spaced points may be used as centers of blending circular arcs so as to obtain a uniform-diameter figure of perimeter exactly given by $p_u = \pi(R + r)$.

Simplicio: Surely this is a very remarkable result! I cannot help being reminded of the formula $p_e = \pi(a + b)$ for the perimeter of an ellipse. But that formula was inexact; consequently for an ellipse and its circumscribing rectangle with sides $2a$, $2b$, the ratio of the perimeters was shown to be *not* equal to the ratio of the areas:

$$p_e : p_r \neq \pi(a + b) : 4(a + b) = \pi ab : 4ab = A_e : A_r.$$

On the other hand, you have now convinced me that there do exist figures that are not circles and yet are not only uniform in diameter but also are of perimeter equal to the circumference of the circle having the same diameter. Could it possibly be that we have here a form or a whole class of forms that are so related to their circumscribing squares as to have with them equal area and perimeter ratios?

Salviati: For these uniform-diameter figures, $p_u = \pi d$, as Sagredo has shown, while for their circumscribing squares, $p_s = 4d$; hence your question hinges upon whether or not the areas of these ovals[99] are given by $A_u = (\pi/4)d^2$. If they are, it would mean that a true circle could be reshaped to give a form such as that indicated in Fig. 31, so that the *area as well as the perimeter* would remain *unchanged.* Do you think that this is possible?

Simplicio: I do not know how to answer. I have no recollection of ever having encountered these figures before and they still seem a little mysterious to me, in spite of Sagredo's analysis. I think I referred to them as circles that are not circles a few minutes ago; now the suggestion arises in my mind that they might be called "Dionysian circles" but do not ask me why.[100]

Sagredo: I trust, Simplicio, that your initial enthusiasm for these ovals will not be quite dissipated by the wholly rational examination to which we are subjecting them. And now, Salviati, to the question you have just put to us I can only answer as I did in the last hour. In my pedestrian fashion I once compared the areas of a number of figures, such as an equilateral triangle, a square, and other regular polygons, with the area of a circle of equal perimeter, and I found that the more the shape deviated from the truly circular the greater the loss of area. Both as a result of these early calculations and through a very slight acquaintance that I later made with the Greek work on isoperimetry, I became convinced that the circle has the greatest area of all isoperimetric figures.[101] But as I have heard that the Greek geometers achieved only partial success in this particular line of investigation, and since I doubt that they ever suspected the existence of uniform-diameter ovals, it appears to be appropriate for us to determine the areas of these figures forthwith.

Simplicio: I agree. As you yourself said at the beginning of today's discussions, let us not rely on intuition when certainty can be achieved by calculation.

Sagredo: Quite so. Referring, then, once again to Fig. 31, we see the following: Total area = n times the shaded area, where $n = 7$ in the case illustrated, and the shaded area is the sum of sectors of radii R, r, respectively, each with central angle $180°/n$ *minus* twice isosceles triangle AOD, with base $AD = R - r$ and altitude $ON = [(R - r)/2) \tan (180°/2n)$. Hence the total area covered by the whole figure comes to the area of a semicircle of radius R *plus* the area of a semicircle of radius r *minus* $2n$ times the area of the isosceles triangle AOD.

$$\text{Therefore, required area} = \frac{1}{2}\pi R^2 + \frac{1}{2}\pi r^2 - 2n\left\{\frac{1}{2}(R - r) \times \frac{1}{2}(R - r) \tan \frac{180°}{2n}\right\}$$

$$\left.\begin{aligned} &= \frac{\pi}{2}(R^2 + r^2) - \frac{n}{2}(R - r)^2 \tan \frac{180°}{2n} \\ &< \frac{\pi}{2}(R^2 + r^2) - \frac{n}{2}(R - r)^2 \frac{\pi}{2n} \end{aligned}\right]*$$

$$= \frac{\pi}{4}(2R^2 + 2r^2 - R^2 + 2Rr - r^2)$$

$$= \frac{\pi}{4}(R + r)^2$$

Thus the required area, A_u, $<$ area of the circle equal to it in diameter (and in circumference).

Simplicio: I see that you have already marked with an asterisk the step about which I should like elucidation.

Sagredo: This step is a perfectly standard one; I was only about to say that we might show how slight is the lack of equality for n not too small. But since you have raised the question of justification, I will start by reminding you that we are dealing with acute angles of $180°/2n$, with $n = 3, 5, 7$, or any greater odd number.

Simplicio: That is a further question I should like to consider: Why cannot an *even* number of centers be used?

Sagredo: All in good time, Simplicio. For the present, notice how in Fig. 32, arc $BP = 1/2n$ of a semicircular arc of radius AB. That is, arc $BP = \pi \cdot AB/2n$, whereas tgt $BC = AB \tan(180°/2n)$. Consequently, since tgt $BC >$ arc BP, $\tan(180°/2n) > \pi/2n$.

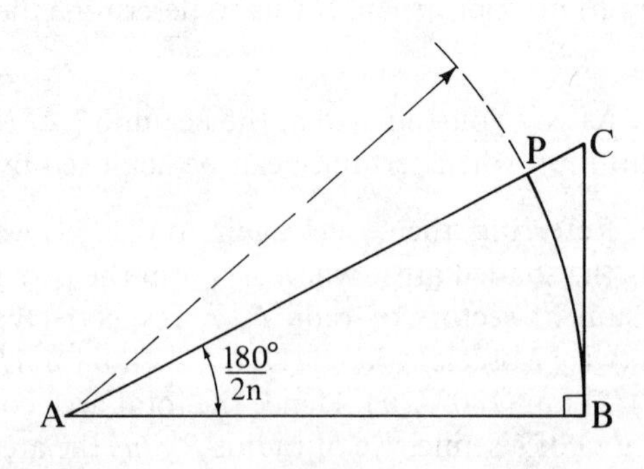

Fig. 32

Simplicio: Am I supposed to accept that the straight line BC exceeds the curved line BP, for any acute value of angle BAC, merely on the strength of a diagram?

Sagredo: No Simplicio; I sought only to remind you of a relation that I thought would have been familiar to you already. A more detailed argument is as follows:

$$\text{area of } \triangle ABC = \frac{1}{2}AB \cdot BC = \frac{1}{2}AB^2 \tan \frac{180°}{2n},$$

whereas, sector $ABP = 1/2n$ of the area of a semicircle of radius AB equal to $(1/2n)(\pi/2) \cdot AB^2$. Then, since triangle ABC exceeds sector ABP (the excess being very small for angle BAC small),

$$\frac{1}{2}AB^2 \tan \frac{180°}{2n} > \frac{\pi}{2n} \times \frac{1}{2}AB^2;$$

hence,

$$\tan \frac{180°}{2n} > \frac{\pi}{2n} \quad \text{with } \tan \frac{180°}{2n} \approx \frac{\pi}{2n} \text{ for } n \text{ large).}$$

Simplicio: Thank you; the demonstration is entirely convincing and I now recall having seen something of the sort before.

Sagredo: Then you will agree that for the oval of uniform diameter $(= R + r)$, we have proved in sufficient detail that the area $A_u < (\pi/4)(R + r)^2$, that is, $A_u < A_c$, where A_c is the measure of the area of the isoperimetric circle. And just how little A_u falls short of A_c can readily be determined in any particular case. For example, taking $n = 7$, and working correct to five decimal places, we have the following:

$$\text{Required area,} \quad A_u = \frac{\pi}{2}(R^2 + r^2) - \frac{n}{2}(R - r)^2 \tan \frac{180°}{2n},$$

$$\text{where } \frac{n}{2} \tan \frac{180°}{2n} = \frac{7}{2} \tan 12° \; 51\tfrac{3}{7}' = 0.79884.$$

$$\begin{aligned} A_u &= 1.57080(R^2 + r^2) - 0.79884(R^2 - 2Rr + r^2) \\ &= 0.77196(R^2 + r^2) + 0.79884(2Rr) \end{aligned}$$

Since $\pi/4 = 0.78540$ correct to five decimal places, it suits our purpose to write this as:

$$\begin{aligned} A_u &= (0.78540 - 0.01344)(R^2 + r^2) + (0.78540 + 0.01344)(2Rr) \\ &= 0.78540(R^2 + 2Rr + r^2) - 0.01344(R^2 - 2Rr + r^2) \\ &= \frac{\pi}{4}(R + r)^2 - 0.01344(R - r)^2 \end{aligned}$$

Thus the area expression approaches $(\pi/4)(R + r)^2$ as $R - r$ approaches zero —that is, as the centers $A, B, C, \ldots$ come together and the figure approaches a true circle. Fig. 33(i) illustrates a case where the centers are sufficiently close together for the form of the outline to be barely distinguishable by eye from the circle equal to it in diameter and circumference. At the other extreme, where $r = 0$, illustrated in Fig. 33(ii), the area expression reduces to $(\pi/4)R^2 - 0.01344R^2$, or $(\pi/4)(R^2 - 0.0171R^2)$ approximately, showing that the area is about 98.29 percent of the isoperimetric circle. As for your other question, about the number of equally spaced centers that might be used, I suggest that you consider the cases represented in Fig. 34, and such others as you might feel the need to draw for yourself.

Simplicio: You are not suggesting that we can conclude from a handful of instances that uniformity of diameter always occurs when the number of centers is odd and never when it is even?

Sagredo: Now, Simplicio, I shouldn't have to remind you that in such circumstances individual examples may well be used as goads and props enabling us to grasp the "commensurate universal" or "essence," as your philosopher and his followers were wont to describe it. The particular cases illustrated in Fig. 34 are assuredly representative of their classes and can be seen to be so representative, with a certainty, incidentally, that does not attend our judgments as to whether the individual objects or events of the external world are truly representative of the general categories they are customarily supposed to exemplify. I was not about to propose a blind induction, but rather to consider whether we could see *that—and* see *why*—uniformity and variability of diameter are necessarily associated respectively with the odd and the even and not merely with the seven and the non-seven, as you seemed prepared to conclude a little while ago!

Simplicio: I was feeling my way then, and I admit I was speaking somewhat fancifully. My question just now was prompted by the sudden thought that perhaps we were sinking to natural history methods when it is mathematical knowledge to which we aspire.

Sagredo: I think an indication of the way in which we may ascend to the general cases will not displease you. We have to see and to show how it is that constancy of $R_1 + r$ (or of R_0 in the case of the inner figures) guarantees uniformity of the diameter in all cases where an odd number of centers is used, but fails to do so—in fact guarantees *variability* of the diameter—if the number of centers is even. Observe first what the broken lines in these diagrams of Fig. 34 serve to indicate. Passing through the centers two at a time they mark out the points of transition between the adjacent arcs of different radius that compose the outer ovals. Only when the number of centers, n, is odd does any (and indeed every) pair of broken lines concurrent at a center demark a pair of arcs of radii R_1 and r lying directly opposite each other. With n odd, then, any chord passing through a center and lying between the pair of broken lines through that center will consist of a pair of collinear radius lines perpendicular to the tangents to the curve at its ends. Conversely, the radii from the points of contact of any pair of parallel tangents pass through their common center collinearly, forming a chord of fixed

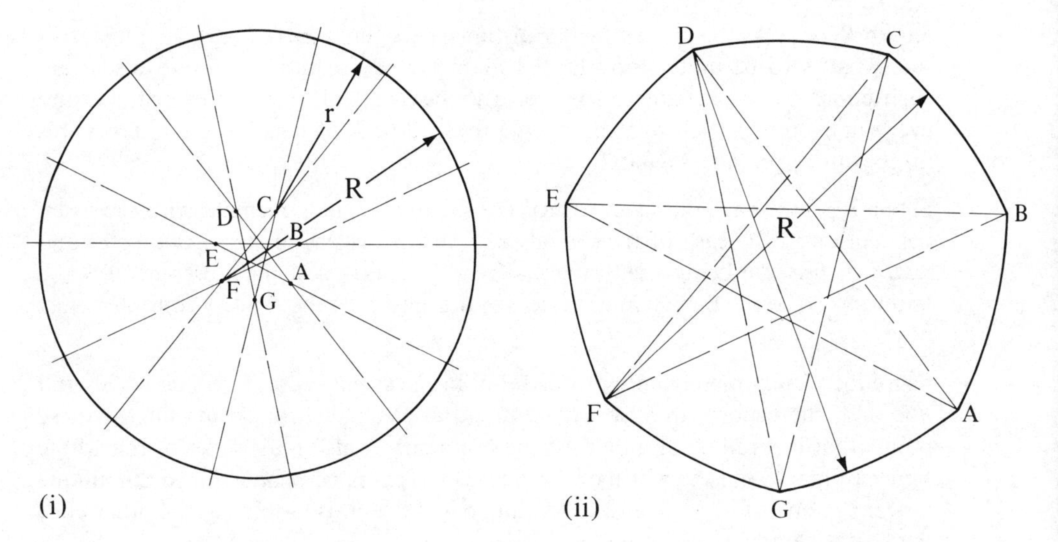
r
R
C
D
B
E
A
F
G
(i)
D
C
E
B
R
F
A
G
(ii)

Fig. 33

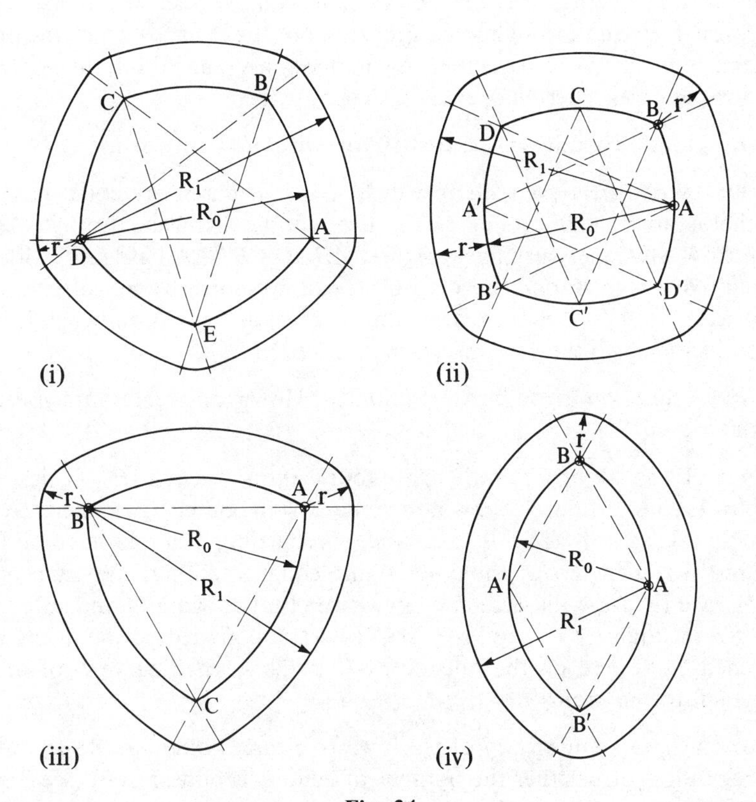
C
B
R1
R0
A
r
D
E
(i)
C
B
r
D
R1
A'
A
R0
r
B'
D'
C'
(ii)
r
A
r
B
R0
R1
C
(iii)
r
B
R0
A'
A
R1
B'
(iv)

Fig. 34

length $R_1 + r$. As the perpendicular distance between a pair of parallel tangents is just what is to be understood by the diameter or breadth of these ovals, every such chord forms a diameter of the ovals having odd numbers of centers, and every diameter of such ovals is one of these chords of length $R_1 + r$. Does this meet your approval, Salviati?

Salviati: The way you have defined the diameter is in agreement with the usual convention in the case of these ovals of uniform width, the terms *diameter* and *width* or *breadth* being used interchangeably. I have a slight preference for the latter terms, as I will explain later; let me not interrupt the flow of your ideas just now.

Sagredo: I was merely going to add that in the special cases in which $r = 0$, and n is odd, the centers are also vertices, and any chord through any one of these points that is a radius of an arc of the boundary is also a diameter of the whole figure (it marks out the width of the figure) since it is perpendicular to the unique tangent at one of its ends and to the line parallel to this tangent at the other end, the vertex center.

Simplicio: You are a cunning fellow, Sagredo; I think you knew that I wished to consider the class of these figures that have an *even* number of centers. Yet you have chosen to give a thoroughgoing exposition of the class for which the number of centers, n, is odd. And so well have you done this that I think the analysis required for the cases where n is even is already evident.

Sagredo: Good. Then it will be best if you will outline it for us.

Simplicio: When n is even, there will be just $\frac{1}{2}n$ diameters equal to $R_1 + r$, equal, that is, to $R_0 + 2r$. In Fig. 34(ii), these are the diameters through AA' and through CC'. The inner curvilinear figure, $ABCDA'B'C'D'$, has least width when measured from A to A' or from C to C' as can be shown by considering the inscribed circle $ACA'C'$. Let me redraw the diagram as Fig. 35, in which it will be useful to include also the circumscribed circle $BDB'D'$.

Sagredo: A question if you please, Simplicio! How many centers are you counting in this figure?

Simplicio: I was taking it that there are four centers—A,C,A',C'—which is only correct for the inner figure; I now notice that for the outer figure there are eight centers. So taking $n = 8$ in this case, and generalizing, I should have said there will be just $\frac{1}{4}n$ diameters of the outer figure equal to $R_1 + r$. My main point is that these are the *least* diameters: everywhere else the width of the outer figure will be greater than $R_1 + r$, or $R_0 + 2r$. The track bounded on the inside by the four-arched curve and on the outside by the eight-arched curve is of uniform width, equal to the radius r.

Sagredo: Yes, this uniformity of track width clearly holds for all such pairs of curves regardless of whether the number of centers is odd or even (see Fig. 34), so, in general, $R_1 = R_0 + r$.

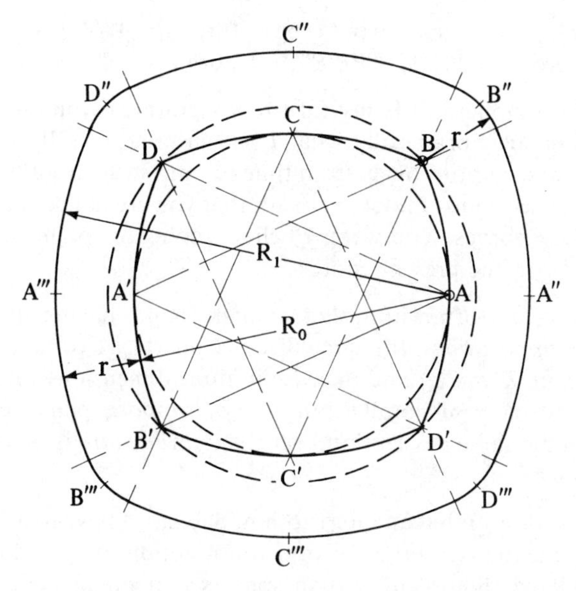

Fig. 35

Simplicio: Now let me see if I can match the thoroughness of the exposition you gave us for the odd-numbered cases. Referring first to the case illustrated in Fig. 35, we observe that the arc *DA′B′*, with center *A* and radius *AA′*, touches the circle *ACA′C′* at *A′* but otherwise lies wholly outside this circle. As you will agree, Sagredo, this can be seen not merely with the sensual eye but "with the eye of the soul," as Socrates would have expressed it.

Sagredo: None the less he would have expected that the geometer provide a rational account of such a seeing.

Simplicio: And I have it for you. The reason why arc *DA′B′* remains confined within the annulus bounded by the inscribed and circumscribed circles is that its curvature is less (because its radius is greater) than that of either of the arcs of these circles. This means that the arc *DA′B′* occupies an intermediate position between the arc of the inscribed circle near *A′* and the tangent touching this arc at *A′*; it means also that this arc *DA′B′* is positioned between arc *DB′* of the circumscribed circle and chord *DB′* of the same circle. (There is no need to clutter up the diagram with representations of all the lines mentioned.) An argument of the same form shows that corresponding relations must hold for the arc *B′C′D′*, and in general for any arcs making up a curvilinear figure of this kind having any even number of centers. A pair of concentric circles may be provided for any such ovals, the diameters of the circles being respectively equal to the least and the greatest widths of the oval. It should not be difficult to calculate the ratio of these diameters in any particular case.[102] Clearly, for the outer oval of the diagram of Fig. 35, the circles would pass through $A''C''A'''C'''$ and through $B''D''B'''D'''$, where $A''AA'A'''$, $B''BB'B'''$, etc., are axes of symmetry.

Sagredo: Well, Simplicio, it is my turn to congratulate you on the aptness of your presentation. It is practically what I had in mind myself and I have been thinking about these various ovals from time to time ever since I happened to see one of those 50 pence coins. I hope, Salviati, that you are not dissatisfied with our discussion of these curves. You were, I believe, going to comment on the use of the term *diameter* in the present context?

Salviati: Ovals of uniform width, that is, plane figures having all pairs of parallel tangents, or *support lines,* the same distance apart, are commonly designated *figures of constant diameter;* and likewise in three dimensions solids of uniform width—having all pairs of parallel tangent, or *support,* planes equally spaced —are also given the same description. Yet the term *breadth* is perhaps to be preferred to *diameter.*

Sagredo: If you don't mind an interruption, Salviati, I have to admit that it had not occurred to me to extend the uniform width notion to three dimensions, so I have not considered figures other than spheres that might have this property. Also, I have not heard of support lines or planes before.

Salviati: The simplest explanations of both these matters may be given by referring to the inner figure, the curvilinear equilateral triangle, in your diagram of Fig. 34(iii). If this figure, which is known as a *Reuleaux triangle,*[103] is revolved about an axis of symmetry, there is obtained the simplest solid of constant breadth apart from the sphere. Returning to two dimensions, such ovals as those illustrated in Fig. 33(ii) and by *ABCDE* in the first diagram of Fig. 34 are called *Reuleaux polygons.* In your own characterization of these figures, Sagredo, you had occasion to refer to pairs of straight lines that touched the ovals and between which the ovals were confined. In these special regular cases one of each pair of such lines is a tangent and the other is the parallel through a vertex center. Lines of both these kinds are called "lines of support" or simply "support lines." In general, *a support line meets an oval at one or more points and has all the other points of the oval on one side of itself;* support planes are analogously defined with respect to solid figures.

Turning now to a consideration of the term *diameter,* you, Simplicio, were led very naturally to speak of the least diameter of variable diameter figures; yet this usage clashes with a common definition according to which the diameter is taken as the greatest distance separating two points of a figure. The term was and sometimes still is applied to rectilinear as well as to curvilinear figures, the diameter of a parallelogram, for example, being according to this definition the longer of the diagonals. The terms *diameter* and *diagonal* are both from the Greek, διάμετρος and διαγώνιος, but no trace of the latter term is known before Euclid, who used it, if at all, only very rarely, preferring the older and more general διάμετρος, where we would write *diagonal.*[104]

Sagredo: But the Greeks also spoke of the diameters of an ellipse, did they not, and were these not variable for any particular ellipse?

Salviati: Quite true; that is a different usage again. In the theory of conic sections as presented by Apollonius, a *diameter* is defined as the straight line bisecting each of a series of parallel chords of a section of a cone,[105] or, as it would perhaps be expressed nowadays, a diameter of a conic section is the locus of the midpoints of any set of parallel chords. (See Fig. 36.) In the case of an ellipse, this means that the diameter is the join of the points of contact of a pair of parallel tangents or support lines.[106] This is clearly not the same, or not usually the same, as the breadth, if the breadth is defined as the perpendicular distance between a pair of parallel tangents. The diameter and the breadth both vary with the direction of measurement, each according to its own distinct law.[107]

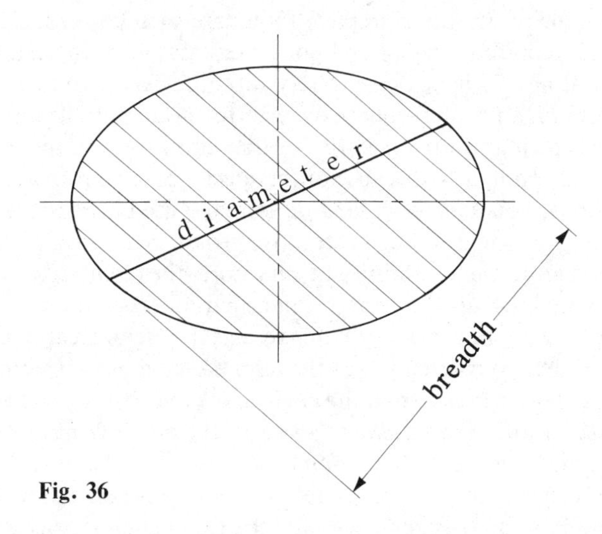

Fig. 36

Simplicio: It appears that the diameter of any ellipse (whether or not it is a circle) may be defined either as the locus of the midpoints of any family of parallel chords or as any line through the center terminated by the curve.

Sagredo: But I take it that only in the case of circles may a diameter be adequately defined either as any axis of symmetry or, say, as any chord perpendicular to the tangents at its ends, or better, perpendicular to *a* tangent at *an* end, since that is a sufficient specification.

Salviati: You are right, Sagredo, in noticing that your definitions would exclude all except two diameters in the case of any noncircular ellipse, but your last definition would apply in the case of noncircular ovals of uniform diameter. I can see that both of you have realized that within a restricted domain there are liable to be several properties, or "symptoms" (συμπτώματα) as the Greek geometers called such characteristics, which may be used in providing logically distinct but factually equivalent definitions of the diameter. Alternative definitions may appear otiose where they are interchangeable over all classes for which they have known application, yet it might well turn out that, outside this domain, differences in denotation as well as of connotation are discoverable.[108]

But now, as our hour is almost over, allow me to draw together two or three matters that I think you will find relevant in the remainder of today's investigations. As everyone knows, the ability to entertain novel associations or correspondences is a mark of the successful thinker, whether a mathematician, an inventor, a comedian, or an advertiser. But, as we had occasion to notice in the fourth hour, something else is characteristic of creativity. I refer to the flexibility of thought that shows itself in an imaginative preparedness to consider whether what has become habitually associated in thought might not be merely artifactual, so to say—the result, for example, of ultimately irrelevant social convention or of individual "conditioning"—and not representative of logical, or even physical, necessity at all. To have come to recognize the existence of a great class of ovals of uniform breadth is already to have performed a notable dissociation.[109] However, partly owing to the regularity of that coin, for you the original exemplar of such ovals, and partly because of the special context provided by our present investigations, you have both held on to two quite unnecessary associations throughout your treatment so far. I refer, first, to the regularity of the spacing of the centers; second, to the uniformity of curvature (the circularity) of each of the individual arcs making up the ovals. Emancipation from these gratuitious assumptions is necessary in order to be able to give a general treatment of the subject. I suspected that your retention of the term *diameter* was symptomatic of the incompleteness of your break from the circle and I felt that if you had been using the term *breadth* rather than *diameter,* you might possibly have been less hindered in reaching a more general point of view.[110] The literal meaning of διάμετρος is, of course, 'through measure' or 'distance across'; as we have seen, it was understood in a fairly general sense by the Greeks. Nevertheless, in popular usage, the term is confined to a central chord of a circle. I suggest that this restricted meaning, with its associations of symmetry, could have influenced as well as reflected the line taken in your own thinking.

Sagredo: The possibility of irregular ovals of constant breadth had not occurred to me; I suspect the oversight might be attributable to self-satisfaction with the progress I had already made.

Salviati: Do not be hard on yourself, Sagredo. You know in the history of our subject how unlikely it is that the pioneer will be the one to grasp the full scope or the full significance of the subject matter. Typically, pioneers fail to imagine the more elaborate possibilities or at least do not see how to deal with them, and for that reason may refrain from mentioning them.

Simplicio: A frank admission of outstanding problems would surely facilitate advancement in any field.

Salviati: Of course you are right. And it is pleasing to find that such ingenuousness is not altogether lacking in the publications of this modern age.

VII

There is something impromptu *about doing mathematics. The history of mathematics is perhaps best characterized by the constant improvization—an untidy patchwork, at least in the process of mathematical* creation. *Some or even most parts may be made to look neat* afterward *in the process of* exposition, *but mathematics as a living organism will remain incomplete, forever untidy.*

—Joong Fang

Human thought, before it is squeezed into its Sunday best, for purposes of publication, is a nebulous and intuitive affair: in place of logic there brews a stew of hunch and partial insight, half submerged.

—Liam Hudson

Scientific research is not itself a science; it is still an art or craft.

—W. H. George

The secret of science is to ask the right question, and it is the choice of problem more than anything else that marks the man of genius in the scientific world.

—Henry Tizard

Falling into trouble, encountering some unexpected difficulty, however harassing at the time, is in fact an opportunity for making a fresh advance and most advances have in fact been made by turning failure into success.

—Henry Guy

Science like life feeds on its own decay. New facts burst old rules; then newly developed concepts bind old and new together into a reconciling law.

—William James

HORA SEPTIMA

Salviati: So you have again spent your pause in busy preparation?

Sagredo: All these sketches are the result of your remarks as to how we might generalize our constructions for constant-breadth ovals. Perhaps we may begin with that large one of Simplicio's.

Salviati: Please tell us about it, Simplicio.

Simplicio: I understood you to have indicated that the triangles, pentagrams, heptagrams, and so on, the vertices of which[111] were taken as the centers of the circular arcs, could just as well be irregular as regular. So I supposed that a rod of fixed length be taken and turned through a half-rotation, not all at once but in any odd number of stages, turning first about one end and then about the other, alternately. This is shown in my diagram (Fig. 37) for a case with five steps, the final position, A_3B_4, coinciding with the initial position A_1B_1, except for the reversed sense. Thus, $\alpha^\circ_1 + \alpha^\circ_2 + \alpha^\circ_3 + \alpha^\circ_4 + \alpha^\circ_5 = 180°$, without the necessity for any of the αs being equal to each other. Clearly, the width of the oval is everywhere equal to the length of the rotated rod or line segment AB. An outer figure composed of blending circular arcs may be generated by B', A', respectively on AB, BA produced so that $BB' = AA'$. What do you think of that, Sagredo?

Sagredo: I can see that if AB has length R_0, and $AA' = BB' = r$, so that $A'B' = R_0 + 2r$ (or $R_1 + r$), then it may be immediately proved that the perimeter of the inner figure $= \pi R_0$, and that of the outer figure $= \pi(R_0 + 2r)$, or $\pi(R_1 + r)$, exactly as for the regular cases (pp. 91–92). I have no doubt also that the areas of these ovals are less than $\pi/4$ times the square of the breadth, though the proof would be more complicated than for the regular cases (cf. pp. 93–94).

What is more, it seems to me that there is a direct way of showing the necessity for there being an odd number of centers for any oval of constant breadth constructed in the manner of your diagram, regardless of the number of positions of the rod. If B is the first end to move, the number of A positions will be one less than the number of B positions (and vice versa if A moves first). If the first end takes up k positions, the second takes up $k - 1$ positions, where k is any integer not less than 3. Hence the total number of centers, n say, is equal to $k + k - 1 - 2$, the last subtraction being required on account of the coincidence of the final position of each end of the rod with the initial position of the other end. Thus, $n = 2k - 3$, which is odd.

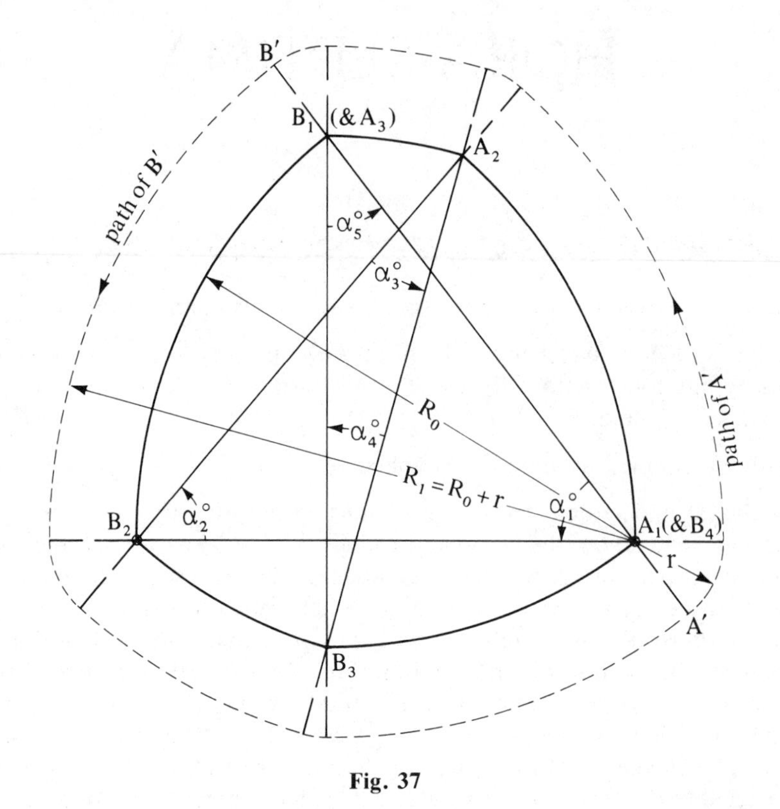

Fig. 37

Simplicio: Your argument appears to be correct, yet what it purports to prove is not yet perfectly transparent to me for the following reason. As there is so much freedom allowed in choosing the relative positions of the vertex centers, I cannot see what is to prevent two of these from coming closer and closer together until they coincide. Referring to Fig. 37, and supposing that A_2 is moved so that it coincides with B_1, thus dispensing with A_3, then are we not left with a four-centered uniform-diameter oval?

Sagredo: In such a case there would be only three centers, namely, A_1, B_2, A_2, because B_3 would cease to be operative as a center simultaneously with the coincidence of A_2 with B_1. The inner oval of Fig. 37 would be transformed into that especially simple and regular oval that Salviati told us is called a "Reuleaux triangle," as shown in Fig. 38.

Simplicio: You have almost satisfied me; yet one puzzle remains. In my diagram (Fig. 37), A_1, B_1, B_2 were not (or at least were not necessarily) at the vertices of an equilateral triangle. But now you have opened up the angle $B_1A_1B_2$ to make it equal to 60 degrees. Was that really necessary?

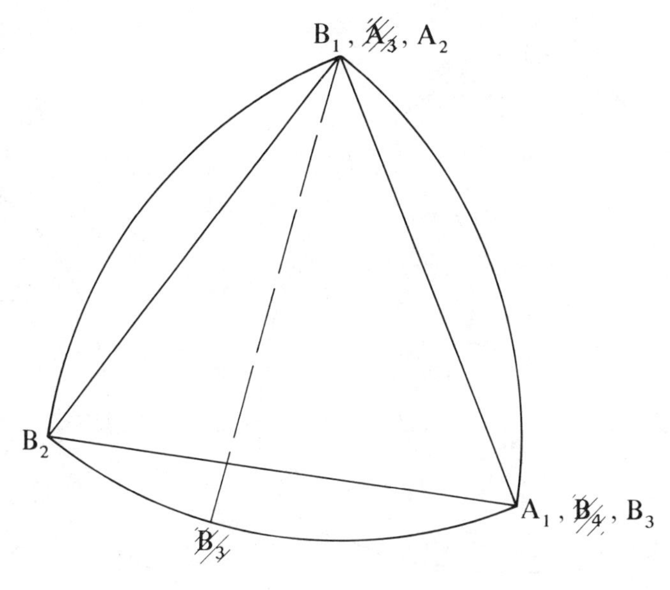

Fig. 38

Sagredo: I am only following your own assumptions. Not only have you been assuming that the "rod" is of fixed length, you have also stated that its final position is to coincide with its initial position, except for the reversed sense. So of course, if A_2 is to coincide with B_1 and arc A_1A_2 is to be centered at B_2, then $B_2A_2 = B_2A_1$. The pentagrams, heptagrams, and so on that you have in mind to use as the skeletons of these ovals, though not regular, were equilateral; reduction from the equilateral pentagram to the equilateral triangle cannot fail to produce regularity and threefold symmetry in the resulting oval.

Simplicio: I see. And are you suggesting that we might have had a rod of variable length?

Sagredo: From the point of view of pure geometry, the rod is a fiction. And the assumption of its rigidity is perhaps indicative of an unnecessary conceptual rigidity. But I am not yet sure how things would work out if we tried to scatter our centers randomly in the plane. Allow me first to show you a pair of diagrams (Fig. 39) that I sketched earlier, while you were preparing your drawing. I simply modified a Reuleaux triangle in the most cautious way I could think of—an unadventurous procedure compared with yours, I admit. Instead of using the vertices of an equilateral triangle as the centers, I used the vertices of an isosceles triangle, with sides a, a, b, say. In the first diagram, as you can see, the breadth is clearly uniform, being measured by $R_0 + r$, which comes to $2a - b$. The drawing of the second diagram made me aware of a complication I had not anticipated. If you

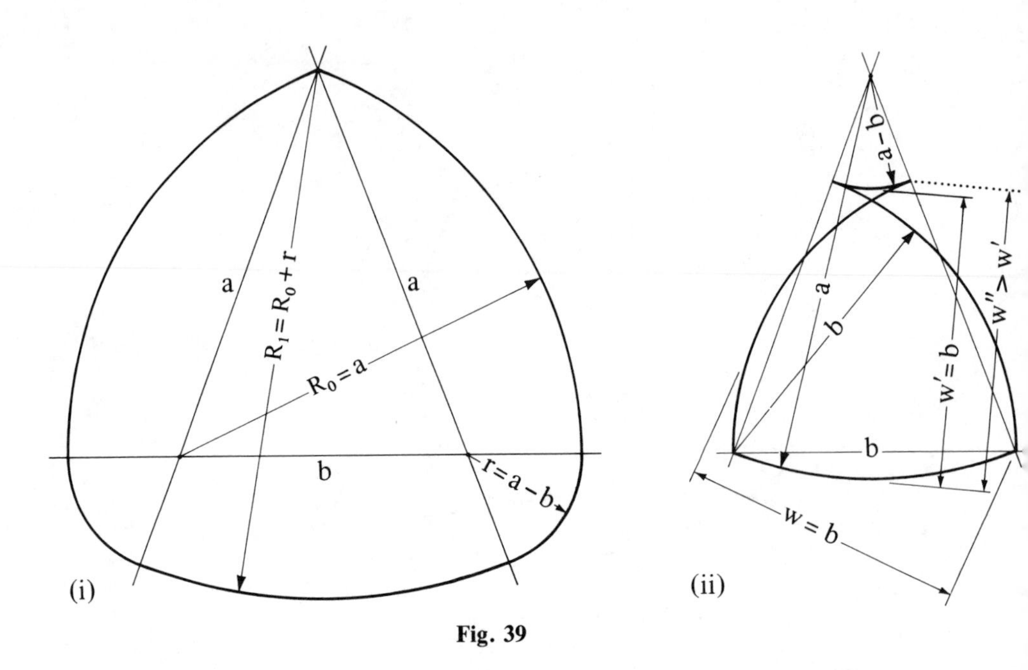

Fig. 39

compare the two diagrams you will see that the fishtail arises because $R_0 < a$. You would have obtained fishtails in your drawing (Fig. 37) if you had drawn the locus of a point on the rod *between A* and *B,* but the way in which an outwardly concave arc causes a breakdown in the constancy of the width is most easily analyzed by referring to my second diagram. There w and w' are each equal to b without any variation resulting from changes in the direction of measurement; on the other hand, w'' *is* variable and becomes considerably greater than b near the cusps. And it is w'', not w', that is the proper overall width—or height—of the figure, is it not?

Salviati: It is a pleasure to witness your reactions to the difficulties as they arise. Your dotted line differs from the other lines parallel to it in that it is not a tangent but a line meeting the curve and separating the infinite plane into two half-planes, one of which contains the figure while the other excludes it. As previously mentioned, such a line is called a "line of support"; this particular line of support meets the curve in a cusp and is actually parallel to two tangents to the curve. In the same diagram, through the lower right-hand vertex you have been led (perhaps without noticing it) to draw another support line in connection with your dimension $w = b$. And now, Simplicio, can you see how to generalize your diagram (Fig. 37) by placing the centers almost anywhere in the plane?

Simplicio: I think I can. First let us choose five points, say, at random: *A, B, C, D, E* like this should do (Fig. 40). And let us describe the vertically opposite arcs of radii R_A, r_A placed as shown.

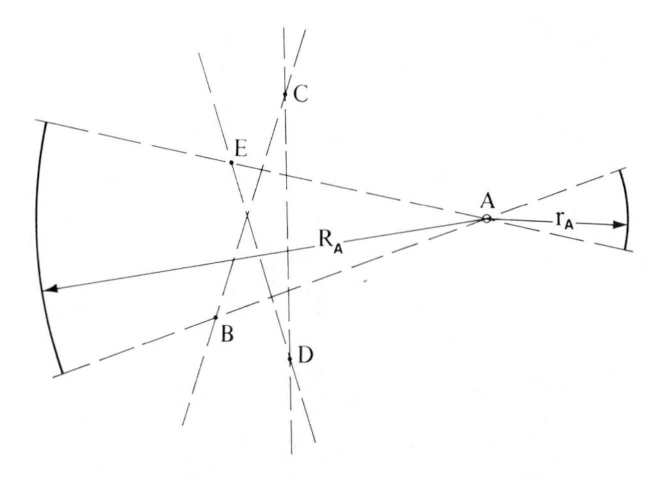

Fig. 40

Sagredo: You say that *A, B, C, D,* and *E* "chosen" like this "should do," and then you describe arcs "of radii R_A, r_A," as shown. Are you suggesting that there is still some restriction upon the relative positions of the centers? And further, what about the relative sizes of R_A and r_A?

Simplicio: I can see no geometrically significant restriction that need be placed upon the relative positions of the five centers. If the constant breadth of the final oval is to be assigned in advance, then $R_A + r_A$ must be made to equal the measure of that breadth. I think you will find that $R_B + r_B$, $R_C + r_C$, and so on, will each equal $R_A + r_A$.

Sagredo: I can see that if R_A is not sufficiently large, one or more fishtails will turn up; but otherwise the situation seems to be as you say. I am sorry if I interrupted unnecessarily.

Simplicio: Well, I have already drawn in the first two arcs. Next, arcs of radii r_B, R_B are described to blend in with those of radii R_A, r_A, respectively. And now the arcs of radii R_C, r_C, and so on (Fig. 41).

Sagredo: And so on, up to the arcs of radii R_E and r_E!

Simplicio: I have to admit it; the oval has failed to close. Just looking at the diagram, it appears that I should have started with a larger value for R_A and a smaller value for r_A.

Sagredo: If the values of R_A and r_A are not individually determined from the preassigned width, they must depend upon the relative positions of the centers.

Simplicio: Let me put *AB* equal to *a, BC* equal to *b* and so on up to *EA* equal to *e*, and then symbolize the relations that the construction provides:

$$W_A = R_A + r_A = (a + r_B) + r_A = r_B + (a + r_A) = r_B + R_B = W_B.$$

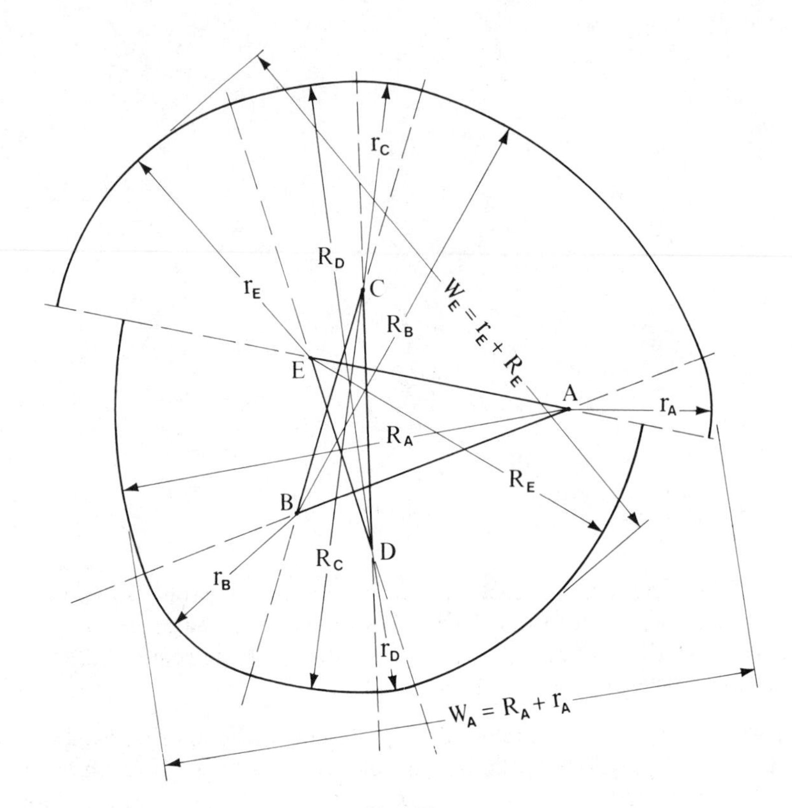

Fig. 41

Next,

$$W_B = r_B + R_B = r_B + (b + r_C) = (r_B + b) + r_C = R_C + r_C = W_C,$$

and,

$$W_C = R_C + r_C = (c + r_D) + r_C = r_D + (c + r_C) = r_D + R_D = W_D;$$

finally,

$$W_D = r_D + R_D = r_D + (d + r_E) = (r_D + d) + r_E = R_E + r_E = W_E.$$

The method of construction thus ensures that $W_A = W_B = W_C = W_D = W_E$, so that the breadth measured across from the upper half oval to the lower (always through the appropriate center) *is* constant. In order to ensure closure, however, I can now see from the diagram that it would have been necessary to arrange in addition that $r_E = R_A - e$, or that $r_A = R_E - e$.

Sagredo: I notice that these last two relations together imply but are not implied by the equality $r_E + R_E = r_A + R_A$—that is, $W_E = W_A$, which, as you have so clearly shown, is provided by the construction. And now, Simplicio, is it not evident that by continuing the construction through a further half-rotation two concentric uniform-width ovals will be completed? In order to construct a single closed uniform-breadth oval, adjacent arcs rather than opposite ones should be consecutively described. Figure 42 shows how your Fig. 41 may be completed. For the inner oval, the arcs of radii R_A, r_B, R_C, r_D, R_E, r'_A, R'_B, r'_C, R'_D, r'_E are described in order. I will not attempt to place dimensions on all of these arcs. There is hardly room to show r'_A, for example, but you will understand, I am sure, to which arcs the various radii refer. In this particular figure r'_A is the measure of the radius of the smallest of all the arcs.

Simplicio: I follow your description. For the outer oval, having chosen the value of R'_A, for the initial arc, the values of r'_B, R'_C, r'_D, R'_E, r_A, R_B, r_C, R_D, r_E for the nine succeeding arcs are determined in this order by the construction.

Sagredo: Exactly.

Simplicio: This appears to work out as you say. Yet I cannot see how you can be so confident that the ovals will close merely because you describe in each case the ten arcs in order the one way round. And further, it seems to me that one will not be in a position to know before the construction is carried out just what the uniform width of each oval is going to be.

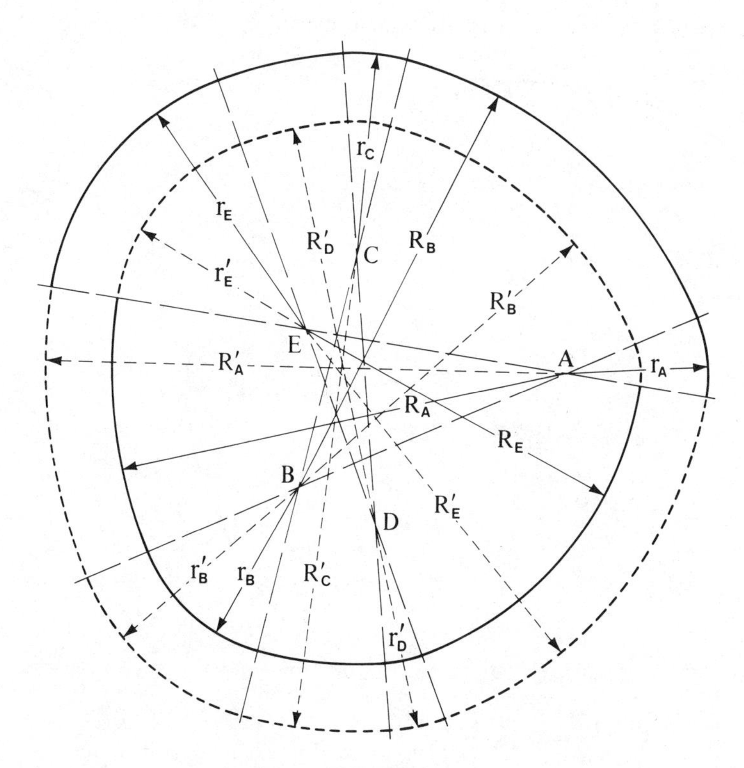

Fig. 42

Sagredo: You are quite right, Simplicio, in requiring satisfaction on these two matters. As to the first difficulty, I freely admit that I have been proceeding intuitively, and so have left myself open to the censure of having been too much influenced by the appearance of the diagram and especially by the evident successes in the cases of the earlier less irregular ovals. Allow me therefore to make amends, if I can, by a careful analysis. This should be facilitated by a somewhat clearer diagram and a simpler symbolism. Let us work with Fig. 43, in which a possible construction sequence is indicated by the subscripts. Now beside this diagram I have written the relations that the construction method ensures. Each of these equations specifies the precise relation between two radii and a center distance corresponding to the choice of an arc so as to meet the previously drawn arc tangentially. Adding all nine of these equations and simplifying, we obtain $\rho_{10} = \rho_1 - e$, and this is precisely the condition guaranteeing closure, since the first and last drawn arcs *are* centered *e* units apart.

Simplicio: Excellent!

Sagredo: And, by adding the appropriate pairs of the listed equations, we obtain immediately:

$$\rho_{10} + \rho_5 = \rho_9 + \rho_4 = \rho_8 + \rho_3 = \rho_7 + \rho_2 = \rho_6 + \rho_1$$

thus establishing that the oval really is of uniform breadth.

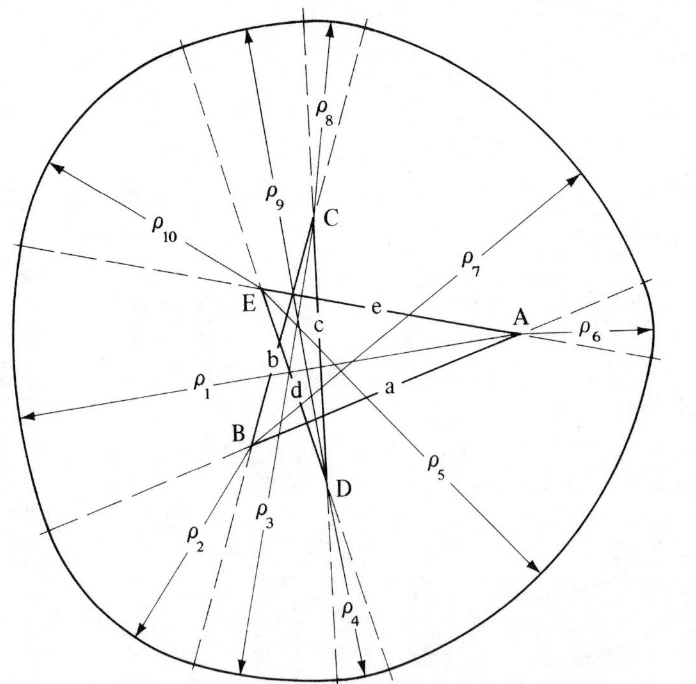

$$\begin{aligned}
\rho_2 &= \rho_1 - a \\
\rho_3 &= \rho_2 + b \\
\rho_4 &= \rho_3 - c \\
\rho_5 &= \rho_4 + d \\
\rho_6 &= \rho_5 - e \\
\rho_7 &= \rho_6 + a \\
\rho_8 &= \rho_7 - b \\
\rho_9 &= \rho_8 + c \\
\rho_{10} &= \rho_9 - d
\end{aligned}$$

Fig. 43

Simplicio: Oh yes!

Sagredo: I fear that the disposal of your second difficulty might not seem quite as satisfactory to you. It is now obvious, from the diagram or from the equations, that once the centers and the first radius have been chosen, all the other radii follow automatically. Since the uniform breadth of the oval depends upon the layout of the centers as well as upon the choice of the first radius and since these points can be scattered at random, the determination of ρ_1 in advance in order to produce an oval of preassigned uniform breadth would appear to be a complicated matter. But if we are already given the position of the centers, we may describe the first arc with any convenient radius (sufficiently large to allow fishtails to be avoided in the subsequent configuration) and then on completing the construction of the corresponding oval we can compare its breadth with that required. If, for example, the constructed oval were found to have a breadth exceeding the required breadth by a distance δ then we could simply construct a second oval with the same centers but with the initial radius less by $\frac{1}{2}\delta$ than that first used. And if the breadth of the trial oval turned out to be too small, the initial radius (and consequently all the subsequent radii) could be just as readily increased by half the discrepancy between the required and the trial breadths. I use the word *trial,* but you will understand that this is no indefinite trial-and-error procedure since the value of the requisite adjustment is determined by the construction itself.

Salviati: I agree with most of what you have said, but you have overlooked a consequence of your own schedule of nine equations. We have:

$$\begin{aligned} \text{Breadth} &= w, \text{ say, } = \rho_1 + \rho_6 = \rho_1 + \rho_5 - e = \rho_1 + \rho_4 + d - e \\ &= \rho_1 + \rho_3 - c + d - e \\ &= \text{etc., giving } w = 2\rho_1 - a + b - c + d - e, \end{aligned}$$

$$\textit{from which } \rho_1 = \frac{1}{2}(w + a - b + c - d + e).$$

Thus the appropriate value of ρ_1 could easily be determined in advance if $w, a, b, c, \ldots$ were given. In practice, however, your method might be just as expeditious.

Sagredo: Not only am I a little disconcerted at missing this implication of the relations I have so recently written out, I am also somewhat puzzled by the meaning of the result, for it seems too simple to do justice to the geometry of the situation. For surely the values of $a, b, c, d,$ and e are not alone sufficient to specify the layout of the centers! A couple of angles would be needed as well, would they not! Let $ABCDE$ (Fig. 44) represent a general pentagram, with sides a, b, c, d, e as shown; then if A and B are considered fixed, E is free to move, to E' say, along a circular arc centered at A, and likewise C can be moved about center B, to C' say; then D is determined by the intersection of arcs of radii c, d respectively centered at C', D'. Thus $ABC'D'E'$ is quite a different figure from $ABCDE,$ even though the side lengths a, b, c, d, e remain unaltered. And consequently, I felt

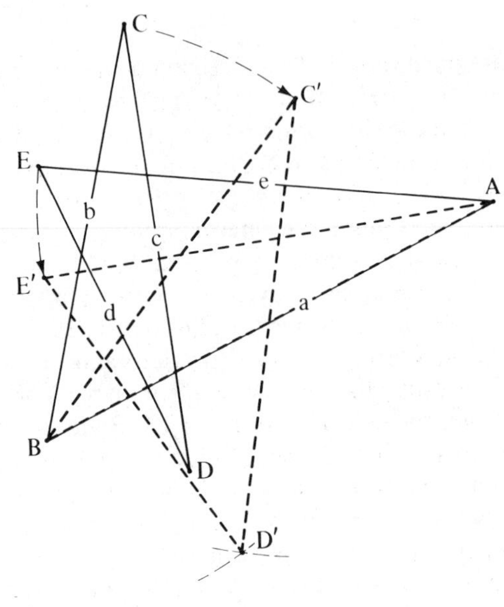

Fig. 44

that quite different ovals would result from them, even though the initial radius be unaltered. It was for this reason that I expected the determination of ρ_1 in advance would be a complex matter.

Salviati: What you say about the pentagram is, of course, perfectly correct. And so clearly have you exposed the grounds for your surprise on finding that the width is given by such a simple relation as $w = 2\rho_1 - a + b - c + d - e$ that I suspect that much of that surprise has already dissipated.

Sagredo: Well yes, that is how it is. I now feel that if you will allow me to produce the sides of both pentagrams, to choose some common value for the initial radius and to complete the construction of the two ovals, then my impression that there is a paradox here will probably no longer remain. Notice, Simplicio, I certainly do not mean that I wish to rely on one particular construction to verify a general proof—I accept the proof as valid—I am wishing rather to modify the discordant and inappropriate intuition that caused me to find the result perplexing.[112] There, you have the first oval with its arcs centered at A, B, C, D, and E fully lined in, and now I will describe with dotted arcs the oval based on the centers A, B, C', D', and E' (Fig. 45).

Simplicio: Oh, now I see; the two distinct sets of centers, though they give rise to two distinct ovals, do have enough in common—namely the values of a, b, c, d, and e—to ensure that the ovals will have equal breadths provided the initial arcs have the same radius.

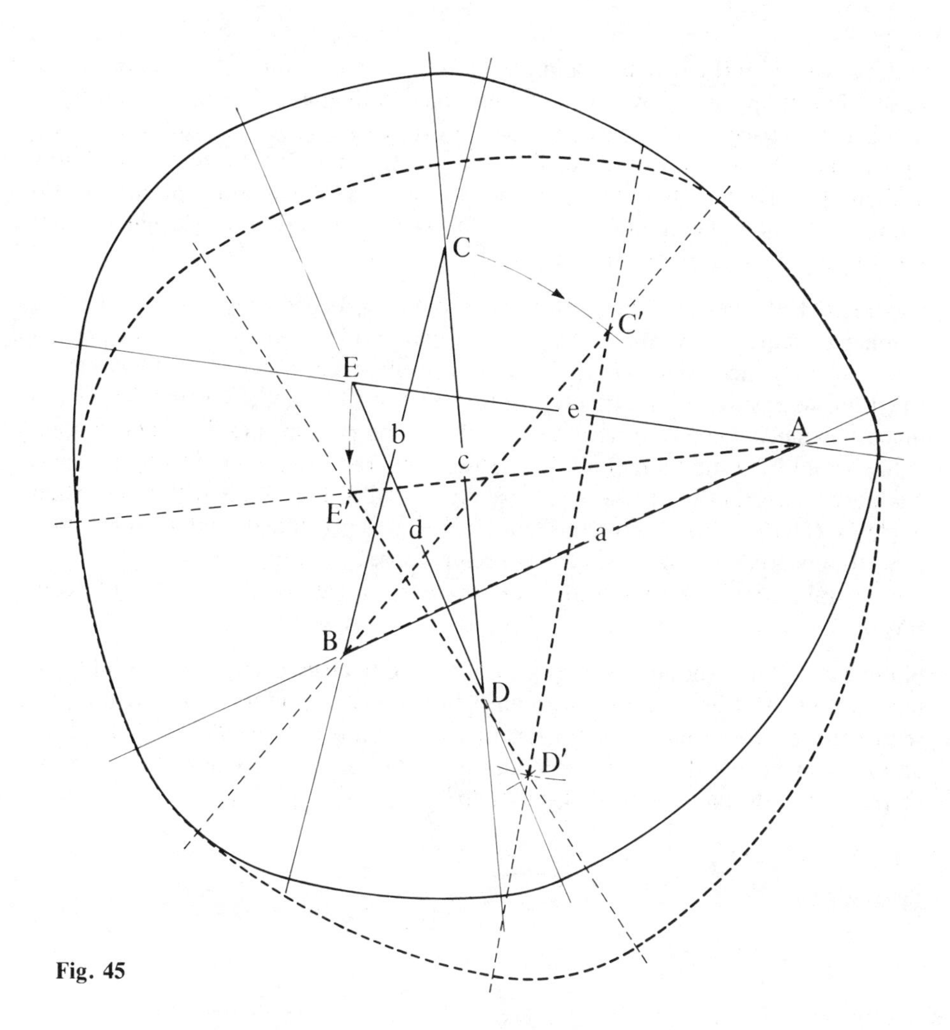

Fig. 45

Sagredo: Exactly. Mind you, I suspect that if E were to be moved to a position on the side of AB remote from C, the resulting oval might be rather different, probably involving fishtails. But I think we will have to refrain at present from investigating every subsidiary matter that occurs to us.

Simplicio: I agree. And as it is apparent that the demonstrations already given could be rewritten for any odd number of centers (from which will spring an even number of arcs), perhaps we can regard this part of our investigation as now brought to a satisfactory conclusion.

Sagredo: I am glad that you are satisfied. I might just remark that in certain cases there will be less than twice as many arcs as centers. Fig. 39(i) provides a simple illustration of this: there are five arcs instead of six from the three centers;

and likewise with the irregular and unsymmetrical cases, one or more arcs might be missing. It appears, however, that their place will always be taken by an equal number of vertices, so that the number of arcs and vertices together total twice the number of centers; I omit consideration of fishtails. But I am reminded now of a query, Salviati. I think you said a while ago (p. 110) something about the odd number of centers being able to take up *almost* any position in the plane. What was the significance of the qualification?

Salviati: To be perfectly accurate, I did not on that occasion specify whether the number of centers was odd or not. What I had in mind was the effect of placing three or more successive centers in the one straight line. That would have to be ruled out to ensure nondegeneracy. I do not wish to belabor the fine details too much, but since you have raised the question it can perhaps best be touched upon with the aid of a diagram (Fig. 46). If the polygram of centers in this figure is compared with that of the general case illustrated in Fig. 43, it will be seen that angle *BCD* has become zero; consequently the arcs previously centered at *C* have now disappeared and so center *C* has been rendered functionless. As you can see, the general figure is thus reduced to a *four-centered,* eight-arched oval of uniform breadth!

Simplicio: But I thought we had convinced ourselves of the impossibility of having a uniform-breadth oval with an even number of centers! Have we not presented two careful arguments purporting to show this (pp. 96–99, 107)? To these arguments you raised no objection, and now you calmly show us this counterexample! What do *you* say to this, Sagredo?

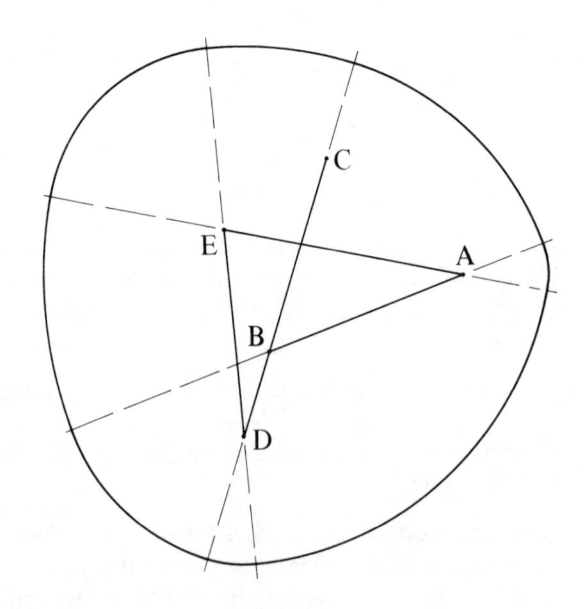

Fig. 46

Sagredo: My immediate reaction was to be taken aback. I had not thought such a thing would be possible. Yet it now occurs to me that our earlier considerations of this matter were undertaken relative to the assumption of a constant distance between successive centers. We have learned through our most recently considered examples that this condition may be dispensed with. Yet if we were now to reimpose it, or at least, in the example of Fig. 46, if we were to have *CD* equal to *BC,* simultaneously with having angle *BCD* equal to zero, then *D* would coincide with *B* and consequently *both C* and *D* would cease to be functioning centers.

Simplicio: I see: back to a three-centered oval again.

Sagredo: Yes, but if *CD* is not equal to *BC,* then the making of angle *BCD* equal to zero leads to the loss only of *C* as a center. So, Simplicio, we must accept that there is no absolute ban upon having uniformity of breadth with an even number of centers. Did I not begin my proof (p. 107) that the number of centers must be odd with the proviso that the oval of constant breadth be constructed in the manner of your Fig. 37? Now I have admitted that I did not foresee examples such as that of Salviati's final diagram; nevertheless I would not, I hope, be so incautious as to claim without further proof that a proposition established for a restricted category of cases would apply to a whole class, of which this is but a proper subclass. But I do confess, Salviati, that this is not the first time in today's discussion that I have been disconcerted by the discovery that our progress toward secure propositions, with a properly defined range of applicability, is full of hazards and is much too much dependent upon what appear to be almost chance insights and corrections.

Salviati: Creative thought is typically hazardous in just this way.

Sagredo: I suppose so. Yet I cannot but feel that there will remain blemishes in our account because some necessary modification or other will have been overlooked.

Salviati: This feeling is perfectly in order and is not at all unexpected in one so cautious and meticulous as yourself. In the initial stages of an investigation, however, we may have to be satisfied with the progressive grasping of true insights and the rejection of pseudoinsights via the testing of conjectures, with the explanation of such experience in ways leading on to further promising conjectures and insights. Our discoveries today may be taken as a kind of microcosmic representation of centuries-long developments in the actual history of mathematics. Restricted glimpses now and again become superseded by more and more adequate views, or sometimes by mistaken ones, though by and large our subject is sufficiently self-corrective for these latter to hold sway for only very short periods. Stages of conjecture and plausible reasoning are followed by stages of informal verification or of falsification, modification, and retesting; the desire for rigorous demonstration can consequently be expected to become stronger and stronger, at least in the "best" periods of the historical development and may issue in formal, axiomatic-deductive systematization. But the latter is a venture that can hardly be undertaken successfully without long preliminary work of the kind in which we are at present engaged. Premature attempts to achieve finality are very liable to be forced and insecure. Our own short proofs are appropriate to our immediate needs.[113] In time they could be utilized or adapted in the building of a general theory of more precisely defined scope and of greater rigor.[114]

Simplicio: Well, Salviati, it is reassuring to hear that you are satisfied with our progress.

Sagredo: It is indeed. And now, am I right in assuming that we are ready for the further generalization that you foreshadowed at the end of the last hour, the generalization to uniform-breadth figures in which none, or not all, of the component arcs are circular?

Salviati: We still have a little time. Have either of you any suggestions?

Sagredo: It occurs to me that in the case of a uniform-breadth oval composed of blending circular arcs, it ought to be possible to choose the centers in such a way that a sequence of successive arcs, going more or less halfway around the figure, would approximate to a semi-ellipse, say. In the diagram of Fig. 46, for example, the line AE cuts off from the oval, on the same side of AB as point C, a four-arched curve that in spite of its evident lack of symmetry cannot be so very far from having a semi-elliptical form. It seems evident that if we were to replace these four circular arcs by an exact semi-ellipse there would then be some curve not unlike the other portion of Fig. 46 that would complement the semiellipse in such a way as to retain the uniform-width property. I am more hesitant about suggesting this than I would otherwise be, owing to my unfortunate experience when I previously tried to bridge the gap between circular-arc approximations and true ellipses. But the mistake I made on that occasion (see Hora Tertia) does not seem to be applicable here. So I am led to believe that not only in the case of a semiellipse but for an arch of the cycloid, or of a sine curve, or perhaps of an arbitrarily sketched curve, it might be possible to determine the complementary curve that would produce an oval of uniform width.

Simplicio: I should very much like to see how this works out in a particular case.

Sagredo: Then let us try it with half of the ellipse $x^2/5^2 + y^2/3^2 = 1$, for which we learned earlier (pp. 49, 57) that the evolute is $(5x)^{2/3} + (3y)^{2/3} = 16^{2/3}$. We have in fact already drawn out parts of these two curves, in Fig. 25, but here they are again in a new diagram (Fig. 47). Corresponding to each point P selected on the semiellipse, let us draw the normal (tangent to the appropriate branch of the evolute) and on it obtain the point Q such that PQ is equal to the breadth defined by the chord cutting off the semielliptical region, in this case the major axis, which is 10 units long. The locus of Q as P moves on the given curve completes the required uniform-breadth oval.

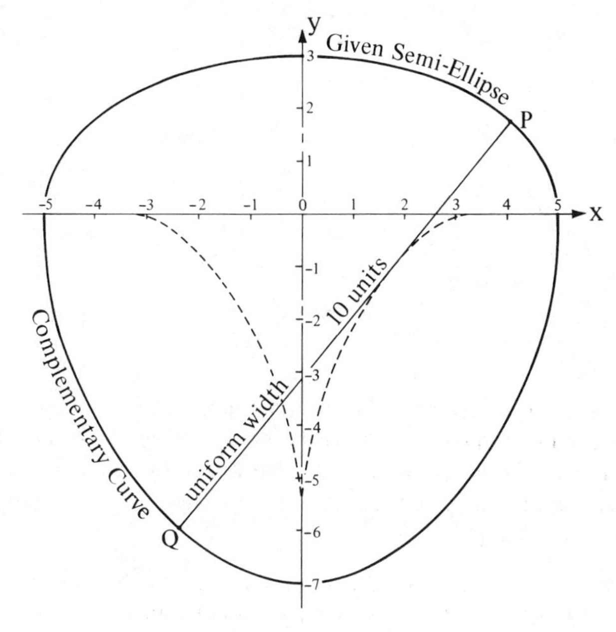

Fig. 47

Simplicio: Excellent! And it looks as though a very wide class of uniform-breadth ovals would be constructible by this method, if we knew how to find the evolute. You mentioned starting with an arch of a cycloid, did you not?

Sagredo: Yes, and in that case I think you would find that the one construction would give both points on the curve and the normals from those points, *without* the need for the evolute.[115] The general shape of the oval obtained would be pretty much like that of Fig. 47. I don't think we need to stop now to draw it in detail.

Simplicio: Not long ago I believed that only circles had uniform diameter or breadth; now it seems that there is no end to the kinds of noncircular ovals having this property. I think you hesitatingly suggested that it might even be possible to complete an oval of uniform breadth given only an arbitrarily drawn arc of it. If we could see how to do that, I would agree that for our purpose we need no longer consider special initial curves.

Sagredo: Suppose, then, that *APB* (Fig. 48) is an arbitrary curve. The uniform breadth of the required oval must be at least equal to *AB*. If it is to be greater, the given curve may be extended, say to *A'*, so that *A'B* is equal to the specified breadth. But let us suppose that the breadth is to be equal to *AB*. Can you see how to obtain the rest of the oval, Simplicio?

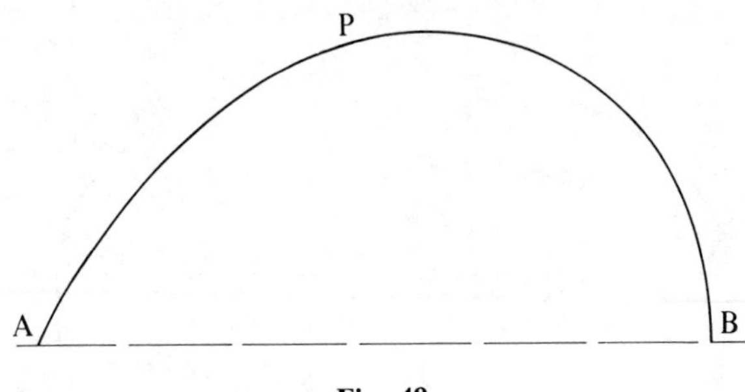

Fig. 48

Simplicio: Indeed I can, now. The mystery has vanished! Since every pair of parallel straight lines tangent to the completed oval is to be the same distance apart (equal to AB), all we need do is draw a dozen or so pairs of appropriately spaced tangents and to carry out freehand the extrapolation, so that the tangents are touched one by one (Fig. 49).

Sagredo: Yes, if you are satisfied that you can place the tangents to the given arc accurately enough, this does provide a practical solution. Doubtless your construction would be facilitated if you were to use a piece of tracing paper upon which two properly spaced parallel lines were drawn. And a common normal

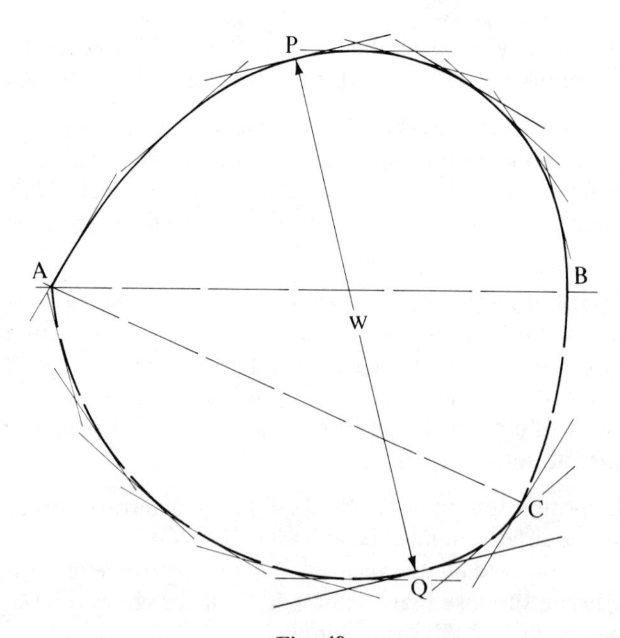

Fig. 49

drawn to these parallels would help as well, since that would assist in the location of the points of contact. It is worth noticing that no difficulty is caused by the fact that the given arc does not meet its chord perpendicularly at one end (A). This simply means that that point will be both a vertex and a center for the opposite arc (BC), the only arc of a circle to appear in the whole diagram. I have a feeling, however, that there will be some rather severe restrictions on the shape of the given curve. But our hour is up, so let us leave it there, if that is satisfactory to you both.

Salviati: Yes gentlemen; and you may rest assured that you have good reason to congratulate yourselves on your treatment of these curves. May I suggest that you reflect upon the general pattern of your progress so far? In particular, you might note how the removal and imposition of conditions opens up whole new classes of problems, of differing generality and intricacy, and how concepts may be extended and refined.

VIII

If you do not expect the unexpected, you will not find it; for it is hard to be sought out and difficult.

—Heraclitus

In the customs and institutions of schools, academies, colleges, and similar bodies destined for the abode of learned men and the cultivation of learning, everything is found adverse to the progress of science; for the lectures and exercises there are so ordered, that to think or speculate on anything out of the common way can hardly occur to any man.

—Francis Bacon

Where all men think alike, no one thinks very much.

—Walter Lippmann

The effect of exclusive attention to those parts of mathematics which offer no scope for the discussion of doubtful points is a distaste for modes of proceedings which are absolutely necessary to the extension of analysis.

—Augustus De Morgan

The element of constructive invention, of directing and motivating intuition, is apt to elude a simple philosophical formulation; but it remains the core of any mathematical achievement, even in the most abstract fields. If the crystallized deductive form is the goal, intuition and construction are at least the driving forces. A serious threat to the very life of science is implied in the assertion that mathematics is nothing but a system of conclusions drawn from definitions and postulates that must be consistent but otherwise may be created by the free will of the mathematician. If this description were accurate, mathematics could not attract any intelligent person. It would be a game with definitions, rules, and syllogisms, without motive or goal.

—R. Courant and H. Robbins

HORA OCTAVA

Sagredo: I know it was our intention to move on from the ovals of uniform breadth, but I have not been able to stop thinking about the restrictions that would need to be placed on the given curve if the completion construction of Fig. 49 is to result in a wholly convex oval, without fishtails. I should like to submit to you both, if I may, two final diagrams that I hope will elucidate the source of departures from (outward) convexity. In Fig. 50 I have again taken for convenience semielliptical initial arcs *APB*. It will be seen that the trouble in the first diagram, Fig. 50(i), in which two fishtails or a pair of devil's horns are exhibited, arises because the distances from *A* and/or from *B* to other points on the given arc *APB* exceed the width *AB;* consequently the complementary curve has to double back on itself. For the complete oval to be convex, I think it is evident that no part of the given curve can be permitted to lie outside the region bounded by *AB* and the pair of circular arcs centered at *A* and *B* with radii *AB, BA* (dotted in the first diagram). In the second diagram, what appears to be another requirement is vio-

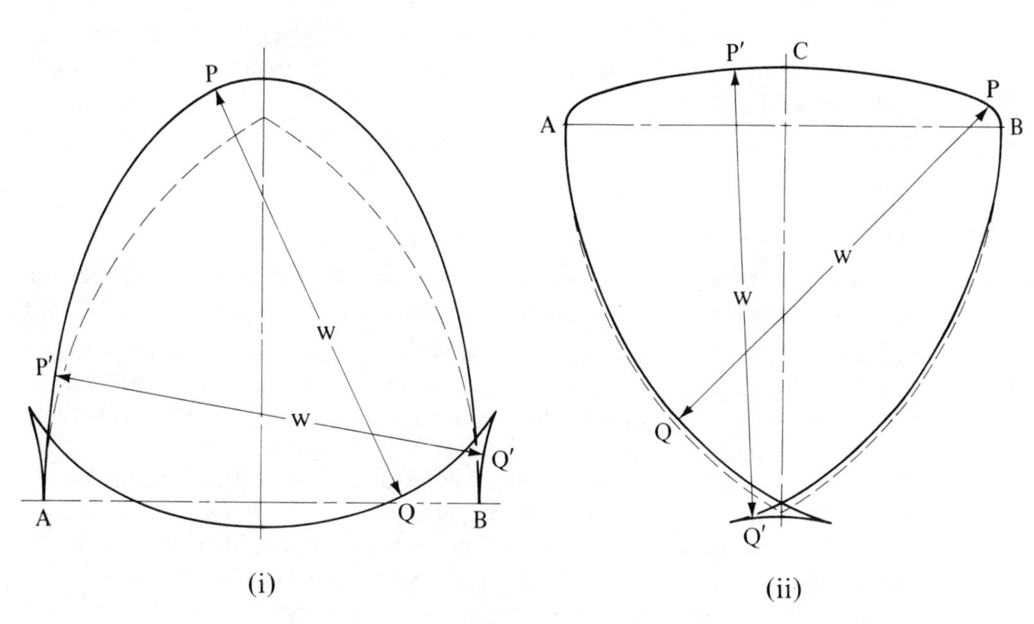

Fig. 50

lated. Here the given curve is so flat that, in places, the radius of curvature exceeds the chord AB. This leads to reversals in the path of Q as P moves along the given arc.

Salviati: Very good. The intimate connection between the two requirements will be appreciated if it is noted that, first, for parts of the semiellipse near A or near B in the first diagram, the radius of curvature exceeds AB, while, second, for positions of P close to C in the second diagram, the corresponding positions of Q are such that AQ and/or BQ exceed AB. The reversals of Q would be clearly understood if the evolutes of the given curves were drawn in and the successive positions of PQ observed in each case as this line rolls around the evolute. You may wish to plot a few cases accurately at your leisure. If one arch of a sine curve were taken as the initial arc, then, though no points on it are further from each other than the endpoints, fishtails would nevertheless arise in the complementary arc owing to the excessive radius of curvature near the end points. (These are, of course, points of inflection[116] for the whole sine curve.)

Considerations of the sort you have discovered in connection with your Fig. 50, Sagredo, lead up to the following theorem:

> *Given a convex arc (an arc which together with its chord bounds a convex region) having its chord of length* b, *if the entire arc lies between the two perpendiculars to the chord at its ends, and if it has the property of being enclosed by every circle of radius* b *tangent to a line of support at its point of contact and lying on the same side of the line as the arc, then the curve can be extended to form a (wholly convex) curve of constant breadth* b.

I do not think we need to stop to demonstrate this[117] or the even more noteworthy theorem that *all ovals having the same constant breadth* b *have the same perimeter* πb.[118]

Simplicio: Whether or not they are composed of circular arcs?

Salviati: Correct.

Sagredo: Well *that* is a remarkable result!

Salviati: Yes indeed, but time is against us, so perhaps we had best refrain from going more deeply into this topic today.

Simplicio: That is my feeling also; I should be glad to return to our original problem. The hope that we might be able to find some kind of oval that in relation to its circumscribing rectangle or square would yield equal area and perimeter ratios sustained us in the earlier part of our prolonged discussion. But after all these hours, I cannot see, Salviati, that we are any closer to being able to accept your claim that the equality of area and perimeter ratios is *not* something unique to the circle and circumscribing square but is merely exemplified by this pair of figures along with a vast number of other pairs. We have not succeeded in discovering any further examples. Yet such is my respect for your knowledge that I confidently await the strange revelations you doubtless have in store for us.

Salviati: I assure you that it will not be necessary to introduce any exotic forms into our repertoire. What will seem paradoxical to you at the moment is that this equality is, if I may so express it, no respecter of forms! I fear that it is partly because you have been trying too hard to determine some special forms that might exhibit the desired relation that you have missed another aspect of the problem, and so have missed the multiplicity of potential examples at hand among the commonest of figures.

Sagredo: I have to admit that I am still in the dark. I know Simplicio joins me in wishing that you would speak in a less puzzling fashion and set us on the right path.

Salviati: I shall be glad to do so, for I am confident that soon the slightest hint will enable you to sort the whole matter out for yourselves.

Simplicio: If the results that have so far eluded us are as simply obtained as you appear to be suggesting, how is it that Archimedes himself missed them, or missed formulating them?

Salviati: We cannot be sure that Archimedes did not discover the generalizations after which you are now striving. If he did remain unaware of them, we should remember that even a genius can be so preoccupied with certain implications of the results he has already discovered as to miss others that we lesser mortals may be able to uncover for ourselves by approaching the matter from a different point of view or because we have the advantage of a certain familiarity with techniques or substantial results belonging to an intervening period. It is quite possible that Archimedes did discover the theorems that I expect you will reach in today's investigation, but they might have appeared rather isolated from the main lines of interest at the time and, in particular, from the specific programs of investigation that he had set himself. So it would be understandable if he did not deal with this topic in a manner likely to lead to the preservation of any discoveries he might have made.

And now I am going to ask you to attempt something I think may lead you to see the crucial methodological step needed to reach your goal. While your successes with the ovals of uniform breadth are fresh in your minds, I would urge you to engage in some introspection and to indicate, if possible, the salient feature of your own emancipation from the belief that it was necessary to use *circular* arcs in the construction of these ovals.

Sagredo: Yes, I think that "emancipation" is not too extravagant a term, but I should add that the importance is not to be underestimated of our having had your assurance (p. 102) that an extension from circular to noncircular arcs was possible. And I know that you will agree that creative thinkers, no less than those who unfortunately devote their lives only to the pursuit of material things are greatly influenced by their largely unconscious estimates of just what moves are likely to bring them nearer to their goals. Please excuse these general remarks; I am having some difficulty in concentrating on the specific matter upon which you have asked us to reflect. But I do now remember that the thought dawned upon me with satisfaction that when these ovals *were* composed of circular arcs, it was easy to see that they must have certain characteristics, and the most obvious but least general proofs of these properties exposed them as consequences of the circularity of the arcs. Of course, this was not to say that other ovals not composed of circular arcs, might not also possess some of the same properties. In retrospect it is clearly seen to be a matter of avoiding the fallacy of (i) the denial of the antecedent and that of (ii) the affirmation of the consequent.[119] But it is not easy to be calmly logical during the excitement of plausible thinking, while one is endeavoring to intuit the likely relations. Our striving for generalization seems to have culminated with Simplicio's construction for so easily obtaining the complementary curve as the envelope of a set of straight lines that are to be its tangents (Fig. 49). Despite complications that would sometimes be encountered, it may be said that we succeeded by taking the property we wished to obtain and directly arranging for it to be present. Instead of hunting for this or that special figure, in the hope that it might have a uniform breadth, we *imposed* this property in an already partially defined situation.

Simplicio: The fundamental construction was yours, Sagredo. You cut off from normals drawn to a given curve the fixed distance equal to the required breadth in order to obtain points on the complementary curve (Fig. 47). I merely adapted your procedure for cases where the normals were readily determinable to suit cases where they were not. Otherwise I agree entirely with what you have said and I have nothing to add.

Salviati: Well then, I should like to suggest that a moral might be drawn from this experience, helpful in enabling us to cut the Gordian knot that has so far prevented us from a direct engagement with our initial problem of finding pairs of figures for which the ratio of the areas is equal to the ratio of the perimeters. You were drawn to the consideration first of ellipses and then of the uniform-breadth ovals in the hope of coming across some case, closely related to that of a circle and its circumscribing square, that might also possess the property that $A_2 : A_1 = p_2 : p_1$. So far this search has been in vain. I say, therefore, offering you back almost your own words in this new context: *Directly arrange for the required property to be present* in some already partially defined case. In a word, *impose* the relation on various pairs of figures of arbitrarily chosen shape.

Simplicio: I am afraid I do not quite see how this could be possible.

Salviati: You will soon see. Let me first remind you of some very obvious facts. For two *squares* of different size, the area ratio $A : a$ differs in value from the perimeter ratio $P : p$. Since $A : a = P^2 : p^2$ it follows that,

$$\text{if } A > a, \text{ then } A : a > P : p > 1 : 1.$$

You will immediately perceive that this same result must hold also for any pair of similar figures of different size, whatever their common shape, simply because the areas of similar figures are to each other as the squares of corresponding linear dimensions. Passing over the obvious case of congruent figures, for which $A_2 : A_1 = p_2 : p_1 = 1 : 1$, the question of the possibility that $A_2 : A_1 = p_2 : p_1 \neq 1 : 1$ leads to a consideration of the area and perimeter ratios holding between figures that *differ in shape* as well as in size.

Simplicio: Might we not first consider differences in shape without differences in size?

Sagredo: If you have in mind two dissimilar figures of equal area, then their perimeters will be different. Conversely, if the perimeters of nonsimilar figures are equal then their areas will differ. For such figures, $A_2 : A_1 \neq p_2 : p_1$.

Salviati: There are exceptions even to what you are claiming, Sagredo,[120] though your remarks well characterize the usual situation. The importance of shape is exemplified in the following easily verified instances:

- **(i)** For a square and an equilateral triangle of equal perimeter, the area of the triangle is approximately 77 percent of the area of the square.
- **(ii)** For a square and an equilateral triangle of equal area, the perimeter of the triangle exceeds that of the square by about 14 percent.

Sagredo: Yes, I once carried out calculations to show relations of this sort myself, and I found that the triangular form of boundary is less effective as an area container than the square form.

Salviati: Then I shall omit the quantitative treatment and merely indicate the constructions for converting an equilateral triangle into a square. In the first diagram of Fig. 51, showing the constant-perimeter construction, the areas are clearly unequal:

square $ADEF >$ rhombus $APRQ >$ triangle ABC.

Hence,

$$A_{\square} : A_{\triangle} > p_{\square} : p_{\triangle} = 1 : 1.$$

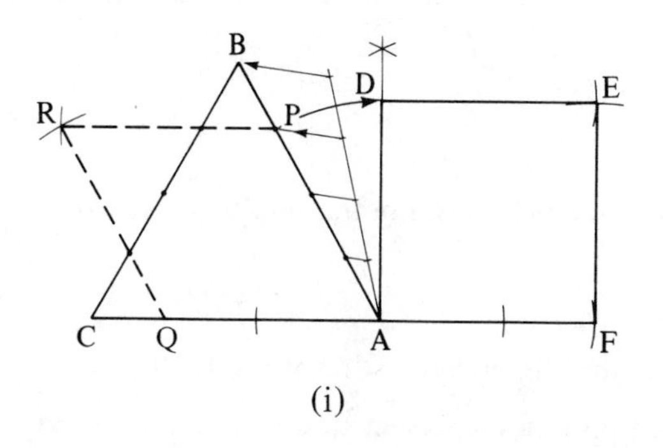

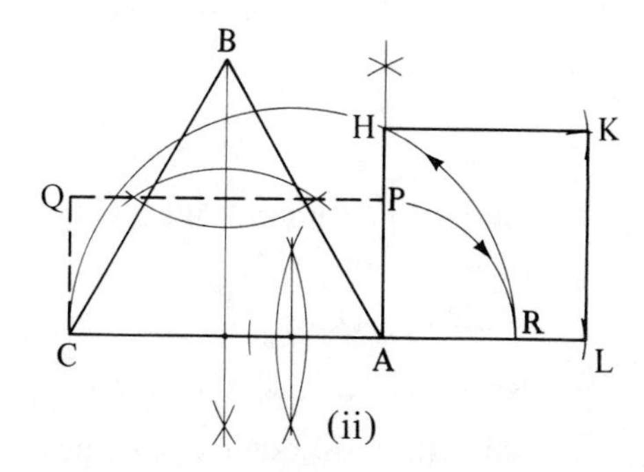

Fig. 51

And in connection with the area-preserving construction of the second diagram, the following perimeter inequality is easily established:

$$\text{perim. } AHKL < \text{perim. } APQC < \text{perim. } ABC.$$

Hence,

$$p_\square : p_\triangle < A_\square : A_\triangle = 1:1.$$

Simplicio: So with neither construction do we obtain $A_\triangle : A_\square$ equal to $p_\triangle : p_\square$. As Sagredo expressed it, the triangular form of boundary is less effective as an area container than the square form.

Salviati: Nevertheless, the area ratio may yet be made equal to the perimeter ratio by arranging things so that a disadvantage from the point of view of shape is compensated for by an advantage in size. By examining particular cases you will soon see not only that conditions are readily determinable for which $A_2 : A_1 = p_2 : p_1$, but that we may form conjectures as to one or two general theorems with application far wider than to equilateral triangles and squares. But let us begin with the examples indicated in Fig. 52. The area of the equilateral triangle in the first diagram is clearly less than half of the square, yet the perimeter of this triangle is three-quarters that of the square. From this, and a simple calculation for the next case, we find that

$$A_\triangle : A_\square < p_\triangle : p_\square.$$

On the other hand, in the last two cases illustrated in Fig. 52, it will be found that

$$A_\triangle : A_\square > p_\triangle : p_\square.$$

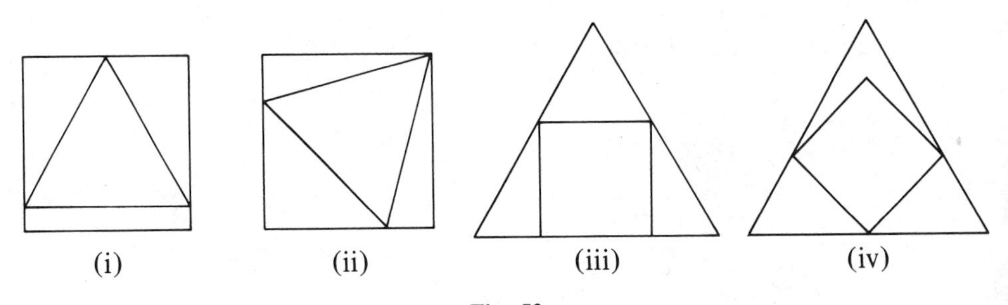

Fig. 52

Simplicio: Indeed? Well, if this is how it works out, I notice that there is a common relation here,

$$A_o : A_i > p_o : p_i,$$

where the subscripts refer to the outer and inner figures, respectively.

Salviati: Although this is true of the four cases represented in Fig. 52 and would be quite widely applicable to other instances, it is not without exceptions. I think you would find, for example, that the sense of your inequality would require reversal for the pair of figures in Fig. 53. (For ease of computation you might assume that the outer polygon is decomposable into four equilateral triangles and a central square.)

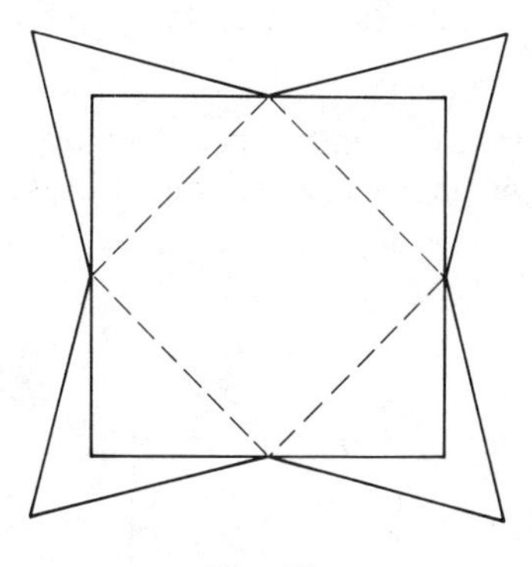

Fig. 53

Simplicio: Perhaps my conjecture is true for *convex* cases, at least if they are regular.

Sagredo: I rather think that we are getting away from the main point, which is that of the compensation for a relatively disadvantageous form by a sufficient increase in its size above that of the more advantageous form. Figure 52 parts (iii) and (iv) apparently represent cases of overcompensation; the triangle has been made too large in relation to the square. It should not be difficult to determine the precise size-compensation appropriate in the case of any two adequately specified shapes.

Simplicio: Ah! So, looking again at the pair of opposed inequalities, the one applying to the first two diagrams of Fig. 52 and the other to the last two diagrams, it is evident that we will not pass from "the lesser" to "the greater" without passing through exact equality.

Sagredo: Yes, there will be some one particular value for the ratio of the side of the equilateral triangle to the side of the square for which

$$A_{\triangle} : A_{\square} = p_{\triangle} : p_{\square}.$$

Salviati: You have got to the heart of the matter. Please carry out the calculation.

Sagredo: Denoting the measures of the sides of the square and triangle by a and b, respectively, we have, for the area ratio, $\frac{1}{2}b(b\sqrt{3}/2) : a^2$, and for the perimeter ratio, $3b : 4a$. Equating, we have

$$\frac{b^2\sqrt{3}}{4a^2} = \frac{3b}{4a},$$

from which $b:a = \sqrt{3}:1$. The steps are reversible so that this particular ratio of side lengths guarantees that

$$A_\triangle : A_\square = p_\triangle : p_\square.$$

Salviati: This is indeed the correct result. An equilateral triangle and superimposed square having the critical relative size that you have determined is shown in Fig. 54.

Sagredo: The solution to our central problem now appears to be absurdly simple. I fear you must consider us remarkably obtuse, Salviati.

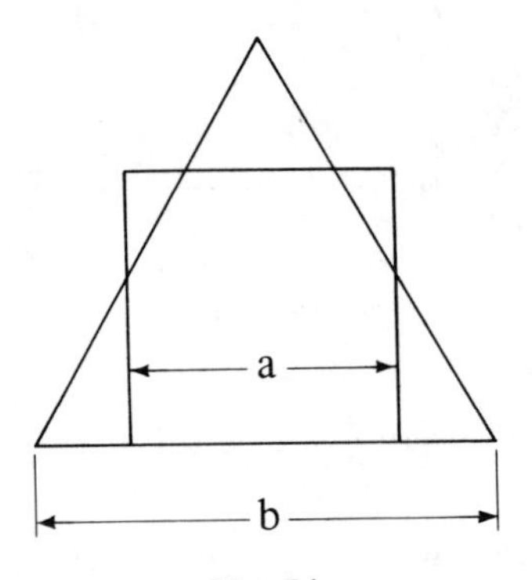

Fig. 54

Simplicio: Our investigations of the elliptical perimeter and of the uniform-breadth ovals appear to have been irrelevant to our initial problem!

Salviati: It is not to be expected that the path of discovery will be ordered like the systematic exposition you will doubtless be capable of giving at the end of your investigations. To have reached a point where one can look back on one's confusions and follies and to see at last how the paradoxes become resolved when unjustified assumptions are exposed and given up, is to have attained a more thorough and abiding knowledge, not only of the subject matter, but often of oneself as well. But, as you know, candid heuristic accounts are almost never published; it is a pretty well established practice among mathematicians to destroy all but their final, most polished version of any particular piece of their work. The contents of their wastepaper baskets would yield a very different story from that manifested in their communications to the learned journals.[121]

Sagredo: You are a most encouraging mentor, Salviati. I do not doubt that what you have just said is perfectly true. And do not think that we have altogether forgotten the various remarks you have made at critical stages of our investigation, for all that their full significance was often unclear to us. At least you can be assured that we have reached, belatedly to be sure, an understanding of your claim that the circle and its circumscribing square are not unique in the way in which I took them to be at the outset of our discussion, but that, on the contrary, there is no end to examples of pairs of figures for which $A_2:A_1 = p_2:p_1$. Yet I feel the need to investigate just how the method of deliberately equating the area ratio to the perimeter ratio will work out for different pairs of specified shapes.

Simplicio: So do I. Perhaps we shall find that among the multitude of pairs of forms matched in this way, the circle and the square are unique in that one figure is *inscribable* within the other.

Sagredo: Well, that is a possible conjecture to bear in mind. But what appears important to me is poles apart from uniqueness. I think that some generalization ought to lie behind the various cases, for I suspect that some other geometrical uniformity will probably be associated with equal area and perimeter ratios.

Simplicio: I have just thought of a rather remarkable pair of figures I once found to be of interest in another connection: a regular hexagon drawn within an equilateral triangle (Fig. 55). I now notice that the area ratio is as 6 is to 9, or 2 : 3, precisely as is the perimeter ratio!

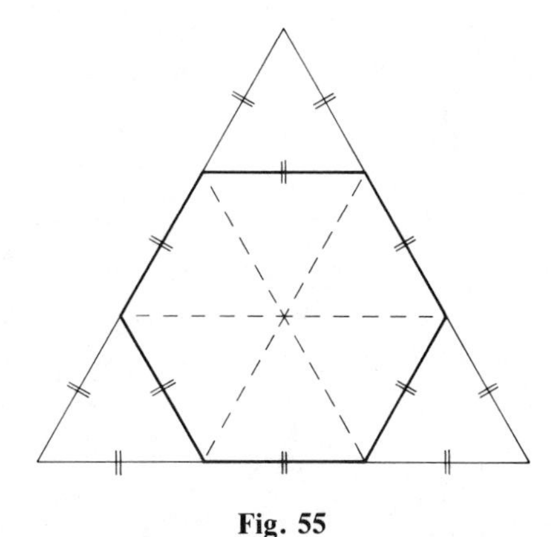

Fig. 55

Sagredo: Well so it is!

Simplicio: Previously we spoke of one figure being inscribed within another only in cases where there was point contact between them. Yet I suppose that here the hexagon may be judged to be inscribed within the triangle, even though whole sides of the inner figure coincide with parts of the boundary of the outer one.

Sagredo: I have no doubt that it will be convenient to adopt a broader definition of *inscribed* to include just such cases. You know, Simplicio, it really was quite astute of you to produce this example, which so aptly disposes of your own conjecture of a minute ago. You have shown that, of figures having equal area and perimeter ratios, the circle and square are *not* unique in forming an "inscribed-circumscribed" pair.

Simplicio: It cannot be denied that this triangle and hexagon do constitute a second pair so related. I wonder what others there are.

Sagredo: I suspect that there will be many others. First, I wonder what will happen if the inscribed hexagon is varied as indicated in the diagrams of Fig. 56. In the first of these diagrams, since the area of each corner triangle is to the area of the whole figure as 2^2 is to 5^2, the area ratio, hexagon to circumscribing triangle, is $\{5^2 - 3(2^2)\} : 5^2$, which comes to 13 : 25, while the perimeter ratio is 9 to 15, that is as 3 : 5. For the second diagram, the corresponding ratios are $\{5^2 - 3(1^2)\} : 5^2$, equal to 22 : 25, and 12 to 15, equal to 4 : 5. In the first case, then, the area ratio, hexagon to triangle, is less than the perimeter ratio; in the second case it is greater.

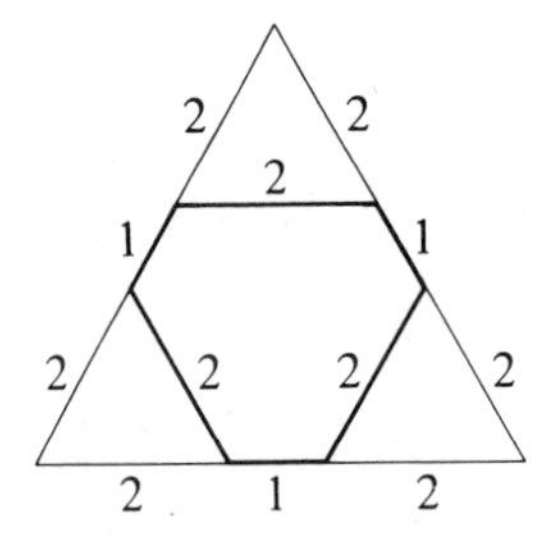

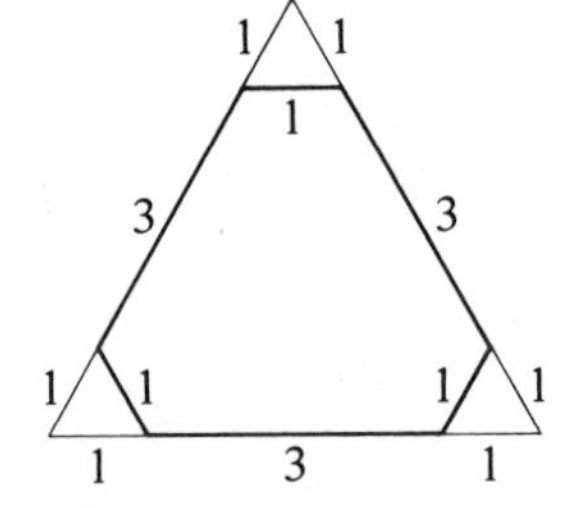

Fig. 56

Simplicio: It would seem that only when the inscribed hexagon is regular will the area and perimeter ratios be equal.

Sagredo: I think you are right. For a hexagon with sides of length $a - 2x$, x, $a - 2x$, x, $a - 2x$, x, inscribed within an equilateral triangle of side a (Fig. 57), the area ratio is $(a^2 - 3x^2) : a^2$, while the perimeter ratio, $\{3(a - 2x) + 3x\}$ to $3a$, simplifies to $(a - x) : a$. Equating these two ratios, we may write $1 - 3(x/a)^2 = 1 - x/a$. Hence $(x/a)\{3(x/a) - 1\} = 0$, which shows that the area and perimeter ratios are equal if and only if $x = 0$ or $x = a/3$, the former root obviously referring to the uninteresting extreme case where the inscribed hexagon passes over into a triangle coinciding with the given triangle. The sole nonzero root, $x = a/3$, provides verification of the conjecture that the *regular* hexagon is the only member of this family of symmetrically inscribed hexagons having the required relation to the given triangle.

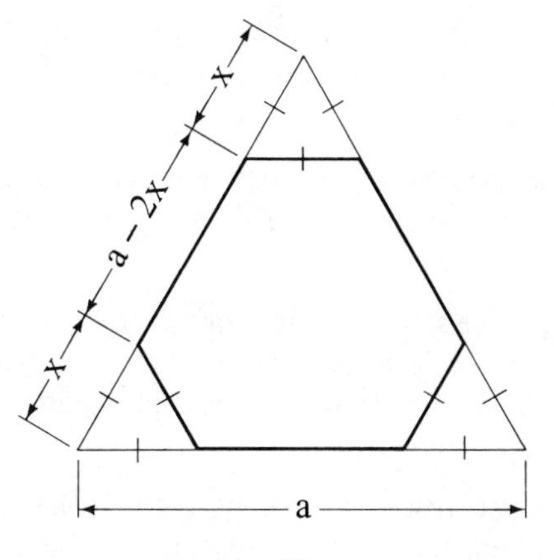

Fig. 57

I am led to suspect that a like relation holds for a square and its inscribed symmetrical octagons; only in the case of the regular inscribed octagon, I believe, would we find $A_8 : A_4 = p_8 : p_4$. A calculation like the one just completed would provide the test. Or we might now be satisfied simply to determine the relation between the side length of a regular octagon and that of a square (c and a in Fig. 58) necessary for the equality of the area and perimeter ratios. Would you care to carry out the calculation, Simplicio?

Simplicio: The area of the octagon is given by $b^2 - 2(c/\sqrt{2})^2$, where $b = c + 2(c/\sqrt{2})$. Hence, area $= (c + c\sqrt{2})^2 - 2(c^2/2)$, which simplifies to $2c^2(\sqrt{2} + 1)$. Setting up the required proportion, $A_8 : A_4 = p_8 : p_4$ as you just

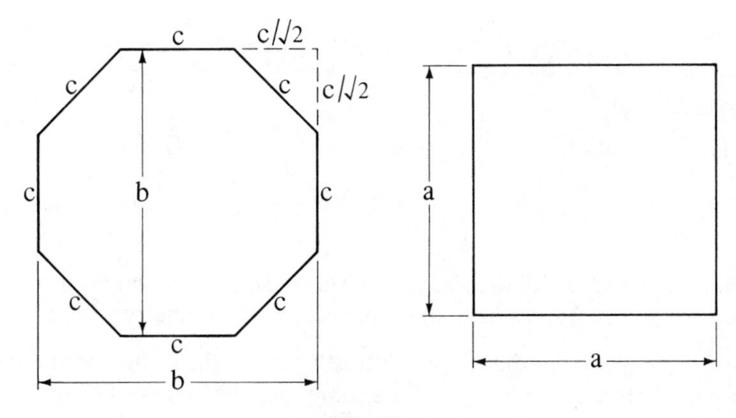

Fig. 58

expressed it, we obtain $2c^2(\sqrt{2}+1)/a^2 = 8c/4a$, from which $(\sqrt{2}+1)(c/a)^2 = (c/a)$, giving for the only nonzero root $c/a = 1/(\sqrt{2}+1)$. Since we know $b/c = \sqrt{2}+1$, it follows that $b = a$.

Sagredo: Thus a square and a regular octagon bear to each other equal area and perimeter ratios only when their relative sizes are such that the octagon could be inscribed within the square.

Simplicio: And yet, as we saw from the example of the square and the equilateral triangle (Fig. 54), the inscription of the "superior" polygon, the one with more sides, within the "inferior" one is *not* the general condition ensuring the equality of the area and the perimeter ratios.

Sagredo: But when the inscribed polygon has *twice as many* sides as the circumscribing one (when, that is, the inner one can be obtained from the outer one by cutting off the corners in such a way that it will be regular), I suspect that we shall then have a pair of figures with equal area and perimeter ratios: $A_{2n}:A_n = p_{2n}:p_n$.

Salviati: This hypothesis does indeed turn out to be correct, as I am sure you would be able to show. But let me invite you to form a still less restricted conjecture, one that will certainly include the case of the equilateral triangle and the square. For a start, consider the pairs of figures we already know have equal area and perimeter ratios (Fig. 59). I would ask you to imagine these diagrams placed one on top of the other.

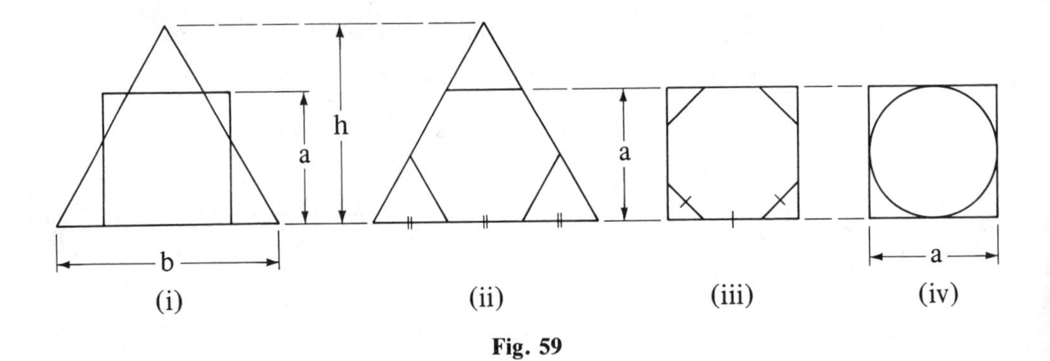

Fig. 59

Simplicio: I can visualize the situation clearly enough; the superposition can be made so that all the figures have the same center of symmetry.[122]

Sagredo: You mean, so that the centroids[123] of all the figures would coincide. But more than this, all would share the same inscribed circle. For notice that in the first diagram (as we found for Fig. 54), $a = b/\sqrt{3}$, or better $a = b\sqrt{3}/3$, whereas the altitude $h = b\sqrt{3}/2$. Hence $a = \frac{2}{3}h$, and this is also the distance between the parallel sides of the inscribed regular hexagon in the second diagram.

I am led to wonder whether perhaps for *any* regular polygon and its inscribed circle, the ratio of the areas is equal to the ratio of the perimeters. If this were true, we would have as an obvious corollary that for any two regular polygons in which equal circles are inscribable, the area ratio is equal to the perimeter ratio.

Simplicio: So you are now suggesting that for *any* pair of regular polygons circumscribable about the same circle, regardless of the numbers (n, m) of their sides, $A_n : A_m = p_n : p_m$.

Sagredo: Well, first of all it is to be noticed that for polygons of this class, not only have we $A_4 : A_3 = p_4 : p_3$ and $A_6 : A_3 = p_6 : p_3$, but, by division, $A_6 : A_4 = p_6 : p_4$. Other obvious combinations of the already established results lead to $A_8 : A_3 = p_8 : p_3$ and to $A_8 : A_6 = p_8 : p_6$.

Simplicio: I had not noticed these implications. Even so, your new general conjecture strikes me as a pretty extravagant extrapolation from the few examples we have before us.

Sagredo: It would perhaps seem so to me also if I did not already feel almost in possession of a proof. Yes, I do believe that we have only to write the area and perimeter ratios in general terms in order to display their equality! Look, Simplicio, at the diagrams in Fig. 60:

Case (i), regular n-sided polygon and its inscribed circle:

Area ratio, $A_n : A_o$ say, $= (n \text{ times area of } \triangle OV_1V_2) : (\pi r^2)$;

$$\therefore A_n : A_o = n \cdot \frac{1}{2} ar : \pi r^2 = na : 2\pi r = \text{perimeter ratio, } p_n : p_o.$$

Case (ii), regular polygons, of n and m sides, with equal inscribed circles:

$$A_m : A_n = m \cdot \frac{1}{2} a'r : n \cdot \frac{1}{2} ar = ma' : na = p_m : p_n$$

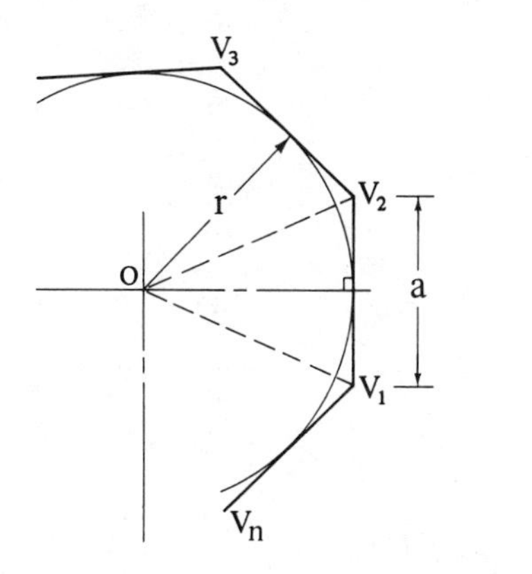

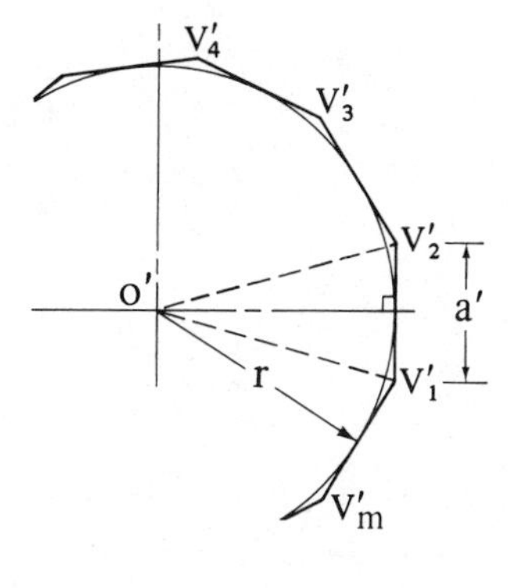

Fig. 60

Simplicio: You have completely satisfied me; you have established your conjecture as a theorem.

Salviati: All the special cases considered so far may be subsumed under this generalization which I shall call Theorem α:

> *For any regular polygon and its inscribed circle* (*or for any two regular polygons circumscribable about the same circle*), *the ratio of the areas is equal to the ratio of the perimeters.*

Your proof of this theorem is so clear, Sagredo, that we may be confident that there can be no counterexamples to it. But let me conclude this hour with a question: Are there not examples of pairs of figures with equal area and perimeter ratios that however are *not* the special regular figures with which your Theorem α is concerned?[124]

IX

The world little knows how many of the thoughts and theories which have passed through the mind of a scientific investigator have been crushed in silence and secrecy by his own severe criticism and adverse examinations; that in the most successful instances not a tenth of the suggestions, the hopes, the wishes, the preliminary conclusions have been realized.

—Michael Faraday

The great tragedy of science is the slaying of a beautiful hypothesis by an ugly fact.

—T. H. Huxley

If error is corrected whenever it is recognized as such, the path of error is the path of truth.

—Hans Reichenbach

In formal logic, a contradiction is the signal of a defeat; but in the evolution of real knowledge it marks the first step in progress toward a victory.

—A. N. Whitehead

Heuristic is concerned with language-dynamics, while logic is concerned with language-statics.

—Imre Lakatos

An enquiry which proceeds like a monologue, without interruption, is not altogether free from danger. One is too easily tempted into pushing aside thoughts which threaten to break into it, and in exchange one is left with a feeling of uncertainty which in the end one tries to keep down by over-decisiveness.

—Sigmund Freud

HORA NONA

Salviati: I think you find these short pauses between the hours as rewarding as our discussions themselves.

Simplicio: They are necessary not because we are subject to any weariness, as we were during our life on Earth, but in order for us to reflect on our progress.

Sagredo: Or upon some hint you have given us or on some question raised.

Simplicio: Quite so, like your question to us, Salviati, at the end of the last hour. I think you were suggesting that we might consider whether a theorem more general than Theorem α might not be found. Before discussing that, I should like to remark that we seem to have been restricting ourselves almost exclusively to the realm of two dimensions. After all, our starting point was Archimedes' own three-dimensional example: For a sphere and its circumscribing cylinder, $V_{\text{sph}} : V_{\text{cyl}} = A_{\text{sph}} : A_{\text{cyl}}$ $(= 2:3)$. For convenience we concentrated our attenion on the two-dimensional analogue: For a circle and its circumscribing square, $A_{\bigcirc} : A_{\square} = p_{\bigcirc} : p_{\square}$ $(= \pi : 4)$. But it has now occurred to me that if we had happened to have started with this two-dimensional case and then asked ourselves what its three-dimensional analogue is, we might just as well have taken this to have been a sphere and its circumscribing cube, and then conjectured that perhaps, for these two solids, $V_{\text{sph}} : V_{\text{cube}} = A_{\text{sph}} : A_{\text{cube}}$.

Sagredo: And do you believe that this relation is true?

Simplicio: The volume ratio, sphere to cube, $V_{\text{sph}} : V_{\text{cube}} = (4/3)\pi r^3 : (2r)^3$, which cancels to $\pi : 6$; the area ratio, $A_{\text{sph}} : A_{\text{cube}} = 4\pi r^2 : 6(2r)^2$, which also comes to $\pi : 6$.

Sagredo: So! You are absolutely right, Simplicio, $V_{\text{sph}} : V_{\text{cube}} = A_{\text{sph}} : A_{\text{cube}}$. I now realize this is something that we should have found out in the very first hour of today's discussions. We might have arrived at it in the following way: For a circle and its circumscribing square, $A_{\bigcirc} : A_{\square} = p_{\bigcirc} : p_{\square}$ $(= \pi : 4)$. Consequently, for a cylinder and its circumscribing square-based prism, height *h,*

> ratio of volumes $= A_{\bigcirc} h : A_{\square} h = p_{\bigcirc} h : p_{\square} h =$ ratio of areas of *side* surfaces.

In particular, when height $h =$ diameter,

> ratio of *total* surface areas = ratio of areas of side surfaces,

since for this form of cylinder, as well as for the cube,

total surface area = $1\frac{1}{2}$ times area of side surfaces.

Consequently, $V_{cyl}/V_{cube} = A_{cyl}/A_{cube}$ $(= \pi/4)$. Multiplying this by the Archimedean result, $V_{sph}/V_{cyl} = A_{sph}/A_{cyl}$ $(= 2/3)$, we obtain

$$V_{sph}/V_{cube} = A_{sph}/A_{cube} \qquad \left(= \frac{\pi}{4} \times \frac{2}{3} = \frac{\pi}{6}\right).$$

Simplicio: Now that this relation is so clear to us, it does seem that Archimedes could hardly have remained unaware of it.

Salviati: If he did notice it, he is unlikely to have stopped there.

Simplicio: You have in mind the other regular solids, each of which may be circumscribed about a sphere?

Sagredo: We ought to investigate the three-dimensional proposition analogous to Theorem α (p. 140):

For any regular polyhedron and its inscribed sphere (or for any two regular polyhedra circumscribable about the same sphere), the ratio of the volumes is equal to the ratio of the surface areas.

Simplicio: This proposition has a prima facie plausibility. Yet I am somewhat suspicious of it, since the analogy between the tetrahedron and the equilateral triangle appears to be less direct than that holding between a cube and a square.

Sagredo: You may have a point. The cube is the only regular solid having square faces and its volume is given simply by the product of the area of a square face and the length of an edge. On the other hand, there are three quite distinct regular solids having equilateral triangular faces: the regular tetrahedron, octahedron, and icosahedron. It would perhaps hardly be expected, then, that the relation $A_{\bigcirc} : A_{\triangle} = p_{\bigcirc} : p_{\triangle}$, for a circle and its circumscribing equilateral triangle, could be shown to lead to

$$V_{sph} : V_r = A_{sph} : A_r$$

where the subscripts refer to a sphere and to the regular solid of four triangular faces, or of eight, or of twenty such faces.

Simplicio: And finally, we should have to consider the regular dodecahedron, with its twelve regular pentagonal faces, circumscribed about a sphere.

Sagredo: Well, we do have a conjectured proposition that must be investigated, in spite of our doubts as to its truth.

Simplicio: Of course. It ought to be easy enough to evaluate each of $V_{sph} : V_{tet}$ and $A_{sph} : A_{tet}$, for a sphere and its circumscribing regular tetrahedron, and again, $V_{sph} : V_{oct}$ and $A_{sph} : A_{oct}$ for the sphere and the octahedron, but I suspect that the corresponding calculations for the circumscribing regular icosahedron and dodecahedron would be somewhat formidable. (See Problem 52.)

Sagredo: I am inclined to agree with you, in view of my memories of Book XIII of Euclid's *Elements*.

Salviati: I suggest to you both that you avoid the piecemeal tests that you are considering and attempt instead to prove the conjecture with a single demonstration, analogous to that employed in Theorem α.

Sagredo: There is a definite suggestion, Simplicio! Why did we not think of that ourselves?

Simplicio: Doubtless we were diverted by our reservations as to the truth of the conjecture, not that it would necessarily be unprofitable to *attempt* to prove a false proposition, I suppose!

Sagredo: I suspect now that the proposition is not false at all. Proceeding analogously to the demonstration of Theorem α, we should try to reduce the volume ratio, regular n-faced solid to inscribed sphere $V_n : V_{\text{sph}}$ to the ratio of the surface areas, $A_n : A_{\text{sph}}$. Considering the regular solid as composed of n congruent pyramids having the faces of the solid as bases and with the apices of the pyramids coinciding at the center, we obtain

$$V_n : V_{\text{sph}} = \left(n \cdot \frac{1}{3} A r\right) : \left(\frac{4}{3}\pi r^3\right),$$

where A measures the area of a single face, so $n \cdot A = A_n$. Therefore,

$$V_n : V_{\text{sph}} = \left(\frac{r}{3} \cdot A_n\right) : \left(\frac{r}{3} \cdot 4\pi r^2\right) = A_n : 4\pi r^2 = A_n : A_{\text{sph}}.$$

And, for any two of the five regular solids, of n, m faces, respectively, circumscribable about equal spheres,

$$V_m : V_n = m \cdot \frac{1}{3} A' r : n \cdot \frac{1}{3} A r = mA' : nA = A_m : A_n,$$

where A' denotes the area measure for each face of the m-faced solid.

Simplicio: So in spite of our initial doubts, we do have a new theorem; I suppose we may call it Theorem β. As you enunciated it a little while ago:

> *For any regular polyhedron and its inscribed sphere (or for any two regular polyhedra circumscribable about the same sphere), the ratio of the volumes is equal to the ratio of the surface areas.*

Sagredo: Naturally, this is a pleasing result, but let us learn from our former experience and so avoid the self-satisfaction so liable to follow in the wake of a new discovery. I cannot forget Salviati's query following our proof of Theorem α. Before he raises the corresponding question with regard to Theorem β, I should like to consider for a minute the doubt we had about the strength of the analogy between the equilateral triangle and its inscribed circle on the one hand the various regular solids with equilateral triangular faces and their inscribed spheres on the other. It now occurs to me that a closer analogy would be with a *circumscribable prism* and its inscribed sphere or cylinder. Indeed, I am led to conjecture that for such a pair of solids (whether the prism has equilateral triangular ends or regular polygonal ones of any kind), the ratio of the volumes is equal to the ratio of the total surface areas. Let us examine this as follows. For a regular-polygonal circumscribable prism and its inscribed cylinder,

$$\begin{aligned} V_{\text{prism}} : V_{\text{cyl}} &= \text{ratio of base areas, } A_n : A_o \\ &= \text{ratio of base perimeters, } p_n : p_o \text{, by Theorem } \alpha, \\ &= k : 1 \text{, say, where } A_n = k \cdot A_o \text{, and } p_n = k \cdot p_o. \end{aligned}$$

Total surface area ratio,

$$\begin{aligned} A_{\text{prism}} : A_{\text{cyl}} &= (2A_n + p_n d) : (2A_o + p_o d) \\ &= (2kA_o + kp_o d) : (2A_o + p_o d) \text{, simplifying to } k : 1. \end{aligned}$$

$$\therefore V_{\text{prism}} : V_{\text{cyl}} = A_{\text{prism}} : A_{\text{cyl}}.$$

Further, multiplying this by

$$V_{\text{cyl}} : V_{\text{sph}} = A_{\text{cyl}} : A_{\text{sph}},$$

we obtain

$$V_{\text{prism}} : V_{\text{sph}} = A_{\text{prism}} : A_{\text{sph}}.$$

It follows then that,

> *Given any prisms, P_n, P_m, with regular polygonal ends (having n, m sides), circumscribable about a sphere S and a cylinder C, then for any two of P_n, P_m, S and C, the ratio of the volumes is equal to the ratio of the surface areas.*

So there you have it, Theorem γ.

Simplicio: Good! At last we are getting some positive results. I notice that the cube is the one polyhedron that belongs to both families of solids, I mean those that are the subject of Theorem β and those that you have dealt with here in Theorem γ.

Sagredo: Yes, and the circumscribable cylinder is now more understandably accommodated. When we knew only Theorem β, this cylinder was on its own, being isolated from the regular solids, whereas now it takes its place as the limit approached by the regular-polygonal circumscribable prisms as the number of sides increases without limit.

Salviati: You have done well to find Theorems β and γ as analogues of Theorem α. But now I suggest that we return to two dimensions.

Sagredo: During our last pause I reflected upon the range of applicability of Theorem α, since I felt you were suggesting that it was unnecessarily restricted. I tried to recall our earlier discussion on generalization (pp. 62–68), but somehow I felt that I was starting all over again. I found myself worrying about necessary and sufficient conditions and the choice of definitions most appropriate in the light of known theorems. I should like to clarify my ideas on these matters if there is time.

Salviati: By all means.

Sagredo: The proposition that *the bisectors of the angles of a triangle are concurrent* came immediately to mind. This seemed to me to be a perfectly respectable theorem, even though concurrent lines in a triangle are not necessarily angle-bisectors; this simply means that the converse, "Concurrent lines of a triangle bisect the angles" is *not* (generally) true, even where the concurrent lines pass through the vertices. Usually we seem content to have additional theorems of restricted generality that supplement each other, as, for example, "The medians of a triangle are concurrent," "The perpendicular bisectors of the sides are concurrent," and so on. Yet the perpendicular bisectors of the sides, not only of a triangle, but of any cyclic polygon are concurrent. And if we found other figures besides triangles for which the bisectors of the angles are concurrent, I am not sure that we should feel dissatisfied with the theorem as it is usually stated, simply for triangles.

Salviati: Well, judge for yourself: For any polygon (whether regular or irregular) that is circumscribable about a circle, the bisectors of the angles are concurrent.[125] Incidentally, I recommend this class of polygon as especially worthy of your attention.

Sagredo: I note your recommendation. As for your general proposition, I can see that it is true—the point of concurrency being the incenter—but I have never come across this given as a theorem in a textbook.

Salviati: What is presented in textbooks is written with particular classes of student in mind; it is surely not inappropriate if we find that the temptation to say everything at once has been resisted.

Sagredo: I suppose not. And doubtless, because the needs of the learner change as he progresses, so will the need arise for redefinition of concepts to match the more general points of view that it becomes possible to adopt.

Salviati: Quite so, and as we are all learners, and since new, more general propositions cannot be known so long as they remain undiscovered, it would seem to be unrealistic to demand that definitions never be altered. The development of our subject or of an individual's knowledge of it makes it expedient to consider redefinition from time to time. Admittedly, piecemeal definition, typically involving the association of a term first with a limited domain, then with a wider one, may be found disconcerting since it obliges one to go back and reconsider the various propositions employing the word or symbol concerned. Since our time, however, mathematicians, ever aiming to achieve greater generalization, economy, and elegance, seem not to have scrupled to make frequent use of such redefinition.[126]

Simplicio: It is, I suppose, because most theorems about polygons are true whatever the number of sides, that triangles and quadrilaterals came to be counted as polygons in mathematics, though in popular usage what are called polygons have at least five sides. I can appreciate that just which definitions (and postulates) will turn out to be the most convenient cannot always be seen very far in advance. As our horizons enlarge, so do the opportunities for more effective definition and more adequate theorems present themselves.

Salviati: For pedagogical purposes the natural restriction of horizons is accepted as a matter of course and may even be deliberately sustained in the interests of simplicity. Whether, for example, that great prince of textbook writers, Euclid himself[127] gave expression to the full range of his geometrical imagination in his written works, or whether he deliberately restricted the domains under consideration for the sake of an enhanced amenability to mathematical treatment, is a difficult and probably unanswerable problem for the historian. Where a restriction is not explicitly noted, at least in a work as thoroughgoing as the *Elements,* some prima facie evidence is provided for the conclusion that, while writing, the author was not conscious of the restriction. Doubtless you remember the note—I do not know whether or not it was Euclid's own—that appears following the final proposition of Book XIII of the *Elements.* Here it is "proved" that "no other figure, besides the said five figures [the regular solids], can be constructed which is contained by equilateral and equiangular figures equal to one another."[128]

Simplicio: I do indeed remember this concluding item, and a perfectly transparent argument I took it to contain. Are you going to find some defect in it?

Salviati: The defect is the common one of failing to make explicit all the restrictions required in order for the argument to be valid. As enunciated the proposition is patently false. You will agree, as soon as your attention is drawn to it, that there is no end to the number of different polyhedra that may be formed so as to be bounded by congruent equilateral triangles alone. Of the examples represented in Fig. 61, only the octahedron (ii) is recognized in the demonstration provided by Euclid or his scholiast, yet the hexahedron (i) and the decahedron (iii) satisfy the stated requirement equally well, as indeed do endless other polyhedra obtained by joining two or more of the illustrated ones together, face to face, or attaching them or their halves, as required, to other polyhedra having only triangular, square, and/or regular pentagonal faces with the same uniform edge length.

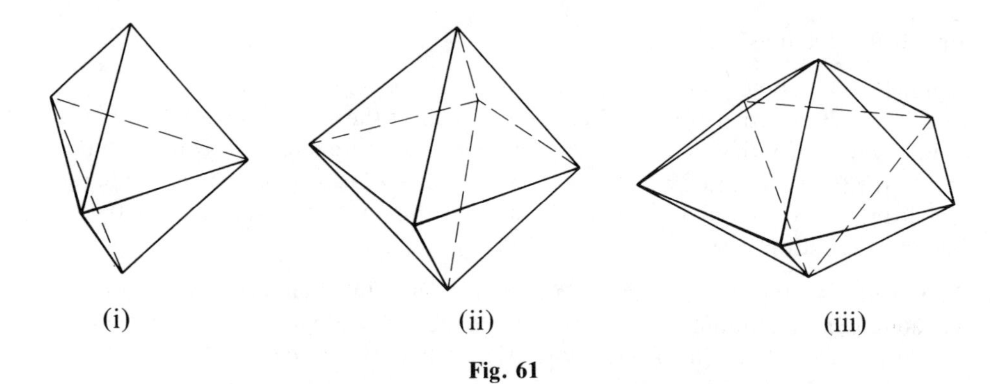

Fig. 61

Sagredo: I suppose that Euclid was concerned only with *convex* polyhedra, but I see that, even with this restriction included, the claim of the scholium is incorrect: the polyhedra illustrated in your first and third diagrams, being convex, still serve as counterexamples. I notice, incidentally, that these two polyhedra are circumscribable about a sphere; they are therefore analogous to the polygons circumscribable about a circle that you recommended for our attention.

Simplicio: They are not, however, inscribable within a sphere.

Sagredo: Nor have they all their solid angles alike. In the solid of the first diagram, for example, some angles are bounded by three faces each and others by four faces; three edges only meet at two of the vertices, four edges at the three remaining vertices. And the angles at which pairs of adjacent faces are inclined to each other are not all equal.

Simplicio: So there are several ways of indicating the irregularity of a nearly regular polyhedron. What then is the criterion, or criteria, of regularity? And how did Euclid define the regular solids?

Sagredo: Imperfectly, I suspect, at least in the case of the icosahedron that he simply defined as "a solid figure contained by twenty equal and equilateral triangles," did he not?

Salviati: That was his definition.[129]

Simplicio: Did you say that that was his definition of a *regular* icosahedron, or of an icosahedron in general?

Sagredo: For Euclid, icosahedra and the other solids with which he was concerned in Book XIII were understood to be both convex and regular. If he had defined a *regular* icosahedron as a "a *convex* solid figure contained by twenty equal and equilateral triangles," then he might perhaps have succeeded in excluding any twenty-faced solids that would have satisfied the definition as he actually gave it, and yet that clearly he would have wished to exclude.

Simplicio: Such as?

Sagredo: Oh, some pretty monstrous combinations I suppose. Yes, look, if we were to attach together, face to face, one of each of the solids illustrated in Fig. 61 in any way such that at each attachment two faces were removed from the surface of the final solid, this would have $6 + 8 + 10 - 4$ faces. A very inelegant icosahedron indeed would be obtained but one that would nevertheless satisfy Euclid's definition.

Simplicio: An ingenious counterexample, I'm sure. Or, I suppose, not a counterexample but an unintended example, since the definition of an icosahedron as given by Euclid, being insufficiently restricted, admits too many instances.

Sagredo: Quite so. And clearly the whole class of polyhedra, extending to all solids bounded by plane faces, may be restricted in several ways. We are free, for instance, to specify that all the edges are to be equal: then the faces may or may not be regular.[130] The added requirement that all the faces be *regular* polygons may itself be imposed in a more or in a less restricted manner, according to whether the faces are to be all alike or whether they may be of two or more kinds of regular polygon.

Simplicio: Could you give an example of the latter?

Sagredo: If you imagine the regular octahedron, Fig. 61(ii), truncated by the removal of six congruent square-based pyramids with apices at the six vertices of the octahedron and with the base vertices of each pyramid at the points of trisection of the edges of the octahedron nearest to the apex, then you will see that the resulting solid will have six square and eight regular hexagonal faces.[131]

Salviati: Alternatively, the six removed pyramids may have their bases at the *mid*points of each set of four adjoining edges and the polyhedron obtained will then have six square and eight equilateral triangular faces.[132]

Sagredo: I notice that the original octahedron is circumscribable about a sphere, the radius of which may readily be expressed in terms of the edge length, but since the two truncations just described, though so alike, are of different severity, it seems evident that the two resulting solids cannot both be circumscribable. This is because the square faces would have to be tangent planes to the sphere that could be inscribed in the original octahedron, parts of the faces of which remain whichever of the truncations is carried out.

Salviati: Aha, Sagredo, it is good to see that you are bearing this notion of circumscribability in mind; I have no doubt that it will be of service to us yet. And you are right in your observation. Indeed, neither of these 14-faced polyhedra turns out to be circumscribable, though each is centrally symmetrical and inscribable within a sphere. I am sure that you will not find any difficulty in showing this on your own. These polyhedra are two of the thirteen so-called semiregular ones ascribed to Archimedes by Pappus.[133] But let us return to the traditional five regular polyhedra to which Euclid restricted himself, the "Platonic" solids.[134] You would now see, if you were to read again the note at the end of the *Elements* (addendum to XIII, 18) that, besides taking the faces to be regular and congruent, the author implicitly assumed that the solid angles were to be all alike, that is, congruent with each other.[135] Euclid's failure to give an explicit general definition of a regular polyhedron and to provide a more complete argument than that in the addendum, were shortcomings that very probably went unnoticed by most of his readers. While they would have been well acquainted with the five Platonic solids, they would more than likely have had little or no knowledge of other, less regular solids bounded by regular polygonal faces.[136]

Sagredo: I can well understand how Euclid could have carried his readers along with him, the traditional line of interest and the immediate context being what they were. Euclid's attention in the final book was exclusively directed to the regular polyhedra, and apparently it did not occur to him to acknowledge that there were other kinds, even in the indirect way of specifying as regular those with which he was concerned.[137]

Salviati: That appears to be a fair statement of the matter. Considerations both of elegance and of priority would seem to be sufficient to account for Euclid's particular concern for the regular solids, and to have made them appear an appropriate subject for special treatment in a book of "elements." There is no evidence that Euclid himself would have believed in the supposed cosmic significance of the forms.[138] And now, let us acknowledge that to establish propositions for a restricted class of entity is in itself, logically, to make no assertion about a less restricted class. But, psychologically, there is much more to be said. Investigators, though meticulous within the limits of restricted treatments, are too often liable to assume that the restrictions adopted (perhaps only for reasons of simplicity or convention) necessarily coincide with and define the full range of applicability of the propositions arrived at. How often do we say, "In any triangle . . . ," "In any regular polygon . . . ," "In any circle . . . ," in introducing statements that, formally, make no claims respectively about quadrilaterals or pentagons, about irregular polygons, or about ellipses or conic sections in general!

Sagredo: And yet they may be applicable in these wider domains. I know from my own experience how easy it is to assume, prima facie, that a proposition established for triangles will have no application to polygons in general. Nevertheless, sometimes the description of a property possessed by all triangles can be so expressed in terms of the number of sides or angles that it applies as well to all polygons. "The sum of the interior angles of any n-sided polygon is $2n - 4$ right angles" is a well-known example.

Salviati: Quite so. Further, sometimes a proposition true for triangles is dismissed as inapplicable in the case of polygons in general because of obvious counterexamples, yet it might be true for all polygons of some special kind (including triangles)—polygons inscribable within a circle, say, or those circumscribable about a circle, or those, perhaps, belonging to the intersection of these two species. When the assumption of certain relations leads to consequences of some importance or interest, it is likely that the recognition of the necessary and sufficient conditions will be formalized in a definition of a general mathematical "object." In this way new concepts and categories of thought arise and are named, enter the literature, and influence the course of subsequent work. Thus there came to be defined so-called poristic polygons, inscribable in one circle and circumscribable about another. These figures attracted the attention of Fuss, Jacobi, Poncelet, and Cayley, the last named generalizing the notion to the class of polygons inscribable in one *ellipse* and circumscribable about another.[139]

Simplicio: It certainly appears that definition and classification have the greatest chance of facilitating investigation when they evolve in conjunction with investigation itself. Aristotle himself said as much. I am reminded of this passage from the *Topics,* which I should like to read:

> [J]ust as in the assemblies the ordinary practice is to move an emendation of the existing law and, if the emendation is better, they repeal the existing law, so one ought to do in the case of definitions as well: one ought oneself to propose a second definition: for if it is seen to be better, and more indicative of the object defined, clearly the definition already laid down will have been demolished, on the principle that there cannot be more than one definition of the same thing.[140]

Salviati: What Aristotle says here may be all very well as regards legal matters: it is proper to change the laws if justice can thereby be better attained. But the case is certainly different in mathematics, where we are concerned to facilitate not justice but the comprehension of necessary relationships. Your philosopher's notion that each object or class of objects has a unique essence that the definition has to capture has caused more than enough confusion in the past. In mathematics, of all subjects, it ought to be patently clear by now that definitions are selected at the discretion of mathematicians on the basis of convenience. Of course, as previously noted, changes in mathematical definitions may sometimes be quite disturbing because of the effect upon already established propositions. Possible extensions of a given concept are likely to be numerous, and different extensions may well be appropriate with respect to different propositions already proved for a less extended range. The logical basis for a proper discrimination between possible extensions has been concisely expressed as follows:

> When the set of properties of a mathematical entity used in the proof of a proposition about this entity does not determine this entity, the proposition can be extended to apply to a more general entity.[141]

Simplicio: I am not quite sure that I fully understand this last remark.

Salviati: It can be illustrated by an example referred to by Aristotle himself. You remember the proof (preserved in Euclid I, 32) for the angle sum for any triangle?

Simplicio: Of course. One simply produces one side of the general triangle beyond a vertex and draws a line through that vertex parallel to the opposite side and then makes use of the equality of corresponding and of alternate angles formed whenever parallel lines are cut by a transversal.

Salviati: Right. Now just suppose that this proof had been given in an earlier age in respect not of the general triangle, but in relation only to right-angled triangles, say, or only for isosceles triangles.[142] Could it not then be pointed out that the specific property "right-angled," or "isosceles," might and should be dispensed with, since it is extraneous to the proof?

Simplicio: It certainly could, for to conclude that the angle sum of only right-angled (or only isosceles) triangles is equal to two right angles would be to fail to appreciate the full implication of the proof and to hold to an unnecessarily restricted enunciation.

Salviati: Exactly so. It is all too easy to overlook the need for evaluating a proof at the time when the result understood to have been established is to be summarized in a general conclusion. A more reflective attitude in this matter could be expected to bring about many an improvement in an initially proposed enunciation. But, to return to the example just considered, let me suggest that our hypothetical early geometer, still preoccupied with right-angled or isosceles triangles, would be quite *unlikely* to give a proof for the angle sum like the one that appears in Euclid. Instead, he would probably see the right-angled triangle as half a rectangle, and an isosceles triangle as decomposable into two matching right-angled triangles, and so obtain simpler but less general proofs,[143] indeed proofs so evident that it might not have seemed necessary to set them down in writing.

Sagredo: I suspect, Salviati, that you intend us to apply the lessons of this discussion to the case of our Theorem α (p. 140).

Salviati: Please tell us what you take these lessons to be.

Sagredo: First, that a particular proof may be perceived in an unnecessarily restricted way, its more general implications simply being overlooked. Second, that the proofs that actually come to mind may very easily do so in unnecessarily restricted forms because of their dependence upon necessarily restricted concepts. Proofs, concepts and their definitions interact with each other, or, if you prefer it, they are related aspects of the one creative process.

Salviati: Admirably put, Sagredo. And now I think that Simplicio has another passage to read us from his copy of Aristotle.

Simplicio: Yes indeed, it is from the *Posterior Analytics,* and after all that you have been saying, I shall be very disappointed if it does not meet with your complete approval:

> full knowledge is attained when an attribute no longer inheres because of something else. Thus, when we learn that exterior angles are equal to four right angles because they are exterior angles of an isosceles [triangle], there still remains the question

> "Why has isosceles this attribute?" and its answer "Because it is a triangle, and a triangle has it because a triangle is a rectilinear figure." If rectilinear figure possesses the property for no further reason [i.e., for no reason other than its own nature], at this point we have full knowledge—but at this point our knowledge has become commensurately universal, and so we conclude that commensurately universal demonstration is superior. . . . He who possesses commensurately universal demonstration knows the particular as well, but he who possesses [only] particular demonstration does not know the universal. . . . For example, if one knows that the angles of all triangles are equal to two right angles, one knows in a sense—potentially—that the isosceles' angles also are equal to two right angles, even if one does not know that the isosceles is a triangle; but to grasp this posterior proposition is by no means to know the commensurate universal either potentially or actually.[144]

Salviati: You have found the very passage I had in mind.

Sagredo: It does seem to be very much in harmony with what you have just been saying.

Salviati: I am bound to add, however, that while many of the characterizations of philosophers of science, from Greek times to the present, seem unobjectionable when illustrated by well-known mathematical examples, they become suspect in one way or another in more complex cases, and especially in fields that are less self-corrective, less cumulative, and less able to be formalized than is mathematics. The view that you have just quoted and with which the observations I have been making doubtless appear to be in such close agreement is part of a very much larger doctrine that has long since ceased to be acceptable, for all that it provided a plausible alternative to the Theory of Forms and of the Recollection of the Forms, and for all that it or, rather, a later version had for many scholastic followers the status of a ruling theory or a whole system of ruling theories.[145] Whatever might have been Aristotle's motivation in devising his doctrine of syllogistic logic, it appears that the hope that natural science might be presented deductively from indubitable primary premises figured prominently in his thinking.[146] It was the attempt to create a general philosophy of "science" (ἐπιστήμη) which required, or seemed to Aristotle to require, that he undertake his hazardous struggle with such controversial notions as prime matter, essence, cause, potentiality, actuality, and coming-into-being. It must have been a relief to be able occasionally to dispense with so much of this complex apparatus and to see in the most austere "limiting case" of mathematics[147] how apparently secure primary premises could yield new unsuspected propositions lending themselves to systematic arrangement and to further and further elaboration. But returning to the sciences in general, Aristotle recognized the need to build up and work from an empirical base more thoroughgoing than that available to any of his predecessors; this dual task of increasing the store of knowledge and of reflecting upon it from fundamental points of view, he assiduously undertook. But his assumption that things had "essences" that disciplined (or "conditioned"!) human minds could apprehend, and his suggestion that such minds (sufficiently educated in the light of relevant experience) might somehow arrive at reliable primary premises and grasp the "commensurate universals"[148] were burdens under which his latter-day followers felt obliged to labor. Working in an intellectual climate that tended to keep them remote from a first-hand acquaintance with the requisite experience, these followers not unnaturally were tempted to rely much too heavily upon the intuitions of "the philosopher" himself or of "authorities" in the tradition that stemmed from his works.[149]

Oh, but how I have been going on! If we restrict ourselves to the passage that you have just read out, Simplicio, it is to be admitted that it well illustrates the logic of generalization that mathematicians do indeed strive on occasion to satisfy. However, in new situations, where unfamiliar entities are being discovered or invented, the mathematical innovator may rightly be more concerned with finding *secure* results, even though these may be found at a later stage to be subject to extensions and generalizations.

Sagredo: I have been wondering about the two examples in this extract from Aristotle. It occurs to me that the restrictions to rectilinear figures and to triangles, respectively, are not absolutely necessary. For surely we could extend our investigation to *curvilinear figures* having an angle sum of two right angles or any other value we might like to specify. For example, we might choose to study biangles.

Simplicio: Biangles?

Sagredo: Yes, biangular figures bounded by two arcs—the lunes studied by Hippocrates are crescent-shaped biangles, but we could just as well study gibbous ones.[150]

Simplicio: I take it, then, that a semicircle (with its bounding diameter) would be an example of a biangle having each of its angles right angles!

Sagredo: Exactly. Notice how we approach a new concept through simple special cases.

Salviati: And in so doing we are often recapitulating to some extent the historical development of our subject.

Sagredo: Historically too, I suppose, it is true to say that special cases of one kind or another have frequently been studied for a long time, and often quite thoroughly, before it seemed worth extending the domain under investigation, or even before it occurred to anyone to try doing so. Following the example of a moment ago, it would seem that the extension to biangles bounded by arcs of *noncircular* curves intersecting each other orthogonally, or in some other specified way, is hardly something the mind would have focused upon in the absence of coordinate geometry and at least the beginnings of the differential calculus. Of course, this and further generalizations might have been ignored because they seemed insignificant in the absence of any interesting context. Be this as it may, these examples have led me to see that sufficient conditions asserted or implied in a proposition are not infrequently also the necessary conditions relative to some conventional restriction. Thus, within the domain of plane rectilinear figures, it is necessary as well as sufficient that a figure be a triangle in order that the sum of its interior angles be two right angles. But, that this is not necessary absolutely is seen as soon as we notice two-dimensional curvilinear figures (or three-dimensional figures either rectilinear or curvilinear) having an angle sum of two right angles.

Simplicio: Yes, I can see this now; also I appreciate what you have stressed, Salviati—that it is often wise (and certainly natural) to study special cases before attempting to master the difficulties of less restricted treatments.[151] On the other hand, there comes a time when the possibility of generalization should no longer be ignored. It may doubtless present itself as a spontaneously recognized challenge, or the learner may be appropriately stimulated by the teacher.

Salviati: What you say does seem to well characterize many instances. However, other cases are now easily recognized by us, with hindsight, as involving generalizations but could at the time have only appeared as constituting quite *new departures.* The radical extensions to the number system have already been alluded to as examples of this (p. 67). Perhaps there is time for one further remark. Mathematics considered "longitudinally"—I mean developmentally—displays both the blindnesses and master strokes characteristic of inventiveness in any field of high human endeavor. It is easy to deplore in general terms the presence of unnecessarily restricted definitions and theorems, of imperfectly understood domains of validity, though it may not be so easy to detect actual instances of these! Just *which* generalizations are possible, and worth pursuing, and what are the best means of articulating whole areas of subject matter, are hardly matters that an isolated investigation is likely to reveal; rather, they are the outcome of a dynamic historically engendered experience, typically involving many minds.

X

The aims of scientific thought are to see the general in the particular and the eternal in the transitory.

—A. N. Whitehead

[In mathematics] every substantial advance takes place together with the discovery of sharper means and simpler methods which simultaneously facilitate the understanding of earlier theories and abolish old long-winded developments.

—David Hilbert

Just as perceptual attention concentrated on one object is thereby relatively switched off for all others, so the associations relating to a problem bar the way to all others. The cat [getting ready to seize its prey] fails to notice the approaching hunter, Socrates immersed in speculation absentmindedly ignores Xanthippe's questions and Archimedes constructing his circle pays with his life for his defective biological adaptation to immediate circumstances.

—Ernst Mach

We suffer from an incapacity to make our thoughts mobile. In order that we may have some guarantee of arriving at the same opinion about an idea, the minimal requirement is that we should have had different opinions about it in the first place. If two men want to agree they really have to contradict one another first. Truth is the daughter of discussion, not of sympathy.

—Gaston Bachelard

No man thoroughly understands a truth until he has contended against it.

—Emerson

HORA DECIMA

Salviati: Well now, are you ready to enunciate a generalization of Theorem α, which we reached at the end of Hora Octava?

Simplicio: I fear that your faith in our powers may not always be justified, Salviati. For myself, I am not able to do better than to explain why I am bewildered by the suggestion of a generalization in this case.

Salviati: Please go ahead. The clear articulation of a difficulty is often the best way of preparing for a successful solution.

Simplicio: You enunciated Theorem α as follows:

> *For any regular polygon and its inscribed circle (or for any two regular polygons circumscribable about the same circle), the ratio of the areas is equal to the ratio of the perimeters.*

Now it seems to me from the indications of the early part of our investigations that the equality of area and perimeter ratios will not be maintained when the polygons are made irregular instead of regular. In this respect the situation will surely be as it was in the case of the elongation of a circle and its circumscribing square so as to obtain an ellipse and its enclosing rectangle with sides parallel to the axes. Before the transformation, the area and perimeter ratios were both equal to $\pi : 4$; after the transformation only the area ratio remained unchanged. $A_e = \pi ab$, and $A_r = 4ab$, but as was shown, $p_e > \pi(a + b)$, whereas $p_r = 4(a + b)$, so that the perimeter ratio, ellipse to rectangle, is greater than $\pi : 4$, greater, that is, than the area ratio. Or, to take an even more transparent example that I have just been sketching out, let us consider the simple rectilinear case illustrated in these diagrams (Fig. 62). For the pair of congruent squares, the area ratio is equal to the perimeter ratio, both being $1 : 1$. For the pair of elongated figures, the area ratio, rhombus to rectangle, is still $1 : 1$, since the area of each is given by $k \cdot a^2$. But there is a difference in the perimeters of these two figures. The perimeter of the rhombus is measured by $4\surd\{(\frac{1}{2}ka\surd 2)^2 + (\frac{1}{2}a\surd 2)^2\}$, which comes to $2a\surd\{2(k^2 + 1)\}$, and this differs from $2a + 2ka$, the measure of the rectangle's perimeter.

Sagredo: Yes, I can see the the ratio,

$$\text{perim. of rhombus} : \text{perim. of rectangle} = 2a\surd\{2(k^2 + 1)\} : 2a(k + 1),$$

and this is not equal to the area ratio of $1 : 1$ unless $2a\surd\{2(k^2 + 1)\} = 2a(k + 1)$, that is, unless $k = 1$. As k gets larger and larger, this perimeter ratio approaches $\surd 2 : 1$.

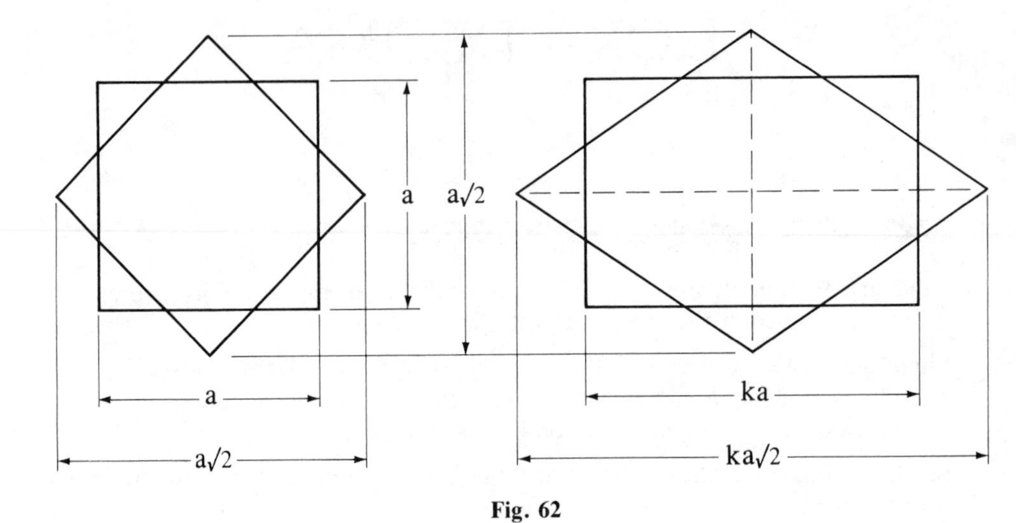

Fig. 62

Simplicio: I am sure you will agree, then, that the slightest deviation from the original regularity of the polygons upsets the equality of the area and perimeter ratios! I have no doubt that this could be shown in as many cases as might be presented.

Salviati: Then let me offer for your consideration the pairs of figures previously illustrated in the four diagrams of Fig. 59, but now imagined to be simply elongated as shown here in Fig. 63 or elongated or compressed uniformly in any other direction you like.

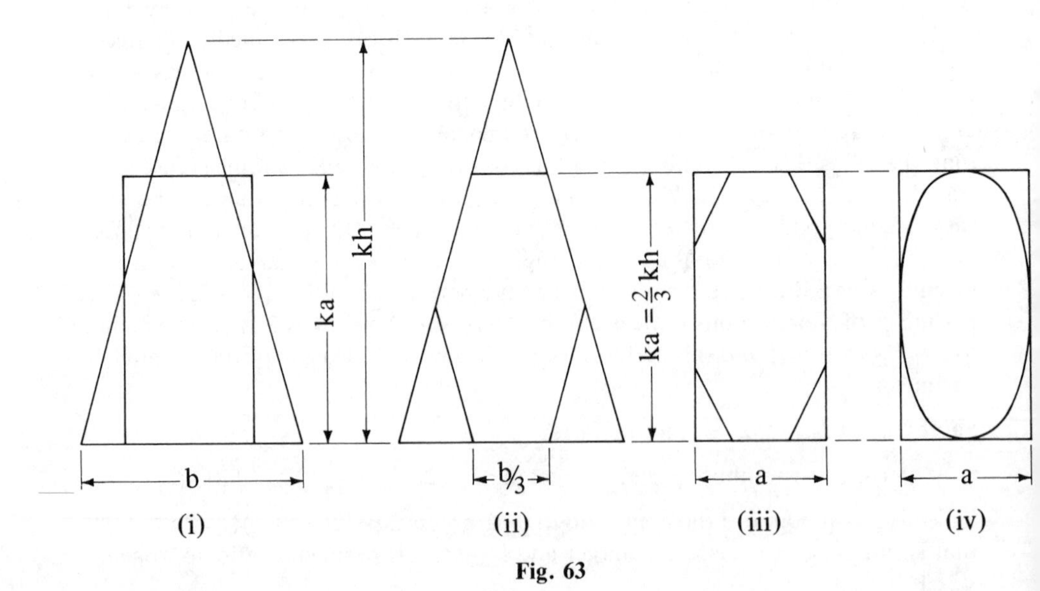

Fig. 63

Sagredo: Shades of El Greco![152]

Salviati: Yes, artists caricatured their subjects before they ever thought of deliberately defining constructions whereby distortions might be methodically produced and, of course, before mathematicians theorized upon such constructions.[153] But now tell me, do you consider that the equality of area and perimeter ratios will be upset in every one of the cases shown here, or do you perceive that there is an exception to this?

Sagredo: Could it perhaps be . . . ? Yes indeed! The second of these diagrams makes me realize that not only for an equilateral triangle, indeed not only for an isosceles triangle, but for *any* triangle and the inscribed hexagon having its vertices at the points of trisection of the sides of the triangle (Fig. 64), the areas and perimeters will *each* still be as 9 : 6, that is as 3 : 2, exactly as shown earlier for the equilateral case (Fig. 55).

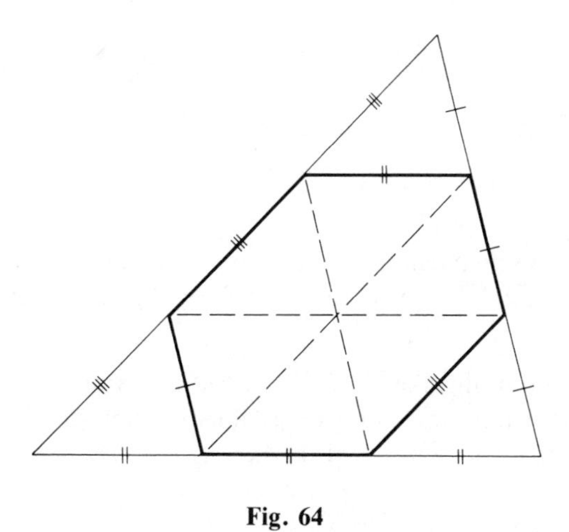

Fig. 64

Simplicio: What you say is undeniable, and we did overlook it before. It seems, however, a rather atypical example and I think that the calculation performed in connection with Fig. 62 is representative of most deviations from regular cases, at least if they consist of simple elongations.

Sagredo: That was an apt example of yours, and, incidentally, the calculation indicated that if *one* side of a right-angled triangle be multiplied by a factor, k say, then the area is multiplied by k, but the hypotenuse and the perimeter are changed in other ratios that could easily be expressed in terms of k. I admit all this, but a different way of passing from regular to irregular polygons occurred to me during the last pause. After our long discussion on extensions and generalizations, I tried to see in my mind's eye how the range of Theorem α might be extended. Since any regular polygon and its inscribed circle bear equal area and perimeter ratios to each other, it does appear that we might combine parts of different regular polygons together, in such a way that their sides still touch the one circle, without upsetting the equality of these ratios. Two examples that came to mind are shown in Fig. 65. The isosceles trapezium shown in the first diagram is to be supposed to be made up of parts of an equilateral triangle and a regular hexagon—two-thirds of the former and one-third of the latter, circumscribed about the same circle. And the second diagram illustrates a symmetrical hexagon composed of half a square and half of a regular octagon likewise circumscribed about a common circle.

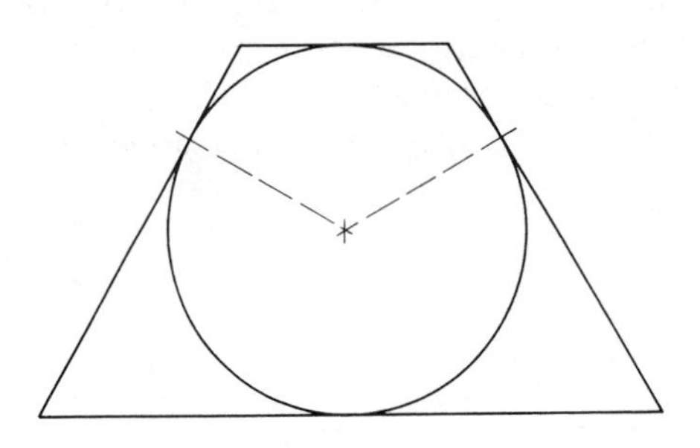

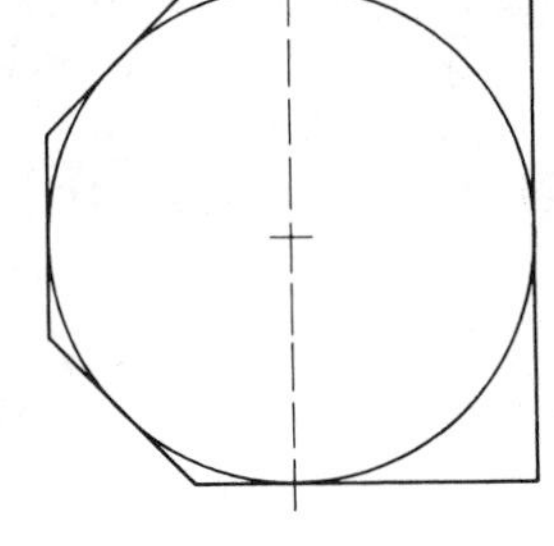

Fig. 65

Simplicio: Aha! So you mean that Theorem α, though explicitly concerned with regular polygons, implicitly covers this new class of irregular polygons and their incircles?

Sagredo: Unless I am mistaken, this is so. But let us proceed with appropriate formality. In obvious notation, for the case illustrated in the first of the diagrams of Fig. 65, area ratio $= (\frac{2}{3}A_T + \frac{1}{3}A_H) : A_o$, or $\frac{2}{3}A_T/A_o + \frac{1}{3}A_H/A_o$, which, by Theorem α, is equal to $\frac{2}{3}p_T/p_o + \frac{1}{3}p_H/p_o$, or $(\frac{2}{3}p_T + \frac{1}{3}p_H) : p_o$, precisely the perimeter ratio for the same pair of figures!

Simplicio: I am convinced, and I can see that exactly the same kind of demonstration will apply for the second pair of figures, or for any other case composed in a like manner.

Salviati: What about the cases illustrated in Fig. 66 then?

Simplicio: Assuming the polygon of the first diagram to consist of half a square and half a regular hexagon, we could readily show that the area ratio is equal to the perimeter ratio as before. But it will be otherwise with the case illustrated in the second diagram. The interior angles of the regular polygons are

60° for the regular triangle,
90° for the regular quadrilateral,
108° for the regular pentagon,
120° for the regular hexagon,

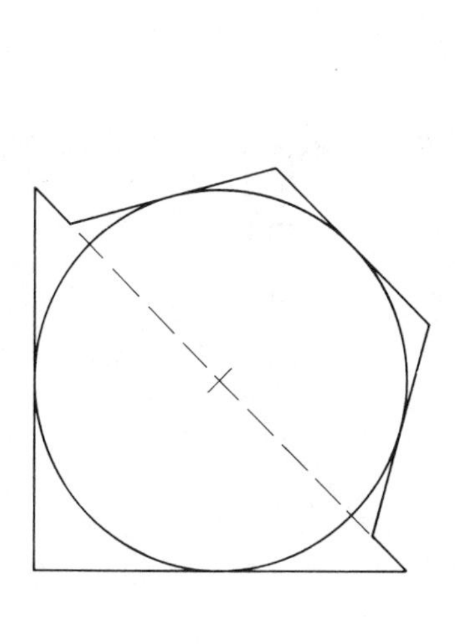

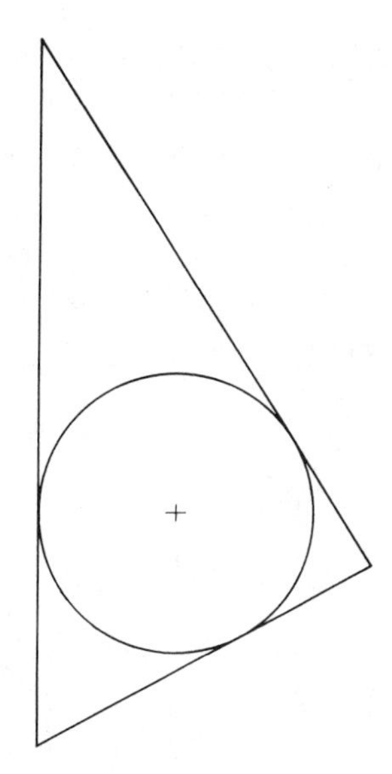

Fig. 66

and so on. Now, no triangle other than the equilateral can have all its angles chosen from this list. This means that, although all triangles are circumscribable about a circle, none, apart from the equilateral, can be entirely decomposed into parts of regular polygons in the way required in Sagredo's analysis.

Salviati: And what do you say, Sagredo?

Sagredo: Simplicio seems to be right about your second example, but I notice that he has overlooked something in the first one. As the diagram shows, the circumscription is improper since not all the sides of the outer figure are tangent to the inside circle: the shortest sides of the polygon, if produced, would cut, not touch, the circle. What is more, these shortest sides belong neither to the half-square nor to the half-hexagon. The perimeter of the polygon—the irregular but symmetrical heptagon—therefore exceeds the sum of the half-perimeters of the square and hexagon, though its area is simply the sum of the two component areas, the half-square and the half-hexagon. Consequently, the area ratio, polygon to enclosed circle, is less than the corresponding perimeter ratio.

Salviati: Thank you, Sagredo, for a clear explanation. But I must ask both of you to reconsider the case of the triangle and its inscribed circle. To show that a particular analysis (into elements of regular polygons) is not applicable is not to show that the result (equality of area and perimeter ratios) could not be true. It is interesting that neither of you thought it worthwhile to compare by calculation these ratios for a triangle and its inscribed circle! I should like to see you less attached to the "requirement" of regularity, which after all entered into our investigations only for reasons of convenience. Simplicio, do you think that a 3-4-5 triangle bears equal area and perimeter ratios to its inscribed circle?

Simplicio: I should not have thought so.

Salviati: Then let us perform the calculation. First the radius of the incircle is to be determined. Referring to Fig. 67, you will observe that, by the equal-tangents theorem, $(3 - r)$ and $(4 - r)$ are together to be equated to 5, the measure of the hypotenuse.

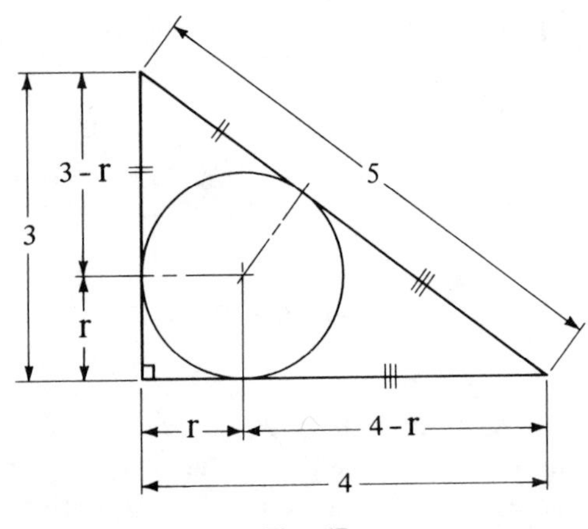

Fig. 67

Simplicio: So, $7 - 2r = 5$, giving $r = 1$; the incircle of a right-angled triangle with sides 3, 4, 5 units has a radius of exactly 1 unit!

Sagredo: Clearly, this example too was chosen for its amenability to mathematical treatment.

Salviati: It will be up to you to consider how representative the outcome is. And now, what does the ratio of the areas come to?

Simplicio: $$\frac{\text{area of triangle}}{\text{area of incircle}} = \frac{\frac{1}{2} \times 4 \times 3}{\pi \times 1 \times 1} = \frac{6}{\pi}.$$

Salviati: And the ratio of the perimeters?

Simplicio: $$\frac{\text{perim. of triangle}}{\text{circum. of circle}} = \frac{3 + 4 + 5}{2\pi \times 1},$$

and this also comes to $6/\pi$. Well, I had not expected this result! So the area and perimeter ratios *are* equal in this special case. I doubt if it holds for right-angled triangles in general.

Salviati: What do you think, Sagredo?

Sagredo: I am not sure, but I am almost prepared to believe that it could be true for any right triangle and its incircle and possibly also in the case of any triangle whatsoever. You now have me wondering whether it is not perhaps circumscribability about a circle, instead of regularity, that is the essential requirement. Is not the radius of the incircle of any triangle given by $r = \Delta/s$, where Δ, s, denote the measures of the area and the semiperimeter respectively? Have I recalled the result correctly, Salviati?

Salviati: You have, Sagredo, and I think it would be useful if you could establish this result without losing your present train of thought.

Sagredo: I am still feeling my way, but I suspect that this relationship, $r = \Delta/s$, lies at the heart of the matter. As I recall it, the demonstration is very simple. Any triangle ABC, with incentre I, may be subdivided into three triangles, IBC, ICA, IAB, of areas $\frac{1}{2}ar$, $\frac{1}{2}br$, $\frac{1}{2}cr$, respectively. Hence, if Δ denotes the measure of the area of triangle ABC,

$$\begin{aligned}\Delta &= \frac{1}{2}ar + \frac{1}{2}br + \frac{1}{2}cr\\ &= \frac{1}{2}r(a + b + c)\\ &= rs, \text{ where } s = \frac{1}{2}(a + b + c).\end{aligned}$$

I realize that this is a well-known proposition.[154] But only now do I see its relevance to our problem of equal area and perimeter ratios:

$$\frac{\text{area of triangle}}{\text{area of incircle}} = \frac{rs}{\pi r^2} = \frac{s}{\pi r} = \frac{2s}{2\pi r} = \frac{\text{perim. of triangle}}{\text{circum. of circle}}.$$

Salviati: So you have proved that for any three-sided polygon, regular or irregular, and the circle about which it is circumscribable, the ratio of the areas is equal to the ratio of the perimeters. What next?

Sagredo: When you express it *that* way, how can we fail to inquire whether the number of sides is relevant or not? You have made me realize that, for any polygon within which a circle can be inscribed, the area of the polygon will be given by the product: half perimeter times incircle radius.

Simplicio: Do you mean to say that Theorem α can be revised to apply to any convex polygon, no matter how irregular, given only that it is circumscribable about some circle?

Sagredo: I do not think it is necessary to specify that the polygon be convex. Look, here are a couple of typical polygonal circumscriptions about a given circle (Fig. 68). Allowing for a slightly extended sense of circumscribability, you will observe that there is no need to exclude concave polygons as such. Let us agree

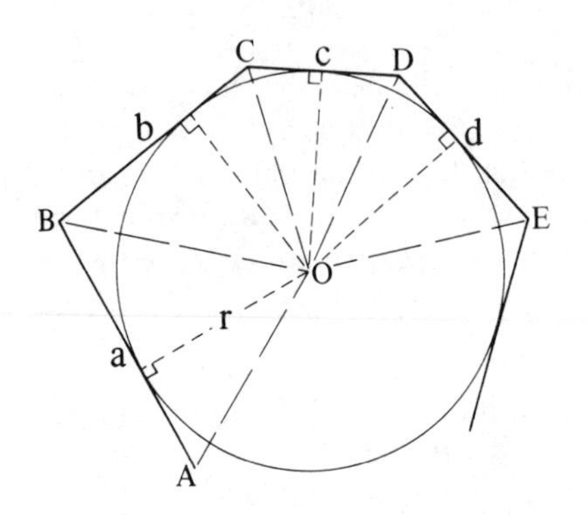

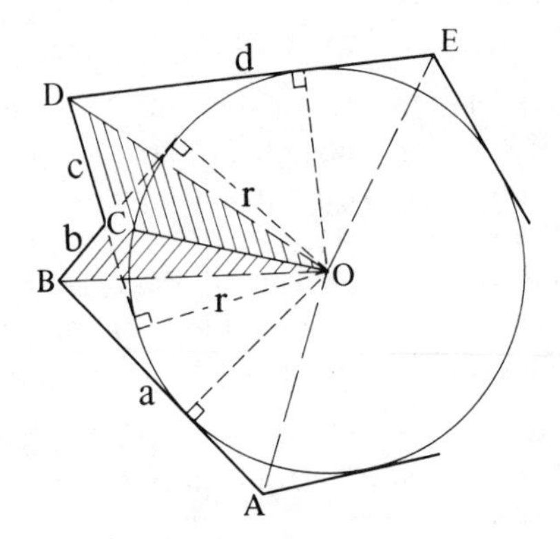

Fig. 68

that any plane polygon will be counted as circumscribable about a circle provided only that all of its sides, produced if necessary, are tangent to the circle. Referring, then, to either diagram of Fig. 68:

$$\begin{aligned}\text{Area of circumscribing polygon} &= \triangle OAB + \triangle OBC + \triangle OCD + \ldots \\ &= \frac{1}{2}ar + \frac{1}{2}br + \frac{1}{2}cr + \ldots \\ &= \frac{1}{2}pr.\end{aligned}$$

(Thus for any such polygons, just as for any triangle, $A = \frac{1}{2}pr$ or $A = rs$.)[155]

$$\begin{aligned}\text{Area ratio, polygon to incircle, } &= \frac{1}{2}pr : \pi r^2 \\ &= p : 2\pi r = \text{perimeter ratio.}\end{aligned}$$

Salviati: This generalization of Theorem α may be appropriately designated Theorem A:

For any circumscribed polygon and its inscribed circle, the ratio of the areas is equal to the ratio of the perimeters.

Corollary: If two polygons are circumscribable about equal circles, then the ratio of the areas of the polygons is equal to the ratio of their perimeters.

Proof: $$\frac{A_2}{A_1} = \frac{A_2/A_o}{A_1/A_o} = \frac{p_2/p_o}{p_1/p_o} = \frac{p_2}{p_1}$$

Converse: If two circumscribable polygons (or a circumscribable polygon and a circle) have equal area and perimeter ratios, then they have equal inscribed circles (or the circle is inscribable in the polygon).

Proof: Since, for any circumscribable polygon or for any circle, $A = \frac{1}{2}pr$, it follows that if it is given that $A_2 : A_1 = p_2 : p_1$, then $\frac{1}{2}p_2r_2/\frac{1}{2}p_1r_1 = p_2/p_1$, and hence $r_2 = r_1$.

Sagredo: I now realize that particular cases of this converse were found via the calculations accompanying Figs. 54, 57, and 58.

Simplicio: Well, really—I must say you are both pretty casual about this new theorem! You who have discovered it appear far less elated than I who have been a mere observer. A Pythagorean, on discovering such a theorem, would have counted himself possessed by a god, divinely inspired, and would doubtless have experienced a feeling of reverence, as well as of great excitement, calling for some offering of thanksgiving.[156]

Salviati: In this age, as you will know, instead of speaking of a god within, the enthusiasm[157] to which you refer is very differently regarded. The creative process is nowadays often characterized (not, of course, explained) in terms of the largely unconscious separation of relevant from irrelevant ideas. This process can lead to a conscious manifestation—an intuition or insight—provided the receptivity of the conscious mind is not too much diminished through preoccupation with other matters. The experience of *illumination* can sometimes be extraordinarily satisfying, counting as a "peak experience," to use a popular designation.[158]

Simplicio: And in the case of this theorem?

Sagredo: The discovery of this theorem was for me much too closely associated with the discovery that I had been blind for far too long to a generalization that Salviati had been so patiently inviting us to make. Realizing how tardy our response had been, I felt more subdued than elated.

Salviati: Sagredo, you ought not to undervalue your achievements.

Sagredo: I am sure that Simplicio will join me in acknowledging how very easy it is for convenient special cases to dominate one's thinking. What I have especially in mind is our initial belief that the area and perimeter ratios for a nonequilateral triangle and its incircle would be *unequal* just because such a case could not be completely reduced to parts of regular figures in the manner illustrated in the diagrams of Fig. 65.[159]

Simplicio: It is, I suppose, one thing to recognize, abstractly, that what is easily amenable to currently known methods is not necessarily representative of significant general classes of entity or event; it is another matter to know how one should proceed in accordance with this recognition when one is engaged in an actual investigation.

Salviati: It is indeed, and I suppose that every mathematician has cause to remember to his chagrin missed opportunities for generalization, opportunities that were overlooked because mastery of a special case had induced the feeling of a task completed. But I think, Sagredo, that you have already made this very point.

Sagredo: I know from experience how reduction in my own receptivity and flexibility of thought accompanies such complacency. And I agree with you too, Simplicio; an awareness of the various psychological lapses or constraints that can interfere with the progress of creative thought does not necessarily afford us protection against being so hindered in our own efforts to gain new and more adequate insights. It is remarkable how what is not taken to concern us can remain below the threshold of perception, even though it present itself a dozen times.

Simplicio: I know how highly selective perception is in all its forms.

Salviati: It should not be too surprising, then, if the apparently well-prepared mind shows itself very far from being consistently acute when not focused upon matters of its own special concern. Doubtless successful mathematicians so cultivate the striving for generalization that it seems to become second nature to them. And yet, even the greatest have left results their successors are able to see can be generalized further.

Simplicio: Perhaps our Theorem A might be generalizable!

Sagredo: That is something I should like to consider at length. But first, Simplicio, let me mention a further reason for my lack of elation over our discovery of this theorem. Because of its very close connection with the long-known formula $A = \frac{1}{2}pr$, it is hardly to be supposed that Theorem A could have remained undiscovered until now. Realizing this, I could not seriously believe that the result was new, for all that I had never come across it in my studies.

Salviati: Most of what we are discovering in today's investigations, though not original in the sense of having been previously undiscovered, nevertheless can be claimed to be original in the etymological sense, on account of the way in which *origins*, heuristic origins, are exhibited. With our approach, too, I believe, comes an enhanced awareness of the ways in which various results can be connected together and illuminate each other. Connectedness is crucial in discovery, in dialectic and demonstration, in the appreciation of history and in all forms of human understanding; without it there is *ἀβλεψία*.[160]

Simplicio: I agree. And now that we have Theorem A, how do we proceed?

Sagredo: It occurs to me that a polygon may fail to be circumscribable about a given circle either because it is the wrong size (Fig. 69(i) illustrates such a failure, the square is simply too small) *or* because it is the wrong shape; see, for example, Fig. 69(ii). In the first of these diagrams, in which the square is inscribed within, instead of circumscribed about, the circle, the ratio of the area of the square to that of the circle comes to $2:\pi$, and this is less than the corresponding perimeter ratio, $2\sqrt{2}:\pi$. In general, for a circle of diameter d and a square of side kd, the area ratio, square to circle, is $k^2d^2:(\pi/4)d^2$, that is, $4k^2:\pi$, whereas the perim-

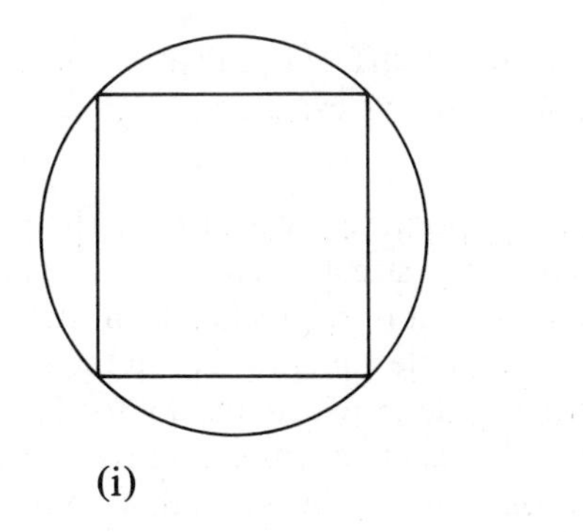
(i)

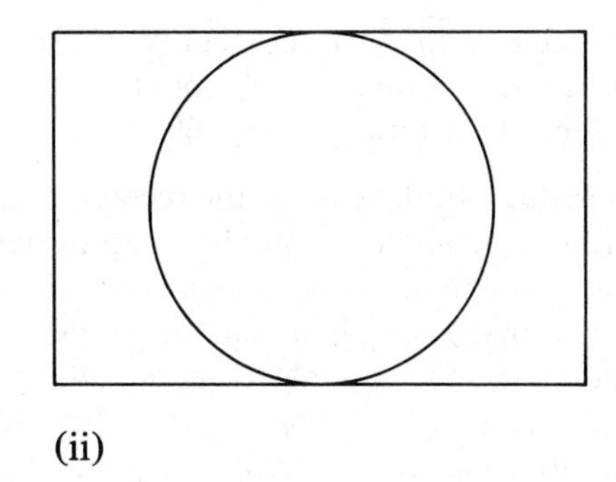
(ii)

Fig. 69

eter ratio is $4kd : \pi d$, or $4k : \pi$. Hence: area ratio $\gtreqless$ perimeter ratio, according as $k \gtreqless 1$, that is, according as the square is greater than, equal to, or less than a square circumscribed about the circle. And corresponding relations must hold in the case of any other circumscribable polygons compared with a specified circle or with each other. I think it can be seen, then, that we may enunciate the following:

Extended Version of Theorem A and Its Corollary and Converse

For any two circumscribable polygons (or for a circumscribable polygon and a circle) the ratio of the areas, $A_2 : A_1$, is greater than, equal to, or less than the ratio of the perimeters, $p_2 : p_1$, according as the ratio of the radii of the respective incircles (or of the incircle and the given circle), $r_2 : r_1$, is greater than, equal to, or less than unity. And, conversely, the ratio of the incircle radii of the two circumscribable polygons (or the ratio of the incircle radius of a circumscribable polygon to the radius of any given circle), $r_2 : r_1$, is greater than, equal to, or less than unity according as $A_2 : A_1 \gtreqless p_2 : p_1$.

Simplicio: I am afraid that I cannot fully grasp this.

Sagredo: Let me try to show it symbolically. If A_1 and A_2 denote the measures of the areas of two circumscribable polygons (of any size) or of any circumscribable polygon and any circle (or, indeed, of any two circles), then since for such polygons and for any circle, $A = \frac{1}{2}pr$, we have,

$$\frac{A_2}{A_1} = \frac{p_2\, r_2}{p_1\, r_1}, \left\{ \begin{array}{rll} \text{so if } r_2/r_1 > 1, & \text{then} & A_2/A_1 > p_2/p_1, \\ \text{if } r_2/r_1 = 1, & \text{then} & A_2/A_1 = p_2/p_1, \\ \text{and if } r_2/r_1 < 1, & \text{then} & A_2/A_1 < p_2/p_1, \end{array} \right.$$

and, conversely, if $A_2/A_1 \gtreqless p_2/p_1$, then $r_2/r_1 \gtreqless 1$, respectively.

Simplicio: Thank you. I accept that you are correct, though I still find this extended theorem a little difficult to take in. I note that it is not a generalization of Theorem A in the sense that it says anything about polygons of noncircumscribable form. I suspect that we are going to find it difficult to deal with these. Have you any more to say about the noncircumscribable example you gave us in the second diagram of Fig. 69?

Sagredo: In the case of the rectangle, circumscription fails because the shape is unsuitable. There is nothing inappropriate about the size as such, for there is no limit to the number of polygons, of area equal to that of the rectangle illustrated in Fig. 69(ii), which *are* circumscribable about a circle equal to that in the same figure—or, for that matter, about any specified circle provided it is not too large. The triangle and the pentagon of Fig. 70 may be taken as representative of this infinitely numerous family of equiareal polygons. (It is to be noted that both the area and the perimeter of any circumscribable polygon increase without limit as a pair of adjacent sides approach parallelism.) It follows from the relationship, area $= \frac{1}{2}pr$, for any circumscribable polygon, that *if for a set of such polygons any two of the area, the perimeter, and the incircle radius are kept constant, the third must also remain constant.* Therefore the members of a set of equiareal polygons circumscribable about the same circle must also be isoperimetric with each other, though, of course, they will not ordinarily be isoperimetric with an equiareal noncircumscribable polygon.

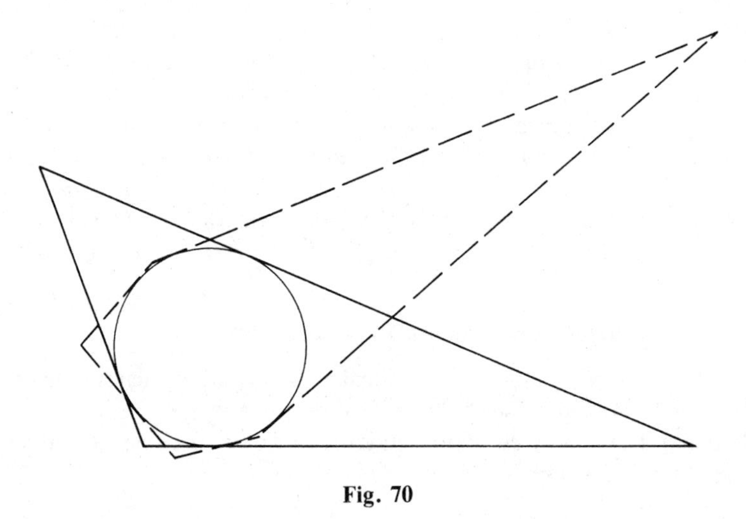

Fig. 70

Simplicio: I see. It looks as though it is going to be difficult to extend our considerations to noncircumscribable polygons. Or is there something more to be learned from your example of Fig. 69(ii)?

Sagredo: I am still feeling my way. The rectangle fails to be circumscribable since *the only quadrilaterals that are circumscribable are those having the sum of one pair of opposite sides equal to the sum of the other pair of opposite sides; hence the only circumscribable rectangles are squares.*

Simplicio: Believe me, Sagredo, I am not wishing to find any fault with this statement, although I shall have a question concerning it, but I can well imagine that many would be perplexed, and some might even be tempted to ridicule you for giving a reason such as this. They might say, "This man purports to explain a most obvious and commonplace truth by asserting another proposition that claims far more, is unfamiliar, not at all obvious, and is even suspect."

Sagredo: I have no doubt we could answer such people perfectly well.

Simplicio: I have not forgotten Aristotle's teaching to the effect that if we want scientific knowledge of a thing, we cannot rest content with mere description but must instead endeavor to uncover the causes or reasons for what is the case. This was illustrated in the treatise *De Anima* with the example of the quadrature of a rectangle. To say that this is the construction of an equilateral rectangle equal to a given oblong rectangle is mere description, whereas to tell us that it is "the finding of a mean [the mean proportional] between the two unequal sides of the given rectangle discloses the ground of what is defined".[161] And I would justify your statement about circumscribable quadrilaterals similarly.

Sagredo: Well yes, though of course the bare descriptive definition does have its important place; but the assertion that an *additional* property, not given in the definition, is assignable to the objects defined, is just what requires proof, as you very well know. Doubtless Aristotle was influenced by the realization that the mean proportional is the key not only to the problem of squaring a rectangle, but to several other problems as well. But I can see that you have something on your mind, Simplicio.

Simplicio: First, as to the incontrovertible fact that the only rectangle that will circumscribe a circle is a square, it has only to be said that if the two longer sides of an oblong rectangle touch the circle then the shorter ones are too far apart, whereas if the two shorter ones touch it the longer ones are too close together (Fig. 71). But, I can imagine a critic saying, instead of being satisfied with such self-evidence, these geometers bring in the proposition that if and only if $BC + DA = AB + CD$ will a quadrilateral $ABCD$ be circumscribable.

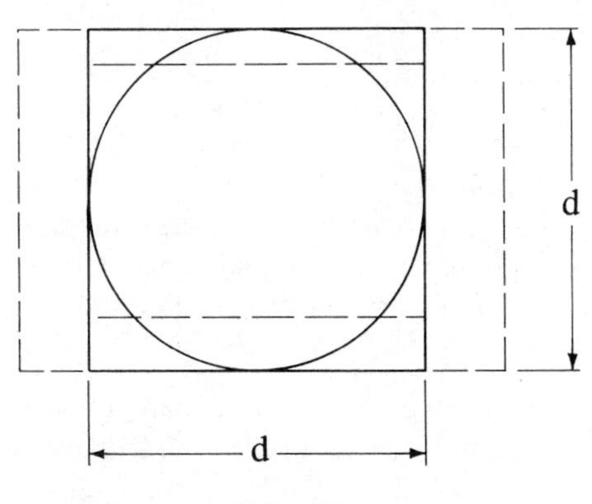

Fig. 71

Salviati: You would no doubt tell such a critic that quite a number of restricted propositions, such as that *all circumscribable parallelograms are rhombuses,* could be subsumed under, and to that extent explained by, the same general theorem.

Simplicio: I should hope I would remember to do that.

Sagredo: Is something still bothering you?

Simplicio: I have to admit that I am a little suspicious of the general proposition to which you have appealed. I do know that *if* a quadrilateral is circumscribable *then* the sums of the opposite sides are equal, each sum being half the perimeter. Referring to Fig. 72, if the sides are all tangent to the circle, then

$$w_2 = w_1, \qquad x_2 = x_1, \qquad y_2 = y_1, \qquad z_2 = z_1;$$

hence

$$w_2 + x_1 + y_2 + z_1 = x_2 + y_1 + z_2 + w_1.$$

That is, *if* a quadrilateral $ABCD$ is circumscribable about a circle, *then* $a + c = b + d$.

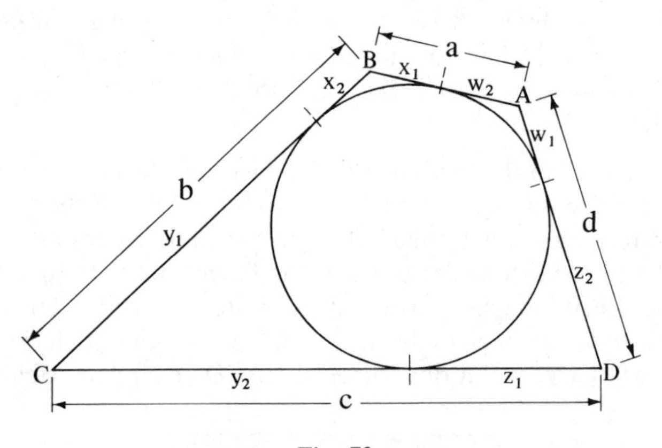

Fig. 72

Sagredo: Very good Simplicio. You know, until now I had not noticed that this theorem is immediately generalizable to circumscribable polygons having *any even number of sides.* The pairing of equal tangents will then enable it to be similarly shown that the sum of one set of alternate sides is equal to the other set of alternate sides: $a + c + e + \ldots = b + d + f + \ldots$

Salviati: This is a noteworthy extension. And can you also see how the theorem concerning the opposite angles of a cyclic quadrilateral may be analogously extended?

Sagredo: Could it be that $A + C + E + \ldots = B + D + F + \ldots$? Let me draw a figure for the case of a cyclic hexagon (Fig. 73). These marked angles at the center clearly total twice four right angles. Hence the angles at the circumference subtending the same arcs total just four right angles, that is, $A + C + E = 4$ right angles. Similarly (or by subtraction from $2n - 4$ right angles with $n = 6$) we find that $B + D + F = 4$ right angles also.

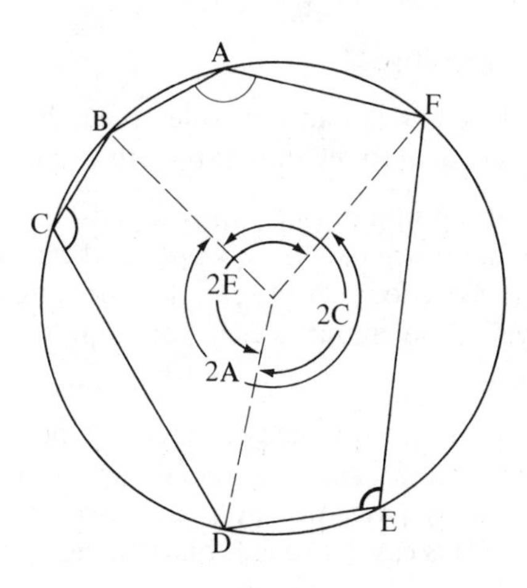

Fig. 73

Simplicio: These are interesting extensions, but my original doubt concerning the criterion for the circumscribability of a quadrilateral remains. I am, of course, familiar with the well-known theorem about the opposite angles of a cyclic quadrilateral and with the converse of this. I am not questioning the proposition that a quadrilateral is inscribable within a circle if and only if its opposite angles are supplementary. But I am still *uncertain* about its counterpart: *A quadrilateral is circumscribable about a circle if and only if the sum of one pair of opposite sides is equal to the sum of the other pair.*

Sagredo: It strikes me, Simplicio, that the relation of proposition and converse to such an if-and-only-if statement is not quite as crystal clear to me as it ought to be. The proof you sketched with reference to Fig. 72 shows that

> If a quadrilateral is circumscribable, then the sum of one pair of opposite sides is equal to the sum of the other pair. (1)

It immediately follows that

> If the sum of one pair of opposite sides is *not* equal to the sum of the other pair, then the quadrilateral is not circumscribable. (2)

This is the contrapositive proposition that is logically implied without the need for any additional geometrical proof—(2) follows from (1) in accordance with the formula (if p then q) implies (if not q then not p).[162]

Simplicio: Exactly. But what has *not* yet been established is the converse of the first proposition,

> If the sum of one pair of opposite sides is equal to the sum of the other pair, then the quadrilateral is circumscribable. (3)

Sagredo: Nor *its* contrapositive,

> If a quadrilateral is not circumscribable, then the sum of one pair of opposite sides is not equal to the sum of the other pair. (4)

So that the assertion that the sum of one pair of opposite sides of a quadrilateral is equal to the sum of the other pair *if and only if* the quadrilateral is circumscribable is equivalent to assertions (1) and (4); the "iff statement" that you made a minute ago (italicized), and about which you expressed uncertainty, is the equivalent of (2) and (3).

Simplicio: I can now see that, from the geometer's point of view, it is enough to consider propositions (1) and (3), since the remaining pair of numbered propositions are, so to speak merely parasitic upon these two, and convey no new insights. As we have seen, (1) is easily proved, but I cannot see how to prove (3), if indeed it be true.

Sagredo: Well, Simplicio, you are *now* showing admirable caution, if I may say so, caution of the kind we had need of earlier. From (i): All circles have uniform diameter, we may immediately infer the contrapositive (ii): All nonuniform-diameter figures are noncircles, but *not* the converse (iii): All uniform-diameter figures are circles, *nor its* contrapositive (iv): All noncircles are nonuniform-diameter figures. The falsity of this last pair of propositions we exposed by numerous counterexamples. But to return to our quadrilaterals, I can hardly doubt that the converse proposition, (3) will be found to be true. But our hour is up so let us make the proof of this the first item of our next discussion.

Simplicio: Well, thank you anyway for the logical exposition, Sagredo.

Sagredo: I assure you that I have been saying what I have in order to clarify my own ideas. We seem to be like ships becalmed, so slow is our progress. I fear that our stumbling efforts must appear most tedious to you, Salviati.

Salviati: You may certainly set your mind at rest on that score, Sagredo. You ought to know how delighted I am to take part in these discussions.

XI

He who first gave names gave them according to his conception of the things which they signified . . . [but] if his conception was erroneous . . . shall we not be deceived by him? . . . [And] if he did begin in error, he may have forced the remainder into agreement with the original error and with himself; there would be nothing strange in this, any more than in geometrical diagrams, which have often a slight and invisible flaw in the first part of the process, and are consistently mistaken in the long deductions that follow. And this is the reason why every man should expend his chief thought and attention on the consideration of his first principles—are they or are they not rightly laid down?

—Plato

I have heard many people say, "Give me the Ideas. It is no matter what Words you put them into." . . . [But I reply:] "Ideas cannot be given but in their minutely appropriate words."

—William Blake

Something we see quite clearly, and which nonetheless is very difficult to express, is always worth the trouble of trying to put into words.

—Paul Valéry

It requires a very unusual mind to make an analysis of the obvious.

—A. N. Whitehead

Concepts, like individuals, have their histories, and are just as incapable of withstanding the ravages of time as are individuals.

—Søren Kierkegaard

Let us remember that science is a pursuit of living men, and that its most marked characteristic is that when it is genuine, it is in an incessant state of metabolism and growth.

—C. S. Peirce

HORA UNDECIMA

Sagredo: Well, Simplicio, have you made any progress toward obtaining a proof of the proposition that any quadrilateral having the sum of one pair of opposite sides equal to the sum of the other pair is circumscribable about a circle?

Simplicio: I am afraid I have not, and until you convince me otherwise, I must regard this "proposition" as a doubtful conjecture. My difficulty is that, while we can obviously assign the individual side lengths a, b, c, d of a quadrilateral, so as to have $b + d$ equal to $a + c$, such an assignment does not fix the shape or size of the quadrilateral. The angles and the area may be altered merely by lengthening or shortening a diagonal at pleasure.[163] I find it difficult to see how the requirement $b + d = a + c$ could be sufficient to ensure circumscribability when it is insufficient to determine the proportions of the figure itself.

Sagredo: I admit that the *size* of the inscribed circle will be changed if the angles are altered. We have only to think of the simplest case, namely, that in which the equality of $b + d$ and $a + c$ is ensured by making a, b, c, d all equal. The quadrilateral will then be a rhombus (Fig. 74). As the smaller angles increase from zero up to right angles, the radius of the incircle will increase from zero to half the side length a.

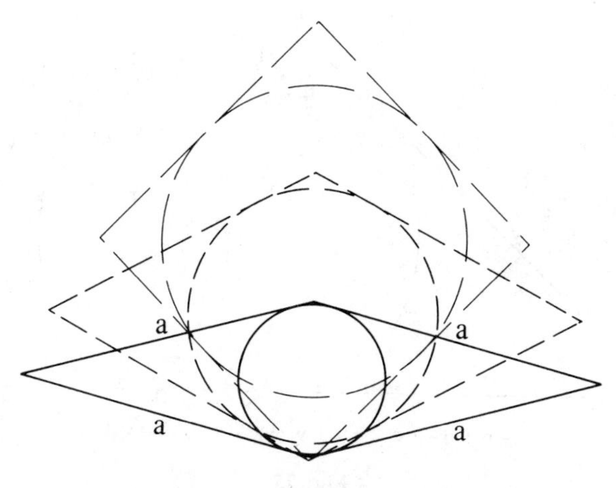

Fig. 74

Salviati: Yes, the radius is easily shown to be given by the equation $r = \frac{1}{2}a \sin A$, where A is the measure of *any* angle of the rhombus. And now, Sagredo, I think that you are ready to turn to the general case, to judge from the diagrams that I see that you have been preparing.

Sagredo: If I am not mistaken, I have found a satisfactory proof. May I submit it to you both for your judgment?

Salviati: Please do.

Sagredo: I begin by noticing that it is easy to find a circle to touch any *three* sides of any quadrilateral. Whether this circle will also touch the fourth side remains to be determined. If J be the point of intersection of the bisectors of angles A and B of a quadrilateral $ABCD$, then J is equidistant from the sides DA, AB, BC, as shown in Fig. 75. Let r units be the common measure of the equal perpendiculars JH, JE, JF. We have now to show that the distance of J from the fourth side, CD, is also measured by r. It has taken me most of our pause period to see how this can be demonstrated. If JG is constructed perpendicular to CD, the congruency of triangles CJG and CJF cannot be proved. And to suppose that $JG = r$, and to attempt to prove that $JG \perp CD$, also gets us nowhere. If we knew that any *two* of the following were true, then it would be easy to show by congruent triangles that the third property is present also:

(i) $JG \perp CD$, (ii) $JG = r$, (iii) $CG = y$.

Simplicio: But we cannot assume—that is, we cannot claim to set up by construction—more than *one* of these relations.

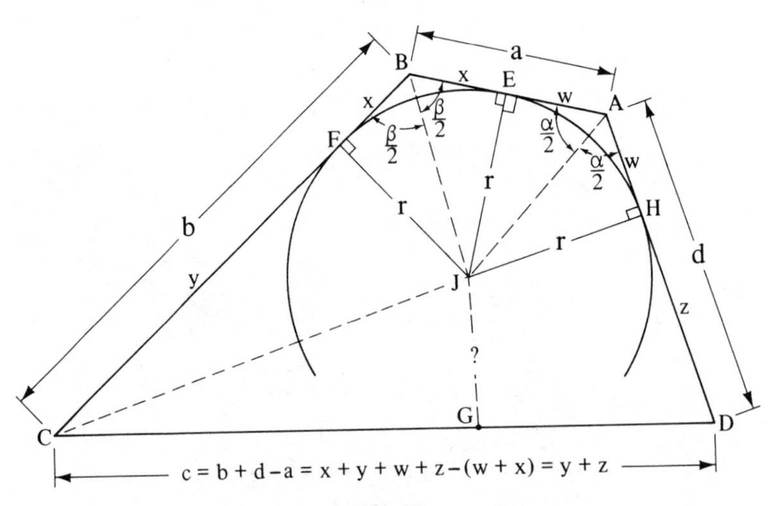

Fig. 75

Sagredo: True; that was just my difficulty. But then I noticed that I had made no use of the data demarcating this particular species of quadrilateral from all others. Referring to Fig. 75, you will notice that if we make CG equal to y units, then it follows from the equality of $a + c$ and $b + d$ that $GD = z$, and so, by joining *HE, EF, FG, GH*, the whole figure becomes divided into four isosceles triangles and an inner quadrilateral *EFGH* whose angles may readily be dimensioned in terms of the angles, α, β, γ, δ, of quadrilateral *ABCD*. As soon as this is done, it is found that each pair of opposite angles of *EFGH* sums to $\frac{1}{2}(\alpha + \beta + \gamma + \delta)$, that is, to half of four right angles (Fig. 76). Hence EFGH has its opposite angles supplementary; therefore it is a cyclic quadrilateral.

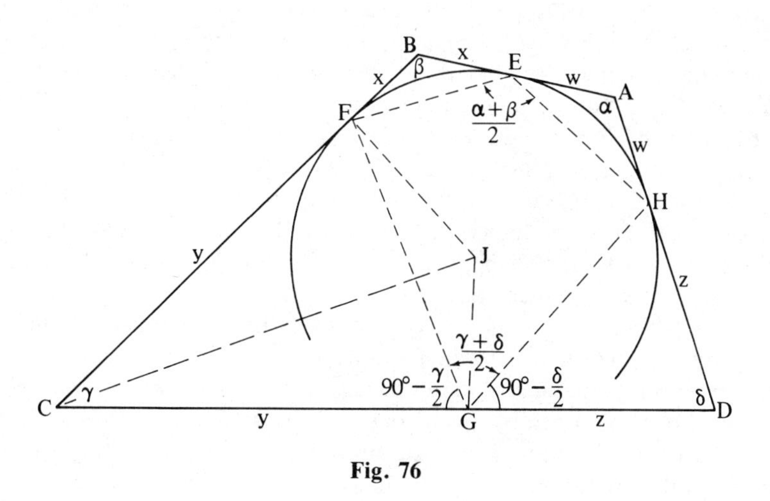

Fig. 76

Simplicio: And this means that the circle passing through *H, E,* and *F*, will also pass through *G*. Oh, well done, Sagredo!

Sagredo: It remains to show, however, that this circle only *touches*, and does not cut, side *CD*. Now that we have $JG = JF = r$, the congruency of triangles *CJG* and *CJF* is no longer in doubt; three sides of the one are known to be respectively equal to three sides of the other. Consequently, $\angle JGC = \angle JFC = 1$ right angle, showing that the circle touches the fourth side as well as the other three, and so is truly inscribed. Thus it is established that $b + d$ equal to $a + c$ is a sufficient condition for the circumscribability of a quadrilateral *ABCD*.[164] That it is also a necessary condition is known from the original proposition, which you yourself proved, Simplicio (p. 172), or from the contrapositive of that proposition.

Simplicio: I am indeed obliged to you, Sagredo, for demonstrating so convincingly that, in spite of all the changes of shape possible without disturbing the assumed equality of $b + d$ and $a + c$, the circumscribability property is still preserved.

Salviati: You might also, at your leisure, go over the slightly modified proof appropriate to *concave* quadilaterals, satisfying yourselves that essentially the same argument is applicable when you work from a diagram in which one of the interior angles of $ABCD$ is reflex in the manner of the second diagram of Fig. 68. But I must not fail to congratulate you, Sagredo, for the illuminating proof you have shown us. This circumscribability condition suggests to me further, related investigations that we might pursue sometime. I notice, for example, that the requirement that $BC + DA$ be equal to $AB + CD$ implies (and is implied by) the relationship $DA - AB = CD - BC$, and this makes me think of the locus of a point moving in a plane so that the difference of its distances from two fixed points is constant. You doubtless remember that, whereas if PD plus PB is constant, the locus of P is an ellipse—for D, B fixed points, called the foci—when PD *minus* PB is constant, the locus of P is a *hyperbola*. More precisely, the plane locus of P, for $PD - PB$ equal to k, where D, B are fixed points and $k > 0$, is the branch nearer to B of the hyperbola with transverse axis of length k, the other branch of the same hyperbola being the locus of P for $PB - PD$ equal to the same positive constant k, as indicated here (Fig. 77). The foci B, D and any two positions of P on the one branch of the hyperbola will therefore constitute the vertices of a circumscribable quadrilateral. I shall only add that, associated with every convex circumscribable quadrilateral, at least if it has no parallel sides, is a concave quadrilateral obtained by producing the sides so that they meet in two more vertices (Fig. 78). The same inscribed circle is then shared by the two quadrilaterals, and also by the two triangles that overlap within the region of the original quadrilateral. I think you can see that to examine these and related matters could well involve us in yet another lengthy investigation.

Sagredo: Yes, especially if we were to try to extend the results from quadrilaterals to other polygons.

Simplicio: Perhaps, then, we may leave these things for another day.

Sagredo: Yes, but let me just remark that I notice that in Fig. 78 there are actually five figures—two quadrilaterals, two triangles, and the circle—any two of which have their areas in the same ratio as their perimeters.

Simplicio: Oh yes! I do see that this follows from Theorem A.

Sagredo: You are right, Simplicio, and this reminds me that we have still to return to the consideration of whether there can be found figures other than those covered by Theorem A, which nevertheless have equal area and perimeter ratios.

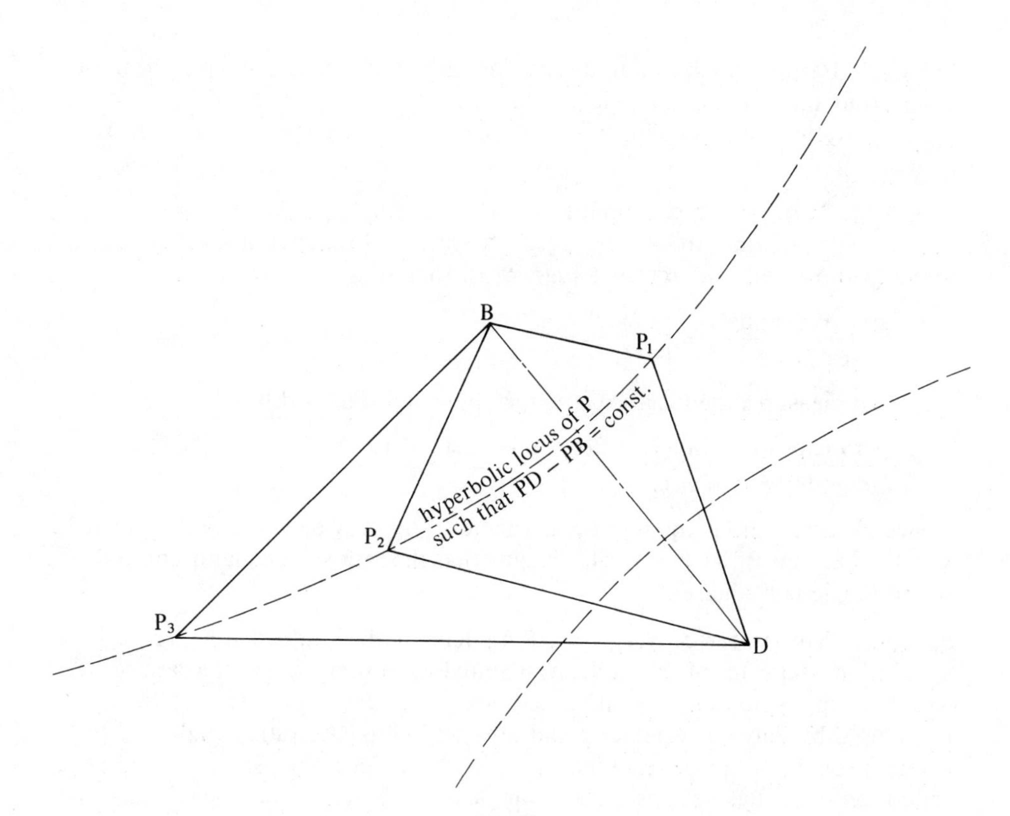

Fig. 77

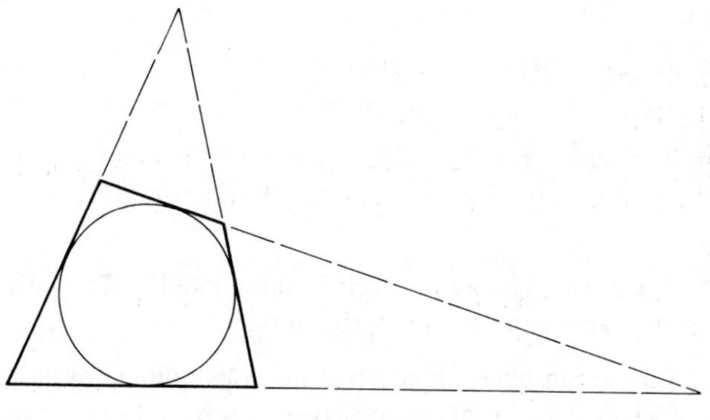

Fig. 78

Salviati: Earlier, you began to discuss the case of a rectangle and a circle in this connection, did you not, Sagredo? Tell me now, do you think that equal area and perimeter ratios could possibly hold between a rectangle (other than a square) and a circle?

Simplicio: You are very thoughtful, Sagredo; remember how you pointed out that the only circumscribable rectangles are squares. I thought then that you were going to show, with reference to Fig. 69(ii), that

$$\frac{\text{area of rectangle}}{\text{area of circle}} = \frac{(kd)\cdot d}{(\pi/4)d^2} = \frac{4k}{\pi},$$

where kd measures the length of the rectangle and d its width. Also that

$$\frac{\text{perimeter of rectangle}}{\text{circumference of circle}} = \frac{2d(k+1)}{\pi d} = \frac{2(k+1)}{\pi}.$$

Hence the area ratio is equal to the perimeter ratio provided, and only provided, $k + 1 = 2k$, that is, iff $k = 1$. This means that the ratios are equal if and only if the rectangle is a square.

Sagredo: This is perfectly correct, Simplicio, if the width of the rectangle is equal to the diameter of the circle, as it suited us to suppose was the case earlier. But that supposition must be discarded now, if we are to do justice to the question raised by Salviati: whether equal area and perimeter ratios might not hold between some nonsquare rectangle and a given circle. Look, Simplicio, suppose for convenience that we consider a rectangle of fixed shape, one twice as long as it is wide will do. If w denotes the measure of the width, then the length, the perimeter, and the area will be measured by $2w$, $6w$, and $2w^2$, respectively. Now let us see what happens when we *impose* the equality of the area and perimeter ratios between this rectangle and a circle of radius r. We make no assumption in advance concerning the relative sizes of w and r, but now that we are dealing with fixed shapes we can expect to determine the value of the ratio $w : r$ in order that it will be true that

$$\frac{\text{area of rectangle}}{\text{area of circle}} = \frac{\text{perim. of rectangle}}{\text{circum. of circle}}.$$

We have, then: $2w^2/\pi r^2 = 6w/2\pi r$, that is $2w/r = 3/1$, if we ignore the uninteresting possibility that $w = 0$. Thus we are left with the result that w must equal $1\frac{1}{2}$ times r.

Simplicio: So, for a rectangle and a circle looking like Fig. 79, with $w : r = 3 : 2$, the areas *are* in the same ratio as the perimeters!

Sagredo: You know, Simplicio, it dawned upon us quite a long time ago, even before we got to Theorem α, that, once two shapes have been specified, we may simply *equate* the area and perimeter ratios in order to determine the *relative sizes*

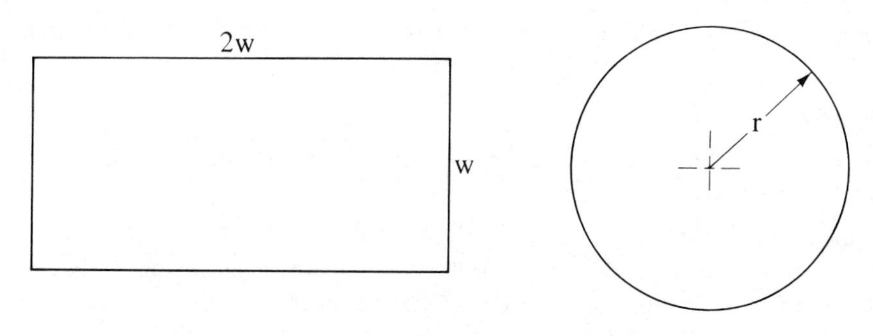

Fig. 79

of figures of these shapes necessary and sufficient to ensure that $A_2 : A_1 = p_2 : p_1$. (See pp. 130–134.) But we had practically to discover this all over again, so preoccupied had we become with circumscribable polygons.

Simplicio: Yes, it does seem that Theorem A (p. 166), the generalization of Theorem α (p. 140), temporarily became something of a ruling theory, and our flexibility of thought suffered accordingly. It is good that you have recovered the lost insight.

Sagredo: It was, I should say, under the stimulus of Salviati's direct question concerning the rectangle and the circle that this understanding was revived. I should now like to apply the principle more generally. Suppose, for example, that we have a figure like Fig. 80, with an area of 3 square units and a perimeter of 5 units, say.

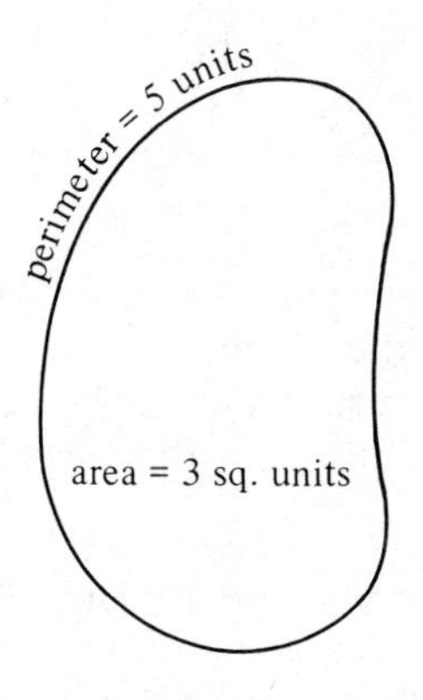

Fig. 80

Salviati: Better make it a perimeter of say 7 units, Sagredo, at least if your area and perimeter units are to be consistent.

Sagredo: Why so?

Salviati: Just because it would be impossible to have a figure of the area and perimeter you suggest.

Sagredo: Oh, I think I see what you mean: a string of length 5 units is not long enough to enclose an area as large as 3 square units. I suppose that is the objection. Let me see, even for a perfectly circular boundary of length 5 units, the area would be . . . Well: $A = \pi r^2$, and $2\pi r = C$, thus $A = \pi(C/2\pi)^2$, that is, $A = C^2/4\pi$. With $C = 5$, this gives $A = 25/4\pi$, indicating an enclosed area of only about 2 square units. Or, if we keep to the supposition of an area of 3 square units, then, from $C = \sqrt{(4\pi A)}$, the perimeter is found to exceed 6 units.

Simplicio: I do not doubt that it is the circular form that encloses the greatest area for a given perimeter—or encloses a given area with the least perimeter—but I should like to see this proved.

Salviati: A satisfactory proof has not been at all easy to come by, but by all means, let us try to find time later to discuss the matter.

Sagredo: Shall I continue?

Simplicio: Please do.

Sagredo: Suppose then, as in Fig. 81, that the first of two figures has an area of 3 and a perimeter of 7, while the second has an area of, say, only 2, for a perimeter of 14. It is quite obvious that the area and perimeter ratios are very different; we have $A_2 : A_1 = 2:3$, whereas $p_2 : p_1 = 14:7 = 2:1$. Nevertheless, simply by altering their relative size, figures of these same two shapes can be found for which the ratio of the areas is equal to the ratio of the perimeters. If we leave the first figure unchanged and enlarge the second figure lineally in the ratio $c:1$, then its area will be increased in the ratio $c^2:1$, since the areas of similar figures are to one another as the squares of corresponding linear dimensions. Let us express this by $p_2' = 14c$, $A_2' = 2c^2$, then we may readily determine the required value of c for which $A_2' : A_1 = p_2' : p_1$.

Simplicio: Now I see clearly what you are about. As in the case of the rectangle and the circle, you are taking the equality of the area and perimeter ratios as a datum, and then you determine the value of the factor c required in order that this equality holds true. Thus in this case the area ratio $2c^2/3$ is to be equated to the perimeter ratio $14c/7$.

Sagredo: Quite so. Rejecting the zero root (we are not interested in having either of the figures shrinking to a point!), the remaining value of c, for which $A_2' : A_1 = p_2' : p_1$, is found to be $c = 3$. Then each of the area and the perimeter ratios has the value $6:1$. I seem to remember, Salviati, your telling us a long while

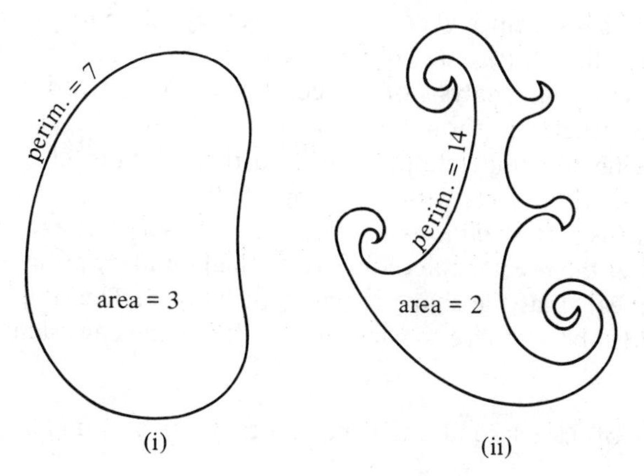

Fig. 81

ago (p. 132) that we could arrange for the ratio of the areas of two figures of any given shapes to be equal to the corresponding ratio of the perimeters by appropriately enlarging the figure with the less advantageous shape or by reducing the size of the other. But only now do I fully appreciate the significance of your earlier remarks.

Simplicio: Well, at least now we understand that whatever shape is assigned to each of two figures, there is one and only one relative size for which the ratio of the areas is equal to ratio of the perimeters.

Sagredo: I think it will be worthwhile trying to show this in a general way, and perhaps, at the same time we will be able to give precision to the notion of a more or less advantageous shape. For figures of any given shape (distinguished by subscript $_1$, say) the area varies directly as the square of any chosen linear dimension, for example the perimeter. Thus, $A_1 = k_1 p_1^2$. And likewise for figures of any other fixed shape, $A_2 = k_2 p_2^2$. Since the constants k_1 and k_2 measure the areas of figures of the respective shapes having *perimeters of unit length*, they are characteristic of these shapes, and so we might regard them as "shape parameters." For the two forms illustrated in Fig. 81, these shape parameters would be $k_1 = 3/49 \approx 0.0612$, and $k_2 = 2/196 \approx 0.0102$. In general, the area ratio

$$\frac{A_2}{A_1} = \frac{k_2 p_2^2}{k_1 p_1^2}, \quad \text{and this may be written} \quad \frac{A_2}{A_1} = \left\{\frac{k_2 p_2}{k_1 p_1}\right\}\frac{p_2}{p_1},$$

showing that the area ratio is equal to the perimeter ratio if and only if $k_2 p_2 / k_1 p_1$ is equal to unity, that is, iff $p_2/p_1 = k_1/k_2$.

Simplicio: So you have shown that the area ratio is equal to the perimeter ratio only where each of these ratios is equal to the inverse ratio of the shape parameters. Let me check this in the case of the shapes represented in Fig. 81.

Yes, I see that this is so, since $k_1 : k_2 = 6 : 1$, precisely the common value of $A_2 : A_1$ and $p_2 : p_1$ when these ratios are equal, as you have shown. I believe that I have correctly followed your analysis of the combined effect of differences in shape and in size. Nevertheless, I should like to consider these effects separately for the sake of clarifying my understanding still further. Please tell me whether the following formulation meets with your approval:

(1) *The effect of variations in size when the shape is held constant.* The larger the figure, the greater the economy of the boundary as an area container, assuming, I repeat, that the shape remains unchanged. The simplest illustration of this is provided by a square of varying size. Assuming consistent units, we find that

A square of area 1 square unit has a perimeter of 4 units,
" " 4 " " 8 " ,
" " 16 " " 16 " ,
" " 64 " " 32 " ,

and so on. Working our way back to smaller and smaller squares:

A square of area $\frac{1}{4}$ square unit has a perimeter of 2 units (side $\frac{1}{2}$ unit),
" " $\frac{1}{16}$ " " 1 unit (side $\frac{1}{4}$ unit),

and so on. The area of a square of unit perimeter being $\frac{1}{16}$ sq. unit, we may record the value of the "shape parameter" for the square form as 0.0625. A comparison of this value with the values corresponding to other shapes leads naturally to the following.

(2) *A measure of the effect of variations in shape when the perimeter is held fixed.* This is most conveniently accomplished if all figures are standardized so as to have unit perimeter; then a comparison of their areas reflects only the differences in their shapes, unconfounded by alterations of scale. For example, from the formula $A = C^2/4\pi$, for the area of a circle of circumference C, we find that the area is $1/4\pi$ square unit when the perimeter is of unit length.

Sagredo: You have summed up the matter admirably, Simplicio. Do you not agree, Salviati?

Salviati: I am as satisfied as you both appear to be. And if you wish to have the approximate value of the shape parameter for a circle, I have already calculated it as 0.0796.

Simplicio: Perhaps it is not wasting words on too trivial a matter to remark upon the fact that the number of area units measuring the content of a square is equal to the number of length units in the perimeter only when the side is 4 units long.

Sagredo: The uniqueness of this result is best appreciated by noting that the only nonzero value of s satisfying the equation $s^2 = 4s$ is $s = 4$. But to avoid any misunderstanding, it should be added that while the square of side 4 units is unique among squares in having its area and perimeter measures equal, it is certainly not absolutely unique in this regard, since for any shape whatsoever, there will be some particular size for which these measures are equal.[165] For example, since a 3-4-5 triangle has an area of 6 and a perimeter of 12, it is obvious that by doubling the linear dimensions to obtain a triangle with sides 6, 8, 10 units, we shall have an area of 24 as well as a perimeter of 24.

Salviati: And a triangle with sides 5, 12, 13 units also has equal area and perimeter measures.

Simplicio: I can see, now that you point out the case, that the area of such a (right-angled) triangle is 30 square units and the perimeter is 30 units.

Sagredo: I think, Salviati, that you have something more in mind in mentioning this example. Having already calculated that the incircle of a 3-4-5 triangle has a radius of 1 unit (p. 164), we are in a position to observe that what is common to a square of side 4 and a triangle with sides 6, 8, 10, is that each has an incircle of radius 2 units. And I feel confident, without yet having performed the calculation, that the incircle of a 5-12-13 triangle will also have a radius of 2 units. The circle of this particular size constitutes a link between the members of an indefinitely large class of rectilinear figures that have equal measures of area and perimeter.

Simplicio: I realize that you are referring to all the circumscribable polygons having incircles of radius 2 units. Yet, I feel the need of a fuller explanation than you have given.

Sagredo: From our previously obtained result that the area of any circumscribable polygon is given by the formula $A = \frac{1}{2}pr$, it immediately follows that when $r = 2$, $A = p$, and conversely, when $A = p$, $r = 2$.

Salviati: You have responded well, Sagredo. We can, of course, go beyond circumscribable polygons and beyond rectilinear figures altogether. As you have already correctly asserted, for any shape whatsoever, there will be some critical size for which $A = p$. I might mention just one nonrectilinear example: for a figure bounded by the circumferences of circles of radii 6 and 8 units, respectively, with centers not more than 2 units apart, the measure of the area is equal to the measure of the perimeter.

Simplicio: Allow me to draw the diagram (Fig. 82). The area is given by $A = \pi(8^2 - 6^2)$, which comes to 28π square units. The perimeter is given by $p = 2\pi(8 + 6)$, which comes to 28π units.

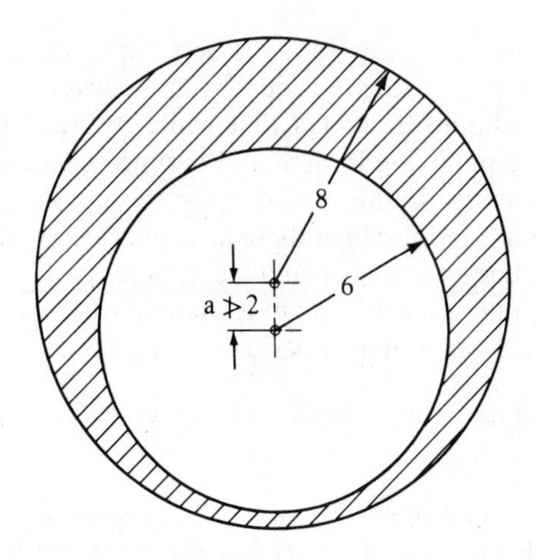

Fig. 82

Sagredo: I cannot help being reminded by this example of the construction that was implicit in one of Valerio's diagrams, the one that our first author discussed in the *Discorsi*.[166] This depicted a cone, a hemisphere, and a cylinder all having a common axis and with bases equal to each other. The solids were disposed as indicated in Fig. 83. In the right-angled triangle ABC, $CA = CD = r$, the radius of the circular section of the cones; $CB = R_i$, the radius of the section of the hemisphere at the same height; and $AB = AE = CF = R_o$, the constant radius of the cylinder. Now the horizontal cross-section shown in the lower view may be regarded quite apart from the three-dimensional context represented in the elevation or axial section drawn above it. Triangle HJK in the lower view is congruent to triangle ABC in the elevation, having sides $HK = r$, $JK = R_i$, $JH = R_o$. Hence the initial construction of triangle HJK, right-angled at K, will enable the three circles to be drawn with radii r, R_i, R_o, such that $R_o^2 - R_i^2 = r^2$, or $\pi(R_o^2 - R_i^2) = \pi r^2$, thus ensuring that the area of the outer annulus is equal to the area of the central circle.[167]

Simplicio: So the two areas shaded in the lower view are equal to each other, for all that their shapes and their perimeters are so different.

Sagredo: Exactly so. If, for example, the right-angled triangle were taken with $HK:KJ:JH$ as $5:12:13$, then the perimeter of the annulus would be to the perimeter of the equiareal circle as $(12 + 13):5$, that is, as $5:1$.

Simplicio: Ah, yes! I now recall how you, Salviati, presented us with a paradox arising out of such relations, a paradox that you called a marvel, if not a miracle. Since the circular section of the cone and the coplanar annular "ribbon" between the hemisphere and the cylinder are of equal area whatever the height of the sectioning plane, you argued that we should therefore not deny their equality in the

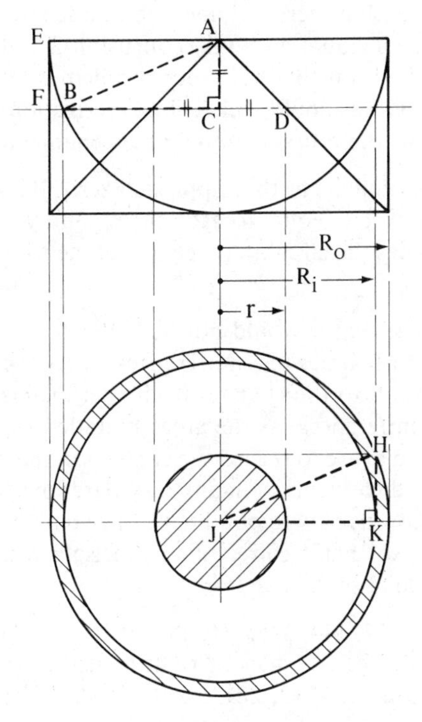

Fig. 83

end when the section of the cone diminishes to a point (the apex) and the annular section simultaneously becomes a circular razor's edge. And so you were led to call the circumference of the circle of radius AE, and the point A, *equal,* on the ground that they were "the last remnants and vestiges left by equal magnitudes."[168]

Salviati: I have to admit that not everything I said on the occasion of our original discussion has given satisfaction to those of our successors whose opinions we should respect, but at least they have been generally appreciative of our efforts. The view prevailing among mathematicians of the present age is that the paradoxes of the infinite are resolvable, whereas, as you well know, Galileo was convinced that in these matters there will always remain something "incomprehensible to our finite understanding."[169]

Sagredo: I am not proposing that we enter into an extensive discussion of this whole subtle subject, but it does occur to me that the context provided by our present investigation allows us to view the particular paradox of the apex of the cone and the razor edge of the bowl in a way that will remove mystery instead of fostering it. Admittedly, a point and a line each have zero area—and a point, a line, and a surface are each of zero volume—so in one sense these are equal, but in other senses they are unequal and only confusion is induced if the necessary distinctions are not drawn. Furthermore, our problem is not merely one of static equalities or inequalities; we should seek a characterization that will do justice to the relations involved as the cross-sectional areas change together.

Simplicio: And, in particular, as they approach zero. But we have seen that the two areas ever remain exactly equal to each other, and yet when they disappear together, the one becomes the circumference of a circle while the other becomes a point.

Sagredo: I am not sure that it would not be better to say that the annulus becomes a coincident *pair* of circles (circumferences). In any case, I think that if we were to give consideration to the changes in the perimeters, say by tabulating a few values of the circumferences, as the areas bounded by them approach zero, we should no longer find cause to regard the changes in the variables, or the limiting values to be assigned to them, as in any way paradoxical.[170] Salviati, would you be so good as to show us how to obtain one or two pythagorean triangles having small acute angles so that I can explain precisely what I have in mind by means of a simple calculation.

Salviati: Such triads as 17, 144, 145; 31, 480, 481; 39, 760, 761; 55, 1512, 1513 provide integral measures of the sides of right-angled triangles. Are these sufficient? Or shall I mention 75, 2812, 2813; 101, 5100, 5101; . . .?

Sagredo: Stop, stop, Salviati, please! You know that I have always had the highest regard for your profound knowledge and wisdom, but it would astonish me to discover either that you had memorized such information or that you were a lightning calculator.

Simplicio: Yes, Salviati, by what secret are you enabled to recite this impressive list of values almost as though you were just reading them out?

Salviati: The "secret" as you call it is something that it would be better for you to discover for yourselves. But I suggest that we delay consideration of this further problem, at least until you have shown us, Sagredo, what it is you wish to do with these triangles.

Sagredo: Very well. I did manage to copy down the triads as you gave them to us. Let us suppose that the sides of triangle HJK are as 2,813 : 2,812 : 75. Then, since HJ, of length equal to R_o units, is supposed to be constant,

$$JK = R_i = \frac{2{,}812}{2{,}813}R_o, \quad \text{and} \quad KH = r = \frac{75}{2{,}813}R_o.$$

Consequently, the area ratio, annular ribbon to central circle, given by $\pi(R_o^2 - R_i^2) : \pi r^2$, may be written

$$\left(R_o + \frac{2{,}812}{2{,}813}R_o\right)\left(R_o - \frac{2{,}812}{2{,}813}R_o\right) : \left(\frac{75}{2{,}813}R_o\right)^2,$$

which comes to $(2{,}813 + 2{,}812)(2{,}813 - 2{,}812) : 75^2$, that is, $(5{,}625)(1) : 5{,}625$, or $1 : 1$, the value we knew it must have. The corresponding perimeter ratio is

$$
\begin{aligned}
&2\pi(R_o + R_i) : 2\pi r \\
&= \left(R_o + \frac{2{,}812}{2{,}813}R_o\right) : \frac{75}{2{,}813}R_o \\
&= (2{,}813 + 2{,}812) : 75 \\
&= 5{,}625 : 75 \\
&= 75 : 1.
\end{aligned}
$$

Simplicio: So, although the areas of the two figures are equal, the less compact one has 75 times the perimeter of the more compact one![171]

Sagredo: You are right, Simplicio. And this perimeter ratio can be *increased* as much as we like, simply by choosing the plane containing the two cross-sections sufficiently close to the apex of the cone or, what amounts to the same thing, simply by taking r, R_i, R_o to be proportional to the three sides of a sufficiently acute right triangle. So why should we want to claim that the last traces of these *perimeters*—the apex of the cone and the razor's edge—are equal?

Simplicio: When you put it that way, we should not be tempted to make such a claim. And I have to admit that we do not have to think of the apex and the razor's edge as "the last remnants and vestiges" left by the pair of cross-sectional *areas* that had remained equal to each other while they approached zero together.[172] Indeed, it now seems to me that to do so would be no more appropriate than to regard the memory of the last survivor of a battle as the last remnant of the fighting itself.

Sagredo: We may speak of things in different ways as suited to our various purposes; the things themselves are not changed by our choice of words, though apposite choices enable us to see now one aspect of them and now another.

Simplicio: You remind me of my difficulties with Zeno's paradoxes. Even Aristotle's explanations have not quite sufficed to set my mind permanently at rest.

Salviati: Well then, after that confession, we are perhaps ready to proceed to the subsidiary problem of the generation of triads providing integral side measures for increasingly acute right triangles.

Sagredo: Yes, and I have the suspicion that in some way the calculations just carried out might indicate what we are looking for. Yes indeed, looking back I now see that $2{,}813 + 2{,}812 = 75^2$. Let us examine the other triads: 17, 144, 145 was your first one. Why yes! $17^2 = 289 = 144 + 145$. Next you gave us 31, 480, 481, and here again the corresponding relation holds: $31^2 = 961 = 480 + 481$.

Simplicio: Also, $3^2 = 9 = 4 + 5$, and $5^2 = 25 = 12 + 13$.

Sagredo: And $7^2 = 49 = 24 + 25$.

Simplicio: Is a triangle with sides 7, 24, 25 right angled then?

Salviati: Just before you answer that for yourself by testing the relation between the squares of the sides, let me acknowledge that you have indeed found the procedure that I used in order to find so readily the triads you asked for. Notice that this method could have been discovered historically even before the relation between the squares of the sides was known. In trying to understand how the *original* discovery of the "Pythagorean" relation between the sides of a right triangle might have been made, the modern student too easily begins by drawing squares on the sides of such a triangle. But suggestions requiring that this would have been done initially by a pioneer investigator are inherently implausible and could only appear satisfactory to those relying upon a hindsight conditioned by proofs devised long afterwards. Even at the *arithmetical* level, the generalization that we write as $a^2 + b^2 = c^2$ does not seem to me to be quite the most likely to have been *first* conjectured by an ancient mathematician reflecting upon such empirically discoverable special cases as 3, 4, 5 and 5, 12, 13.[173] If these instances had become known as a result of measurement of the diagonals of rectangles 3 by 4 and 5 by 12, our hypothetical innovator might well have proceeded to conjecture that since $3^2 = 4 + 5$, and $5^2 = 12 + 13$, it would be worth considering $7^2 = 49 = 24 + 25$, and $9^2 = 81 = 40 + 41$, and so on, as possibly indicative of further rectangles having integral sides and diagonals. And since these would appear to be verified by actual measurement, the general procedure might have come to be accepted as a rule of thumb in a corpus of predemonstrative "mathematics," especially if the further relation, involving the squares of the measures of the side lengths, were also noticed.

Sagredo: Yes, I can see that the discovery of an apparently unfailing correspondence between one or more arithmetical recipes and such a noteworthy characteristic of geometrical form could have sustained the faith that a valid recipe had been found. Yet I wonder whether we ought to regard rules of thumb of this kind as constituting mathematics at all.

Salviati: Such rules were probably associated with other items that we would be happier to accept as genuinely mathematical. Perhaps the crucial thing is whether the rules were simply dogmatically asserted and accepted on faith or whether, on the other hand, they were recognized as being in need of justification and whether this need was in some degree met. I mean insightful justification that goes some way toward exposing necessary connections between various apparently true assumptions. There would have been numerous levels of insight possible and it may be supposed that many of these occurred, only to be lost again; they would have been more difficult to preserve than the rules themselves. Perhaps at historical stages now regarded as predemonstrative, some partial proofs *were* devised and transmitted in the form of oral glosses on the rules and diagrams.[174] But I am talking too much; the time is late and you have something to say, Simplicio?

Simplicio: Well, to return to the particular generating procedure that you used, I notice that it is always an *odd* number that is squared and that this square is divided into two parts *differing by just one unit.*

Sagredo: Yes, Salviati, what if your primitive mathematician had become acquainted with the 8 by 15 rectangle and its diagonal of 17 units?[175]

Salviati: A good question! $8^2 = 64$, whereas 15 and 17, which differ by 2, add up to 32, which is but *half* of 64. Now a further case would be known, namely the 6 by 8 rectangle with its diagonal of 10, and here $6^2 = 36$, while $8 + 10 = 18$, which again is half of the square of the measure of the shortest side.

Sagredo: The general procedure appears obvious. Let me start with 12: split half of 12^2 into two parts differing by 2, namely, ($\frac{1}{2}$ of 72) $\pm$ 1, to obtain 12, 35, 37 as our triad.[176]

Salviati: Yes, though even if we were to proceed just as we did with the squares of the odd numbers we should also be led to valid triads. For example: $8^2 = 64$, and this, if divided into two parts differing by a unit, gives $31\frac{1}{2} + 32\frac{1}{2}$; hence 8, $31\frac{1}{2}$, $32\frac{1}{2}$, or as we should prefer to give it, 16, 63, 65, measure the sides and diagonal of a rectangle, or the sides of a right triangle. You may care to consider the systematic development of these ideas at your leisure.

Simplicio: The procedures you have indicated appear to be convenient in practice, but they certainly require validating.

Salviati: Of course you are right, Simplicio. I interrupted you a few minutes ago when you were going to check on one individual case, namely, the triad 7, 24, 25 generated by writing $7^2 = 49 = 24 + 25$.

Simplicio: $7^2 = 49$, and $24^2 = 576$, while $25^2 = 625$; therefore, since $49 + 576 = 625$, a triangle with sides 7, 24, 25 will be right angled (by Euclid I, 48).

Sagredo: Allow me to set out this verification differently:

$$25^2 - 24^2 = (25 + 24)(25 - 24) = (49)(1) = 7^2.$$

Not only does this simplify the calculation; it also points, I think, to a more illuminating expression of the general method and will perhaps lead us to a general demonstration of its validity. The square of an *odd* integer is itself odd, and it may be partitioned into a pair of consecutive integers of which the larger is odd. We see this by representing the form of the first procedure by $(2n + 1)^2 \equiv 4n^2 + 4n + 1 \equiv (2n^2 + 2n + 1) + (2n^2 + 2n)$. The validation will be achieved if we can show that the difference of the squares of these last two integers does in fact simplify to the square of the originally chosen integer:

$$\begin{aligned} &(2n^2 + 2n + 1)^2 - (2n^2 + 2n)^2 \\ &\equiv (2n^2 + 2n + 1 + 2n^2 + 2n)(2n^2 + 2n + 1 - 2n^2 - 2n) \\ &\equiv (4n^2 + 4n + 1)(1) \\ &\equiv (2n + 1)^2. \end{aligned}$$

Simplicio: A very succinct proof, I am sure.

Sagredo: I have little doubt that the other forms of the procedure—for example, where we begin with an even number, a equal to $2p$, say, and obtain b, c differing by 2—may be validated just as readily. Yet it seems to me that we would need many forms of the method in order to obtain all the pythagorean triads.

Salviati: The method is an excellent one for obtaining more and more acute right triangles, which is the problem which you proposed initially. It will be found to provide a piecemeal and overelaborate approach to the problem of obtaining "all" pythagorean triads. Nevertheless it is a method that contains within itself the seeds of its own improvement. I will be very brief in indicating what I mean. If we were to seek all the pythagorean triads a, b, c, say, in which b, c differ by d, we would partition a^2/d into two parts differing by d, obtaining $a^2/2d \pm d/2$. Suppose, for example, we wished to draw up a schedule of pythagorean triangles for which $d = 8$, then a would be chosen so that a^2 would be divisible by 16; that is, a would be selected from the natural numbers divisible by 4. For example, if $a = 20$, the procedure gives $400/8 = 50$, which is partitioned to give 25 ± 4, giving 20, 21, 29 as a pythagorean triad. But the efficiency of this procedure is less than appears at first sight, because of the high frequency of triads of numbers obtained that are not prime to one another and that on reducing to their lowest terms are found to be repetitions of triads given with lower values of d. For example, with $a = 24$ and $d = 8$, we have $b = 24^2/16 - 8/2$, and $c = 24^2/16 + 8/2$, that is, $a = 24$, $b = 32$, $c = 36$, giving $a:b:c = 3:4:5$. Viewed overall, the procedure is correct but labored.

Sagredo: If I am not mistaken, the proof of the correctness of the method in its full generality is just as simple as the proof for any one of the forms corresponding to a particular value of d. For we have in general, as the measures of the sides, a, $b = a^2/2d - d/2$, $c = a^2/2d + d/2$, therefore

$$c^2 - b^2, \text{ equal to } (c + b)(c - b), = (2 \cdot a^2/2d)(2 \cdot d/2) = a^2.$$

Salviati: Yes, and notice that the method makes

$$\begin{aligned} a:b:c &= a:\frac{1}{2}(a^2/d - d):\frac{1}{2}(a^2/d + d) \\ &= 2ad:(a^2 - d^2):(a^2 + d^2). \end{aligned}$$

Expressed in the lowest terms, we write, $a = 2mn$, $b = m^2 - n^2$, $c = m^2 + n^2$, and specify that m, n be integers having no common factor and not be both odd, since such values of m, n can be shown to be necessary and sufficient for a, b, c to be integral and prime to one another.[177] This is the improved (and well-known) general method for generating pythagorean triads.[178]

XII

Let no one say that I have said nothing new; the arrangement of the subject is new. When we play tennis, we both play with the same ball, but one of us places it better. I had as soon it said that I used words employed before.

—Pascal

The value of men's works is not in the works themselves but in their later development by others, in other circumstances. We never know in advance whether a work will live. *It is a seed endowed with more or less vitality and it needs special conditions; even the frailest may be favoured by circumstances.*

—Paul Valéry

Mathematical writings may carry entailments which lay outside the mind of their author, and nevertheless exert a strong influence at a later date. Did not Renaissance mathematics sit under the Greeks for a century? . . . Was not Dedekind stimulated on his own avowal, and led to his number constructions, by Eudoxus of Cnidus?

—D. A. Steele

The unwritten rules of the game [of science] provide recognition and approbation for work which is imaginative and accurate, and apathy or criticism for the trivial or inaccurate . . . . Thus, it is the communication process which is at the core of the vitality and integrity of science. . . . The system of rewards and punishments tends to make honest, vigorous, conscientious, hardworking scholars out of people who have human tendencies of slothfulness and no more rectitude than the law requires.

—P. H. Abelson

HORA DUODECIMA

Salviati: Since we are so soon to be returned again to the shades, let us review our progress and see just which things most need to be discussed in our remaining hours.

Simplicio: As a matter of fact, Sagredo and I have been drawing up an outline of our main results and this has led us to notice that though we have established Theorem α (p. 140) and its generalization, Theorem A (p. 166), we have not so far generalized the corresponding three-dimensional theorem, Theorem β (p. 145), which concerned the regular polyhedra and their inscribed spheres.

Sagredo: Yes, instead of proceeding naturally to a Theorem B, we were diverted into a consideration of Theorem γ (pp. 146) concerning circumscribable prisms, and I now realize that *that* theorem was stated in an unnecessarily restricted form. In spite of our previous experience of unnecessarily restricted theorems and our explicit recognition that these occur to us because of habits of thought too much influenced by the attractions of symmetry and simplicity and the desire for convenience, we ourselves remained under the spell of the regular polyhedra (Theorem β) and the regular polygonal circumscribable prisms (Theorem γ). But now, reflecting on the matter from a little distance, it becomes clear that Theorem β may be generalized as follows:

> **Theorem B**
> *For any circumscribed polyhedron (regular or irregular) and its inscribed sphere (or for any two polyhedra circumscribable about the same sphere), the ratio of the volumes is equal to the ratio of the surface areas.*

Salviati: You have just mentioned Theorems β and γ and asserted that the first of these generalizes to Theorem B. Have you not noticed that Theorem B serves as a generalization of Theorem γ also, given only that it is allowable to assimilate the circumscribing cylinder to the circumscribing polyhedra, in a limiting sense?

Sagredo: Of course you are right, Salviati! Whereas previously, we enunciated Theorem γ in the following form:

> *Given any prisms, P_n, P_m, with regular polygonal ends (having n, m sides), circumscribable about a sphere S and a cylinder C, then for any two of P_n, P_m, S, and C, the ratio of the volumes is equal to the ratio of the surface areas,*

it became clear to me a few minutes ago that the restriction to regular polygonal-ended prisms is quite unnecessary and ought to be waived. Yet somehow I had not made the connection in my mind between the prisms of a generalized Theorem γ and the polyhedra of the generalized Theorem β.

Simplicio: If I remember correctly, the key insight used in proving Theorem β was the reducibility, or decomposability, of each of the regular polyhedra to sets of congruent pyramids having their bases coinciding with the faces and their apices meeting at the center of the polyhedron.

Sagredo: You are right, and the adaptation of the earlier proof to the more general Theorem B is doubtless obvious to you.

Simplicio: Let *a, b, c* . . . and so on denote the *areas* of the faces of any polyhedron circumscribable about a sphere of radius r. The component pyramids are no longer necessarily congruent, but the perpendicular height of each is still measured by r. Hence:

$$\text{Volume ratio, polyhedron to sphere} = \left(\frac{1}{3}ar + \frac{1}{3}br + \frac{1}{3}cr + \ldots\right) : \frac{4}{3}\pi r^3$$
$$= \frac{1}{3}r\left(a + b + c + \ldots\right) : \frac{1}{3}r(4\pi r^2)$$
$$= (a + b + c + \ldots) : 4\pi r^2,$$

which is the corresponding surface area ratio.

Sagredo: And for two polyhedra circumscribable about the same or equal spheres?

Simplicio: In obvious notation:

$$\frac{V_2}{V_1} = \frac{V_2/V_s}{V_1/V_s} = \frac{A_2/A_s}{A_1/A_s} = \frac{A_2}{A_1}.$$

I think, Sagredo, that in the case of Theorem A, you went further with an extended version.

Sagredo: So we did (p. 169). And likewise we may enunciate

An Extended Version of Theorem B (and Converse)

For any two circumscribable polyhedra, the ratio of the volumes, $V_2 : V_1$, is greater than, equal to, or less than the ratio of the surface areas, $A_2 : A_1$, according as the ratio of the radii of the respective inspheres, $r_2 : r_1$, is greater than, equal to, or less than unity. And conversely, the ratio of the insphere radii is greater than, equal to, or less than unity, according as $V_2 : V_1 \gtreqless A_2 : A_1$.

Simplicio: And the proof of that, Sagredo? Or are you now content to regard it as obvious?

Sagredo: It will be a good exercise to compose a proof, whether or not we regard the proposition as obvious. Suppose we denote by P_1 and P_2 two circumscribable polyhedra having as the measures of their respective volumes, surface areas, and insphere radii, V_1, V_2; A_1, A_2; r_1, r_2. Let P_2' denote a solid similar to P_2 and having corresponding measures V_2', A_2', r_2', such that $r_2' = r_1$.

Then, by Theorem B, as just proved, $V_2'/V_1 = A_2'/A_1$. (1)

And by the fundamental theorems for similar figures,

$$V_2/V_2' = r_2^3/r_2'^3 = r_2^3/r_1^3; \quad (2)$$
$$A_2/A_2' = r_2^2/r_2'^2 = r_2^2/r_1^2. \quad (3)$$

From (1) and (2),

$$(V_2'/V_1)(V_2/V_2') = (A_2'/A_1)(r_2^3/r_1^3). \quad (4)$$

From (3),

$$r_2^2/r_1^2 = A_2/A_2', \quad \text{hence,} \quad r_2^3/r_1^3 = (A_2/A_2')(r_2/r_1). \quad (5)$$

From (4) and (5),

$$V_2/V_1 = (A_2'/A_1)(A_2/A_2')(r_2/r_1) = (r_2/r_1)(A_2/A_1).$$

Thus $V_2/V_1 \gtreqless A_2/A_1$ according as $r_2 \gtreqless r_1$; and conversely, $r_2 \gtreqless r_1$ according as $V_2V_1 \gtreqless A_2A_1$.

Simplicio: I follow what you have done. Let me ask you, though, whether the omission of reference to the pair "a circumscribable polyhedron *and a sphere*" in your enunciation of the extended version of this theorem was deliberate.

Sagredo: I refrained from explicitly mentioning not only the sphere but also the circumscribable cylinder and the cone. Your query reminds me to state that these three kinds of solid are to be understood as included among the circumscribable polyhedra in so far as the applicability of Theorem B is concerned. This can be seen by considering the number of faces of suitably defined polyhedra to increase without limit.

Simplicio: I suppose you are regarding the cylinder as the limit of circumscribable prisms with regular polygonal ends, but I am not sure that I am clear about the other two cases.

Sagredo: Well, of course, the number of faces of a *regular* circumscribable polyhedron cannot be made as large as we choose; as you know, that number cannot exceed twenty![179] But we may start with *any* circumscribable polyhedron and truncate it by as many cutting planes tangent to the insphere as the polyhedron has vertices, repeating the operation over and over again in such a way that the solid approaches the sphere as a limit.

Simplicio: And the cone?

Sagredo: Circumscribable pyramids are a species of circumscribable polyhedra. If we consider as a subspecies, the right pyramids of height h, circumscribable about a sphere of radius r, and having regular polygonal bases circumscribable about a circle of radius ρ, then, given any two of h, r, ρ, the remaining dimension can be readily found. As the number of sides of the base is increased without limit the pyramid becomes, or approaches as a limit, a cone of height h and base radius ρ, circumscribable about the same sphere. I assume, Simplicio, that you have not been confused as a result of our initial discussion concerning the sphere, cone, and cylinder (Hora Prima). In *that* instance, the cone was *not circumscribable* about the sphere.

Simplicio: This is all clear to me now, and I have no doubt that the same relation that holds for the pyramids will hold also in the limiting case of the cone. Nevertheless, I should like to work through a particular example.

Sagredo: As you wish.

Simplicio: Then let us begin with a sphere of radius r equal to 1 unit, circumscribed by a cone of base radius ρ equal to 2 units. After first determining the height h of the cone, we will be able to calculate the ratio of the volumes of the two solids and to compare this with the separately calculated ratio of their surface areas. Working from the similar triangles indicated in the diagram, Fig. 84:

$$\{\sqrt{(h^2 + 2^2)}\}/(h - 1) = 2/1, \text{ from which } h^2 + 4 = 4h^2 - 8h + 4.$$

So $3h^2 - 8h = 0$; hence (rejecting the inadmissible zero root), $h = 2\frac{2}{3}$.

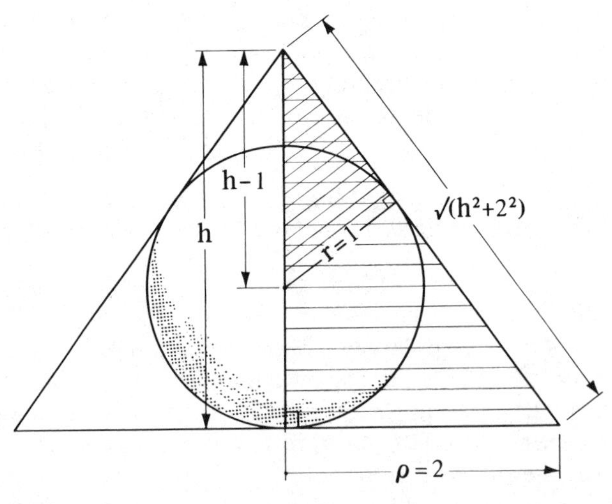

Fig. 84

$$\text{Volume ratio, cone to sphere,} = \frac{1}{3}\pi\rho^2 h \Big/ \frac{4}{3}\pi r^3 = 4 \times \frac{8}{3} / 4 \times 1 = 8:3.$$

$$\text{Slant height of cone} = \sqrt{\left\{2^2 + \left(\frac{8}{3}\right)^2\right\}} = \sqrt{\left(4 + \frac{64}{9}\right)} = \sqrt{\frac{100}{9}} = \frac{10}{3}.$$

Hence,

$$\text{Curved surface of cone}^{[180]} = \frac{1}{2} \times 2\pi \times 2 \times \frac{10}{3},$$

$$\text{Total surface} = \frac{20}{3}\pi + (\pi \times 2^2),$$

so that

$$\text{Area ratio, cone to sphere} = \pi(\tfrac{20}{3} + 4)/(4\pi \times 1^2) = 8:3.$$

Thus the volume ratio and the area ratio do have the same value.

Sagredo: Quite so, Simplicio. How clearly you have shown it. Such particular examples do enhance our familiarity with the generalizations that they illustrate.

Salviati: Well my friends, you have effectively removed the chief dispensable restrictions in the theorems you have discovered, or rediscovered. Perhaps now is the time to read to you a passage from a book that our author showed me during the last pause. As you know, his main labors consist in searching for the historical origins and in trying to follow the lines of development of various mathematical ideas. He is especially concerned to indicate where our present discoveries are really rediscoveries. Consequently he wishes me to read a passage to you from the 1851 English edition of Jacques Ozanam's "Recreations" (*Recréations mathématiques et physiques*); this passage is a direct translation of the corresponding item in the 1778 French edition:

What bodies are those, the surfaces of which have the same ratio to each other as their solidities?

This problem was proposed, in the form of an enigma, in one of the French Journals, entitled the Mercury, of the year 1773,

"Reponds-moi, d'Alembert, qui découvre les traces
Des plus sublimes vérités;
Quels sont les corps dont les surfaces
Sont en même rapport que leurs solidités?"

We do not find that d'Alembert condescended to answer this problem, for it may be readily seen, by those in the least acquainted with geometry, that two bodies well known, the sphere and the circumscribed cylinder, will solve it. Archimedes demonstrated long ago, that the sphere is equal to two thirds of that cylinder, both in surface and solidity, provided the two bases of the cylinder are comprehended in the former; and this is the answer which was given to the enigma in the following Mercury.

But we may go a little further, and say, that there are a great number of bodies which, when compared with each other, and with the sphere, will answer the problem also: such are all solids formed by the circumvolution of a plane figure circumscribed about the same sphere, and even all plane-faced solids, regular or irregular, that can be circumscribed about the same sphere; for the solidity of all these bodies is the product of their surface by the third of its radius.

Thus, the equilateral cone is to the inscribed sphere, both in surface and solidity, as 9 to 4.

The case is the same in regard to the sphere and the circumscribed isosceles cone; except that the ratio, instead of 4 to 9, will be different according to the elongation or oblate form of the cone.

If the sphere and the circumscribed cylinder possess this property, it is because the latter is a body produced by the circumvolution of the square about the great circle of the sphere, on an axis perpendicular to two of the parallel sides.

If the square and inscribed circle revolved around the diagonal of the square, the surface and solidity of the body, thus produced, would be to each other as $\sqrt{2}$ is to 1.

We shall here propose a similar problem:

What are those figures, the surfaces and perimeters of which are to each other in the same ratio?

The answer is easy: the circle and all polygons, regular or irregular, circumscriptible of it.[181]

Sagredo: Well, I suppose it would have been surprising if no one had found Theorems A and B before us. I notice that here there is a slightly different point of view. It is as though the solids of revolution were conceived of as prior to the polyhedra, but I think we should prefer to think of this the other way around since the underlying relation, volume = one-third surface area times radius of in-sphere, appears to be more fundamentally associated with the plane-faced circumscribable solids. Further, I am not altogether satisfied with the wording in one or two places. Just what are we to conclude from the reference to "all solids formed by the circumvolution of a plane figure circumscribed about the same sphere"? I suspect that if, for example, a semicircle of radius $2r$ were circumscribed about a circle of radius r (Fig. 85), and the whole configuration rotated about the axis of symmetry, the resulting hemisphere and sphere will *not* bear to each other equal volume and area ratios.

Simplicio: You say that you suspect that these ratios will not be equal. Allow me to carry out the calculation. In obvious notation:

$$V_H/V_S = \frac{2}{3}\pi(2r)^3/\frac{4}{3}\pi r^3 = 4:1;$$

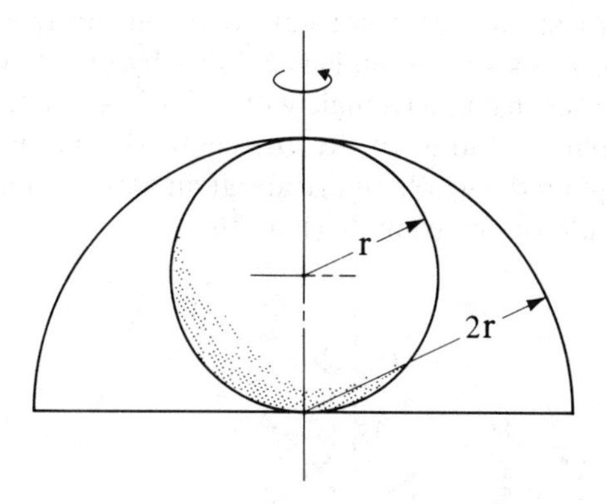

Fig. 85

whereas

$$A_H/A_S = \{2\pi(2r)^2 + \pi(2r)^2\}/4\pi r^2 = 3:1.$$

So your suspicion was justified.

Sagredo: Yes, I thought of the hemisphere as . . . No, that is not what I did! I simply intuitively resisted the idea that $V_H : V_S$ would be equal to $A_H : A_S$, without consciously analyzing the situation. But now, on reflection, I see that we might think of the hemisphere as the limit of a number of adjacent conical frusta with parallel bases. A typical set of such frusta would be generated by the rotation of a symmetrical polygon about its axis of symmetry, but such a polygon, which would lie close to the semicircle of radius $2r$, would *not* be circumscribable about the circle that generates the sphere of radius r.

Simplicio: I think I understand the significance of your remarks. The way in which you previously dealt with the cone in relation to the circumscribable pyramids comes to mind. I recall that it was because these pyramids were recognized as circumscribable polyhedra that they were seen to be subsumable under Theorem B.

Sagredo: Yes, that is what I meant by giving priority to the *plane*-faced solids circumscribable about the sphere.[182] And now, Simplicio, it is I who feel the need for a particular worked example, the better to understand the claim made in the foregoing passage concerning all those "solids formed by the circumvolution of a plane figure circumscribed about the same sphere." I notice that the examples mentioned all involved rotations about an *axis of symmetry* of the circumscribing figure that coincided with a diameter of the circle generating the sphere. I feel somewhat uneasy about the claim, if such were intended, that the volume ratio will be equal to the area ratio in cases where the circumscribing figure has no axis of symmetry, or at least does not have one coinciding with the axis of rotation. An example that occurs to me is suggested by one of our earlier diagrams showing the circle inscribed in a triangle with sides 3, 4, 5 units (Fig. 67). You will recall, Simplicio, that the radius of the incircle works out as 1 unit exactly. Suppose now that we rotate the whole figure about an axis through the incenter parallel to the shortest side of the triangle (Fig. 86).

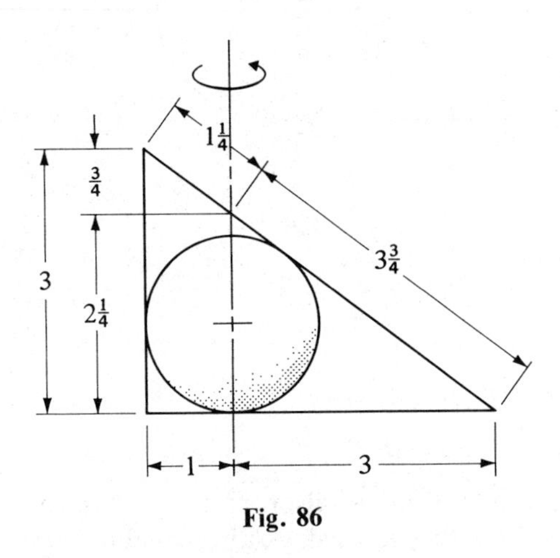

Fig. 86

Simplicio: I find the generated figure difficult to visualize. Let me see: the part of the triangle to the right of the axis of rotation will sweep out a right circular cone of base radius 3 units, but the solid generated by the remaining part of the triangle will clash with this.

Sagredo: Whenever rotation is not about an axis of symmetry, the solids generated by the two parts of the rotated figure must fail to coincide, but one can always consider the composite figure formed by the simultaneous *half*-rotations of these two parts. The measure of the volume of the half-cone generated by the half-rotation of the right triangle with sides 3, $2\frac{1}{4}$, $3\frac{3}{4}$, is given by $\frac{1}{2} \times \frac{1}{3}\pi \times 3^2 \times (9/4)$, which comes to $(27/8)\pi$. The measure of the surface area, semicircular base plus half curved surface of cone

$$= \frac{1}{2}\pi \times 3^2 + \frac{1}{2}\pi \times 3 \times \frac{15}{4}$$

(since the area of the curved surface of a complete cone of base radius r and slant height ℓ is given by $\pi r \ell$). The surface of this part of the solid of rotation thus works out to $(81/8)\pi$ square units. As a check on these calculations, we may test whether the relation

$$\text{Volume} = \frac{1}{3} \text{ surface times insphere radius}$$

is satisfied by the values obtained.

Simplicio: The right side comes to $\frac{1}{3} \times (81/8)\pi \times 1$, that is $(27/8)\pi$, as it ought.

Sagredo: The portion of the triangle to the left of the axis of rotation generates a half cylinder of radius 1 unit and height 3, from which is scooped out a half cone to a depth of $\frac{3}{4}$ unit. Its volume is therefore given by

$$\frac{1}{2}\pi \times 1^2 \times 3 - \frac{1}{2} \times \frac{1}{3}\pi \times 1^2 \times \frac{3}{4} = \frac{3}{2}\pi - \frac{1}{8}\pi = \frac{11}{8}\pi \text{ cubic units.}$$

The surface area is given by

$$\frac{1}{2} \text{ of } 2\pi \times 1 \times 3 + \frac{1}{2}\pi \times 1^2 + \frac{1}{2}\pi \times 1 \times \frac{5}{4}$$
$$= 3\pi + \frac{1}{2}\pi + \frac{5}{8}\pi = \frac{33}{8}\pi \text{ square units.}$$

And one-third of this times the insphere radius agrees with the volume as just calculated. Total volume of the solid circumscribing the sphere

$$= \frac{27}{8}\pi + \frac{11}{8}\pi = \frac{19}{4}\pi \text{ cubic units.}$$

Next, adding the surface areas of the two parts, we obtain

$$\frac{81}{8}\pi + \frac{33}{8}\pi = \frac{57}{4}\pi \text{ square units.}$$

And for the inscribed sphere, of unit radius, the respective measures of volume and surface area are $(4/3)\pi$ and 4π.

Simplicio: You are near your goal, Sagredo; yet you hesitate. The volume ratio, outer solid to insphere $= (19/4)\pi/(4/3)\pi = 57 : 16$. And the surface area ratio $= (57/4)\pi/4\pi = 57 : 16$ also! Does not this result please you?

Sagredo: I think that we have arrived at this apparently agreeable outcome only by inadvertently neglecting a part of the surface of the outer solid. Let me carefully sketch a plan and two elevations of this solid. (See Fig. 87.) Yes, Simplicio, it *is* as I suspected! I did overlook the two triangular faces F and F'. Including these, the total surface area has to be corrected to $(57/4)\pi + 3$ square units; consequently we find that the volume ratio, outer solid to sphere, is exceeded by the corresponding surface area ratio. This is analogous to the two-dimensional example illustrated in the first diagram of Fig. 66; in each case the circumscription is improper because there are some parts of the boundary of the outer figure that do run tangent to the inner figure.

Simplicio: It appears that in three dimensions there will be such an impropriety whenever the rotation is not about an axis of symmetry of the outer plane figure that passes through the centre of the incircle.

Sagredo: I agree, and although such cases were probably not intended to be covered by the enunciation quoted, it would have been safer if they had been explicitly excluded.

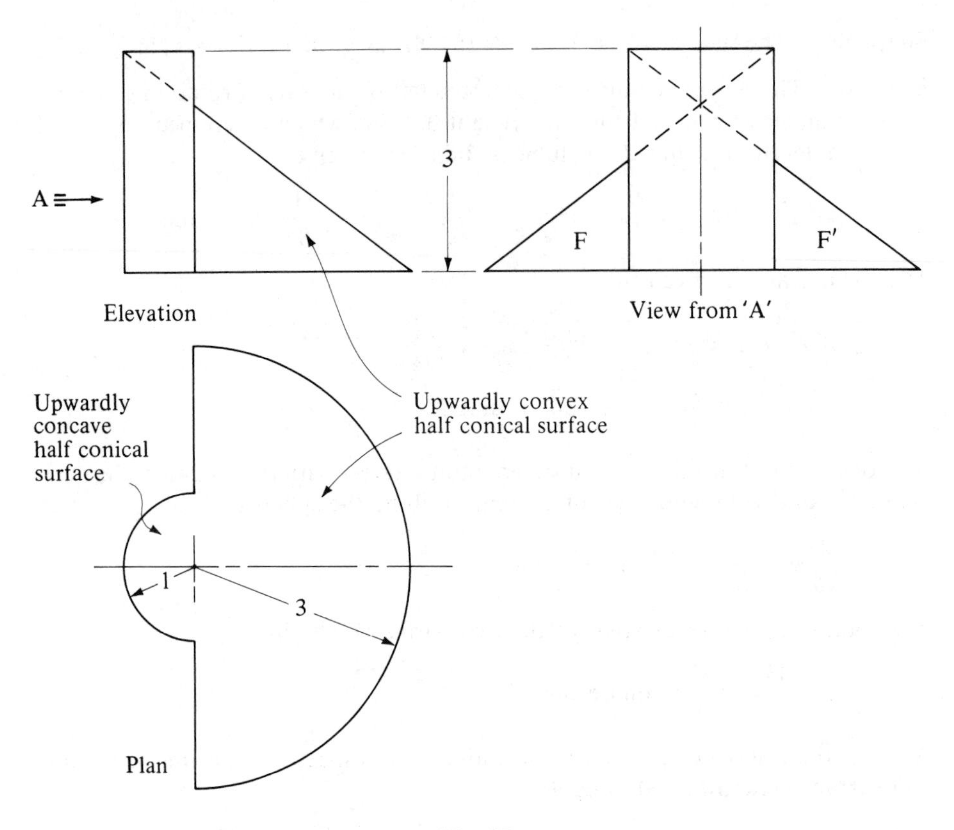

Fig. 87

Salviati: I hesitate to interrupt your deliberations, but perhaps you will allow me to remark that in adhering to our traditional convention concerning the disposition of the three views represented in Fig. 87, we may leave ourselves open to criticism.

Simplicio: How is that?

Sagredo: I suspect, Simplicio, that what Salviati has in mind is the lack of agreement between draughting practices in different countries.

Salviati: You know, my friends, how it is American English that is rapidly becoming the real world language, not some artificial one like Esperanto. Conventions of nonverbal communication are spreading from the United States more quickly still! Whatever reservation we may have about other examples of Americanization, we should doubtless applaud the particular form of standardization that would require us to replace the arrangement of the views of Fig. 87 with that shown in Fig. 88. For a reason that it is unnecessary for us to go into, this American system is known as "third angle projection," while the system we have been used to is called "first angle projection."

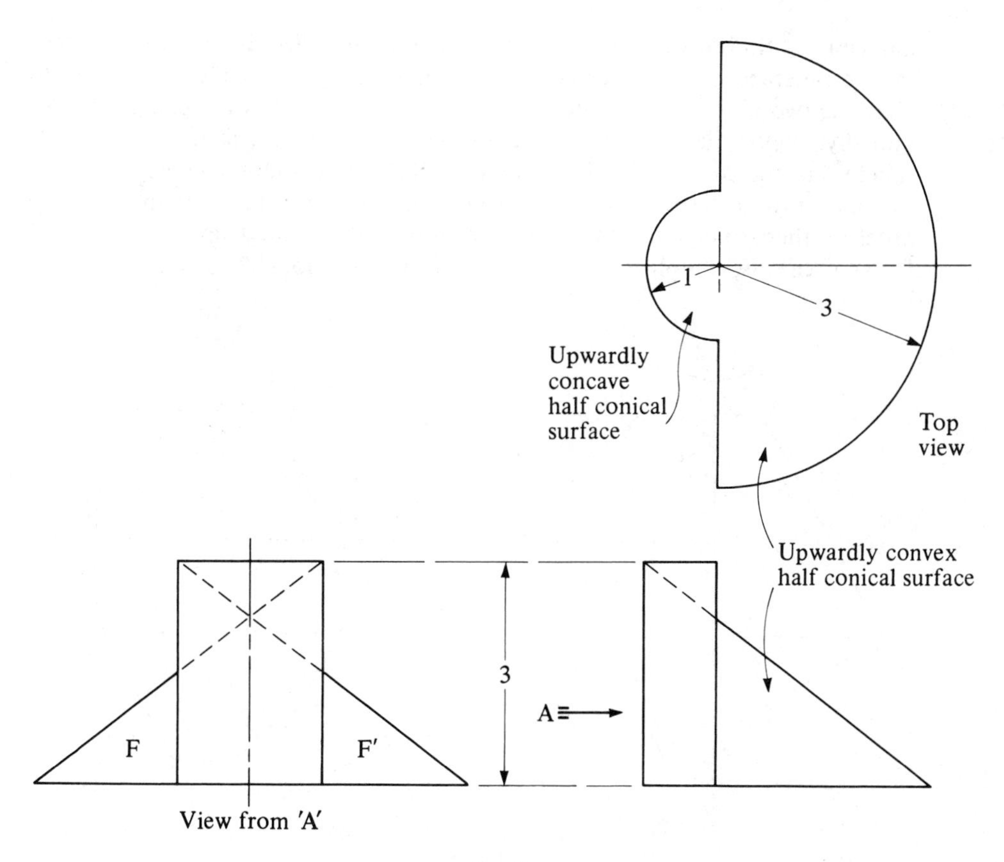

Fig. 88

Simplicio: While I find this new arrangement strange, it would be foolish to resist what is apparently an inevitable trend.

Sagredo: Let me see: having drawn one view, the others are placed on the same side as the viewer, not remote from the viewer as we have been accustomed to place them. You know, Salviati, I quite like this method of arrangement, which puts the top view *on top,* and the view from the left *on the left.* I am convinced that we should welcome the change. I propose, Simplicio, that while we have Salviati to guide us in case of error, we follow this convention henceforth.

Simplicio: I have no objection. And now, if we may continue our previous discussion, I have an example I should like you to consider. It is a centrally symmetrical case. I have in mind a circle and a circumscribed square rotated about any coplanar axis through their common center that does *not* run parallel to a pair of sides or pass through a pair of opposite vertices. Are we correct in assuming that the resulting solid will have two nonmatching halves?

Sagredo: I find this case just a little difficult to visualize. Let me see whether it can be satisfactorily represented in a couple of views. Yes, Fig. 89 should do. Here the two halves will "match" in one sense; indeed they are congruent. Admittedly, the whole figure will have a center of symmetry, at the center of the sphere, but the two halves of the circumscribed solid, simultaneously generated by a half turn of the original square, will be out of alignment where they meet. In order for them to match in this sense, one of the halves would have to be inverted. Do you see this, Simplicio? Are my diagrams clear enough?

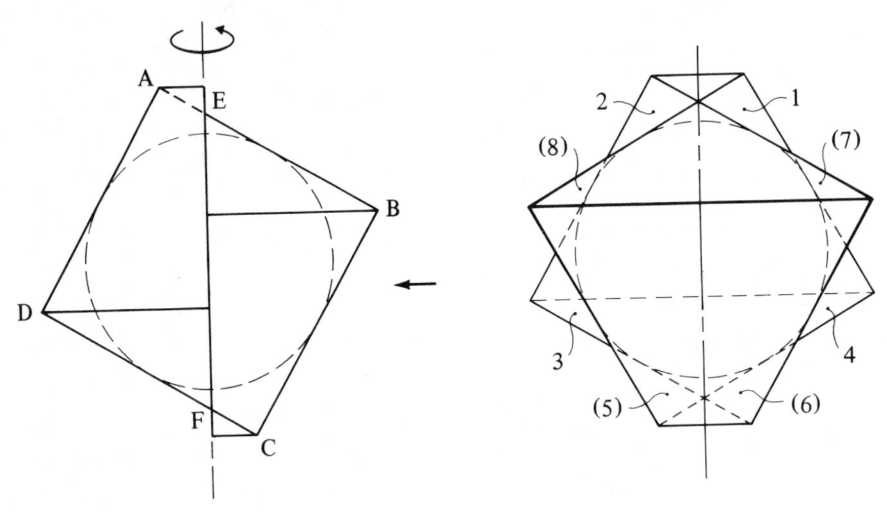

Fig. 89

Simplicio: I think so. As side *AD* sweeps out half the curved surface of a conical frustum, *BC* sweeps out a congruent but inverted surface. The semicircle generated by *B* is the base edge of a half cone whose surface is swept out by *EB*. *E,* the apex of this half cone, is also the apex of the inverted semiconical surface generated by *EA*. Likewise, *F* is the common apex of two semicones whose curved surfaces are generated by *FD* and *FC*.

Sagredo: Right; then you will appreciate that in addition to the conical surfaces, there are eight small plane faces. Four of these, which I will number 1 to 4, face an observer viewing in the direction of the arrow, the remainder, numbered (5) to (8), face the other way. It is precisely on account of these eight faces, which are not tangent to the sphere, that the surface of the outer solid is increased so that the ratios $V_o : V_i$ and $A_o : A_i$ will be found to be unequal. (In writing these ratios, the subscripts *o* and *i* refer, of course, to the outer and inner solids.)

Simplicio: But if we were to rotate only one half of the square, *AEFD* say, about axis *EF,* turning it through one complete revolution, then, I suppose, the ratios $V_o : V_i$ and $A_o : A_i$ would be equal?

Sagredo: Undoubtedly.

Simplicio: You may be quite certain of this, but I should like to consider the case in some detail.

Sagredo: Well then, please make a drawing of the figure for us to work from.

Simplicio: It is just a matter of duplicating half of the first elevation of your previous drawing. (See Fig. 90.) If we let the side AD of the original square be s units and the arbitrary distance AE be a units, then $DF = s - a$ units. It should not be too difficult to find expressions for the volume and surface area of this outer solid of rotation as well as for the sphere about which it is circumscribed.

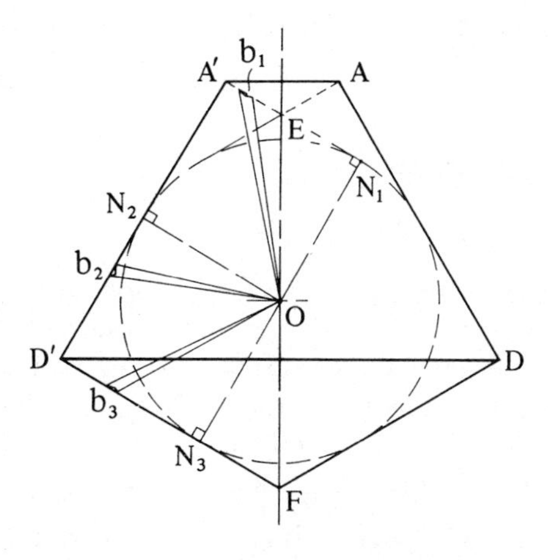

Fig. 90

Sagredo: I suggest that it will be easier and more generally illuminating if we proceed rather differently. Allow me to add a few lines to your diagram.

Simplicio: As you wish.

Sagredo: I first ask you to imagine the whole surface of the outer solid to be enclosed by a surface composed of a very large number of very small plane faces, each of which touches the surface of the solid of revolution tangentially. I will indicate by b_1, b_2, b_3 just three of these tiny faces, or the measures of their areas, and by these lines joined to the center O, I intend to mark out representative elementary pyramids into which the whole enclosing solid—a many-faced polyhedron—may be supposed to be divided. Since the perpendicular heights, ON_1, ON_2, ON_3, and so on for all the component pyramids, will in every case be equal to the radius r of the insphere, the volume of the whole outer (plane-faced) solid is given by the sum of the pyramidal volumes, which comes to $\frac{1}{3}(b_1 + b_2 + \ldots + b_n) \cdot r$, or $\frac{1}{3}P \cdot r$, where $P\ (= \Sigma b)$ measures the complete polyhedral surface area. And, by continually increasing the number of elements of volume so that the individual bases approach zero, P may be made to approach A_o as a limit.[183]

Simplicio: Now I see what you are up to! Your conclusion is that for the solid of revolution itself $V_o = \frac{1}{3}A_o \cdot r$.

Sagredo: Then you can see that the same argument will apply also in the case of the inner solid, the sphere itself.

Simplicio: Well so it will! Yes, the relation *is* implicit in the pair of formulae $V_i = (4/3)\pi r^3$ and $A_i = 4\pi r^2$; these show that $V_i = \frac{1}{3}A_i r$. Or the argument involving the pyramidal elements of volume would enable us to obtain the formula for the volume of a sphere from that for the surface area!

Sagredo: Or vice versa.[184]

Simplicio: That is true too. And now, for the pair of solids illustrated in Fig. 90,

$$V_o : V_i = \frac{1}{3}A_o r : \frac{1}{3}A_i r = A_o : A_i.$$

Sagredo: Exactly so.

Simplicio: I am especially grateful to you, Sagredo, for having shown how this case is to be dealt with in such a general way that it may be seen as an exemplar for all the solids of revolution having inscribed spheres such that the volumes will be in the same ratio as the surfaces. I am now fairly confident that I understand the range of applicability of the theorem under consideration, thanks to your treatment of this example, and of the counterexamples that preceded it.

Sagredo: You will realize, of course, that the argument just indicated combines insights that we reached earlier and used in proving Theorem B (p. 198) and in discussing its extension to cones as well as to pyramids circumscribed about a sphere (p. 200). Notice too that, in your particular example (Fig. 90), the conical depression *AA'E* is to be counted as circumscribed with respect to the sphere since the generators of this surface, when produced, are tangents to the sphere. This is in conformity with the view that we found it fitting to adopt in the corresponding two-dimensional case (Fig. 68). Well, Salviati, you have heard our response to the passage that you brought to our attention.

Salviati: You have done well, and I assure you that you need not be disappointed to find that Theorems A and B are not new. As you have already observed, Sagredo, it would have been surprising if they had not been discovered by someone before today. In spite of the popularity of Ozanam's work, these theorems seem to have been little noticed. Perhaps they were judged to be relatively isolated from a sufficiently interesting mathematical context. For although some unification with other geometrical items has emerged from our discussion, and although these theorems have a prima facie impressiveness on account of their generality, they may have appeared on reflection somewhat as oddities or "fancy" results when it was realized that *whatever form is assigned to each of two figures, there is one (and only one) relative size for which the ratio of the contents is equal to the ratio of the boundaries.* And for any pair of figures, so long as they are readily amenable to mensurational treatment, the appropriate relative size is easily determined independently of these theorems that Ozanam published two centuries ago.

Simplicio: We have had a couple of examples of this in two dimensions (pp. 182–185); perhaps we can conclude this hour with a three-dimensional example.

Sagredo: Are we to be permitted to continue our discussion beyond sunset on this occasion?

Salviati: I believe so. Certainly I can assure you that our author does not share the common prejudice against the number thirteen. You will remember that Euclid and Ptolemy did not disdain to divide their greatest works into thirteen parts.[185] At least we have time for a three-dimensional illustration now. You may care to work through the following example, Simplicio:

> For the spherical segment and the cuboid shown in Fig. 91, find the ratio of R to a in order that $V_s : V_c = A_s : A_c$.

You will perhaps recall that the curved surface of a spherical segment is equal to the area of a band of equal height around the cylinder that would circumscribe the whole sphere. For the volume of the spherical segment, you could use the formula $V = (h/6)\{A_A + A_B + 4A_C\}$, where $A_A = 0$, $A_B = \pi r_B^2$, $A_C = \pi r_C^2$. But the most elegant method would be to make use of the results established by Archimedes: (i) the curved surface of a segment of a sphere is equal in area to a circle whose radius is equal to a certain chord (in our diagram, chord AD); (ii) the volume of the segment is given by the difference of the volumes of the corresponding

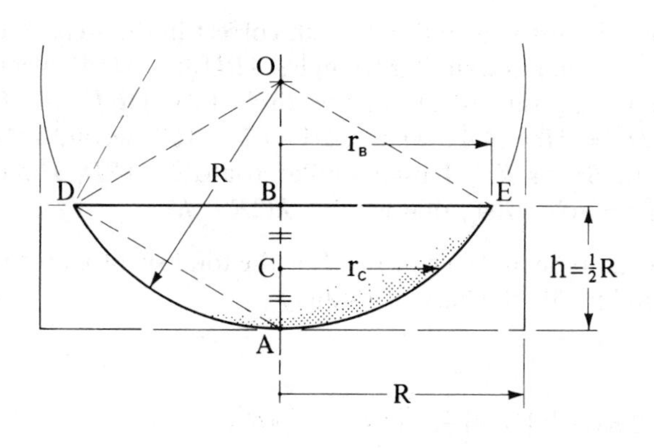

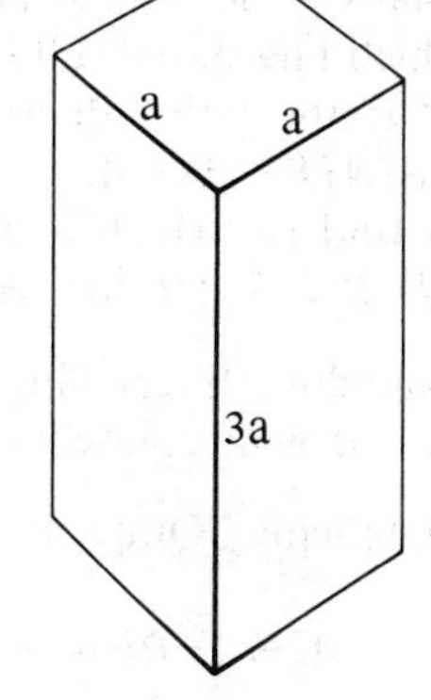

Fig. 91

sector and cone (in the diagram, segment DAE = sector $ODAE$ − cone ODE, where the volume of the sector is equal to that of a cone whose base is equal to the curved surface of the segment and whose height is equal to the radius of the sphere).[186]

Sagredo: I shall certainly have a question or two concerning what you have just said, but for the moment it is best if I do no more than acknowledge that the very last proposition, that concerning the volume of the spherical sector, corresponds exactly to the result we arrived at a couple of minutes ago for the volume of the whole sphere: $V_s = \frac{1}{3}A_s \cdot r$. I fear, Simplicio, that we must appear as children to anyone who knows his Archimedes.

Simplicio: Probably so. At least see if I can deal satisfactorily with the example Salviati has put to us:

$$\begin{aligned}
\text{Curved surface of segment} &= \pi \cdot AD^2 \\
&= \pi(AB)(2 \cdot AO)^* \\
&= 2\pi Rh. \\
\text{For the volume, segment } DAE &= \text{sector } ODAE - \text{cone } ODE \\
&= \frac{1}{3}(2\pi Rh)R - \frac{1}{3}\pi r_B{}^2(R - h) \\
\text{where } h = \frac{1}{2}R \text{ and } r_B{}^2 &= h(2R - h).^{**} \\
\text{Hence, volume, } V_s &= \frac{1}{3}\pi R^3 - \frac{1}{3}\pi\left(\frac{1}{2}R\right)\left(2R - \frac{1}{2}R\right)\left(R - \frac{1}{2}R\right) \\
&= \frac{1}{3}\pi R^3 - \frac{1}{6}\pi R \cdot \frac{3}{2}R \cdot \frac{1}{2}R \\
&= \frac{1}{3}\pi R^3 - \frac{1}{8}\pi R^3 \\
&= \frac{5}{24}\pi R^3.
\end{aligned}$$

Thank you for your patience. I trust you find that I am correct in the steps over which I hesitated. First (*), I had in mind similar triangles ABD and ADA', where A' is the point diametrically opposite to A on the circle through D, A, E. $AB/AD = AD/AA' \Rightarrow AD^2 = AB \cdot AA'$, where $AA' = 2 \cdot AO$. Second (**), to find r_B, which is equal to BE or BD, I used similar triangles ABD, DBA'. $BD/BA' = AB/BD \Rightarrow BD^2 = AB \cdot BA'$, that is $r_B{}^2 = h(2R - h)$.

Sagredo: Everything appears to be in order so far. For the total surface area of the segment you will need to include the base of radius r_B.

Simplicio: Quite so.

$$A_s = 2\pi Rh + \pi r_B{}^2 = 2\pi R\left(\frac{1}{2}R\right) + \pi\left(\frac{1}{2}R\right)\left(\frac{3}{2}R\right) = \frac{7}{4}\pi R^2.$$

Turning now to the prism or square-ended cuboid,

$$V_c = 3a^3, \quad \text{and} \quad A_c = 2a^2 + 4(3a^2) = 14a^2.$$

Hence $V_s/V_c = A_s/A_c$ implies and is implied by $(5/24)\pi R^3/3a^3 = (7/4)\pi R^2/14a^2$. Solving for (the nonzero value of) R/a, we find $(5/72) \cdot R/a = 1/8$. Thus the ratio of the volumes is equal to the ratio of the surface areas if and only if $R : a = 9 : 5$.

Sagredo: Thank you, Simplicio. And now, Salviati, you surprised me earlier by informing us that the volume of the spherical segment could be found by the formula $V = (h/6)\{A_A + A_B + 4A_C\}$. If this is so, we may check the first part of Simplicio's working by an alternative method, by evaluating V when

$$h = \frac{1}{2}R, \qquad A_A = 0, \qquad A_B = \pi r_B{}^2 = \pi\left(\frac{1}{2}R\right)\left(\frac{3}{2}R\right) = \frac{3}{4}\pi R^2,$$
$$A_C = \pi r_C{}^2 = \pi\left(\frac{1}{4}R\right)\left(\frac{7}{4}R\right) = \frac{7}{16}\pi R^2.$$

Then,

$$V = \frac{R}{12}\left\{\frac{3}{4}\pi R^2 + 0 + 4 \times \frac{7}{16}\pi R^2\right\} = \frac{\pi R^3}{12}\left\{\frac{3}{4} + \frac{7}{4}\right\} = \frac{5}{24}\pi R^3,$$

as before.

Salviati: You might like to consider the following third method, suggested by Fig. 83 and the accompanying discussion. Referring to this new diagram, Fig. 92:

> Any circular section (radius r_s) of the spherical segment
> + the corresponding annular section of the scooped-out bowl
> = the circular section of the outer cylinder of radius R.

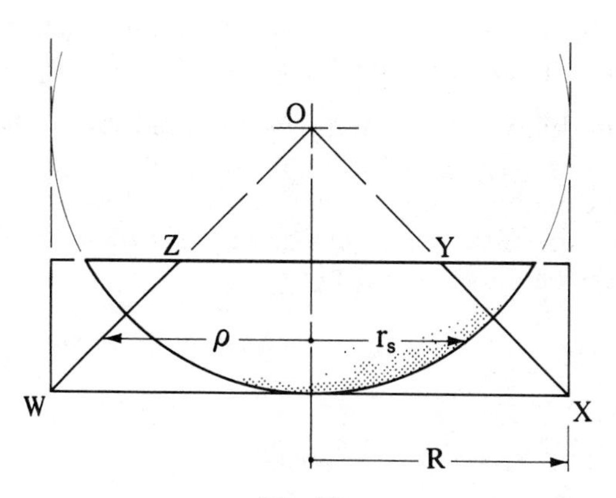

Fig. 92

That is, $\pi r_s^2 + \pi(R^2 - r_s^2) = \pi R^2$. And since, as shown in our earlier discussion (after Fig. 83 and in note 167), the area of the annulus is equal to the area of the circular section (radius ρ) of the cone (OWX), we have $\pi r_s^2 + \pi\rho^2 = \pi R^2$. Then, by Cavalieri's principle,[187]

Volume of the spherical segment (V_s) + volume of frustum of cone (V_{FC})
= volume of outer cylinder (V_{OC}),

where the heights of the three solids are understood to be equal, in this case $h = \frac{1}{2}R$. Since the cone OZY is similar to and half the linear scale of cone OWX, the volumes are as 1 : 8, and so the frustum $ZWXY$ has seven-eighths of the volume of cone OWX, that is

$$\frac{7}{8} \times \frac{1}{3}\pi R^2 \cdot R.$$

Thus $V_s + V_{FC} = V_{OC}$ becomes

$$V_s + \frac{7}{24}\pi R^3 = \pi R^2 h, \quad \text{giving} \quad V_s = \frac{1}{2}\pi R^3 - \frac{7}{24}\pi R^3 = \frac{5}{24}\pi R^3.$$

Simplicio: Oh how ingenious! Does this not please you, Sagredo?

Sagredo: Of course. Nevertheless, I am still puzzled over the applicability of the formula $V = (h/6)\{A_A + A_B + 4A_C\}$, which is nothing else but the formula that I have always known as the "prismoidal formula," $V = (h/6)\{A_T + A_B + 4A_{MS}\}$, where the subscripts refer to the top, bottom, and mid-section, respectively. I now have to admit that this formula does indeed give the correct measure of the volumes of spherical segments, at least if they are of the form proposed in this example, but since I had always connected it with plane-faced prismoids, such as frusta of pyramids, you will understand that I was surprised to learn that it could be used here.

Simplicio: Your remarks remind me of another volume formula that I have seen used for the frusta of pyramids, namely, $V = (h/3)\{A_T + A_B + \sqrt{(A_T A_B)}\}$.

Sagredo: Yes, but that would certainly *not* be applicable in the case of the segment of a sphere since, with $A_T = 0$, the formula reduces to $V = \frac{1}{3}A_B h$, the correct measure for the volume of a *cone,* and so clearly it would give a gross underestimate for the volume of the spherical segment. But I fear the time has got away from us again. Yes, we must leave it there.

XIII

Every thought is an exception to a general rule—which is not-thinking.

—Paul Valéry.

An intellectual is someone whose mind watches itself.

—Albert Camus

The larger the island of knowledge, the longer the shoreline of wonder.

—R. W. Sockman

Is not mathematics the alembic in which the most quintessential abstractions are distilled?

—F. Le Lionnais

If he [the mathematician] be asked why he persists on the high glaciers whither no one but his own kind can follow him, he will answer, with Jacobi: For the honour of the human spirit.

—André Weil

In growing mathematical theories . . . the most exciting developments come from exploring the boundary regions of concepts, from stretching them, and from differentiating formerly undifferentiated concepts. In these growing theories intuition is inexperienced, it stumbles and errs. There is no theory which has not passed through such a period of growth; moreover, this period is the most exciting from the historical point of view and should be the most important from the teaching point of view. . . . [However,] mathematics is presented as an ever-increasing set of eternal, immutable truths. . . . [This] deductivist style hides the struggle, hides the adventure. The whole story vanishes, the successive tentative formulations of the theorem in the course of the proof-procedure are doomed to oblivion while the end result is exalted into sacred infallibility.

—Imre Lakatos

HORA TERTIA DECIMA

Sagredo: Though our day is nearly over, and I know there is still much to discuss, I should like to follow up the question of the range of applicability of the volume formulae

$$V = \frac{h}{3}\{A_T + A_B + \sqrt{A_T A_B}\} \quad \text{and} \quad V = \frac{h}{6}\{A_T + A_B + 4A_{MS}\}$$

if you agree to this.

Salviati: I have just spoken with our author concerning the possibility of extending our day so as to include a fourteenth hour. I even mentioned that if Euclid had continued his *Elements* as far as a Book XIV, he might have forestalled the efforts of certain of his successors to supplement his mighty treatise.[188] Our writer claimed at first that he was merely the recorder of our discussion and that he had little influence over our present very temporary situation. Nevertheless, a moment later, he did encourage me in the belief that we will be permitted to continue here until darkness overtakes us; so let us proceed on that assumption. If Simplicio is willing, let us consider the matter you have raised.

Simplicio: Of course; please go ahead, Sagredo.

Sagredo: Well, I think that I am able to derive the two volume formulae just mentioned using a difference-of-two-pyramids method. I assume

(i) Volume of a complete pyramid = $\frac{1}{3}$base area × perpendicular height;
(ii) Areas of similar polygons are to one another as the squares on corresponding linear dimensions.

Then, briefly, referring to Fig. 93, it can be seen that

$$\text{Volume of frustum} = \frac{1}{3}A_B(h+k) - \frac{1}{3}A_T k,$$
$$\text{where } A_B = c(h+k)^2,\ A_T = ck^2,$$

c being a constant of proportionality. Hence the volume is given by

$$V = \frac{1}{3}\{c(h+k)^3 - ck^3\}$$
$$= \frac{c}{3}\{h^3 + 3h^2k + 3hk^2\}.$$

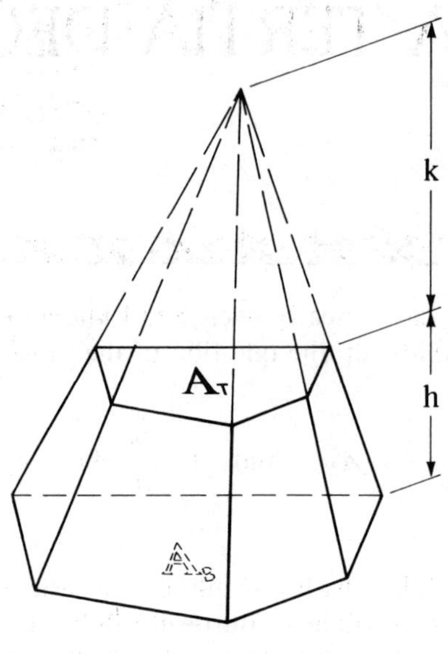

Fig. 93

Having in mind the first formula to be established, it is natural to rearrange this as follows:

$$\begin{aligned} V &= \frac{h}{3}\{ch^2 + 3chk + 3ck^2\} \\ &= \frac{h}{3}\{ck^2 + c(h^2 + 2hk + k^2) + ck^2 + chk\} \\ &= \frac{h}{3}\{A_T + A_B + ck(k + h)\} \\ &= \frac{h}{3}\{A_T + A_B + \sqrt{(A_T A_B)}\}.\text{[189]} \end{aligned} \tag{1}$$

Having in mind the second formula, the last four lines may be replaced by the following:

$$\begin{aligned} V &= \frac{h}{6}\{2ch^2 + 6chk + 6ck^2\} \\ &= \frac{h}{6}\{ck^2 + c(h^2 + 2hk + k^2) + ch^2 + 4chk + 4ck^2\} \\ &= \frac{h}{6}\left\{ck^2 + c(h + k)^2 + 4c\left(k^2 + kh + \frac{1}{4}h^2\right)\right\} \\ &= \frac{h}{6}\{A_T + A_B + 4A_{MS}\}.\text{[190]} \end{aligned} \tag{2}$$

This much I managed to work out during our pause; but I could see that it gives no clue to the fact that formula (2) correctly measures the volume of a frustum or segment of a *sphere*.

Salviati: Or of a segment of a spheroid or of a whole sphere or spheroid or of a paraboloid of revolution and indeed of a vast range of solid forms strikingly different from pyramid frusta.

Simplicio: You mean, do you, that if we take A_T and A_B each equal to zero and A_{MS} equal to the measure of the great circle area, and h equal to $2r$, then we shall obtain the correct expression for the volume of a sphere?

Salviati: Try it for yourself.

Simplicio: $V = \frac{2r}{6}\{0 + 0 + 4\pi r^2\} = \frac{1}{3}r \times 4\pi r^2 = \frac{4}{3}\pi r^3$. Amazing!

Salviati: The range of applicability of formula (1) is much less remarkable. The general methodological approach to the problem of determining the domain in which a formula may be validly applied is more easily illustrated in the case of (1) than (2), and more easily still in the case of the formula to which (1) or for that matter (2) reduces when $A_T = 0$.

Simplicio: You mean $V = \frac{1}{3}A_B h$ for a pyramid?

Salviati: Yes, for a pyramid or cone or certain other related solids.

Simplicio: I must confess that I do not even recall how to establish the formula for the volume of a pyramid.

Salviati: Such basic and easily understood problems are often far from being easily solvable in a rigorous manner. Euclid devoted most of the twelfth book of his *Elements* to pyramids and cones and their relation to the circumscribing prisms and cylinders. That is something you will have to study on your own. For the present, I shall mention the special case of the right square pyramid of height equal to half the base edge length, $h = \frac{1}{2}e$. Since six such congruent pyramids may be juxtaposed with their bases forming the faces of a cube and their apices coinciding at the center of the cube, their combined volume is measured by e^3, equal to $e^2 \times 2h$, that is $2A_B h$. So the volume of each is given by $V = \frac{1}{3}A_B h$. The original motivation for determining the volume of a pyramid might well have arisen in ancient Egypt in connection with the great funerary pyramids. This is not to say that the pyramid builders or designers would themselves have had any need to know the volumes of their structures.[191] Rather it is to suggest that architecture, and technology and art in general, would occasionally have stimulated reflective minds to respond in protomathematical ways in an endeavor to extract some sort of understanding of the abstract features involved.[192] As a simple example, consider 100 equal cubes arranged in square formation, and upon these a layer of 9×9 cubes of the same size placed symmetrically and similarly aligned upon the first layer; then a third layer of 8×8 cubes, and so on up to a single cube at the top. The total number of unit cubes in this step pyramid is the sum $100 + 81 + 64 + \ldots + 1$, which comes to 385.

Simplicio: This is rather far from one-third of the 1,000 blocks in the enclosing prism, in this case the $10 \times 10 \times 10$ stack of unit cubes.

Salviati: But it is a very excellent approximation to one third of $10\frac{1}{2} \times 10\frac{1}{2} \times 10\frac{1}{2}$, which product, as you can see from Fig. 94, gives the number of cubic units in the whole cube with which it is most reasonable to compare the pyramidal stack.

$$\begin{aligned}\left(10\frac{1}{2}\right)^3 &= 10^3 + 3 \cdot 10^2 \cdot \frac{1}{2} + 3 \cdot 10 \cdot \left(\frac{1}{2}\right)^2 + \left(\frac{1}{2}\right)^3 \\ &= 1000 + 150 + 7\frac{1}{2} + \frac{1}{8} \\ &= 1{,}157\frac{5}{8},\end{aligned}$$

and one-third of this is $385\frac{7}{8}$.

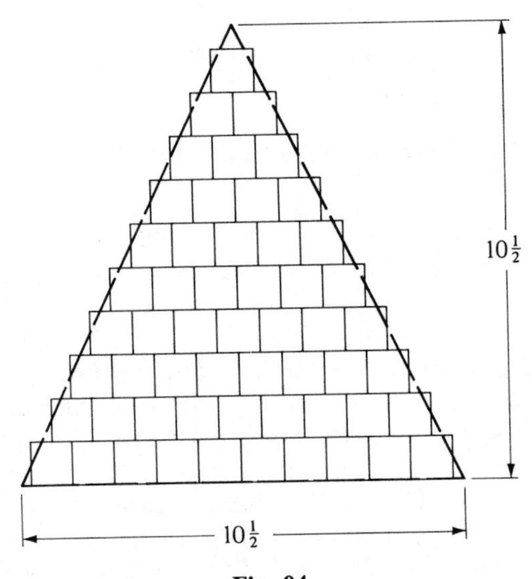

Fig. 94

Simplicio: I see. And if the numbers of blocks in the two stacks, the pyramidal and the cubic, are respectively ν and N, doubtless the ratio $\nu : N$ approaches $1:3$ as the numbers involved increase without limit.

Salviati: The formula for ν would *now* be written[193]

$$\nu = \sum_1^n r^2 = \frac{n}{6}(n + 1)(2n + 1).$$

Hence, taking $N = (n + \frac{1}{2})^3$ (and skipping several millennia of mathematical development from where we were a minute ago!),

$$\nu/N = \frac{n}{6}(n+1)(2n+1)\Big/\left(n+\frac{1}{2}\right)^3$$
$$= \frac{1}{6}\left(\frac{n}{n+1/2}\right)\left(\frac{n+1}{n+1/2}\right)\left(\frac{n+1/2}{2n+1}\right)$$
$$= \frac{1}{6}\left(\frac{n+1/2-1/2}{n+1/2}\right)\left(\frac{n+1/2+1/2}{n+1/2}\right)\left(\frac{2}{1}\right)$$
$$= \frac{1}{3}\left(1-\frac{1}{2n+1}\right)\left(1+\frac{1}{2n+1}\right),$$

which, as you just concluded, approaches $\frac{1}{3}$ as a limit, as n increases without limit. Even if we were to use the cruder correspondence involving the cubical stack with n^3 instead of $(n + \frac{1}{2})^3$ blocks, the same limit would be obtained, though the convergence would be much less rapid:

$$\nu/N = \frac{n}{6}(n+1)(2n+1)/n^3$$
$$= \frac{1}{6}\times\frac{n}{n}\times\frac{n+1}{n}\times\frac{2n+1}{n}$$
$$= \frac{1}{6}\left(1+\frac{1}{n}\right)\left(2+\frac{1}{n}\right)$$

which $\rightarrow \frac{1}{3}$ as $n \rightarrow \infty$.

Sagredo: The example of the ancient determination of the volume of a pyramid is one of the kind that could well have been "begun with the hands and finished in the head."[194] For clearly, it is unnecessary actually to build a pyramidal stack in order to determine the number of blocks in it. It is unnecessary to count the blocks one by one when they can be dealt with layer by layer. And there would be no need to proceed by addition if the summation formula were known:

$$1^2 + 2^2 + 3^2 + \ldots + 10^2 = \frac{10}{6}(11)(21) = 385.$$

I notice also that the form of the pyramid does not have to be restricted to that given by a stack of *cubes*. The ratio of the height of the pyramid to the base edge length may be varied simply by altering the component cubes to cuboids having their height-to-base-edge ratio whatever is required for the pyramid itself. The arithmetic would remain unchanged. I mention this because it shows why the volume of a pyramid on a given base is directly proportional to the height, a relation that might otherwise be doubted owing to the change of shape.

Salviati: Yes, and this could have been appreciated by the ancient Egyptians. At least they built their pyramids in various proportions.[195] And for obvious practical reasons, they used large stone blocks only near the base of their monumental pyramids, smaller blocks being more convenient in the higher layers. Thus the assumption of a stack of equal cubes is already an abstraction and one that, as you have noticed, can be varied at pleasure. A pyramidal stack might just as well be composed of elementary prisms with bases that are rectangles or parallelograms, or triangles or hexagons, and these prisms could just as well be oblique as right. Even with right prisms, an approximation to an oblique pyramid may be obtained by offsetting each layer with respect to the one below it. In all such cases the arithmetical comparison between the volumes of the pyramid and the equally based circumscribing prism would be of the one form, varying only with the arbitrarily assigned number of layers.

Simplicio: If we may return to the preceding example, then, I should like to ask about the discrepancy between the volume of the ten-layered stack and that of the pyramid of height and base edge length each $10\frac{1}{2}$ units. It appears from your diagram (Fig. 94) that the excess of the pyramid over the approximating stack is due entirely to the lack of agreement at the very top.

Salviati: Our diagram, being only a single elevation, does not do full justice to the spatial relations involved. But you are right to focus your attention on the top in the first place. A careful comparison of the top layer, consisting of a single unit cube, with the pyramid of base $1\frac{1}{2} \times 1\frac{1}{2}$ and height $1\frac{1}{2}$ ought to be sufficient to show that the true pyramid volume is *exactly,* and not merely approximately, one-third of the volume of the circumscribing prism. As I complete the drawing (see Fig. 95), observe that by making *PQ* parallel and equal to *db,* and *RS* parallel and equal to *ac,* and so on, the excess of the pyramid *OABCD* over the unit cube is shown to be precisely equal to six times the triangle-based pyramid *XPQA,* or exactly three times the square-based pyramid *Oabcd.* You will appreciate that the wedge-shaped pieces *XYHGQR* and *XYFEab* are congruent; one of these prismoids may be made to coincide with the other by turning it through two right angles about *XY.* And, of course, a like relation holds between the parts above and below the middle line of each of the other side faces of the cube.

Sagredo: It follows too that at each layer below this top one, the complete pyramid will exceed the approximating stack by four more corner pyramids congruent with *XPQA,* and these are equivalent to two pyramids like *Oabcd.* Thus, for the ten-layered case of Fig. 94, the pyramid exceeds the stack by a volume equal to $(3 + 9 \times 2)$ times that of *Oabcd.*

Salviati: You are quite correct. But now, let us see how further consideration of Fig. 95 will enable the *exact* volume of pyramid *OABCD* to be determined, without the need to consider any other layers at all, provided it can be assumed that the pyramid *Oabcd,* similar to *OABCD,* has one twenty-seventh of its volume. Clearly, the ratio of the corresponding linear dimensions is as $1:3$, and the correct volume ratio might very early have been assumed by analogy with the relation clearly exemplified by a $3 \times 3 \times 3$ stack of cubical or cuboidal blocks,

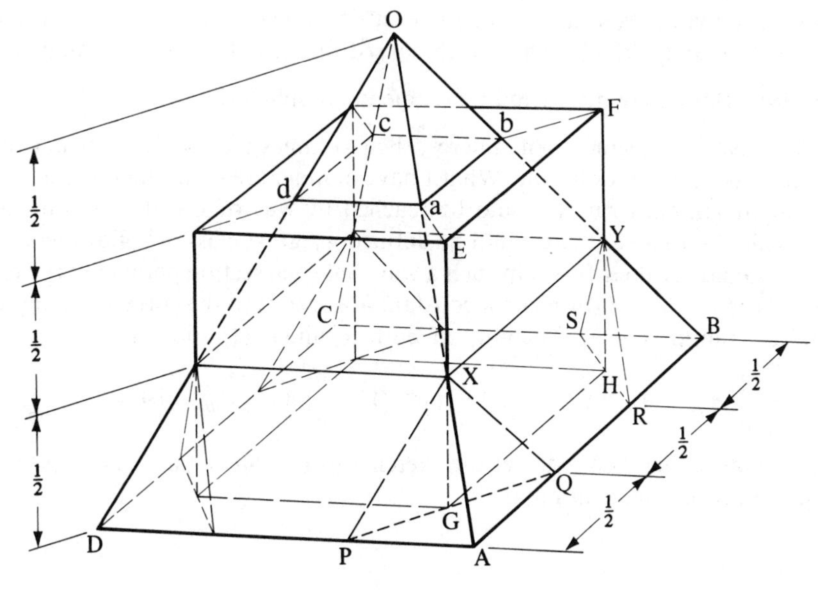

Fig. 95

long before the general principle, $v : V = d^3 : D^3$ was demonstrated, or even explicitly stated.[196] Then, because the whole pyramid *OABCD* exceeds the unit cube by a volume three times that of *Oabcd,* the excess is 3/27 of OABCD. Thus, $OABCD - \frac{1}{9}OABCD$ = unit cube. Hence,[197] the ratio of the volumes, $OABCD$: unit cube = 9 : 8. The volume of pyramid OABCD is therefore $1\frac{1}{8}$ cubic units. And, since the volume of the equally based prism which would circumscribe *OABCD* is given by $1\frac{1}{2} \times 1\frac{1}{2} \times 1\frac{1}{2}$, equal to $3\frac{3}{8}$ cubic units, it follows that the volume of the pyramid is one-third of the product, base area times perpendicular height.

Such a derivation might be considered too elegant for an ancient Egyptian; yet, on the evidence of the surviving mathematical papyri, the arithmetical part would certainly have caused no difficulty. And mastery of spatial relations is well attested in connection with contemporary crafts; the ancient Egyptians were great model makers and their building and cabinetmaking achievements show that they were ingenious designers of various kinds of hidden joints, locking devices, and so forth.[198]

Sagredo: I take it then that you are suggesting that this line of reasoning could have occurred 4,000 years ago. Considered from our standpoint, the approach via stepped pyramidal stacks suggests generalizations that might not have occurred even to the Greeks. We have already agreed that, simply by postulating appropriately shaped elementary prisms, properly arranged, the same reasoning is applicable to pyramids of any form and even to some solids that we should hesitate to call pyramids at all—some with curved or twisted faces for example.

Simplicio: But surely not including segments of spheres!

Sagredo: No, not segments of spheres; their volumes are certainly greater than one-third base area times height. What I have in mind are such generalizations of a regular pyramidal form as could be reached by way of Cavalieri's principle. There "exist"—I mean, there could be defined—a great class of solids related to simple pyramids or their frusta in such a way that each section parallel to the base or bases is equal in area (or bears a constant area ratio) to the corresponding section of a pyramid. The applicability of both formulae (1) and (2):

$$V = \frac{h}{3}\{A_T + A_B + \sqrt{(A_T A_B)}\} \quad \text{and} \quad V = \frac{h}{6}\{A_T + A_B + 4A_{MS}\},$$

to such solids is then assured. Let the sketches in Fig. 96 be taken as representative of the forms I have in mind.

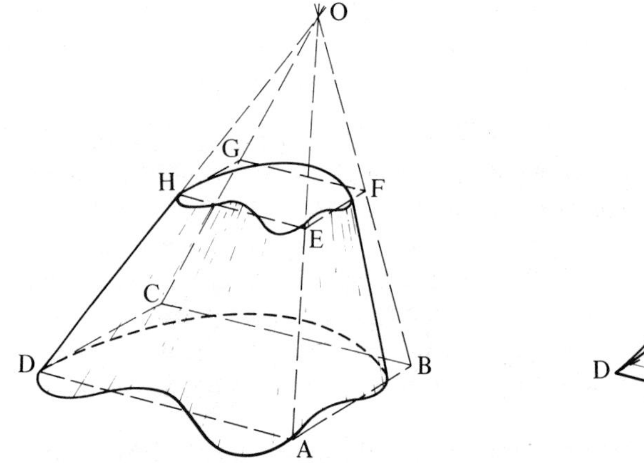

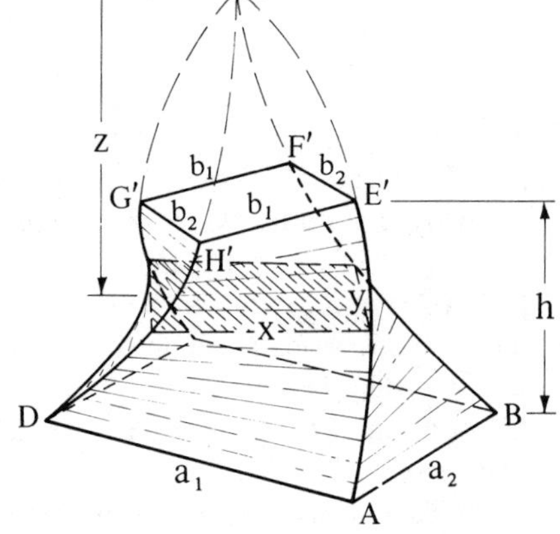

Fig. 96

Salviati: Yes indeed. Corresponding to any given pyramid frustum there is no end to the further solids of equal volume of such a character that not only their total volumes but also the volumes of segments of them bounded by *any* pair of planes parallel to the bases will be correctly measured by (1) and by (2). With $b_2/a_2 = b_1/a_1$, it becomes most convenient to specify the positions of the planes parallel to the bases by their distances from the apex of the complete solid, O or O'. Allow me to add some further dimensions to your second sketch.

Sagredo: Of course.

Salviati: Good. And let us write $x = c_1 z$, $y = c_2 z$, and $A(z) = xy = c_1 c_2 z^2$. So, $A(z) = cz^2$ say. Hence the volume is given by the following expression:

$$V = \int_h^{h+k} cz^2 dz = \frac{c}{3}\{(h+k)^3 - k^3\},$$

with the algebraic work then proceeding exactly as you have already given it, Sagredo (p. 217–218) to obtain formulae (1) and (2)—regardless of what sort of pyramid, or what Cavalierian transformation of a pyramid, we have in mind. What these figures have in common, guaranteeing their interchangeability as alternative "applications" of the same analysis, is simply this: that the cross-sectional areas parallel to their bases obey the same law of variation.[199]

Simplicio: You said just now that $b_2/a_2 = b_1/a_1$. Is this necessary?

Salviati: A good question, though an incomplete one! The short answer is that in the first diagram of Fig. 96, by obvious similar triangles, $EF:AB$ $(= OE:OA) = EH:AD$; and assuming for convenience that the bases, and parallel cross-sections, of the twisted solid of the second diagram are congruent to the corresponding bases and cross-sections of the plane-faced pyramid frustum, so that $E'F' = EF$, $E'H' = EH$, etc., we are left with $b_2/a_2 = b_1/a_1$. I admit that I was lax in stating my assumptions, in particular the assumption that led to the equation $A(z) = xy$. Yes, again for convenience, I did assume that the bases and cross-sections parallel to the bases are all rectangles. But I suspect that your question was not really directed at these details. Allow me, therefore, to respond in a different way, by asking you to consider a simple prismoid such as that shown in Fig. 97 in plan and elevation.

Simplicio: You would like me to calculate the volume?

Salviati: Please—and, if possible by several methods.

Simplicio: Assuming rectangular bases, we have $A_T = 4$, $A_B = 40$, and $h = 6$. Our formula (1) gives $V = (6/3)\{4 + 40 + \sqrt{160}\}$, which comes to $88 + 8\sqrt{10}$ cubic units. Since $\sqrt{10} = \sqrt{(3 \times 3\frac{1}{3})}$, a good approximation for $\sqrt{10}$ will be $3\frac{1}{6}$. Eight times this is $25\frac{1}{3}$; adding 88, we have $113\frac{1}{3}$ very nearly for the measure of the volume. I see that you have been busily calculating, Sagredo; do you agree that this is the correct answer?

Sagredo: I have been applying the same formula to a slightly different solid. Suppose that the side faces of the prismoid are continued upward so that the solid terminates in an edge instead of a rectangular face at the top. Clearly, the length of this edge will be 3 units and the overall height of the solid will be $7\frac{1}{2}$ units. Then formula (1) gives $V = (7\frac{1}{2}/3)\{0 + 40 + \sqrt{0}\}$, which comes to 100 cubic units exactly for the volume.

Salviati: We seem to have stumbled upon a counterexample to the axiom that the whole is greater than the part! I assure you that there are no mistakes in your arithmetic.

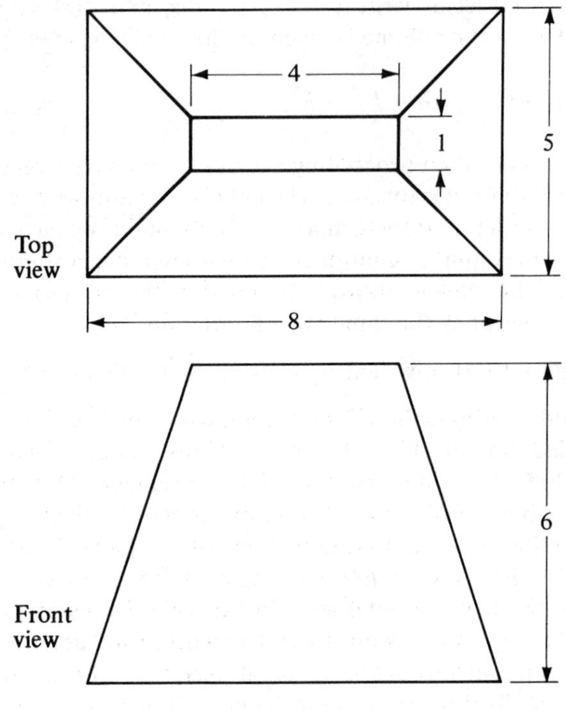

Fig. 97

Sagredo: We would indeed be faced with a paradox if we believed that the formula $V = (h/3)\{A_T + A_B + \sqrt{(A_T A_B)}\}$ provided valid volume measures for such solids. Apparently it does not. Let us try our other formula, Simplicio. If it applies to segments of spheres as well as to pyramid frusta, it might be expected to work in this case.

Simplicio: $V = (h/6)\{A_T + A_B + 4A_{MS}\}$ gives, for the solid of Fig. 97, $V = (6/6)\{4 + 40 + 4 \cdot 6 \cdot 3\}$, which comes to 116.

Sagredo: And for the complete solid, for which $h = 7\frac{1}{2}$ and $A_T = 0$, the same formula yields

$$V = \frac{7\frac{1}{2}}{6}\left\{0 + 40 + 4 \cdot 5\frac{1}{2} \cdot 2\frac{1}{2}\right\} = \frac{5}{4}\{40 + 11 \times 5\} = 1\frac{1}{4} \times 95 = 118\frac{3}{4}.$$

That looks better. At least we can check that there is no internal inconsistency in the application of this formula (2) to such prismoids that have nonsimilar bases and are therefore not frusta of pyramids. The complementary solid that together with the prismoid of Fig. 97 makes up my complete solid has a rectangular base 4×1, a height of $1\frac{1}{2}$ and a top edge of 3 units. Hence the midsection is a rectangle $3\frac{1}{2} \times \frac{1}{2}$, and so formula (2) gives

$$V = \frac{1\frac{1}{2}}{6}\{0 + 4 + 4(3\tfrac{1}{2})(\tfrac{1}{2})\} = 1/4\{4 + 7\} = 2\frac{3}{4},$$

which is precisely the difference between the previously calculated values of $118\frac{3}{4}$ and 116.

Simplicio: What an excellent verification!

Sagredo: I would hesitate to call it a verification, in the strict sense.

Simplicio: Why so?

Sagredo: Well, it would be possible for a formula to pass such a consistency test even while it is yielding incorrect answers. If, for example, a putative formula gives measures having values M_i for a quantity whose true measures are T_i, such that $M_i = kT_i$, then, even though the constant $k \neq 1$, $M_1 + M_2 = M_3$, iff $T_1 + T_2 = T_3$, and so a formula erroneous in this way would not be shown to be false by such a test. Admittedly in the present case it is hardly credible that a formula that has been shown to provide true volume measures for pyramid frusta should produce *constant* fractional errors as the shape of the prismoid to which it is applied deviates in various ways from the form of a pyramid frustum. One would expect the fractional error to depend upon the form of the solid concerned; and the form of the complementary piece surmounting the given prismoid is very different from that of the prismoid itself. I would say, then, that formula (2) has withstood a severe test, and that on this account its applicability in this new case is well supported, whereas that of formula (1) has been clearly falsified.[200]

Salviati: Very good. And, of course, the applicability of formula (2) in the case under consideration may readily be checked by considering the prismoid to be divided into a central prism and two end pieces by a pair of planes 3 units apart and running perpendicular to the longest edges of the solid (Fig. 98).

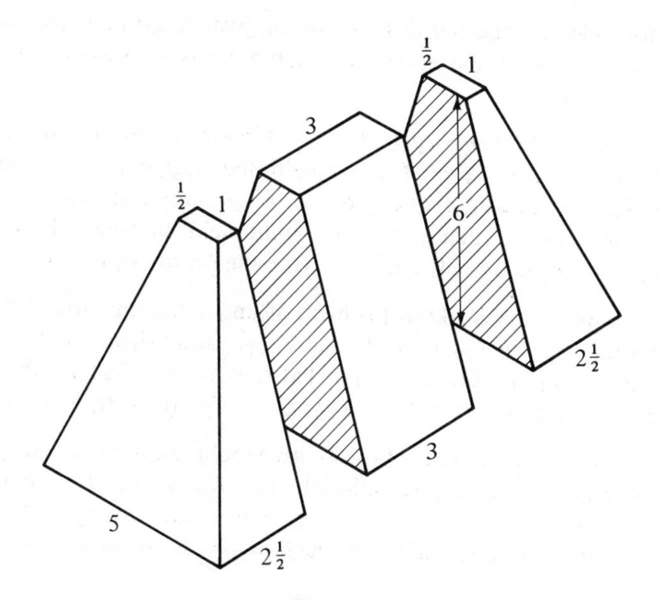

Fig. 98

Simplicio: The central prism has volume, $\{\frac{1}{2}(5+1)6\} \cdot 3 = 54$ cubic units, while the two end pieces may be combined to form a square-based pyramid frustum of volume $\frac{1}{3} \cdot 5^2 \cdot 7\frac{1}{2} - \frac{1}{3} \cdot 1^2 \cdot 1\frac{1}{2} = 25 \times 2\frac{1}{2} - \frac{1}{2} = 62$ cubic units. Hence the total volume is found to be 116 cubic units; so that formula (2) is shown to have given the correct answer in this case, and formula (1) an incorrect answer.

Salviati: Quite so. And if we write, from (1) and (2) respectively,

$$\frac{h}{3}\{A_T + A_B + \sqrt{(A_T A_B)}\} = V_1 \quad \text{and} \quad \frac{h}{6}\{A_T + A_B + 4A_{MS}\} = V_2,$$

we find that

$V_1 = V_2$ iff $2A_T + 2A_B + 2\sqrt{(A_T A_B)} = A_T + A_B + 4A_{MS}$, that is, iff $4A_{MS} = A_T + A_B + 2\sqrt{(A_T A_B)}$. Thus the necessary and sufficient condition in order for V_1 to be equal to V_2 is that A_{MS} be equal to $\{\frac{1}{2}(\sqrt{A_T} + \sqrt{A_B})\}^2$.

Sagredo: I see. I notice that there is a specially simple spatial interpretation for this last relation. If *a, b, m* units denote the side lengths of *squares* respectively equal to the parallel bases and the section of the solid midway between them, then the condition for which $V_1 = V_2$ reduces to $m = \frac{1}{2}(a+b)$. This result is obviously satisfied in the case of pyramidal or conical frusta.[201]

Salviati: Yes, and in the case of the extreme forms of these—prisms and cylinders.

Sagredo: But, clearly, the requisite condition just found in order for formulae (1) and (2) to yield values that agree with each other is *not* satisfied in the case of any segment of a sphere.

Salviati: No, nor should we expect it to be. Already, at the end of our last hour, you noted that in the case of a single-based segment of a sphere, $V = (h/3)\{A_T + A_B + \sqrt{(A_T A_B)}\}$ reduces to $V = \frac{1}{3}A_B h$, and so if this formula were to be applied in the case of such a solid, it would yield the true volume not of the spherical segment itself but of the inscribed cone on the same base.

Simplicio: So we are back to our problem of understanding how it comes about that formula (2): $V = (h/6)\{A_T + A_B + 4A_{MS}\}$, established for the case of any pyramid frustum, can be correctly applied also to a solid of such a different form as a segment of a sphere, or indeed, as you convinced us, to a whole sphere.

Salviati: To clear up this matter thoroughly would require the determination of the most general law of cross-sectional area for which $V = (h/6)\{A_T + A_B + 4A_{MS}\}$ provides correct volume measures. Before attempting that, let me offer for your consideration three solids, for each of which the volume may readily be determined independently of this formula (Fig. 99). You will notice immediately that the volume of each prism is simply given by $A_F \times d$, where $A_F = \frac{1}{2}bh, \frac{1}{3}bh, \frac{1}{4}bh$, respectively. I trust you know that Cavalieri, Torricelli, Fermat, and Roberval found the result

$$A_F \left[= \int_o^h cz^n dz, \text{ in the later Leibnizian notation} \right] = \frac{ch^{n+1}}{n+1} = \frac{1}{n+1}bh.$$

By about 1640, the three last-named mathematicians had discovered the applicability of this result to cases soon to be expressed with fractional and with negative indices ($n \neq -1$).[202]

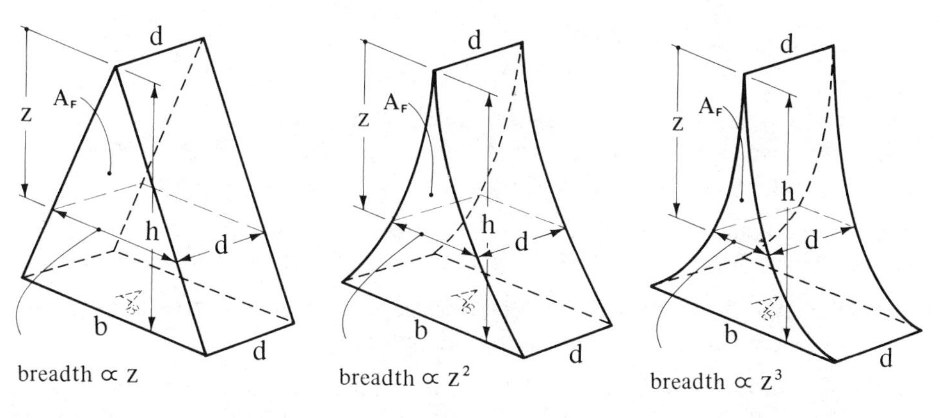

Fig. 99

Sagredo: I cannot help noticing that the second of your solids may be regarded as a Cavalierian transformation of a square-based pyramid, or of any other pyramid or cone for that matter. Consequently, both formulae (1) and (2) will give correct volume measures for this solid.

Simplicio: I can see that formula (1) when applied to the second solid, with $A_T = 0$, gives $V = (h/3)\{A_B + 0 + \sqrt{(A_B \times 0)}\} = \frac{1}{3}A_B h$, which agrees with the result given by $V = A_F d = \frac{1}{3}bdh$.

Sagredo: And it can equally well be seen that the application of formula (1) in this way to the first and third solids yields the identical result $V = \frac{1}{3}A_B h$ instead of the true volume measures, which are respectively $\frac{1}{2}A_B h$ and $\frac{1}{4}A_B h$ (given by $A_F d$ where $A_F = \frac{1}{2}bh, \frac{1}{4}bh$, respectively). We must notice, Simplicio, that formula (1) uses as data only the values of A_T, A_B, and h; hence it cannot be sensitive to changes in the law of cross-sectional area, that is, to changes in the law of increase of volume with respect to z. [$dV/dz = A(z)$.]

Simplicio: Then let us see whether the inclusion of A_{MS} in formula (2) is sufficient to allow the true volumes of the first and third of these solids to be obtained.

$$A_{MS}:A_B = \text{ratio of breadth at } z = \frac{1}{2}h \text{ to breadth at } z = h;$$

$$= 1:2^n, \text{ which } = \frac{1}{2} \text{ for } n = 1, \text{ and } = \frac{1}{8} \text{ for } n = 3.$$

Hence, the estimates provided by the second formula for the volumes of the first and third solids of Fig. 99 are, respectively:

$$V = \frac{h}{6}\left\{0 + A_B + 4\left(\frac{1}{2}A_B\right)\right\}, \text{ which comes to } \frac{1}{2}A_Bh,$$

$$\text{and } V = \frac{h}{6}\left\{0 + A_B + 4\left(\frac{1}{8}A_B\right)\right\}, \text{ giving } \frac{1}{4}A_Bh.$$

Sagredo: So, the second formula is correct for $A(z) = cz^n$, where $n = 1, 2$, or 3, and also, we should note, for $n = 0$, since then $A_T = A_B = A_{MS}$, and $V = (h/6)\{A_T + A_B + 4A_{MS}\}$ reduces to $V = A_Bh$, correct for a prism.

Simplicio: And, presumably, it will be correct for values of n over the whole range from $n = 0$ to $n = 3$. We will capture that segment of a sphere yet!

Salviati: What is true for the greatest and least of a class is not necessarily true for all—or even for any—of the values in between.

Sagredo: In the present case we have found that formula (2) gives correct volume measures when $A(z) = cz^n$, for $n = 0, 1, 2, 3$. Let us consider an intermediate value, such as $n = 1\frac{1}{2}$. $A_{MS}:A_B =$ (breadth at half height):(breadth at base) $= (\frac{1}{2}h)^{3/2}:(h)^{3/2}$. Thus, $A_{MS}:A_B = 1/2\sqrt{2}$, so $4A_{MS} = A_B\sqrt{2}$, and formula (2) gives

$$V = \frac{h}{6}\{0 + A_B + A_B\sqrt{2}\} = \frac{h}{6}(1 + \sqrt{2})A_B \approx 0.4024A_Bh,$$

compared with the exact value given by A_Fd, or by

$$\frac{bh}{n+1} \cdot d, \quad \text{or} \quad \frac{1}{n+1}A_Bh;$$

in this instance

$$\frac{1}{3/2 + 1}A_Bh = 0.4A_Bh.$$

So our formula gives a value approximately 0.6 percent in excess of the true value.

Simplicio: This discrepancy appears undeniable. I wonder what happens beyond the range $0 \le n \le 3$; perhaps we could try letting n equal 4. For the breadth and for $A(z)$ proportional to z^4, $A_{MS}:A_B = 1:2^4$, so $A_{MS} = \frac{1}{16}A_B$, and our formula gives $V = (h/6)\{0 + A_B + \frac{1}{4}A_B\} = \frac{5}{24}A_Bh$, whereas the exact volume expression, $(1/(n + 1))A_Bh$, gives the true value as $\frac{1}{5}A_Bh$.

Sagredo: Yes, the error is considerable. We are clearly in need of a more systematic approach to this problem. Perhaps you are willing to help us, Salviati?

Salviati: Let us first set aside the special case where $n = 0$. In that case the cross-section at depth z is given by Az^0; that is, the cross-section is constant, $A_T = A_{MS} = A_B = A$; and formula (2) and the expression $(1/(n + 1))A_Bh$ both agree in reducing to the correct value Ah for a prism or cylinder as you have noticed. For the other values of n, it is convenient to work in terms of the most obvious generalization of the solids illustrated in Fig. 99. Where the breadth $\propto z^n$, the question then to be asked is simply: For what values of n does the expression $(h/6)\{0 + A_B + 4A_{MS}\}$—in which $A_{MS}/A_B = 1/2^n$—give the same values as $(1/(n + 1))A_Bh$? The values of n that we seek, then, are the values for which,

$$\frac{h}{6}\{A_B + 4(A_B/2^n)\} = \frac{1}{n+1}A_Bh,$$

that is, we have to solve

$$1 + 4/2^n = \frac{6}{n+1}, \quad \text{or} \quad 2^{2-n} = \frac{6-n-1}{n+1}, \quad \text{or} \quad 2^{n-2} = \frac{n+1}{5-n}.$$

Inspection of this last equation shows that for the right side to be equal to the left side, the expression $(n + 1)/(5 - n)$ must not be negative or zero, neither can it be infinitely large; hence the required values of n must certainly lie between -1 and $+5$. Within this range, we have:

$$(n - 2)\log 2 = \log (n + 1) - \log (5 - n),$$

or $F(n) = G(n)$, say, where $F(n)$ is the increasing linear function $(n - 2) \log 2$, and $G(n)$ is the increasing nonlinear function $\log (n + 1) - \log (5 - n)$. As the sketch graph (Fig. 100) indicates (and consideration of the derivatives $G'(n)$ and $G''(n)$ could establish), $G(n)$ increases less and less rapidly as n increases from -1 to $+2$, and more and more rapidly from $n = 2$ to $n = 5$. Hence, since $F(n)$ is increasing uniformly, $F(n) = G(n)$ has at most three real roots. Substitution shows that these are indeed $n = 1, 2, 3$. Incorporating the additional case referred to at the outset, for which $n = 0$, the following alternatives to our original assumption—that $A(z) = cz^2$—are indicated:

$$A(z) = cz, \quad \text{or} \quad A(z) = cz^3, \quad \text{or simply} \quad A(z) = c,$$

where the cs are any arbitrary constants or

$$A(z) = c_0 + c_1z + c_2z^2 + c_3z^3,$$

to give the most general way of specifying a solid for which formula (2) correctly measures the volume corresponding to z equal to any given h, or indeed for which (2) correctly measures the volume of a frustum between parallel planes specified by *any* pair of arbitrary values of z. If you feel that this claim has not been ade-

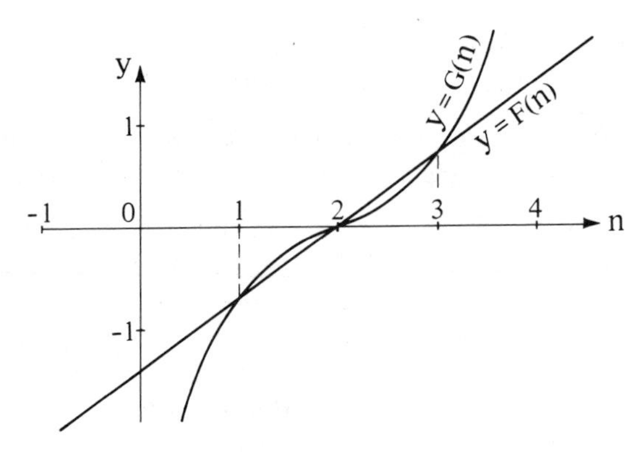

Fig. 100

quately established by the foregoing brief argument, in which, it is true, A_T was taken as zero merely for convenience, allow me to conclude by assuring you that if you write

$$V = \int_k^{k+h} (c_0 + c_1 z + c_2 z^2 + c_3 z^3)\, dz$$
$$= \left[c_0 z + \frac{1}{2} c_1 z^2 + \frac{1}{3} c_2 z^3 + \frac{1}{4} c_3 z^4 \right]_k^{k+h},$$

and if this is simplified and rearranged in the light of the following equivalences,

$$A_T = c_0 + c_1 k + c_2 k^2 + c_3 k^3,$$
$$A_B = c_0 + c_1(k+h) + c_2(k+h)^2 + c_3(k+h)^3,$$
$$A_{MS} = c_0 + c_1\left(k + \frac{1}{2}h\right) + c_2\left(k + \frac{1}{2}h\right)^2 + c_3\left(k + \frac{1}{2}h\right)^3,$$

you will assuredly be able to obtain formula (2): $V = (h/6)\{A_T + A_B + 4A_{MS}\}$. As this is a somewhat lengthy piece of work, it may be better if I leave it for you to do on your own. It occurs to me, however, that we might soon have occasion to work through a special case of it.

Sagredo: Well, in the light of all this, it is instructive to review the proof of formula (2) that I set out at the beginning of this hour. It is doubtless often enough true that a particular demonstration has been given without any thought arising as to possible further ranges of applicability. In our example, we might establish the following proposition:

If $A(z) = cz^2$, then $V = \int_k^{k+h} A(z)\, dz$ implies

$$V = \frac{h}{6}\{A_T + A_B + 4A_{MS}\},$$

yet all too easily ignore the question of whether *other* antecedents are sufficient to guarantee that same consequent.

Simplicio: So the diagram of the pyramid frustum, which it is natural to have accompanying your proof, can now be seen as symbolizing an unnecessarily restricted vision, and, in its turn, as reinforcing that vision.

Sagredo: Very true; as the diagrams of Fig. 96 already served to show, a very much wider range of solids than have traditionally been understood by the terms *pyramid* and *cone* have the essential property, $A(z) = cz^2$, guaranteeing the applicability of the proof under consideration (pp. 217–218). But further than this, as I am sure we now appreciate, the question as to whether or not *different* assumptions can be substituted for $A(z) = cz^2$ without the predication of formula (2) being rendered invalid was a matter requiring separate investigation.

Salviati: Quite so, and as you have indicated, the settling of his question—even the consciousness of it as a question to be asked—is not something that is likely to occur merely by reviewing an existing restricted proof. We have already noted both the logical and the psychological reasons for this. (See especially pp. 68, 167.) An increase in the universality of a proposition,[203] occurring as a result of the recognition that the antecedent has been expressed or understood in too restricted a fashion, *may* require little or no change in the proof. The generalization from Theorem α, involving regular circumscribable polygons, to Theorem A, involving irregular ones as well, is a case in point. On the other hand, an increase in universality sometimes comes about not so much by such direct extension as by the discovery of alternative antecedents that cannot be incorporated into the same proof structure. As for reviewing proofs, old and new, there certainly ought to be more of that. Naturally, the first aim will be to ensure that a proof is as free as possible both from significant omissions and from inelegant superfluities. But, having uncovered and made explicit assumptions that were previously only implicit, and having exposed as simply as possible a secure foundation for a conclusion,[204] we ought not really regard our analysis as finished without considering what is likely to be the more difficult matter: that different premises might also be found to entail the same (or a closely related) conclusion.[205]

Simplicio: I can well see that promising alternative assumptions may long remain hidden, and that, for these to be discovered, more than a casual inspection of an existing proof may be required. I suppose that a general answer to the question of how one comes upon likely alternative assumptions is hardly to be expected. But, to return to our present example, I can well believe that the chance that anyone would undertake such a "damn-fool" experiment as to try applying formula (2) to see whether it would measure the volume of a *sphere* is obviously extremely remote so long as this formula had always been associated only with the "pyramid assumption," $A(z) = cz^2$. If you are willing, Salviati, I should like to see an explicit consideration of the applicability of formula (2) to the case of the sphere and to the segment of a sphere.

Salviati: This case is, of course, subsumable under the general theory already sketched; nevertheless it will be instructive to consider it separately. Admittedly, the sphere is clearly *not* what we have referred to as a Cavalierian transformation, or, to use Cavalieri's own term, an "analogue" of any of the solids illustrated in Fig. 99, though it can be appropriately related to two of them. But first, Simplicio, if z measures the distance along a diameter of a sphere from one of its ends, tell us what is the function $A(z)$, which measures the area of any cross-section perpendicular to the diameter?

Simplicio: The area of the circular cross-section is given by πr^2 where r increases from zero to a maximum value, R say, the radius of the sphere itself, and then decreases back to zero, as z increases through the range from zero to $2R$. The precise relation may be read off from the sides of a right-angled triangle; it is $r^2 = R^2 - (R - z)^2$. Hence, $A(z) = \pi(2Rz - z^2)$.

Salviati: Right; so the volume is given by

$$\pi \int_{z=0}^{z=2R} (2Rz - z^2)\, dz = \pi\left[Rz^2 - \frac{1}{3}z^3 \right]_0^{2R} = \pi\left(4R^3 - \frac{8}{3}R^3 \right) = \frac{4}{3}\pi R^3.$$

But in order to connect this with our formula (2), which will have its real usefulness in the case of *segments* of a sphere having one or two (parallel) bases, let us write the volume integral as

$$V = \int_k^{k+h} (az - bz^2)dz, \quad \text{where} \quad 0 \le k < k + h \le 2R, \quad \text{and}$$

$$a = 2\pi R,\ b = \pi,$$

$$= \frac{1}{2}a(k + h)^2 - \frac{1}{2}ak^2 - \frac{1}{3}b(k + h)^3 + \frac{1}{3}bk^3.$$

Now, Simplicio, please verify that this can be rearranged and simplified in order to obtain the formula $V = (h/6)\{A_T + A_B + 4A_{MS}\}$ exactly. First, it will be a good idea for you to express A_T, A_B, and A_{MS} in terms of a, b, h, and k.

Simplicio: $A(z)$, with $z = k$, gives $A_T = ak - bk^2$,
with $z = k + h$, gives $A_B = a(k + h) - b(k + h)^2$,

and with $z = k + \frac{1}{2}h$, gives $A_{MS} = a\left(k + \frac{1}{2}h\right) - b\left(k + \frac{1}{2}h\right)^2$,

so $4A_{MS} = 4ak + 2ah - 4bk^2 - 4bkh - bh^2$.

Guided by these equivalences, the volume expression may be rearranged, I suppose, without difficulty:

$$V = \frac{1}{2}ak^2 + akh + \frac{1}{2}ah^2 - \frac{1}{2}ak^2 - \frac{1}{3}bk^3 - bk^2h - bkh^2 - \frac{1}{3}bh^3 + \frac{1}{3}bk^3.$$

Eliminating the terms that collect to zero and taking out $h/6$ as a common factor:

$$V = \frac{h}{6}\{6ak + 3ah - 6bk^2 - 6bkh - 2bh^2\}$$
$$= \frac{h}{6}\{(ak - bk^2) + (ak + ah - bk^2 - 2bkh - bh^2)$$
$$+ (4ak + 2ah - 4bk^2 - 4bkh - bh^2)\}$$
$$= \frac{h}{6}\{A_T + A_B + 4A_{MS}\} \quad \text{Q.E.D.}$$

Sagredo: A satisfying reduction, I am sure, at least from a formal point of view. But to delve a little further, I think you remarked, Salviati, that the sphere, or the segment of a sphere, could be related to two of the solids of Fig. 99. I take it that there is an indication of this to be found in the form of the integrand, $az - bz^2$, which is but $A(z) = c_0 + c_1z + c_2z^2 + c_3z^3$ with $c_0 = 0$, $c_1 = a$, $c_2 = -b$, $c_3 = 0$.

Salviati: Yes, and $a = 2\pi R$, $b = \pi$. So when we write for the volume of the whole sphere:

$$\int_0^{2R}(2\pi Rz - \pi z^2)\,dz, \quad \text{or} \quad \int_0^{2R}2\pi Rz\,dz - \int_0^{2R}\pi z^2\,dz,$$

we could just as well be measuring the difference of the volumes of two solids of the same general form as depicted in the first two diagrams of Fig. 99, with $A_B = 4\pi R^2$ and $h = 2R$. If the second solid were cut away from the interior of the first, the cross-sections (parallel to the base edges) of the remainder would all be equal to the corresponding cross-sections of the sphere; that is to say, the remainder solid would be an analogue of the sphere. But, for a more interesting analogue, let us return to the unsimplified expression for the area of cross-section of the sphere. As Simplicio found, this is πr^2, where $r^2 = R^2 - (R - r)^2$. Hence, for the whole sphere,

$$V = \int_0^{2R}\pi\{R^2 - (R - z)^2\}\,dz, \quad \text{or} \quad \int_0^{2R}\pi R^2 dz - \int_0^{2R}\pi(R - z)^2\,dz.$$

The first of these last two integrals measures the volume of the cylinder that would circumscribe the sphere, while the second gives the volume of a double cone that could be inscribed coaxially within the circumcylinder, vertex at the center. If this double conical form is regarded as having been removed from the solid cylinder, the remainder is an analogue of the sphere. The equality of the corresponding cross-sections has already been established in connection with Fig. 83 (cf. also Fig. 92).

Simplicio: You have certainly given us plenty to think about, Salviati. It appears that the kind of solid commonly covered by the term *prismoid* constitutes but a small subclass of the solids whose volumes are correctly measured by the prismoidal formula.[206]

Sagredo: Our hour is up, but not our day's discussion I believe.

Simplicio: Yes, apart from your intimation at the beginning of this hour, Salviati, I have just now received a strange yet distinct impression that we are to be granted another hour.

Salviati: I too think we can count on that.

XIV

Without constant stimulation by the spur of the unknown . . . it might be feared that men of science would become system ridden in their acquirements and their knowledge. Then science would halt through intellectual inertness, just as minerals, in saturated solution, become chemically inert and crystallise.

—Claude Bernard

It should always be required that a mathematical subject not be considered exhausted until it has become intuitively evident.

—Felix Klein

The outcome of any serious research can only be to make two questions grow where only one grew before.

—Thorstein Veblen

I have yet to see any problem, however complicated, which, when you looked at it in the right way, did not become still more complicated.

—Poul Anderson

If, by examining the simplest cases, we can bring to light what mankind has there done by instinct, and can extract from such procedures what is universally valid in them, may we not thus arrive at general methods for forming concepts and establishing principles which will be applicable also in more complicated cases?

—Gottlob Frege

To make a theory of mathematical creativity is almost a contradiction in terms, for nothing can be less easily described in terms of techniques or recipes than creative originality. As soon as one uses a textbook, one establishes a didacticism, an academicism, even if the book be so written as to promote individual research. What is one to conclude other than that efforts at improving pedagogy will always be unfinished?

—René Thom

HORA QUARTA DECIMA

Sagredo: Ah, Salviati! How good it is to be able to continue a little longer. Simplicio and I have been reviewing our day's discussion. Circumstances have certainly turned us into compulsive investigators. It is always the same at these reunions: we continue our discussions until the last moment, only to finish with more problems than we started with!

Simplicio: I admit that more and more problems open up as we proceed, though I feel that these are mostly disposed of, thanks to Salviati's guidance.

Sagredo: I am fully aware of how indebted we are to Salviati; yet each time I reflect on what we have achieved, still further questions come to mind. For example, consider the following table of values:

Figure	Shape Parameter
Equilateral triangle	$\sqrt{3}/36 \approx 0.0481$
Square	$1/16 = 0.0625$
Regular hexagon	$\sqrt{3}/24 \approx 0.0722$
⋮	⋮
Circle	$1/4\pi \approx 0.0796$

(You will remember, Simplicio, that the shape parameter is the area of figure of given form when the perimeter is equal to unity.) Two problems are immediately suggested by this table: (1) How does one *prove* in general that for a regular n-sided polygon, the shape parameter continually increases up to $1/4\pi$ as a limit, as the number of sides is allowed to increase without limit? (2) How can it be shown that for two polygons with the same number of sides, one polygon being regular and the other irregular, the shape parameter of the regular one is always greater than that for the irregular one?

Simplicio: Perhaps we can learn something of value in relation to the second of these problems by showing that in the case of rectangles, the square has the greatest area for a given perimeter. I suppose a Greek geometer would have proved this by applying the relations discussed in our second hour in connection with the semicircle construction for obtaining a mean proportional (Fig. 7). Adapting that diagram to our present needs, we have (Fig. 101), $g^2 = \ell w$, that is, the square of side g is equal in area to the rectangle of area $\ell \times w$. But, for $\ell \neq w$, we have $r > g$; hence the square of side r is greater in area than the rectangle of equal perimeter.

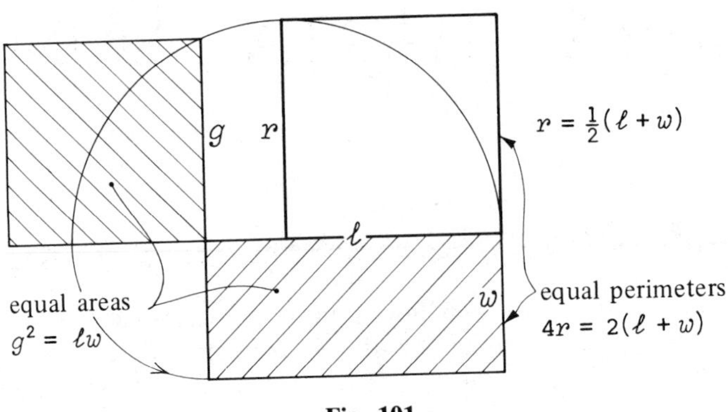

Fig. 101

Sagredo: Yes, or else we may compare any rectangle with its isoperimetric square as in Fig. 102. The area relation may be read from this diagram as $\ell w + d^2 = \{\frac{1}{2}(\ell + w)\}^2$, where $d + w = \ell - d$; that is, $d = \frac{1}{2}(\ell - w)$. Thus the rectangle falls short of the isoperimetric square by the square of area d^2.

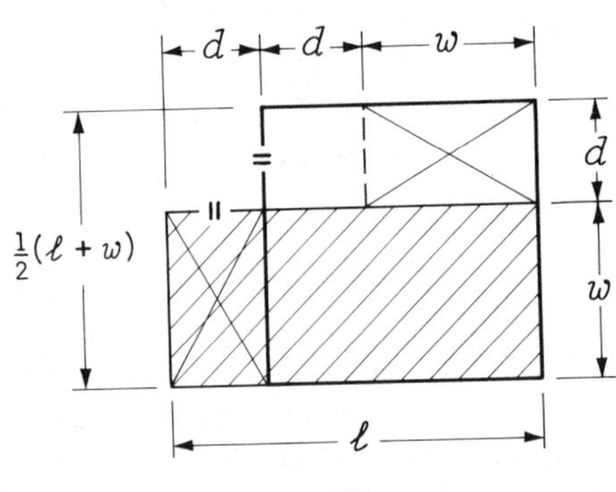

Fig. 102

Simplicio: I notice that the rectangle of area ℓw is equal to the difference of two squares, namely, $\{\frac{1}{2}(\ell + w)\}^2 - d^2$. This suggests a construction based on a right-angled triangle for obtaining a square equal in area to a given rectangle.[207]

Salviati: The construction you have in mind may be superimposed on Fig. 102 so as to yield the very same result as indicated in your own diagram of Fig. 101. The construction implicit in your Fig. 101, which we have already discussed (in connection with Fig. 7; cf. also Fig. 51(ii)), is also given by Euclid: see his II, 14 and cf. VI, 13. But let us illustrate the correspondence between Figs. 101 and 102 in a new diagram, Fig. 103. Here, XY is equal to the semiperimeter of the rectangle and C and A are points that divide this line interval into equal and unequal

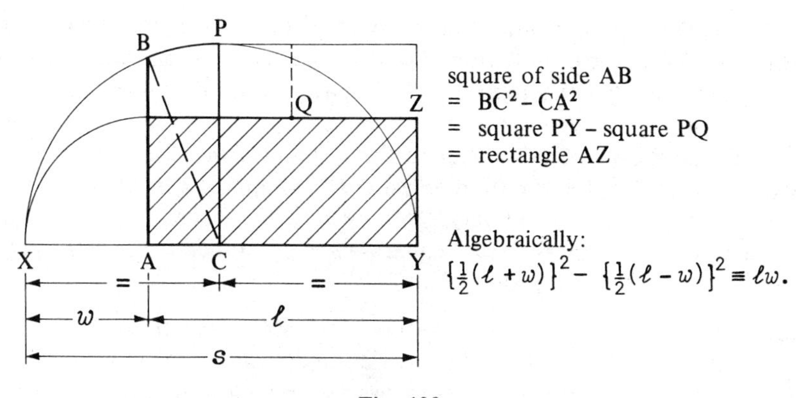

Fig. 103

segments. You will see how this new figure illustrates Euclid II, 5, which is enunciated:

> *If a straight line be cut into equal and unequal segments, the rectangle contained by the unequal segments of the whole together with the square on the straight line between the points of section is equal to the square on the half.*[208]

Observe also that the difference between the area of the square and the area of the isoperimetric rectangle is vanishingly small as the rectangle approaches the form of the square. As Kepler noted, "Near a maximum the decrements on both sides are in the beginning only imperceptible."[209]

Sagredo: A corresponding remark is applicable in the case of typical minimum values and certain inflexions.

Salviati: This is why such points are termed *stationary points.*

Simplicio: I notice that $9.9 \times 10.1 = 99.99$, which differs negligibly, or almost "imperceptibly," from 10 squared.

Salviati: Quite so. Kepler's remark could be regarded merely as a rather obvious comment on a table of values such as the following:[210]

$10 \times 10 = 100$
$9 \times 11 = 99 = 10^2 - 1^2$
$8 \times 12 = 96 = 10^2 - 2^2$
$7 \times 13 = 91 = 10^2 - 3^2$
$6 \times 14 = 84 = 10^2 - 4^2$
. . . (etc.)

On the other hand, we might have here the seed for Fermat's brilliant anticipation of the differential calculus. Perhaps you recall how Fermat dealt with the present problem, or its equivalent, namely, *to divide an interval XY internally at A so*

that $XA \cdot AY$ shall be a maximum. In our symbolism, if the given interval XY has length s of which one part is ℓ, then the other part is $s - \ell$, and their product is $s\ell - \ell^2$. Fermat then lets the segments of XY have the values $\ell + e$ and $s - \ell - e$, their product being $s\ell - \ell^2 + se - 2\ell e - e^2$.
Writing this as approximately equal to the original product $s\ell - \ell^2$ (for which operation Fermat adopted the term *adégalant,* "adequating"), there is obtained after simplifying $s \approx 2\ell + e$. Finally, "suppressing e," Fermat obtains from this (approximate) adequation the exact equation $s = 2\ell$, indicating that the product $XA \cdot AY$ will be a maximum when $\ell = \frac{1}{2}s$.[211]

Sagredo: There appears to be a close resemblance between this method of Fermat's and the later method of differentiation "from first principles," but I understand that Fermat was not able to supply a background theory that would have justified his procedure. Perhaps he felt no need for this.

Salviati: Well, those of his successors who recognized the need were unable to satisfy it for a very long time to come.[212]

Simplicio: Is there not yet another method of showing that the square exceeds in area any rectangle of equal perimeter? I am thinking of the algebraic method called "completing the square."[213] I should like to see how this works out.

Salviati: If s measures the semiperimeter and ℓ the length of a rectangle, then

$$\text{Area} = \ell(s - \ell) = -\left\{\ell^2 - s\ell + \left(\frac{1}{2}s\right)^2\right\} + \left(\frac{1}{2}s\right)^2 = \left(\frac{1}{2}s\right)^2 - \left(\ell - \frac{1}{2}s\right)^2,$$

which shows that the maximum area is given by $(\frac{1}{2}s)^2$ and that this is obtained when $\ell = \frac{1}{2}s$. The relation $\ell(s - \ell) = (\frac{1}{2}s)^2 - (\ell - \frac{1}{2}s)^2$ is nothing else but $\ell w = \{\frac{1}{2}(\ell + w)\}^2 - \{\frac{1}{2}(\ell - w)\}^2$, implicit in our version of Euclid II, 5 (Fig. 103). But now I am sure that we have discussed at sufficient length what is but a very special case of one of the problems that Sagredo posed at the begining of this hour.

Sagredo: You have indeed. I was hoping to prove that for two polygons of equal perimeter, having the same number of sides, one being regular and the other irregular, the area of the regular one exceeds that of the irregular one. In considering this problem during our pause I got as far as showing by way of a construction, that this must be so in the case of four-sided polygons. Figure 104 shows that the conversion of any irregular quadrilateral into a regular one of equal perimeter is necessarily accompanied by an increase in area. $ABCD$ represents any given irregular quadrilateral; it may be converted into a kite $AB'CD'$ of equal perimeter by making $AB' + B'C$ equal to $AB + BC$ and $AD' + CD'$ equal to $AD + CD$. If B is thought of as moving to B' along the elliptical arc through B having its foci at A and C, it can be seen that B' is further from diagonal AC than B is. Likewise D' is further than D from AC. Hence kite $AB'CD' >$ quad. $ABCD$. A similar argument shows that rhombus $A'B'C'D' >$ kite $AB'CD'$. Finally, the square has a greater area than any nonsquare rhombus of equal side length since its opposite sides are further apart.

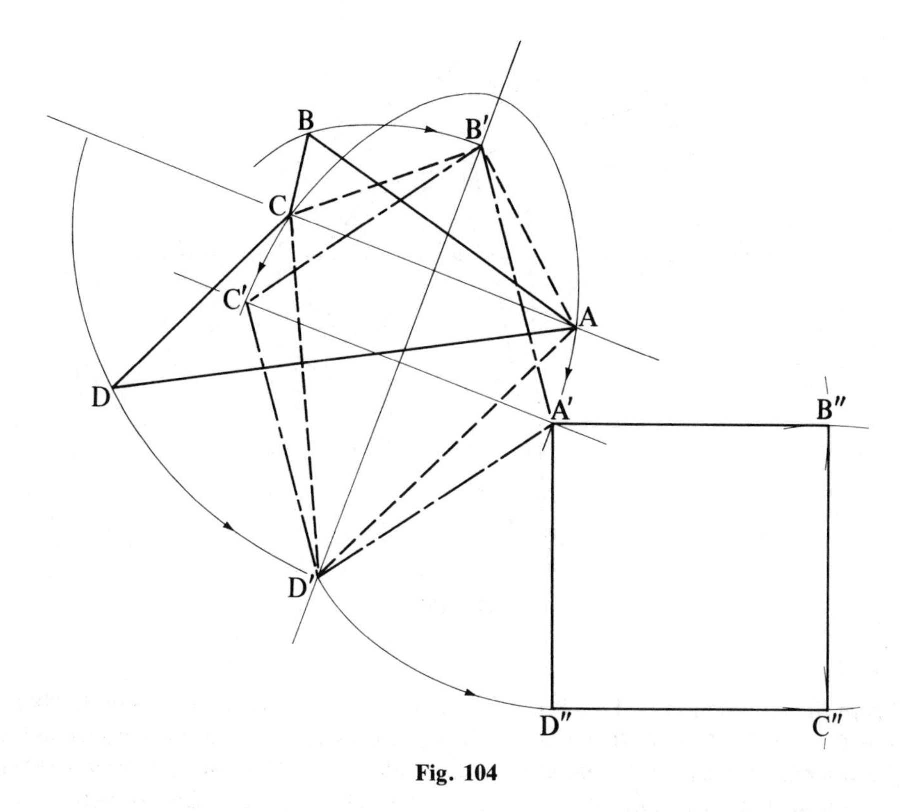

Fig. 104

Simplicio: Oh fine! At each stage of the construction there is an increase in area: square > rhombus > kite > original irregular quadrilateral. But I do not know what a purist would say about your use in elliptical arcs.

Sagredo: They are not an essential part of the construction, Simplicio. Clearly, a straight line interval AP may be marked out equal to $AB + BC$ and bisected at M, say; then B' is located by making AB' and CB' each equal to AM. I did not think it necessary to complicate my diagram by including such details. And, of course, having established what we set out to establish with the aid of the construction as I have given it, we can henceforth substitute for it a very much simpler, straight-edge-and-compasses construction, consisting merely of laying out the perimeter of the given quadrilateral along a straight line and using one-quarter of this as the side of the required square constructed in the usual way. I suspect that like considerations would apply in the case of any polygon. Consider, for example, a triangle ABC (Fig. 105). Yes, we may use the same type of construction as that already incorporated into Fig. 104. With $BC = a$, $CA = b$, $AB = c$, we have, in $\Delta A'BC$, $A'B = A'C = \frac{1}{2}(b + c)$, and in $\Delta A'B'C$, $A'B' = B'C = \frac{1}{2}\{a + \frac{1}{2}(b + c)\}$, and so on. I have no doubt that it would be easy to show that for the sequence of triangles ABC, $A'BC$, $A'B'C$, $A'B'C'$, $A''B'C'$, . . . *ad infinitum,* the side lengths approach $\frac{1}{3}(a + b + c)$ as a limit.

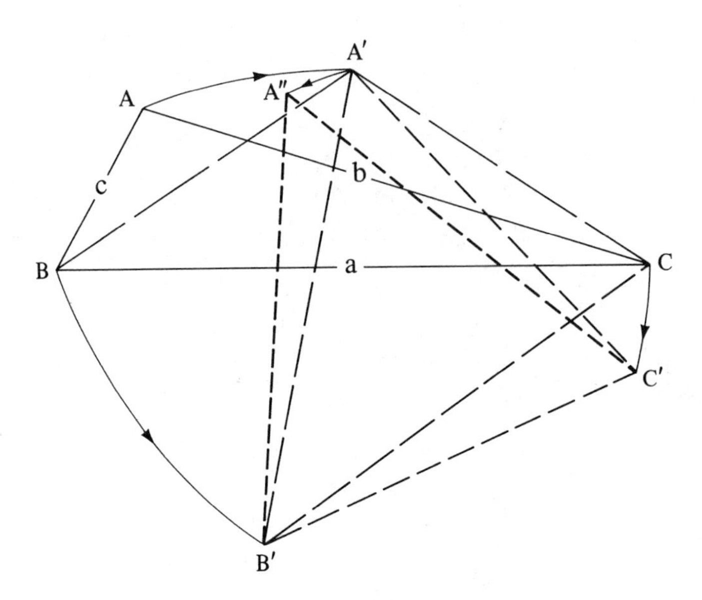

Fig. 105

From the elliptical arcs AA', BB', CC', $A'A''$, . . . (with their foci respectively at B & C, C & A', A' & B', B' & C', . . .) it is evident that the areas of these triangles form a continually increasing sequence of values the closer the equilateral form is approached. It occurs to me now that it ought to be possible, somehow, to show from the general formula for the area of a triangle, namely, $\Delta = \sqrt{\{s(s-a)(s-b)(s-c)\}}$,[214] that the area is a maximum for a given perimeter when $a = b = c$.

Salviati: This is so. Observe that since s, equal to $\frac{1}{2}(a+b+c)$, is assumed fixed, Δ is a maximum when $(s-a)(s-b)(s-c)$ is a maximum. Now $(s-a)+(s-b)+(s-c) = 3s-(a+b+c) = s$, which is constant. Writing for simplicity, $s-a=f_1$, $s-b=f_2$, $s-c=f_3$, we would like to show that the product of these factors, $f_1f_2f_3$, is a maximum when $f_1=f_2=f_3$.

Simplicio: If we can prove this, then it seems to me that we show not only that the triangle of maximum area for a given perimeter is equilateral, but also that the rectangular prism of maximum volume for a given surface area will be a cube.

Sagredo: Well, we have shown that, in two dimensions, for $\ell + w$ constant, the product ℓw is a maximum when $\ell = w$. In three dimensions, we may, I suppose, reasonably conjecture that the product ℓbh $(= V)$ will be a maximum when $\ell = b = h$, given either (1) that $\ell + b + h$ is constant, or (2) that A, equal to $2(\ell b + bh + h\ell)$, is constant. The geometrical interpretations are obvious; I am just wanting to point out that the two conjectures are different and would appear to require distinct verifications.

Simplicio: What you say is undeniable; I see now that I conflated two problems. And I must say that I am still not altogether clear about this since it seems that the surface area (as well as the volume) will actually be a maximum, for $\ell + b + h$ constant, when $\ell = b = h$.

Sagredo: We may call this Conjecture (3).

Simplicio: In the surface area expression, $2(\ell b + bh + h\ell)$, ℓb is a maximum for $\ell = b$, and bh is a maximum for $b = h$, and $h\ell$ is a maximum for $h = \ell$.

Sagredo: What you say does not seem to me to constitute a proof. ℓb is a maximum when $\ell = b$, provided $\ell + b$ is constant. But this is not what we are assuming; our assumption is that $\ell + b + h$ is constant. However, given the symmetry of the surface area expression, I expect that your conclusion is correct, even though I am dissatisfied with your argument for it. I think that we should now ask you, Salviati, to show us that for three positive numbers, f_1, f_2, f_3, having a fixed sum, the product is a maximum when $f_1 = f_2 = f_3$.

Salviati: Very well, but let me first remind you of some related results that we obtained in the second hour. We showed that for positive a and b, $\sqrt{(ab)} \leq \frac{1}{2}(a + b)$, and also that $\frac{1}{2}(a + b) \leq \sqrt{\{\frac{1}{2}(a^2 + b^2)\}}$(See pp. 19 and 22.) So, of course, $ab \leq \frac{1}{2}(a^2 + b^2)$, the equality holding only for a equal to b. And this last relation could have been established very easily on its own in the first place, by writing

$$\frac{1}{2}(a^2 + b^2) - ab \equiv \frac{1}{2}(a^2 - 2ab + b^2) \equiv \frac{1}{2}(a - b)^2 \geq 0.$$

I know you do not require all the details spelled out. But now observe how we may obtain $ab \leq \{\frac{1}{2}(a + b)\}^2$ from $ab \leq \frac{1}{2}(a^2 + b^2)$. In the latter relation, write $a^2 = A$, $b^2 = B$, $ab = \sqrt{(AB)}$, so obtaining $\sqrt{(AB)} \leq \frac{1}{2}(A + B)$. Here A and B like a and b may take any nonnegative real values, and so we may just as well write

$$\sqrt{(ab)} \leq \frac{1}{2}(a + b), \quad \text{or} \quad ab \leq \left\{\frac{1}{2}(a + b)\right\}^2,$$

for any nonnegative a, b.

Sagredo: I find it a little surprising that it is possible to start with

$$ab \leq \frac{1}{2}(a^2 + b^2), \quad \text{or with} \quad \sqrt{(ab)} \leq \sqrt{\left\{\frac{1}{2}(a^2 + b^2)\right\}} \tag{1}$$

and to obtain from this the closer relation

$$\sqrt{(ab)} \leq \frac{1}{2}(a + b). \tag{2}$$

By merely changing the symbols, you have refined a crude relation to obtain a sharper, more informative one; I am, of course, thinking of cases where $b \neq a$.

For example, when $a = 4$, $b = 9$, (1) gives $\sqrt{36} < \sqrt{48\frac{1}{2}}$, or 6 < almost 7; whereas (2) gives $\sqrt{36} < \frac{1}{2}(4 + 9)$—that is, $6 < 6\frac{1}{2}$.

Salviati: It is as you say. With familiarity, the process will doubtless come to seem less strange. Let us proceed now to the analogous situation where *three* factors, or terms, are involved. Just as $\sqrt{(ab)} \leq \frac{1}{2}(a + b)$, for positive a and b, was obtained from $ab \leq \frac{1}{2}(a^2 + b^2)$, so we may obtain $\sqrt[3]{(abc)} \leq \frac{1}{3}(a + b + c)$ from $abc \leq \frac{1}{3}(a^3 + b^3 + c^3)$.

Simplicio: We will first have to establish this latter relation.

Sagredo: Presumably this will be done by showing that $a^3 + b^3 + c^3 - 3abc$ is positive, except when it is zero—when $a = b = c$.

Salviati: That this is true, provided that $a + b + c > 0$, can be shown as follows. The factorization might not be immediately obvious, but I think you would reach it after a few trials:

$$a^3 + b^3 + c^3 - 3abc \equiv (a + b + c)(a^2 + b^2 + c^2 - ab - bc - ca)$$
$$\equiv \frac{1}{2}(a + b + c)\{(a - b)^2 + (b - c)^2 + (c - a)^2\}.$$

Inspection of this last expression shows it to be positive or zero, provided the first factor is not negative. And if $a + b + c > 0$, the whole expression is zero only if $a = b = c$. So, $a^3 + b^3 + c^3 \geq 3abc$, or $abc \leq \frac{1}{3}(a^3 + b^3 + c^3)$, which was the first relation to be established. In our easier example we substituted A for a^2 and B for b^2 at the corresponding stage of the working. But we have already posed the problem in terms of three positive factors, f_1, f_2, f_3, so I will use these symbols to stand, respectively, for a^3, b^3, c^3, thus obtaining

$$\sqrt[3]{(f_1 f_2 f_3)} \leq \frac{1}{3}(f_1 + f_2 + f_3),$$

with the equality holding only if $f_1 = f_2 = f_3$.[215] And so, if we write $f_1 + f_2 + f_3$ equal to $3m$, assumed constant, then the product $f_1 f_2 f_3$ has its maximum value of m^3 occurring when $f_1 = f_2 = f_3$.

Sagredo: In particular, for our rectangular prism with $\ell + b + h$ constant, the volume measure ℓbh has its maximum value when $\ell = b = h$, as we expected on analogy with the two-dimensional case. I cannot yet see, however, a way of showing that ℓbh will be a maximum when $\ell = b = h$ on the different assumption that it is the surface area expression, $2(\ell b + bh + h\ell)$ that is to be held constant.

Salviati: It is only necessary to notice that ℓbh will be a maximum when $(\ell b)(bh)(h\ell)$ is a maximum. And we know that the product of three factors of constant sum is a maximum when the factors are equal—so, Sagredo?

Sagredo: Ah yes, I see it now; regarding ℓb, bh, $h\ell$ as the three factors of constant sum, for a fixed surface area, then the product of these factors (and hence the volume expression ℓbh itself) will be a maximum when $\ell b = bh = h\ell$, that is, when $\ell = b = h$.

Salviati: Likewise, the product $(s - a)(s - b)(s - c)$ must be a maximum when $s - a = s - b = s - c$, if we assume that $s - a + s - b + s - c$ is constant, as it is where s, equal to $\frac{1}{2}(a + b + c)$, is fixed.

Sagredo: Then, since $s - a = s - b = s - c$ iff $a = b = c$, the general area expression for a triangle, $\sqrt{\{s(s - a)(s - b)(s - c)\}}$ is itself a maximum, for s constant, when $a = b = c$.

Simplicio: So, of all triangles having a given perimeter, the equilateral has the greatest area, and, as was shown earlier, of all quadrilaterals with a given perimeter, the square has the greatest area. Also, in our eighth hour, Salviati, you showed us that the square has a greater area than the isoperimetric equilateral triangle. Presumably, corresponding relations consistently hold for what we may call the higher polygons. I see that we are returned to the very questions that you raised at the beginning of this hour, Sagredo. But now I have a vague impression that perhaps what we have need of here is a particular, very ingenious theorem with which you so impressed us, Salviati, long ago. It concerned, I believe, relations between the area of a circle and the areas of certain regular polygons related to it.

Salviati: You can only be referring to the following four-part theorem that our first author issued through me in the last of his books:[216]

> [I] *The circle is the mean proportional between any two similar regular polygons of which one is circumscribed about it and the other is isoperimetric to it. Also* [II], *the circle being less than all circumscribed [figures], it is nevertheless the greatest of all isoperimetric [figures]. And* [III] *among the circumscribed [polygons], those that have more angles are smaller than those that have fewer; on the other hand* [IV], *among isoperimetric [polygons] those having more angles are the greater.*

Sagredo: Oh yes, I have overlooked these proportions, and the possible relevance that they might have for today's investigations. Perhaps the reason is that my appreciation of at least the first part of this theorem was from a different point of view. I connected it in my mind with a problem that must have presented itself to countless mathematicians at least back to early Greek times, namely, that of finding the appropriate mean—that is, the appropriate interpolation procedure—to be used in order to obtain the area of a circle from the determinable areas of two squares, or other similar and regular polygons, one inscribed in and the other circumscribed about the circle.[217] I remember how I at first marveled how this difficult problem was avoided by *replacing* the inscribed regular polygon with another, one isoperimetric with the circle. It was thus possible to establish the beautifully simple relation that the area of the circle is the geometric mean of the area of this isoperimetric polygon and of the similar circumscribing one (Fig. 106). But then, of course, I very soon realized that this no more enables us to *find* the area of a circle than does Archimedes' proposition that this area is given by half the product of the circumference and the radius.[218] Now that Simplicio has raised the question of the possible relevance of this proposition or of the other three just enunciated, I think it would be helpful, Salviati, if you would outline the proofs that you previously gave us.

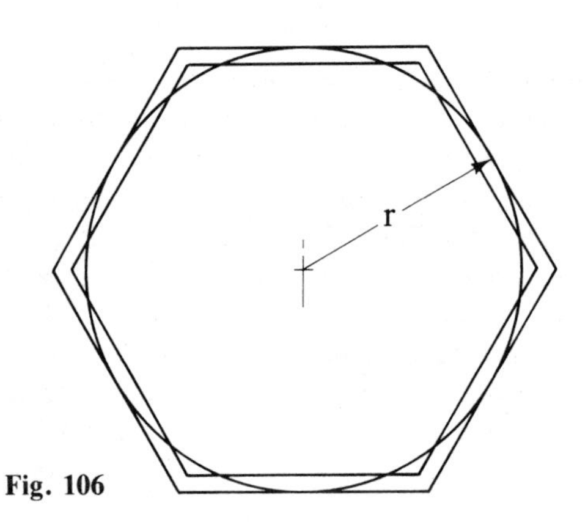

Fig. 106

Salviati: Very well. Although I might as well employ a concise and transparent symbolism of the kind customary with the modern mathematicians, I shall retain exactly the same reasoning as in the original. Let $p_{c,n}$, $p_{i,n}$, $A_{c,n}$, $A_{i,n}$, denote the measures of the perimeters and the areas respectively of the circumscribing polygon of *n* sides and of the similar polygon isoperimetric with the circle; then, if A_o, C, r denote the area, circumference, and radius of the circle,

$$A_{c,n}/A_o = \frac{1}{2}(p_{c,n})r/\frac{1}{2}Cr = p_{c,n}/C. \tag{3}$$

Further, since the areas of similar figures are to one another as the squares of their perimeters,

$$A_{c,n}/A_{i,n} = (p_{c,n}/p_{i,n})^2 = (p_{c,n}/C)^2$$

$$\therefore\ p_{c,n}/C = \sqrt{(A_{c,n}/A_{i,n})} \tag{4}$$

Hence, from (3) and (4), $A_{c,n}/A_o = \sqrt{(A_{c,n}/A_{i,n})}$, from which, $A_o = \sqrt{(A_{c,n}A_{i,n})}$, Q.E.D.[I].

Sagredo: I cannot help noticing now that this proof establishes a relation for a much wider class than is claimed in the enunciation. For if we have in mind, not merely regular polygons but *irregular* circumscribable ones, the whole proof still stands, provided, of course, that the two polygons are kept similar to each other.

Salviati: You are absolutely correct, Sagredo. Yet, while this unnecessary restriction in our first proposition may now appear to be due simply to oversight, I must say in defense, and on behalf of our first author, that this proposition was but one part of a four-part theorem and for the purpose of the overall presentation, the restriction to regular polygons was expedient.

Sagredo: I can appreciate that it may be convenient for a lemma, being ancillary to other propositions, to be expressed in a less general way than might be desirable if it were to be offered as a theorem in its own right.

Salviati: At least in the present case, the third and fourth parts of the theorem are *not* generally true for *all* circumscribed and for *all* isoperimetric polygons respectively, as the slightest consideration will show. The propositions we have designated here are III and IV are certainly to be taken as referring to regular polygons.

Simplicio: The second part of the theorem appears to me to be the most important.

Salviati: Probably so. We regarded this, in effect, as a corollary of the first result, which we may write as $A_o/A_{i,n} = A_{c,n}/A_o$. Then, since $A_{c,n} > A_o$, the whole being greater than the part, it follows that $A_o > A_{i,n}$—Q.E.D.[II]. As I expressed it originally, *the circle "is the greatest of all regular polygons to which it is isoperimetric."* (But since your remark about the range of applicability of the first proposition applies equally here, we may note that "polygons having any circumscribable form" could have been substituted in place of "regular polygons" in this last statement.) And now to the third proposition, according to which $A_{c,n} < A_{c,m}$, for n greater than m. The demonstration was actually given in terms of a particular pair of regular polygons circumscribing the same circle; a pentagon and a heptagon. It was to be understood that the argument could be repeated for any pair of regular polygons circumscribable about the same circle, however many sides each might have. In Fig. 107, *AD, AC* represent half sides of a regular

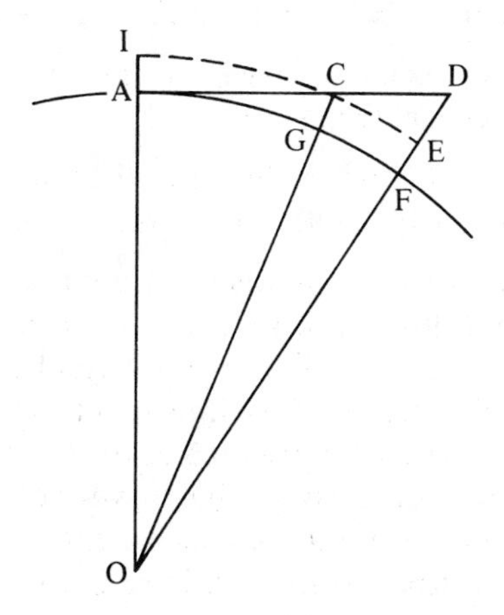

Fig. 107

pentagon and heptagon respectively, circumscribed about the circle with centre O and radius OA. $\triangle DOC >$ sector EOC, whereas sector $COI > \triangle COA$; hence, $\triangle DOC/\triangle COA >$ sector EOC/sector $COI =$ sector FOG/sector GOA.

Componendo: $\triangle DOA/\triangle COA >$ sector FOA/sector GOA,

permutando: $\triangle DOA$/sector $FOA > \triangle COA$/sector GOA,[219]

$\therefore 10 \cdot \triangle DOA/10 \cdot$ sector $FOA > 14 \cdot \triangle COA/14 \cdot$ sector GOA,

i.e., pentagon/circle > heptagon/circle.

Hence, the (circumscribed regular) heptagon is less than the (circumscribed regular) pentagon.

Sagredo: Oh, very clear—and, as you have indicated, the general result presumed to be established by this particular demonstration is that among regular polygons circumscribed about the same circle, "those having more angles are smaller than those having fewer"—Q.E.D.[III].

Salviati: Exactly so. Of course, we should now prefer to write $2m$ in place of the 10 and $2n$ in place of the 14 and so obtain an explicit general demonstration that $A_{c,n} < A_{c,m}$ for $n > m$, without reference to the particular case. The fourth proposition, concerning the relative areas of regular polygons *isoperimetric* with a given circle, was likewise explicitly proved only for the case of the pentagon and heptagon, each assumed regular. By the first proposition, for the pentagon: $A_{i,5}/A_o = A_o/A_{c,5}$, and for the heptagon: $A_{i,7}/A_o = A_o/A_{c,7}$. Therefore, $A_{i,7}/A_{i,5} = A_{c,5}/A_{c,7}$. By the third proposition, this last ratio is greater than unity; hence $A_{i,7} > A_{i,5}$.

Simplicio: I see; once again, we have proof in which the direct substitution of literal symbols—the "pronumerals" n and m for the 5 and the 7, respectively—would convert the particular demonstration into a general one. Thus, for any pair of isoperimetric regular polygons, $A_{i,n} > A_{i,m}$, for n greater than m—Q.E.D.[IV]. Thank you Salviati. It is clear that one or two items of today's investigations have had an affinity with the first of these propositions.

Sagredo: Yes, but especially impressive is the introduction of the set of isoperimetric polygons in the second proposition.

Simplicio: You yourself were led in the same direction with your table of comparative areas of isoperimetric regular polygons.

Sagredo: I was merely engaged in a preliminary maneuver of a pedestrian kind in the hope of reaching a proper theoretical understanding. I suppose I should have written down a general expression for the area of a regular n-sided polygon of fixed perimeter and then attempted to show that this area continually increases as n increases and that it approaches the area of the isoperimetric circle as a limit. I think that very little time remains so that I will not suggest that we try this now.[220] In any case, I do not think that such a procedure would match in elegance Galileo's demonstration, which you have been good enough to expound again, Salviati.

Simplicio: Well, now, Sagredo, are you satisfied with this proposition, that a circle has a greater area than any isoperimetric regular polygon, or are you still going to say that we have more problems than we started with?

Sagredo: I am satisfied with it as a proposition in its own right, but some related matters occur to me. At the beginning of this hour, I asked how it could be shown that a *regular* polygon has a greater area than any isoperimetric nonregular polygon having the same number of sides. Although I think that we were successful in establishing this for triangles and quadrilaterals, it would seem to be important to obtain a proof for polygons in general.

Simplicio: Then we would be able to compare any noncircular plane figure with the isoperimetric circle via inscribed (or circumscribed) approximating polygons (Fig. 108).

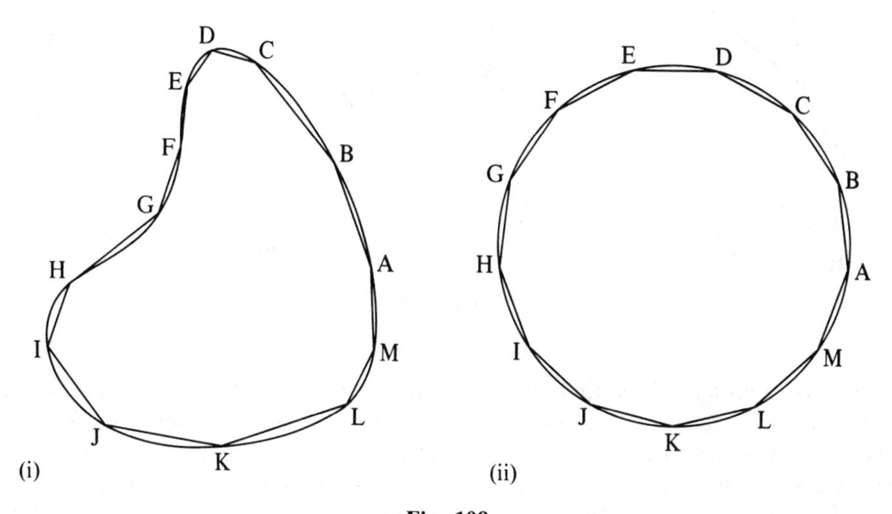

Fig. 108

Sagredo: This seems intuitively satisfying, even though the discrepancies between the two curvilinear figures and their polygons counterparts will not ordinarily be equal. I suppose this would not matter if it could be assumed that the discrepancies, while not equal, could nevertheless be shown to be less than any preassigned magnitude.[221] By the way, it is obvious from your first sketch that any plane figure that is not convex can be converted into a convex figure of equal perimeter but having greater area.

Simplicio: Looking at that part of the boundary that appears concave from the outside, I notice that my polygon is not completely inscribed, as sides *FG* and *GH* lie outside the curved figure.

Sagredo: Notice, too, that by keeping all the vertices fixed except F and G and swinging side EF outward about E as center, and side HG outward about H, obtain the boundary $AB \ldots EF'G'H \ldots MA$, the enclosed area is increased (Fig. 109i).

Simplicio: I see; and having done this, E could be moved outward, keeping D and F' fixed, and a still greater area would then be enclosed for no increase in the perimeter.

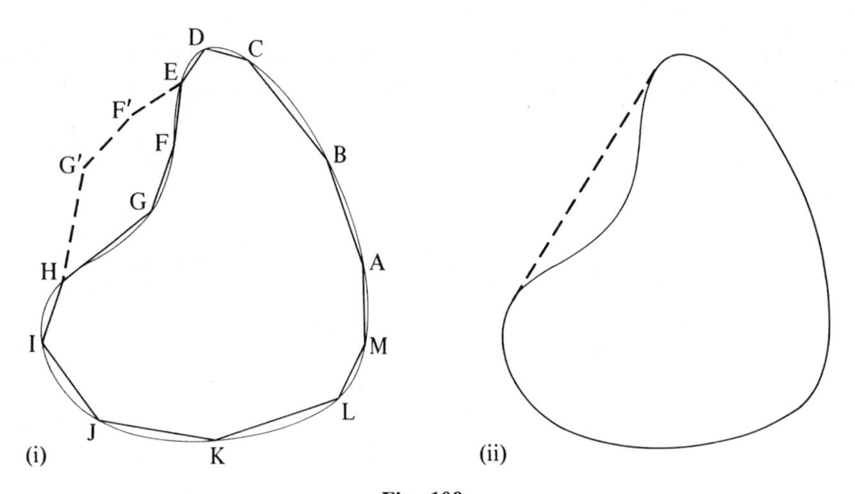

Fig. 109

Sagredo: Quite so. It is obvious, too, that whenever there is an outward facing concavity, this can be bridged by a straight line so as simultaneously to *decrease* the perimeter and *increase* the area enclosed by the boundary (Fig. 109ii). Now that we know that a polygon of maximum area for a given perimeter must be convex, I would like to consider a typical polygon, an irregular octagon, say, such as that in Fig. 110. As in Fig. 104, so here also, the area of the polygon can be increased by averaging adjacent sides; in the present case we may replace scalene triangle ABC by isosceles triangle $AB'C$, in which $AB' + B'C = AB + BC$, and so on around the figure so as to obtain $AB'CDEFGH$, then $AB'CD'EFGH$, then $AB'CD'EF'GH$, and finally $AB'CD'EF'GH'$, each polygon being greater in area than the previous one.

Simplicio: I notice that this last octagon is composed of a central quadrilateral $ACEG$ upon the sides of which the isosceles triangles $AB'C$, etc., are externally constructed. If this quadrilateral joining the alternate vertices of the octagon were arbitrarily fixed in size and shape, then the area of the octagon would have been maximized for the given perimeter. But, obviously, further improvement *is* possible. For one thing, there appears now to be a concavity at G.

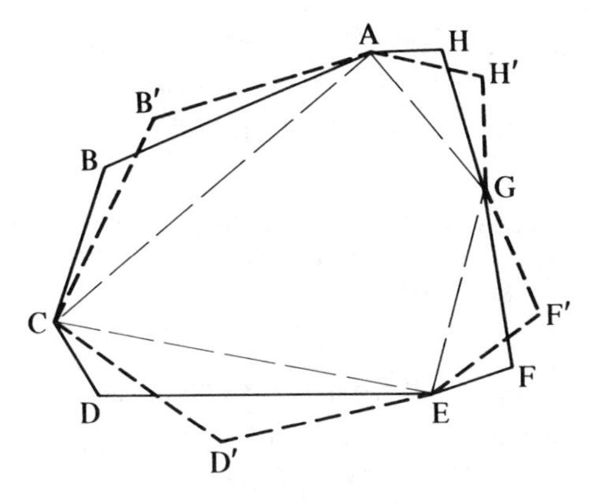

Fig. 110

Sagredo: I am not at all sure that you are correct in claiming that for A, C, E, G arbitrarily fixed, $AB'CD'EF'GH'$ will be the greatest octagon for the given perimeter. I can well imagine that the area might be increased by lengthening each of AB' and $B'C$ by some small amout and shortening each of AH' and $H'G$ by the same amount or perhaps by shortening AB' and $B'C$ and lengthening AH' and $H'G$. But you are certainly correct in noting that we are free to vary quadrilateral ACEG.

Simplicio: Without changing the perimeter of the octagon, the outward concavity formed by sides $F'G$, GH' may be removed, in accordance with your previous indications, by replacing these sides by $F'G'$, $G'H$, where G' is the image of G in $F'H'$ (Fig. 111). But a much greater improvement in shape of this inner quadrilateral is possible. I am imagining it to be like a pin-jointed framework, free to alter its shape.

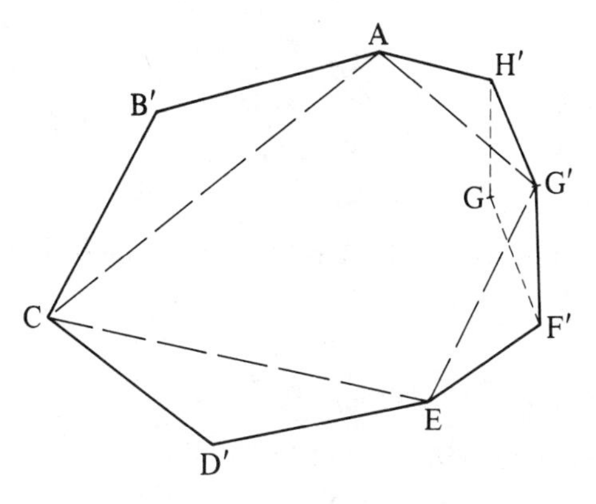

Fig. 111

Sagredo: I would conjecture that its greatest area would be obtained if it were cyclic, though I do not see how this would be proved.[222] But really, Simplicio, we are not obliged to keep the sides of this quadrilateral fixed like the rods of a framework.

Simplicio: I realize that; let us forthwith convert it into an isoperimetric *square*. You have already established that that is the way to maximize its area (see the discussion accompanying Fig. 104).

Sagredo: It is the perimeter of the octagon that has to be kept constant, not necessarily that of the inner quadrilateral. Indeed, in moving vertex G to G' in Fig. 111, you have already increased its perimeter as well as its area. Furthermore, as is shown in Fig. 112, it is not enough to form an octagon $ABCDEFGH$ in which $AB = BC$, $CD = DE$, $EF = FG$, and $GH = HA$, and having A,B,E,G as the vertices of a square. We should still want to "average out" sides HA and AB, BC and CD, DE and EF, FG and EH, and also to ensure that $BDFH$ as well as $ACEG$ was a square. Only if the averaging process were extended to produce an *equilateral* octagon would the fact that $ACEG$ is a square guarantee that $BDFH$ also is a square.

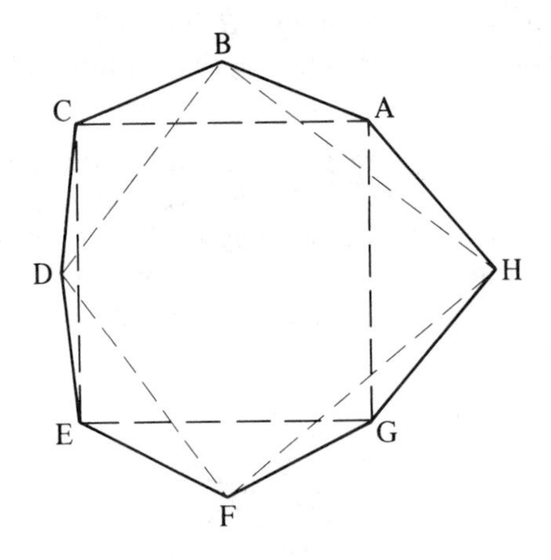

Fig. 112

Simplicio: Only then would we have reached an octagon that could no longer be transformed into another octagon of equal perimeter so as to have a greater (or even equal) area.

Sagredo: I am not satisfied to jump to that conclusion, Simplicio.

Simplicio: Perhaps there was a gap or two in the earlier stage of this discussion, but enough has been said to convince me that the *regular* octagon has a greater area than any other isoperimetric octagon, and doubtless similar arguments could be applied in the case of polygons with any given number of sides.

Salviati: I have enjoyed your discussion, but I must say, Simplicio, that you are a little premature in drawing the conclusion you have. For one thing, as is shown in the next diagram, Fig. 113, it is possible to choose two *unequal* squares, *ACEG* and *BDFH,* with the same center, *O* say, and placed at 45 degrees to each other, such that *ABCDEFGH* is a convex equilateral octagon.

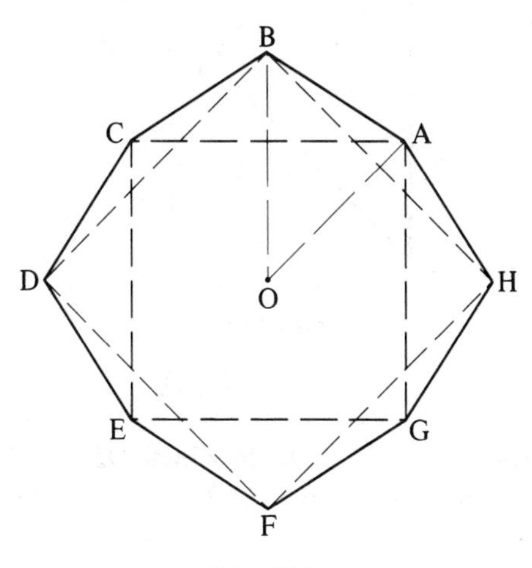

Fig. 113

Simplicio: And yet it is still not regular!

Salviati: You are right. Suppose *BDFH* is a larger square than *ACEG,* then $OB > OA$. If we let $OB - OA$ equal 2δ, then, by moving each of *A, C, E, G* a distance δ away from *O*, and each of *B, D, F, H* a distance δ toward *O,* we would obtain an equilateral octagon with all its vertices on a circle, center *O*.

Simplicio: Ah, a regular octagon at last!

Salviati: Yes, and with *OA* equal to *OB,* triangle *OAB* will have a greater area than before, when *OA* and *OB* were unequal.

Simplicio: And eight times this maximized triangle gives us the whole regular octagon. I see that the new (isosceles) triangle *OAB* is greater than the old (scalene) triangle *OAB*. This is simply an application of what we have shown already: that an isosceles triangle on a given base (here a side of the octagon) has a greater area than any isoperimetric scalene triangle on the same base.

Sagredo: I am not sure that the base *AB* does remain the same.

Salviati: Your doubt is justified, Sagredo. Actually, Simplicio, the procedure I have described for regularizing the octagon not only increases its area, it simultaneously reduces it perimeter!

Simplicio: I am afraid I do not see this.

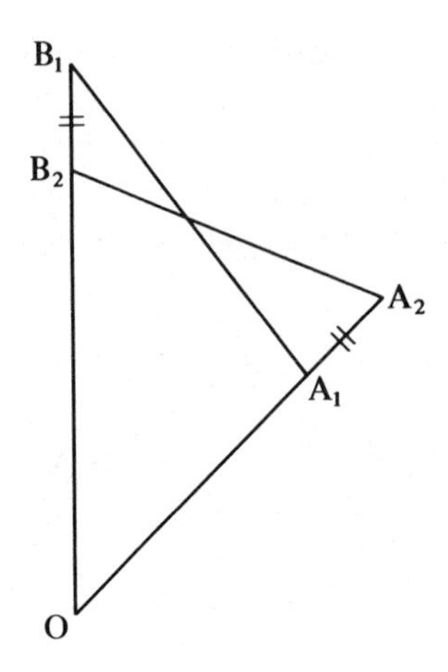

Fig. 114

Salviati: Consider, then, any ΔOA_1B_1 in which $OB_1 > OA_1$ and having point A_2 on OA_1 produced and point B_2 on OB_1, such that $A_1A_2 = B_1B_2$ *and* $OA_2 = OB_2$ (Fig. 114).

I assure you that it can be shown by very simple geometry both that (i) $A_2B_2 < A_1B_1$ and (ii) $\Delta OA_2B_2 > \Delta OA_1B_1$.[223]

Sagredo: So, *a fortiori,* if a regular octagon is constructed with side length equal to one-eighth of the perimeter of any given irregular octagon, the area of the regular figure assuredly exceeds that of the irregular one.

Salviati: Well, between us we have adumbrated a demonstration that this is so. But see how rapidly the light is fading. We can hardly go into more detail now.

Simplicio: The end of our day is surely at hand. We must now appeal to you, Salviati, to bring our discussion to a close.

Salviati: I think we may be satisfied with today's investigations. I would especially refer each of you to what is preserved of the Greek works, by Zenodorous and Pappus, on isoperimetry,[224] as well as to some of the modern papers on the same subject, including those of Jacob Steiner and his successors.[225] You will discover that a rigorous solution to the general isoperimetric problem has not been easy to come by. To show that any plane figure that is not a circle can be "improved" (as an area container) by some kind of transformation while the circle itself cannot, is, if we view the matter strictly, *not* sufficient to establish that the circle contains the greatest possible area for a given perimeter. To illustrate the logic of the situation, consider the transformation of a natural number n into its square, n^2. Now $n^2 > n$ (so the transformation "improves" the magnitude) *except* when $n = 1$. It is unity, and not some extremely larger number, that remains the same after squaring (or cubing, etc.).

Sagredo: Ah yes; but just because unity is the only natural number not subject to increase in this way, it would be absurd to conclude that unity is the largest of the natural numbers.[226]

Simplicio: I seem to remember that our first author, through you Salviati, pointed out a paradox somewhat like this one.[227]

Salviati: We expressed ourselves on that occasion by saying that "*if* any number may be called infinite, it is unity," but we also stated that "no specified number *is* infinite." Most of the short "proofs" of the isoperimetric theorem that you will come across establish only that "*if* the isoperimetric problem has a solution, then that solution is a circle," which is to say that if there is a closed curve for which the maximum possible value for the enclosed area *is* attained, then that curve is a circle. Of course, intuition may convince you that, whereas there can be no largest integer, there must exist a specificable form that encloses the largest possible area for a given perimeter. But I do not need to remind you that intuition has not always turned out to be a safe guide, and even where no evidence suggesting the possible falsity of intuitions has been found, the call for ever more detailed articulations has shown itself to be justified often enough. There is, however, no time to go further into this.[228] Let me conclude by sharing with you what is perhaps the simplest of all ways of showing, on the assumption that a maximum is attained, that the area bounded by a simple closed curve of given length has its maximum when the form is circular. It is due to the Cambridge mathematician A. S. Besicovitch, who was well aware that its brevity could expose it to demands for more detailed justification.[229] The chord joining any two ordinary points,[230] B and C say, of the curve must make equal angles with the tangents at B and C (Fig. 115i); for suppose that this were not the case and let $BA'C$ be the reflection of

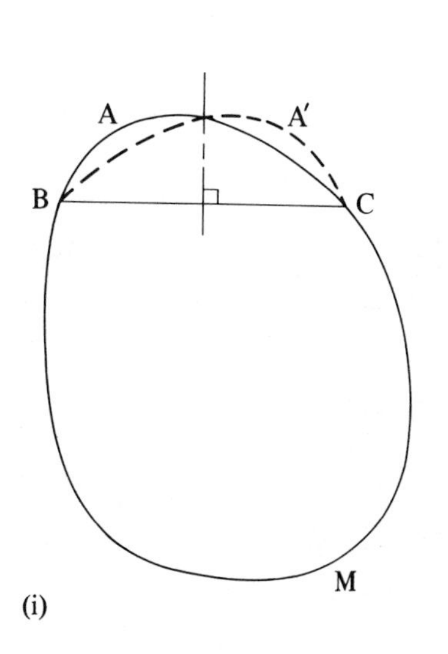

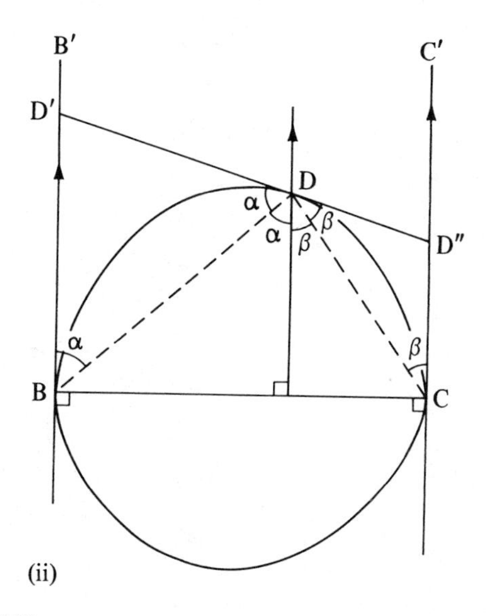

Fig. 115

BAC in the perpendicular bisector of *BC*, then the curve *BA'CMB* would have the same length and enclose the same area as *BACMB*. Hence, if *BACMB* were an extremal curve, then so also would be the curve *BA'CMB:* but this is impossible since it would follow from the supposition that there would be a concavity at *B* or at *C*. Next, suppose that *B* and *C* are points at which the tangents are parallel (Fig. 115ii); then the cointerior angles *B'BC* and *C'CB* will each be right angles, and, taking the tangent *D'DD"* at any ordinary point *D* of the curve, the angles marked α will be equal to each other, and likewise the angles marked β.

Simplicio: How ingeniously simple! Since *D'DD"* is a straight line, $2\alpha + 2\beta = 2$ right angles, hence $\angle BDC = 1$ right angle. Since the same holds for any point *D* that is chosen on the curve, the extremal curve is found to be the locus of *D* such that $\angle BDC$ is a right angle, that is, the circle on *BC* as diameter!

Salviati: You have it well enough, and we must leave it there, for now we are at the end of our day.

Sagredo: I thank you my friends. But excuse me, I am strangely drowsy and am no longer able to concentrate.

Salviati: We are being returned to our dreamworld. I wish you well.

Simplicio: My sight fades. . . . Yet now, through the mists, I discern the shape of a gondola. I see you stepping into it, Salviati, and now you, Sagredo. And there is a third figure following you . . . why, it is myself . . .

XV

The endeavour to show, with regard to almost any history, however true, that it actually occurred, and to produce an intelligent conception regarding it, is one of the most difficult undertakings that can be attempted, and is in some instances an impossibility.

—Origen

To understand a science it is necessary to know its history.

—Auguste Comte

The little present must not be allowed wholly to elbow the great past out of view.

—Andrew Lang

The first condition for the construction of the historical world is . . . the purification, by means of criticism and interpretation, of the confused and often corrupt memories of mankind about itself.

—Wilhelm Dilthey

The historian must serve two masters, the past and the present.

—Fritz Stern

The struggle for existence holds as much in the intellectual as in the physical world. A theory is a species of thinking, and its right to exist is coextensive with its power of resisting extinction by its rivals.

—T. H. Huxley

To find in the works of science of the past, that which is not and cannot be superseded, is perhaps the most important part of our quest. A true humanist must know the life of science as he knows the life of art and the life of religion.

—George Sarton

ANNOTATIONES

1. *Dialogo di Galileo Galilei Linceo . . .*, 1632 (see translation by Stillman Drake, *Dialogue Concerning the Two Chief World Systems—Ptolemaic and Copernican,* Berkeley: Univ. of California Press, 1953), Fourth Day. Galileo's theory of the tides has tended to be too harshly criticized or simply dismissed as a worthless attempt to buttress its author's commitment to the Copernican theory. For the best assessments, see Stillman Drake, *Galileo Studies: Personality, Tradition and Revolution,* Ann Arbor: Univ. of Michigan Press, 1970, pp. 200–213 (revised version of article in *Physis 3* (1961):185–194); H. L. Burstyn, "Galileo's Attempt to Prove that the Earth Moves," *Isis 53* (1962):161–185; Harold I. Brown, "Galileo, the Elements, and the Tides," *Studies in History and Philosophy of Science 7* (1976):337–351.

2. *Discorsi e Dimostrazioni Matematiche intorno à Due Nuove Scienze,* 1638 (see translation by Stillman Drake, *Two New Sciences Including Centres of Gravity and Force of Percussion,* Madison: Univ of Wisconsin Press, 1974), Second Day.

3. See especially, *Dialogue Concerning the Two Chief World Systems,* Second Day, opening pages. When Simplicio asks what better guide could there be than Aristotle, Galileo has Salviati reply: "We need guides in forests and in unknown lands, but on plains and in open places only the blind need guides. It is better for such people to stay at home, but anyone with eyes in his head and his wits about him could serve as a guide for them. In saying this, I do not mean that a person should not listen to Aristotle; indeed, I applaud the reading and careful study of his works, and I reproach only those who give themselves up as slaves to him in such a way as to subscribe blindly to everything he says and take it as an inviolable decree without looking for any other reasons." (Drake trans., pp. 112–113.)

4. Plutarch, "Vita Marcelli," *The Lives of the Noble Grecians and Romans,* Dryden translation, (*Great Books of the Western World,* vol. 14, p. 254). Also see *The Works of Archimedes,* edited by Thomas L. Heath, Reprint of 1897 edition with Heath's 1912 supplement, *The Method of Archimedes,* New York: Dover, n.d.; p. xviii, and E. J. Dijksterhuis, *Archimedes,* Copenhagen: Munksgaard, 1956, p. 32.

5. "[The volume of] every cylinder whose base is the greatest circle in a sphere and whose height is equal to the diameter of the sphere is 3/2 [of the volume] of the sphere, and its surface together with its bases is 3/2 of the surface of the sphere'' (Archimedes, ''On the Sphere and the cylinder,'' Book I, Corollary following Prop. 34, *Works* p. 43, cf. p. 1).

6. The first use of the *symbol* π for the ratio of circumference to diameter was by William Jones in 1706, though in 1647 Oughtred had used δ/π to represent the ratio diameter/periphery. However it was Euler's use of the symbol π from 1736 onward, when, as Cajori remarked, he ''either consciously adopted the notation of Jones or independently fell upon it,'' and especially its appearance in Euler's *Introductio in analysin infinitorum,* 1748, that was to lead to its general adoption. See the facsimile reproduction of Euler, Brussels: Culture et Civilisation, 1967, and Florian Cajori, *A History of Mathematical Notations,* Chicago: Open Court, 1929, vol. 2, pp. 8–13.

7. Simplicio is here referring to the first proposition of the ''Measurement of a Circle,'' *Works,* pp. 91–93. Cf. ''On the Sphere and Cylinder,'' I, 14, in which it is proved that ''The surface of any isosceles cone excluding the base is equal to a circle whose radius is a mean proportional between the side of the cone [a generator] and the radius of the circle which is the base of the cone.'' (Ibid., pp. 19–20). Archimedes' demonstration consists of a double *reductio ad absurdum* involving reference to the surfaces of inscribed and circumscribed pyramids on regular polygonal bases.

8. As in the case of the conical surface, Archimedes' treatment was different from that suggested by Simplicio's statement. Refer, ''On the Sphere and Cylinder,'' I, 13: ''The surface of any right cylinder excluding the bases is equal to a circle whose radius is a mean proportional between the side [i.e., a generator] of the cylinder and the diameter of its base'' (*Works,* pp. 16–18).

9. ''The surface of any sphere is equal to four times the greatest circle in it'' (''On the Sphere and Cylinder,'' I, 33, *Works,* pp. 39–41).

10. ''Measurement of a Circle,'' *Works,* pp. 91–93.

11. I.e., the (major) *auxiliary circle.* The fixed fraction is, of course, the ratio of the minor axis to the major axis. See note 51 for Archimedes, ''On Conoids and Spheroids,'' Prop. 4: ''The area of any ellipse is to that of the auxilliary circle as the minor axis to the major.''

12. The verbal equivalent of $A_e = (11/14)(a+b)^2$ was indeed asserted in a twelfth-century work, the *Liber Embadorum* of Savasorda (Abraham bar Ḥiyya), II, iv, 17 (Hebrew original, A.D. 1116, Plato of Trivoli's trans., A.D. 1145; see pp. 108–109 of M. Curtze, ''Der *Liber Embadorum* des Savosorda in der Übersetzung des Plato von Tivoli,'' *Abhandlungen zur Geschichte der Mathematicischen Wissenschaften* 12 (1902):1–183). Fibonacci (Leonardo of Pisa), in his *Practica Geometriae* (A.D. 1220) gave the verbal equivalents of

$$p_e = \frac{22}{7}(a+b) \quad \text{and} \quad A_e = \frac{1}{2}p_e\left(\frac{a+b}{2}\right).$$

Clearly, these yield $A_e = (11/14)(a+b)^2$.

See *Scritti di Leonardo Pisano*. vol. 2: *Practica Geometriae,* edited by Baldassare Boncompagni, Rome, 1857–62, pp. 101–102. Although these rules bear such an obvious analogy to those for the circle, the precise relation of the ellipse to the circle is not indicated. Neither Savasorda nor Leonardo Fibonacci can be said to have given a definition distinguishing the ellipse from any other oval having a pair of perpendicular axes of symmetry and having a greatest and a least diameter respectively equal to the greatest and least diameter of the ellipse. Fibonacci did not improve upon Savasorda's description, viz.: "a figure which is not a circle but which is formed oblique [*sed obliqua procreatur*], of which the two diameters are unequal. . . ." Curtze rendered the words italicized here as "sondern eine ovale Gestalt hat" and then assumed, in his footnote, that the form is elliptical. (Of course, the equivalence of the various possible definitions of "ellipse" is far from being self-evident and different definitions could be expected to induce different ideas concerning area or perimeter.)

13. Cf. Voltaire, anxious to present Newton's "hypotheses non fingo" in the best light, and generally to promote empirical philosophy in France: "A philosophy, founded only upon accidental Explications advanced at a venture, would not in strictness deserve the least Examination. For the Methods of arriving at Error are innumerable, whilst there is but one Way that leads to Truth; it is therefore the odds of infinite to one, that a Philosopher, who supports his Principles only by Hypotheses, will advance nothing but Chimera's" (*The Elements of Sir Isaac Newton's Philosophy,* translated by J. Hanna, 1738, facsimile, London: Cass, 1967, p. 2.) The important question concerning hypotheses has not to do with their accidental or venturesome origin; it is whether or not they are open to falsification or to a more positive form of proof. And whereas for a sufficiently well-defined problem there may be but one right answer and innumerable wrong ones, there are usually many more *ways* than one of arriving at an answer, true or false. In the abstract world of mathematics, the cultivation of *alternative methods* shows that here too there is not merely "one Way that leads to Truth."

14. "It seems to me that mathematics is, in several respects, the most appropriate experimental material for the study of inductive reasoning. This study involves psychological experiments of a sort: you have to experience how your confidence in a conjecture is swayed by various kinds of evidence. Thanks to their inherent simplicity and clarity, mathematical subjects lend themselves to this sort of psychological experiment much better than subjects in any other field" (George Polya, *Mathematics and Plausible Reasoning,* vol. 1: *Induction and Analogy in Mathematics,* Princeton, N.J.: Princeton Univ. Press, 1954, p. viii.)

15. "The scientist's . . . task is—typically—to accommodate some new discovery to his inherited ideas, without needlessly jeopardizing the intellectual gains of his predecessors. This kind of problem has an order of complexity quite different from that of simple curve-fitting: far from his having an infinite number of possibilities to choose between, it may be a stroke of genius for him to imagine even a single one." (Stephen Toulmin, *Foresight and Understanding: An Enquiry into the Aims of Science,* London: Hutchinson, 1961, pp. 112–113.)

16. It was with unusual candor that Helmholtz referred to his own very indirect path to success:

> In 1891 I have been able to solve a few problems in mathematics and physics, including some that the great mathematicians had puzzled over in vain from Euler onwards: e.g., the question of vortex motion, and the discontinuity of motions in fluids, that of the motions of sound at the open ends of organ pipes, etc. But any pride I might have felt in my conclusions was perceptibly lessened by the fact that I knew that the solution of these problems had almost always come to me as the gradual generalization of favorable examples, by a series of fortunate conjectures, after many errors. I am fain to compare myself with a wanderer on the mountains, who, not knowing the path, climbs slowly and painfully upwards, and often has to retrace his steps because he can go no farther—then, whether by thought or from luck, discovers a new track that leads him on a little, till at length when he reaches the summit he finds to his shame that there is a royal way, by which he might have ascended, he had only the wits to find the right approach to it. In my works I naturally said nothing about my mistakes to the reader, but only described the made track by which he may now reach the same heights without difficulty.

(Quoted by his biographer, Leo Koenigsberger in *Hermann von Helmholtz,* translated by F. A. Welby, Oxford: Clarendon Press, 1906, pp. 180–181.) As this "confession" indicates, the study of only the final successful phases of innovation, to which documentary evidence is so liable to be restricted, is very likely to result in a grossly foreshortened and oversimplified view of human inventiveness. Cf. George Sarton, *The Study of the History of Mathematics,* 1936, reprinted (with *The Study of the History of Science*) New York: Dover, 1957, p. 20.

17. It does need to be said, however, that

> in the course of the step-by-step development of a [deductive] demonstration (devised in response to some previous conjecture), a mathematician could be led on to new results which he had *not* anticipated at the conjectural level. It is because 'conjecture and testing' nowadays forms the all-pervading theme of most discussions on innovation—and 'trial and error' accepted as such a primary paradigm in the learning process—that it is necessary to insist that the experience of deductive proof need not consist merely of an exposé of the skeleton of a previously-encountered organism; it can also very well constitute a scaffolding which provides its constructor, as well as future climbers, with knowledge of hitherto unsuspected relations.

(J. Pottage, "Mensuration of Quadrilaterals . . . ," *Archive for History of Exact Sciences 12,* (1974):299–354; on p. 321). Cf. the Pythagorean motto, "A figure and a stepping-stone, not a figure and three obols." For this, see Proclus, *Commentary on the First Book of Euclid's Elements,* 84, 15–23 translated by Glenn R. Morrow, Princeton, N.J.: Princeton Univ. Press, 1970, p. 69). Cf. also the Greek *porism* in the sense in which it is used in Euclid's *Elements:* "it is what we call a corollary, i.e., an incidental result springing from a proof of a theorem or the solution of a problem, a result not directly sought but appearing as it were by chance without any additional labour, and constituting, as Proclus says, a sort of *windfall* (ἕρμαιον) and *bonus* (κέρδος)" (Thomas L. Heath, *The Thirteen*

Books of Euclid's Elements, Cambridge, England: Cambridge Univ. Press, 1926, vol. 1, p. 278). See further, Evert W. Beth and Jean Piaget, *Mathematical Epistemology and Psychology,* translated by W. Mays, Dordrecht: Reidel, 1966, pp. 93–94, on the "search for a method which is both heuristic and demonstrative."

18. See Eduard von Hartmann, *Philosophy of the Unconscious,* translated from the 9th German edition by W. C. Coupland, London: Trübner, 1884, especially vol. 1, p. 283 ff. Henri Poincaré's celebrated lecture, "L'invention mathématique," was delivered in 1908 before the Société de Psychologie in Paris and first published as "Le raisonnement mathématique" in *Science et méthode,* 1908, which is available in two English translations; refer to the bibliography. Graham Wallas, acknowledging his debt to Helmholtz and to Poincaré, established the terminology of the four stages of the creative process as (1) "preparation," (2) "incubation," (3) "illumination," (4) "verification" (*The Art of Thought,* London: Cape, 1926, p. 80). See Joseph-Marie Montmasson, *Invention and the Unconscious,* translated by H. S. Hatfield, London: Kegan Paul, 1931; Jacques Hadamard, *An Essay on the Psychology of Invention in the Mathematical Field,* Princeton, N.J.: Princeton Univ. Press, 1945, or Dover reprint 1954; Beth and Piaget, *Mathematical Epistemology and Psychology,* pp. 87–93, 199 ff.; Thomas R. Miles, *Eliminating the Unconscious: A Behaviorist View of Psycho-analysis,* Oxford: Pergamon Press, 1966.

19. "Of other lines [i.e., of lines other than straight lines] in a plane and having the same extremities, [any two] such are unequal whenever both are concave in the same direction and one of them is either wholly included between the other and the straight line which has the same extremities with it, or is partly included by, and is partly common with, the other; and that [line] which is included is the lesser [of the two]" (Archimedes, "On the Sphere and Cylinder," Book I, Assumption 2, *Works,* p. 2).

Comment: Figure A1 shows (i) a curve AB, consistently concave with respect to the straight-line AB, (ii) a sequence of circumscribing polygonal figures composed of straight line segments tangent to the arc AB, (iii) a sequence of inscribed polygonal figures composed of successive chords that span AB. It is especially convenient if the vertices of each inscribed figure are taken to coincide with the points of contact of a circumscribed figure. Then, since the sum of any two sides of a triangle is greater than the third side (Euclid, I, 20),

$$AC_0 + C_0B > AC_1 + C_1C_2 + C_2B > AC_1 + C_1C_3 + C_3C_4 + C_4B > \ldots *$$
$$* \ldots > AI_0 + I_0I_1 + I_1B > AI_0 + I_0B > AB$$

But to insert "arc AB" between the asterisks would be, in the absence of any independent theoretical means of determining the length of the arc, to add an *assumption* to the statement of an otherwise simply demonstrable relation. Presumably it was the difficulty of conceiving of any such independent determination that accounts for Archimedes' explicit assumption just quoted and also for the

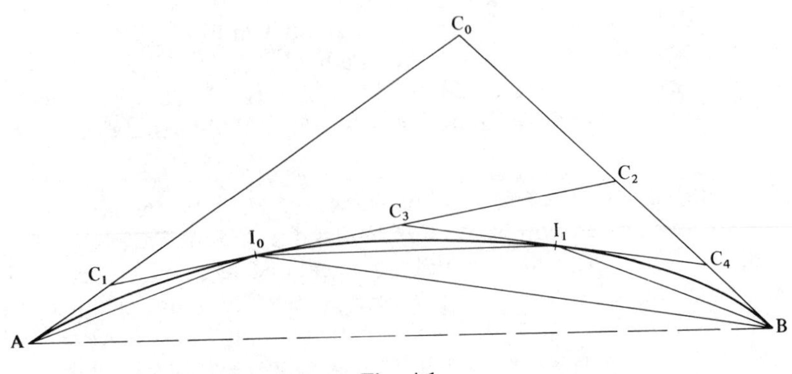

Fig. A1

persistence of the tradition, which continued until the middle of the seventeenth century, concerning the supposed "impossibility" of ever rectifying a curve; see note 47 below. (If it can be shown that the limit approached by the decreasing sequence

$$AC_0 + C_0B, \quad AC_1 + C_1C_2 + C_2B, \quad AC_1 + C_1C_3 + C_3C_4 + C_4B, \quad \ldots$$

is equal to the limit approached by the increasing sequence

$$AB, \qquad AI_0 + I_0B, \qquad AI_0 + I_0I_1 + I_1B, \qquad \ldots,$$

as the lengths of the elementary straight line intervals composing the paths all tend to zero, then the length of the arc AB may be *defined* as being measured by this common limit of the circumscribed and the inscribed sequences.)

20. The inequality signs > and < were introduced by Thomas Harriot (1560–1621). See Cajori, *History of Mathematical Notations,* vol. 1, p. 199, and vol. 2, pp. 115–120; also R. C. H. Tanner, "On the Role of Equality and Inequality in the History of Mathematics," *British Journal for the History of Science 1* (1962):159–169.

21. If, for example, $a = 9$, $b = 4$, then

$$\text{a.m.} = \frac{1}{2}(9 + 4) = 6\tfrac{1}{2}, \qquad \text{g.m.} = \sqrt{(9 \times 4)} = 6, \qquad \text{h.m.} = \frac{2 \times 9 \times 4}{9 + 4}$$

$$= 5\frac{7}{13}, \qquad \text{r.m.s.} = \sqrt{\left\{\frac{1}{2}(9^2 + 4^2)\right\}} = \text{almost } 7.$$

For the harmonic mean, and other means known to the Greeks, see Thomas L. Heath, *A History of Greek Mathematics,* Oxford: Clarendon Press, 1921, reprinted 1960, 1965, vol. 1, pp. 85–87.

22. The ellipse and rectangle may be obtained from the circle and square, respectively, by one-dimensional strains (simple elongations or compressions). In modern terminology, the ellipse is said to be the affine image of the auxiliary circle on the diameter that is the major axis of the ellipse.

23. It will be obvious that the term *analogy* is used here and elsewhere in the dialogue in the general qualitative sense referring to resemblance of attributes or relations of potential use in nonrigorous attempts to facilitate the understanding. The associated claims may amount to anything from mere metaphor on the one hand to an indication of some real identity of structure or function on the other. Since it happens, however, that in the dialogue we have just now been concerned with mean values and proportions, it is worth mentioning the technical senses that the term *analogia* and its variants appear to have had in the Pythagorean theory of means and proportionals, a theory that was naturally of great importance at a time before a more general appreciation of functionality had emerged. Originally, an *analogy* meant a proportion, an equality of ratios. For example, the geometric mean g and the harmonic mean h of two magnitudes a and b may be expressed in terms of these magnitudes by the "analogies"

$$a : g = g : b \quad \text{and} \quad h : 2b = a : (a + b),$$

respectively. (Such proportions suggest obvious geometrical constructions, involving similar triangles, for the determination of g and of h when a and b are given.) See *The Thirteen Books of Euclid's Elements,* Heath ed. vol. 3, "General Index of Greek Words and Forms," where there are listed references to ἀναλογία, 'proportion'; ἀνάλγον = ἀνά λόγον, 'proportional' or 'in proportion'; μέση ἀνάλογον, 'mean proportional' (of lines), μέσος ἀνάλογον (of numbers); τρίτη (τρίτος) ἀνάλογον, 'third proportional'; τετάρτη (τέταρτος) ἀνάλογον, 'fourth proportional'; ἑξῆς ἀνάλογον, 'in continued proportion'. The Greeks, of course, also used and recognized qualitative analogies designating similarities of any kind: "Likeness should be studied, first, in the case of things belonging to different genera, the formula being '$A : B = C : D$' (e.g., as knowledge stands to the object of knowledge, so is sensation related to the object of sensation), and 'As A is in B, so is C in D' (e.g., as sight is in the eye, so is reason in the soul, and as is a calm in the sea, so is windlessness in the air)." (Aristotle, *Topics,* I, 17; 108a 6–11, translated by W. A. Pickard-Cambridge, in *The Basic Works of Aristotle,* edited by Richard McKeon, New York: Random House, 1941.) The Pythagoreans, with their doctrine that "everything is number," would doubtless have regarded the qualitative analogies as somehow ultimately reducible to quantitative ones.

24. *Two New Sciences,* First Day, 134; Drake translation, p. 93.

25. The heuristic precedence of the result over the argument, of the theorem over the proof, has deep roots in mathematical folklore. Let us quote some variations on a familiar theme: Chrysippus is said to have written to Cleanthes: 'Just send me the theorems, then I shall find the proofs' (cf. Diogenes Laertius, *Lives of Eminent Philosophers,* VII, 179). Gauss is said to have complained: 'I have had my results for a long time; but I do not yet know how I am to arrive at them' (cf. A. Arber, *The*

> *Mind and the Eye,* 1954, p. 47), and Riemann: 'If only I had the theorems! Then I should find the proofs easily enough.' (cf. O. Hölder, *Die mathematische Methode,* 1924, p. 487.) Pólya stresses: 'You have to guess a mathematical theorem before you prove it' (*Mathematics and Plausible Reasoning,* 1954, vol. 1, p. vi).

(Imre Lakatos, *Proofs and Refutations: The Logic of Mathematical Discovery,* Cambridge, England: Cambridge Univ. Press, 1976, p. 9, n. 1.)

26. Cf. Norman Campbell, *What Is Science?,* 1921, pp. 101–103 of the reprint, New York: Dover, 1952, and Karl Popper, *Conjectures and Refutations: The Growth of Scientific Knowledge,* London: Routledge & Kegan Paul, 1963, passim, especially pp. 93–95. See also note 149 below.

27. The eccentricity ϵ of a given ellipse is a number in the range from 0 to 1, which may be defined in various ways, just as the curve itself may be variously defined. For example, corresponding to the definition that an ellipse is the plane locus of a point that moves so that the sum of its distances from two fixed points in the plane is constant, the eccentricity may be defined as the ratio of the distance between the fixed points (the "foci") to the length of the major axis. Alternatively, an ellipse may be defined as the locus of a point P moving in a plane containing a fixed point F (a focus) and a fixed line (a directrix), so that the ratio of the distance PF to the distance PN (measured perpendicularly) to the fixed line is a constant less than 1. This constant ratio may then be taken as the definition of the eccentricity, $\epsilon = PF:PN$. Either of the foregoing approaches leads to $\epsilon = \sqrt{(1 - b^2/a^2)}$, where a, b are the measures of the semimajor and semiminor axes, respectively. Clearly, if $b = a$ then $\epsilon = 0$ (and the ellipse is a circle); for a given positive a, and b equal to 0, $\epsilon = 1$ (the ellipse has degenerated into a line interval). Further, if one focus is fixed and b is a given positive number, then as $a \to \infty$, $\epsilon \to 1$ (the second focus recedes to infinity and the ellipse approaches a parabola).

HORA SECUNDA

28. A particularly clear example of a study of extreme behavior that is highly illuminating for an understanding of more normal reactions is provided by Émile Durkheim's classic work *Le Suicide: Étude de sociologie,* 1897 (*Suicide, A Study in Sociology,* translated by J. A. Spaulding and G. Simpson, Glencoe, Calif.: Free Press, 1951). Durkheim studied suicide, not for its own sake, but because he saw that it was able to serve as an index of general social disorganization and disintegration, one cause of suicide being seen as the excessive call for personal initiative when the laws, norms, and support systems of a society can no longer be relied upon. Again, as A. A. Brill noted in the introduction to his translation of Freud's *Psychopathology of Everyday Life* (New York: Macmillan, 1914), "it was while tracing back the abnormal to the normal state that Professor Freud found how faint the line of demarcation was between the normal and the neurotic person, and that the psychopathologic mechanisms so glaringly observed in the psychoneuroses and psychoses could usually be demonstrated in a lesser degree in normal persons."

29. In Plato's *Republic,* 331 c, Socrates is represented as asking: "But are we really to say that doing right consists simply and solely in truthfulness and returning anything we have borrowed? Are those not actions that can be sometimes right and sometimes wrong? For instance, if one borrowed a weapon from a friend who subsequently went out of his mind and then asked for it back, surely it would be generally agreed that one ought not to return it, and that it would not be right to do so, or to consent to tell the strict truth to a madman?" (H. D. P. Lee's translation, Harmondsworth, England: Penguin, 1955.) Cf. Kant, "On a Supposed Right to Lie from Altruistic Motives," *Critique of Practical Reason and Other Writings in Moral Philosophy,* translated and edited by L. W. Beck, Chicago: Univ. of Chicago Press, 1949, pp. 346–350.

30. *Two New Sciences,* First Day, 116–120. Cf. Descartes in his letter to Mersenne, 11 October 1638: "All that he [Galileo] says about the speed of bodies falling in the void is without foundation, for he should have first determined what gravity is; had he known its nature, he would have seen that there is none in the void" (*Oeuvres et lettres,* edited by A. Bridoux, Paris: Gallimard, 1953, pp. 1024–1039, on p. 1028). See also Boyle, *A Continuation of New Experiments Physico-Mechanical Touching the Spring and Weight of the Air, and their Effects,* 1669, Experiment 40, reproduced in J. B. Conant (ed.), *Robert Boyle's Experiments in Pneumatics,* Cambridge, Mass: Harvard Univ. Press, 1950, pp. 54–57. Cf. Newton, *Opticks,* Latin translation, 1706, Query 20; 2nd English edition, 1717–18, and subsequent editions, Query 28: "small Feathers falling in the open Air meet with great Resistance, but in a tall Glass well emptied of Air, they fall as fast as Lead or Gold, as I have seen tried several times" (4th ed., 1730, New York: Dover reprint, 1952, p. 366); and see Newton, *Principia,* Scholium Generale written for the 2nd ed., 1713, (also in the 3rd ed., 1727; Cajori's edn., p. 543); and Voltaire, *Elements of Sir Isaac Newton's Philosophy,* pp. 160–161. The aforementioned letter of Descartes to Mersenne contains numerous strictures against Galileo's *Two New Sciences,* as well as against communications of Fermat, Roberval, and others. An English translation of the half of this letter that relates to Galileo is given by Stillman Drake in *Galileo at Work: His Scientific Biography,* Chicago: Univ. of Chicago Press, 1978, pp. 387–392. Drake comments, "It is evident that Descartes, like all his contemporaries outside Italy and most of them there, considered the discovery of causes to be the whole point of science, in which everything (and not just some things) must be causally explained. Neglect of causes and lack of universality were seen as irremediable and fatal faults of Galileo's books by those who demanded an impossible certainty and applicability of science without restriction" (p. 399). Descartes' remark to Mersenne concerning gravity may be compared with Leibniz's criticism of Newton quoted in Cajori's edition of Newton's *Principia,* 1934 and subsequent reprints, note 52, pp. 668–669; cf. notes 5 and 6, pp. 629–635. See further, Peter Machamer, "Galileo and the Causes," and William R. Shea, "Descartes as Critic of Galileo," in

Robert E. Butts and Joseph C. Pitt (eds.), *New Perspectives on Galileo,* Dordrecht: Reidel, 1978.

31. *Two New Sciences,* Third Day, 244; Fourth Day, 268, 274–276. Galileo was careful to point out that, strictly speaking, any change in distance from the center of the earth would result in a change from uniform speed, though this would be negligible for ordinary distances where the difference between a level plane and a sphere concentric with the earth is negligible. While for most practical purposes "we could treat one minute of a degree [of arc] at the equator as if it were a straight line," the theory is (like Archimedes' theory of the balance) strictly exact only for actions "situated at infinite distance from the centre" (274–275). Cf. Galileo's letter to M. Welser, 14 August 1612: "all external impediments removed, a heavy body on a spherical surface concentric with the earth will be indifferent to rest and to movements toward any part of the horizon. And it will maintain itself in that state in which it has once been placed; that is, if placed in a state of rest, it will conserve that; and if placed in movement toward the west (for example), it will maintain itself in that movement." (Stillman Drake, trans., *Discoveries and Opinions of Galileo,* Garden City, N.Y.: 1957, p. 113.)

32. The problem of inscribing an *equilateral* polygon (of more than four sides) within a given ellipse perhaps first arose early in the seventeenth century after Kepler had considered, rather unsuccessfully, the problem of determining the perimeter of an ellipse. At least we know that in 1673 Michael Dary, perhaps with Huygen's approximations to circular arcs in mind, became interested in the inscribed equilateral polygons of ellipses. In a letter to Collins (17 September 1673), Newton mentioned that Dary had just requested his opinion "about the relation of ye lines one to another which are drawn from ye center of an Ellipsis to the angular points of a polygon inscribed into ye same Ellipsis, wch consists of 24 equall sides." Newton continued: "As it appeares to me, their relation cannot be accurately known without an Equation wch is 4 times decompounded of affected cubic equations & twice of quadratick ones, & by consequence would ascend to 324 dimensions: To compute wch would be a Herculean labour, & when done, it would be unmanageable." (*The Correspondence of Isaac Newton,* vol. 1, edited by H. W. Turnbull, Cambridge, England: Cambridge Univ. Press, 1959, p. 307; see also pp. 308, 332–333.)

33. *Two New Sciences,* Third Day, pp. 208–211. The essential relation is, of course, simply that $(n + 1)^2 - n^2 \equiv 2n + 1$. For n taking consecutive positive integral values, this sequence of "gnomon numbers" was well known to the early Pythagoreans in their arithmetic of figurate numbers.

34. *Eudemian Ethics,* I, 2; 1215 a 35.

35. *Republic,* IX; 580 c, ff., in connection with the tripartite division of the soul.

36. Cicero, *Tusculan Disputations,* V, iii, 9, and Diogenes Laertius, *Lives of Eminent Philosophers,* VIII, 8. Both works available in Loeb Classical Library editions. The early Pythagoreans lived austere lives, and it was prescribed that they not teach for gain or fame.

37. William Wordsworth, "The World Is Too Much With Us," Sonnet (1807).

38. James Wolfe: "I would rather have written that poem [Gray's *Elegy*], gentlemen, than take Quebec tomorrow." The remark and the circumstances in which it was made are variously reported; see, e.g., C. P. Stacey, *Quebec, 1759, The Siege and the Battle,* Toronto: Macmillan, 1959, pp. 122–123.

39. *Phaedo,* 66 c–d, Benjamin Jowett translation. For an alternative translation, see *The Collected Dialogues of Plato,* edited by E. Hamilton and H. Cairns, Princeton, N.J.: Princeton Univ. Press, 1961, p. 49.

40. *Republic,* VII, 525–531. "In these studies some instrument in each man's soul, which was being destroyed and blinded by other pursuits, is here purged and rekindled . . ." (527 d–e, A. D. Lindsay translation, London: Dent, Everyman's Library, 1935, cf. 500 c). Plato actually has Socrates name *five* preparatory studies: arithmetic, geometry, *stereometry,* astronomy, and harmonics, since geometry was understood as being restricted to the investigation of plane figures. The separate naming of stereometry could be dropped as soon as geometry came to include the theory of solid as well as of plane figures. Thus the medieval quadrivium consisted simply of arithmetic, geometry, astronomy, and music. After mentioning Plato, Aristotle, Boethius, and Isidore as authorities for the division of mathematical science into these four parts, Luca Pacioli, in his *Divina proportione* (1509), gave it as his own opinion that *perspective* should be added as a fifth division (or else music omitted as of no more importance than perspective and the classification left as a trichotomy). Refer, Robert Emmet Taylor, *No Royal Road: Luca Pacioli and His Times,* Chapel Hill, N.C.: Univ. of North Carolina Press, 1942, pp. 262–263.

41. Charles Nicolle (Nobel prize for physiology and medicine, 1928), as quoted by René Taton, *Reason and Chance in Scientific Discovery,* translated by A. J. Pomerans, London: Hutchinson, 1957, p. 30. See also H. K. Fulmer, "Motives and Trends in Mathematics," *American Mathematical Monthly 61* (1954): 157–160; Charles S. Fisher, "Some Social Characteristics of Mathematicians and Their Work," *American Journal of Sociology 78* (1973):1094–1118, and Bernard H. Gustin, "Charisma, Recognition, and the Motivation of Scientists," ibid. *78* (1973):1119–1134. For criticism of Fisher, see J. Fang and K. P. Takayama, *Sociology of Mathematics and Mathematicians, A Prolegomenon,* Hauppauge, N.Y.: Paideia Press, 1975, pp. 84–85, 227–235.

42. *Two New Sciences,* First Day, opening pages.

HORA TERTIA

43. Compare "molar" (or so-called macroreductive), as contrasted with "molecular" (or microreductive), explanatory treatments in general. The former involve the comprehension of the explanandum in an appropriate context, as part of a larger system (e.g., any right triangle may be seen as half of a rectangle; consequently *its* angle sum is two right angles). The latter involve instead the resolu-

tion of the explanandum into component parts (e.g., any triangle may be divided into two right triangles by an interior altitude; hence the angle sum for *any* triangle is two right angles. For other purposes, a triangle might be regarded as a set of indefinitely many equally spaced *lines* in arithmetic progression, in the manner of Cavalieri; for this, see note 187).

In the substantive sciences particularly, (micro-)reductionists have sometimes left themselves open to criticism for assuming that complex systems are simply explicable in terms of the properties of the isolated parts. But against this, in all branches of science in recent centuries, explanations have more and more taken the form of specifications, and often quantifications, of *interactions* instead of appeals to essential attributes or natures. Obvious examples are provided by Newtonian gravitation, Lavoisierian combustion, Darwinian selection, and modern ecological science, by the developmental theories of twentieth-century biology and psychology, and, especially in physics, by the recognition of interactions between microphenomena and the instruments of detection, and generally between observers and observed. Perceptions, cognizances, understandings, and misunderstandings, take the forms they do because of interaction between the presented stimulus or data and the characteristics or prejudices of the observer-interpreter.

Returning to the abstract world of mathematics, it is worth noticing that theorems and equations may be seen as giving precise specifications of constrained interactions, or covariations, as may be illustrated by the following extremely simple example: For a variable line ℓ intersecting fixed orthogonal axes as shown in Fig. A2, $\beta = \alpha - 90$ degrees.

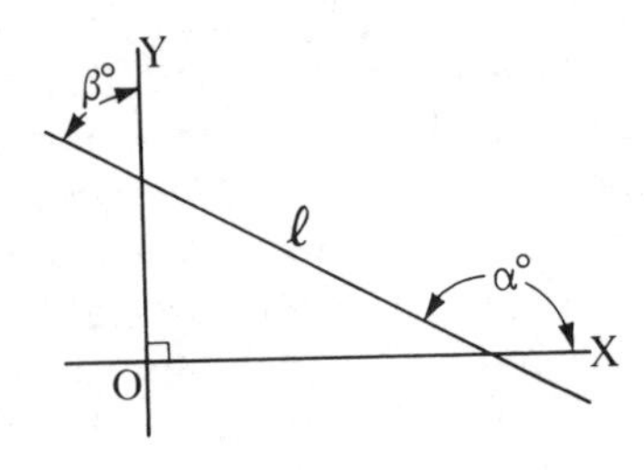

Fig. A2

44. Cf. Poincaré: "It never happens that unconscious work supplies *ready-made* the result of a lengthy calculation in which we have only to apply fixed rules. It might be supposed that the subliminal ego, purely automatic as it is, was peculiarly fitted for this kind of work, which is, in a sense, exclusively mechanical. . . . Observation proves that such is by no means the case. All that we can hope from these inspirations, which are the fruits of unconscious work, is to obtain points of departure for such calculations. As for the calculations themselves . . . they demand discipline, attention, will, and consequently consciousness." ("Mathematical Discovery," in *Science and Method,* translated by F. Maitland, 1914, New York: Dover reprint, pp. 62–63.) (Poincaré was referring to creative mathematicians, not calculating prodigies.)

45. For an appropriate location of the centers used in such circular-arc constructions in order to achieve a "best" fit with the elliptical form, see E. H. Lockwood, *Mathematical Gazette 14* (1928):136–137, Note 891, and cf. ibid., p. 270, Note 912, and *16* (1932):269–270, Note 1045. See also N. T. Gridgeman, "Quadrarcs, St. Peter's, and the Colosseum," *Mathematics Teacher 63* (1970):209–215; this article lists four criteria according to which the agreement with the elliptical form might be optimized.

46. See pp. 10–11 of this book. A general theoretical treatment is given on p. 75.

47. It is hardly possible in the space of a note to discuss the earlier view, widely held until the seventeenth century, that it is impossible ever to rectify any curve in the sense of determining an exactly specifiable ratio between an arc and a given straight line interval, or to more than mention Descartes' distinction between "geometrical" (algebraic) and "mechanical" (transcendental), curves, and his belief that only the former were unrectifiable. Actually two of the latter, the cycloid (unpublished) and the logarithmic spiral, were rectified in Descartes' own lifetime, and soon afterward several of the former also. See Carl B. Boyer, *A History of Mathematics,* New York: Wiley, 1968, pp. 375–376; also the same author's "Early Rectifications of Curves," *Mélanges Alexandre Koyré*, vol. 1, Paris: Hermann, 1964, pp. 30–39; and Joseph E. Hofmann, *Leibniz in Paris 1672–1676, His Growth to Mathematical Maturity,* translated by A. Prag and D. T. Whiteside, London: Cambridge Univ. Press, 1974, chap. 8. (The relatively easy task of determining the *area* under a cycloidal arch had already been successfully carried out by Roberval in 1634 and shortly afterward by Descartes, Fermat, and Torricelli. Mersenne, when he originally asked Roberval to examine the cycloid, erroneously surmised that the locus generated by a point on the circumference of the rolling circle might actually by a semiellipse! See Margaret E. Baron, *The Origins of the Infinitesimal Calculus,* Oxford: Pergamon Press, 1969, pp. 156–164 and 199.)

48. Formula (1), already has been given in the alternative form $p_e = \pi\{(\frac{3}{2})(a + b) - \sqrt{(ab)}\}$, is due to Peano (1887); see *Selected Works of Giuseppe Peano,* translated and edited by H. C. Kennedy, Toronto: Univ. of Toronto Press, 1973, pp. 135–136. Formulae (2) and (3) are due to Ramanujan (1914); see *Collected Papers of Srinivasa Ramanujan,* edited by G. H. Hardy, P. V. Seshu Aiyar, and B. M. Wilson, Cambridge, England: Cambridge Univ. Press, 1927, p. 39. Formula (4) appears to have been first given by R. Goormaghtigh, *American Mathematical Monthly 37* (1930):441–442. To this last was appended the following editorial note: "It may be of interest to emphasize the degree of accuracy of the preceding formula by applying it to an ellipse having the general size and shape of the earth's orbit (which in Newtonian mechanics would be a true ellipse if one neglects, as we shall do, the perturbations due to heavenly bodies other than the sun). Assuming, as at present it seems proper to, that the age of the earth is less than one hundred billion years then the total distance the earth has traveled since its creation would differ from the approximate distance given by the preceding formula by about the thickness of the sheet of paper on which this is written.'' For a theoretical treatment, see Hora Quinta, especially p. 77.

49. *Sophist,* 231a. Salviati's response echoes the *Sophist,* 231e.

50. In his *Physics,* 203 a 10–15, Aristotle alludes to the building up of successive square numbers in Pythagorean arithmetic by placing odd gnomons ''around the one'' (cf. note 33, supra), in which case the successive square arrays are of the same form (εἶδος). Contrary to this is the sequence of *rectangular* numbers obtained ''without the one''; that is, presumably, beginning instead with the *two,* thus obtaining rectangular arrays of ever different form, 1 : 2, 2 : 3, 3 : 4, . . . (corresponding to the concordant intervals of the octave!) See Fig. A3; also Thomas L. Heath, *Mathematics in Aristotle,* Oxford: Clarendon Press, 1949, pp. 101–102.

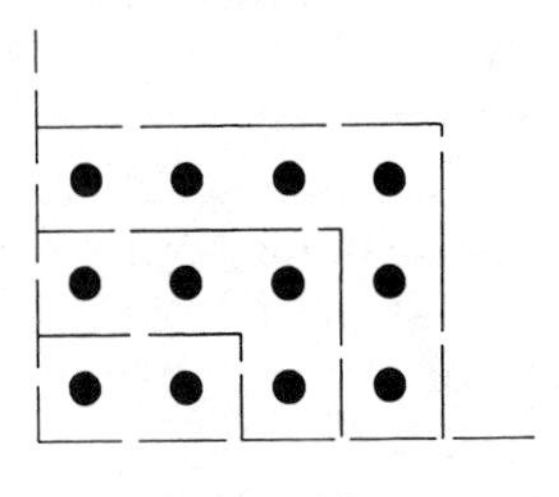

Fig. A3

51. As previously indicated, the formula $A = \pi ab$, for the area of an ellipse, follows immediately from $A = \pi a^2$, for the area of the auxiliary circle, together with Proposition 4 of Archimedes' ''On Conoids and Spheroids.'' The translation of this proposition in Fig. A4 is reproduced from the Heath edition of *The Works of Archimedes,* pp. 113–114:

Fig. A4

Proposition 4.

The area of any ellipse is to that of the auxiliary circle as the minor axis to the major.

Let AA' be the major and BB' the minor axis of the ellipse, and let BB' meet the auxiliary circle in b, b'.

Suppose O to be such a circle that

$$(\text{circle } AbA'b') : O = CA : CB.$$

Then shall O be equal to the area of the ellipse.

For, if not, O must be either greater or less than the ellipse.

I. If possible, let O be greater than the ellipse.

We can then inscribe in the circle O an equilateral polygon of $4n$ sides such that its area is greater than that of the ellipse. [cf. *On the Sphere and Cylinder,* I. 6.]

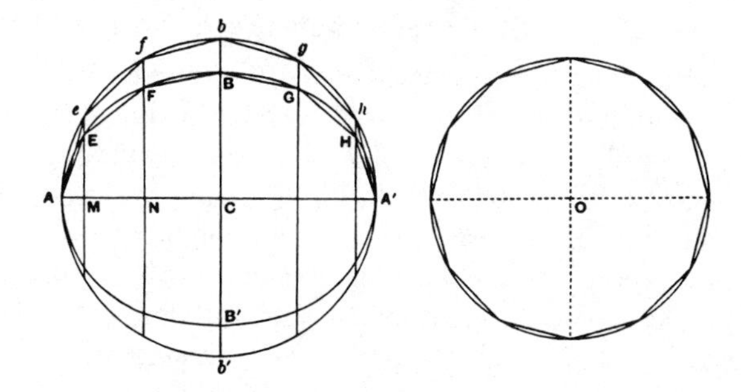

Let this be done, and inscribe in the auxiliary circle of the ellipse the polygon $AefbghA'\ldots$ similar to that inscribed in O. Let the perpendiculars $eM, fN,\ldots$ on AA' meet the ellipse in $E, F,\ldots$ respectively. Join $AE, EF, FB,\ldots$.

Suppose that P' denotes the area of the polygon inscribed in the auxiliary circle, and P that of the polygon inscribed in the ellipse.

Then, since all the lines $eM, fN,\ldots$ are cut in the same proportions at $E, F,\ldots$,

i.e. $$eM : EM = fN : FN = \ldots = bC : BC,$$

the pairs of triangles, as eAM, EAM, and the pairs of trapeziums, as $eMNf$, $EMNF$, are all in the same ratio to one another as bC to BC, or as CA to CB.

Therefore, by addition,

$$P' : P = CA : CB.$$

Now P' : (polygon inscribed in O)

$$= (\text{circle } AbA'b') : O$$

$$= CA : CB, \text{ by hypothesis.}$$

Therefore P is equal to the polygon inscribed in O.

But this is impossible, because the latter polygon is by hypothesis greater than the ellipse, and *a fortiori* greater than P.

Hence O is not greater than the ellipse.

II. If possible, let O be less than the ellipse.

In this case we inscribe in the *ellipse* a polygon P with $4n$ equal sides such that $P > O$.

Let the perpendiculars from the angular points on the axis AA' be produced to meet the auxiliary circle, and let the corresponding polygon (P') in the circle be formed.

Inscribe in O a polygon similar to P'.

Then $$P' : P = CA : CB$$

$$= (\text{circle } AbA'b') : O, \text{ by hypothesis,}$$

$$= P' : (\text{polygon inscribed in } O).$$

Therefore the polygon inscribed in O is equal to the polygon P; which is impossible, because $P > O$.

Hence O, being neither greater nor less than the ellipse, is equal to it; and the required result follows.

52. See the final sentence of note 53, relating to Fig. 20, as well as pp. 10–11, 75.

53. The outer and inner octagons have areas $30\sqrt{2}$ and $6\sqrt{2}$ square units, respectively; hence the double-track area is $24\sqrt{2}$. The perimeter of the intermediate octagon $= 4\{\sqrt{(14 - 4\sqrt{2})} + \sqrt{(26 - 16\sqrt{2})}\}$, $= p$ units, say; and so $p = 18.90$ correct to four significant figures. Defining τ_E, for the elongated tracks, by $2p\tau_E = 24\sqrt{2}$, the value of τ_E is easily shown to be slightly less than 0.9 units. Incidentally, that the expression $\pi(a + b)$ underestimates the perimeter of the ellipse $x^2/4^2 + y^2/2^2 = 1$ can be demonstrated by showing that $\pi(4 + 2)$ is actually less than the expression just given for the perimeter of the intermediate octagon about which this ellipse may be circumscribed.

54. For a historical introduction, see J. L. Coolidge, "The Unsatisfactory Story of Curvature," *American Mathematical Monthly 59* (1952):375–379. According to Coolidge, "The first writer to handle the question of curvature of a plane curve in what we should call today a thoroughly satisfactory manner was Sir Isaac Newton, no less" (p. 377). Newton's pioneering work on normals, curvature, and tangents, as recorded in his "Waste Book," Winter 1664–Spring 1665, is now available for convenient reference in *The Mathematical Papers of Isaac Newton,* edited by D. T. Whiteside, vol. 1, Cambridge, England: Cambridge Univ. Press, 1967, pp. 245–297; see especially pp. 268–271 for the evolute of the ellipse. The first systematic account to be published appeared in *The Method of Fluxions and Infinite Series, with its Application to the Geometry of Curve-lines by the Inventor Sir Isaac Newton,* "translated from the Author's Latin Original not yet made publick" by John Colson, London, 1736, pp. 59–80. For early related work by Huygens, see note 92.

55. Cf. the simpler, easily sketched astroid, $x^{2/3} + y^{2/3} = c^{2/3}$, which turns out to be a hypocycloid of four cusps (generated by a point on the circumference of a circle of radius $\frac{1}{4}c$ rolling inside a circle of radius c).

HORA QUARTA

56. Art historians tend to use the term *elliptical* much too freely, imprecisely (or overprecisely!) characterizing as elliptical various forms of oval ware, arches, vaults, stadia, and the like, when a simple analysis would show that the form is not elliptical at all, in many cases having been laid out by a blending circular-arc construction, or perhaps merely by the sweep of a bent lath.

57. Imre Lakatos argued that proposed proofs, though they may not be valid ("working mathematicians . . . know from experience that proofs are fallible") nevertheless are valuable because the effort to devise them refines the understanding and may indicate necessary modifications of initial beliefs: "proofs, even though they may not *prove,* certainly do help to *improve* our conjecture. . . . *This intrinsic unity between the 'logic of discovery' and the 'logic of justification' is the most important aspect of the method of lemma incorporation.*" (*Proofs and Refutations,* p. 29, n. 1, and p. 37.)

58. Or *disassociation,* if the shortened form is felt to have too many pathological overtones:

> Dissociation. A process by which some thoughts, attitudes, or other psychological activities lose their normal relationship to others, or to the rest of the personality, and split off to function more or less independently. In this way logically incompatible thoughts, feelings and attitudes may be held concurrently and yet conflict between them averted. A chronic state of dissociation is usually regarded as pathological. Some behaviors and phenomena which have been described in terms of dissociation include multiple or 'split' personality, somnambulism (sleep walking), automatic writing, certain delusional symptoms, hypnotism, and hysterical amnesia.

(G. D. Wilson, "Dissociation," *Encyclopedia of Psychology,* edited by H. J. Eysenck, W. Arnold, and R. Meili, 3 vols., London: Search Press, 1972; also in 2 vols., London: Fontana/Collins 1975.)

59. Modern thinkers have been sensitive to the importance for the moral life of just such a preparedness to consider whether restrictions to ways of behaving that are commonly accepted as "necessary" are necessary at all, or even desirable. With this skepticism has been associated, notably by Sartre, the responsibility of the individual to recognize and accept the real freedom of decision that is available, though very often denied. In Sartre's view, we are responsible for the adoption of the very criteria according to which particular choices are referred. The failure to recognize *the very wide range of options open* and the pretence that one "couldn't do anything else" are self-indulgent attitudes constituting the Sartrian "bad faith."

60. For a detailed study, see J. Pottage, "The Mensuration of Quadrilaterals . . . with Special Reference to Brahmagupta's Rules," *Archive for History of Exact Sciences 12* (1974):299–354.

61. See, D. R. Green, "The Historical Development of Complex Numbers," *Mathematical Gazette 60* (1976):99–107; G. Windred, "History of the Theory of Imaginary and Complex Quantities," ibid., *14* (1929):533–541.

62. Book VII, Def. 2: "A number is a multitude composed of units."

63. "Whenever proofs depend upon some only of the marks by which we define the object to be studied, these marks should be isolated and investigated on their own account. For it is a defect, in an argument, to employ more premises than the conclusion demands: what mathematicians call elegance results from employing only the essential principles in virtue of which the thesis is true. . . . every new axiom [assumption or restriction—J. P.] diminishes the generality of the resulting theorems, and the greatest possible generality is before all things to be sought" (from p. 71 of Bertrand Russell's 1902 essay "The Study of Mathematics," reprinted in *Mysticism and Logic and Other Essays,* Harmondsworth, England: Penguin Books, 1953, pp. 60–73.)

64. For example, Waclaw Sierpiński, *Pythagorean Triangles,* translated (from the Polish) by A. Sharma, Scripta Mathematica Studies, no. 9, New York: Yeshiva University, 1962.

65. Daniel Pedoe, *Circles,* London: Pergamon Press, 1957; Julian Lowell Coolidge, *A Treatise on the Circle and the Sphere,* Oxford: Clarendon Press, 1916.

66. Peter B. Medawar, *The Art of the Soluble,* London: Methuen, 1967, p. 87. Cf. Stephen Toulmin: "The gamesplayer improves his sporting techniques most quickly by playing against opponents who are just *one* degree his superior. The scientist, likewise, is on the look-out for events which are not yet *quite* intelligible, but which could probably be mastered as a result of some intellectual step which he has the power to take. So long as everything proceeds according to his prior expectations, he has no opportunity to improve on his theories. He must look out for deviations ["anomalies," or, on the other hand, unexpected coincidences] that are not yet explained, but promise to be explicable." (*Foresight and Understanding: An Enquiry into the Aims of Science,* London: Hutchinson, 1961, p. 45.) Cf. also Marx's overconfident claim that "mankind always sets itself only such tasks as it can solve; since, looking at the matter more closely, it will always be found that the task itself arises only when the material conditions for its solution already exist or are at least in the process of formation" (*A Contribution to the Critique of Political Economy,* p. 363 of the translation from the German edition of 1859, in Karl Marx and Frederick Engels, *Selected Works,* Moscow: Foreign Languages Publishing House, 1962, vol. 1, pp. 361–365.)

HORA QUINTA

67. Newton's analysis is contained in his famous letter to Oldenburg of 13 June 1676 (afterward sent to Leibniz). See *The Correspondence of Isaac Newton,* vol. 2 (1676–1687), edited by H. W. Turnbull, Cambridge, England: Cambridge Univ. Press, 1960, pp. 32–47. John Collins in a letter to James Gregory, dated 24 December 1670, stated that he understood (apparently from a talk with Barrow) that Newton had already found "the length of the portions or intire Elliptick line" (see *James Gregory Tercentenary Memorial Volume,* edited by H. W. Turnbull, London: Bell, 1939, pp. 155, 160). In his letter to Collins, 15 February 1671, Gregory sent the result of his own rectification of the elliptical arc (ibid., pp. 171, 173). The chance survival of a rough note of Gregory's on the back page of a bookseller's letter (bearing the date 30 January 1671) provides evidence of the method used (ibid., p. 366). For eighteenth-century work on the rectification of the ellipse and other curves, see G. N. Watson, "The Marquis and the Land Agent," *Mathematical Gazette 17* (1933):5–17.

68. The well-known construction for an ellipse shown in Fig. 29 is thus justified by showing that the points determined have x, y coordinates, respectively equal to $a \cos \phi$, $b \sin \phi$ (which values satisfy $x^2/a^2 + y^2/b^2 = 1$, as just shown). Clearly, the construction determines points on a simply elongated circle (radius b, elongated in the ratio a/b) or on a simply compressed circle (radius a, contraction ratio b/a).

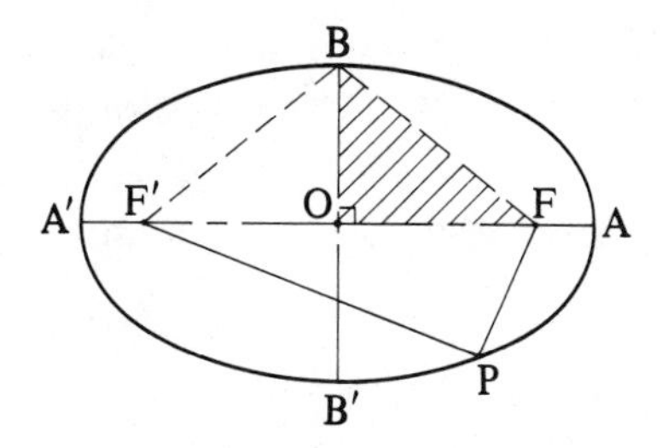

Fig. A5

69. Refer to note 27 and Fig. A5. If the ellipse were defined as the plane locus of a point P such that $PF + PF'$ is a constant, k say, then symmetry about axes through FF' and along the perpendicular bisector of FF' is evident. In particular, if B, B' are the positions of P for which $PF' = PF$, then $BF = BF' = B'F = B'F' = \frac{1}{2}k$; and if A, A' are points on the curve collinear with F, F' then $AF + AF'$ and $A'F + A'F'$ must each equal k. Hence $\frac{1}{2}(AF + AF' + A'F + A'F') = k$, that is $A'A$ $(= 2a) = k$. So in right triangle OBF, $OB = b$, $BF = \frac{1}{2}k = a$, $OF = \sqrt{(a^2 - b^2)}$. If the eccentricity, ϵ, is defined as the ratio $F'F/A'A$, or OF/OA, then $\epsilon^2 = OF^2/OA^2$, giving $\epsilon^2 = 1 - b^2/a^2$.

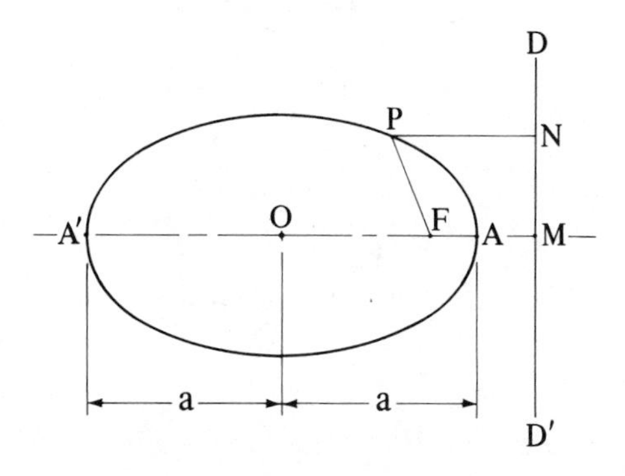

Fig. A6

Alternatively (see Fig. A6), if the ellipse is defined as the locus of a point P in a plane defined by a fixed point F and a fixed line $D'D$ where P moves so that the ratio PF/PN is a constant less than unity, PN being the perpendicular from P to $D'D$, then the eccentricity, ϵ, may be defined by $\epsilon = PF/PN$. (While it is evident from this definition of an ellipse that the locus will be symmetrical about an axis through F perpendicular to $D'D$, it is *not* evident that the locus will be symmetrical about an axis parallel to $D'D$, but this will be shown.) Let A, A' denote the positions taken by P when on the axis through F, M, where FM is the perpendicular from focus F to directrix $D'D$, and let O be the midpoint of $A'A$, so that $A'O = OA, = a$, say. Since A, A' are positions taken up by P, $FA/AM = \epsilon$ and $A'F/A'M = \epsilon$.

Hence, major axis, $A'F + FA = \epsilon(A'M + AM)$, i.e., $2a = \epsilon \cdot 2 \cdot OM$,

$\therefore OM = a/\epsilon$.

$$\begin{aligned} \text{Further, } A'F - FA &= \epsilon(A'M - AM), \\ \text{where } A'F - FA &= A'F + FA - 2FA \\ &= A'A - 2FA = 2(OA - FA) = 2 \cdot OF. \\ &\therefore 2 \cdot OF = \epsilon(AA'), \end{aligned}$$

$\therefore OF = a \cdot \epsilon$.

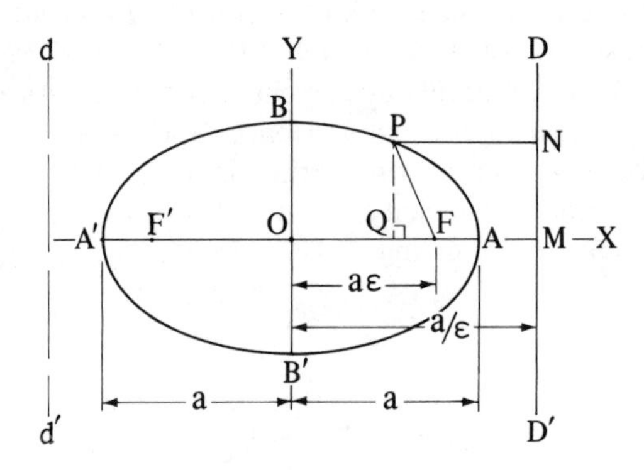

Fig. A7

Now (see Fig. A7) let P be the point (x,y) with respect to coordinate axes through O, with OX along OM, perpendicular to $D'D$, and OY parallel to $D'D$. Then, since

$$\begin{aligned} PF/PN = \epsilon \quad \Rightarrow PF^2 = \epsilon^2 PN^2 \quad \Rightarrow QF^2 + QP^2 = \epsilon^2(QM^2) \\ = \epsilon^2(OM - OQ)^2, \end{aligned}$$

we have,

$$\begin{aligned} &(a\epsilon - x)^2 + y^2 = \epsilon^2(a/\epsilon - x)^2, \\ \Rightarrow\ &a^2\epsilon^2 - 2a\epsilon x + x^2 + y^2 = a^2 - 2a\epsilon x + \epsilon^2 x^2 \\ \Rightarrow\ &\qquad x^2(1 - \epsilon^2) + y^2 = a^2(1 - \epsilon^2) \end{aligned}$$

Thus (letting x equal zero) the curve is found to intersect the y-axis where

$$\begin{aligned} y = \pm a\surd(1 - \epsilon^2), \text{ or at } y = \pm b, \text{ where} \\ b^2 = a^2(1 - \epsilon^2) \quad \Rightarrow \epsilon^2 = 1 - b^2/a^2, \end{aligned}$$

The equation to the curve may now be written as

$$x^2/a^2 + y^2/a^2(1 - \epsilon^2) = 1, \quad \text{or} \quad x^2/a^2 + y^2/b^2 = 1.$$

Since the only powers of x and y are even, the curve must be symmetrical with respect to both coordinate axes. It follows from the symmetry about the y-axis that there will be another focus, F' say, and another directrix, $d'd$ say, symmetrical about OY with F and $D'D$. See also Problems 39 and 42.

70. See p. 39, formula (4), and note 48.

71. See pp. 38–39, formulae (2) and (3), and note 48.

72. Arthur Koestler characterized as a "sleepwalker's performance" Kepler's great discovery, recorded in the *Astronomia Nova,* 1609, IV, lvi. "After six years of incredible labour," Kepler found himself confronting an unexpected, and initially not understood, coincidence involving an *ellipse*—none other than the orbit of Mars (see Koestler, *The Sleepwalkers, A History of Man's Changing Vision of the Universe,* London: Hutchinson, 1959, pp. 331–333, or Harmondsworth, England: Penguin Books, 1964, pp. 336–338, or better: Curtis Wilson, "How did Kepler Discover the First Two Laws?" *Scientific American* (March 1972):92–106, and Curtis Wilson, "Kepler's Derivation of the Elliptical Path," *Isis 59* (1968):4–25).

73. The same procedure was used by Hero of Alexandria in his *Metrica,* I, 8; this is reproduced with English translation in Ivor Thomas, *Selections Illustrating the History of Greek Mathematics,* London: Heinemann, Loeb Classical Library, 1941, vol. 2, p. 471 ff. See also Thomas Heath, *A History of Greek Mathematics,* Oxford: Clarendon Press, 1921, 1960, 1965, vol. 2, pp. 323-326. On an Old Babylonian tablet, now in the Yale Babylonian Collection (YBC 7289), there is drawn a square and along a diagonal is written the sexagesimal number that may be transcribed as 1;24, 51, 10 (i.e., $1 + 24/60 + 51/60^2 + 10/60^3$), working out to 1.41421 . . . , and correctly approximating $\sqrt{2}$ to six significant figures! It is probable that this highly accurate estimate of the multiplier to be applied to the side of a square in order to obtain the diagonal was calculated by the same method: in our fractional notation, starting with

$$\frac{3}{2} \times \frac{4}{3} = 2, \text{ obtain } \frac{1}{2}\left(\frac{3}{2} + \frac{4}{3}\right) = \frac{17}{12}$$

as an approximation for $\sqrt{2}$; then, since

$$\frac{17}{12} \times \frac{24}{17} = 2, \text{ obtain } \frac{1}{2}\left(\frac{17}{12} + \frac{24}{17}\right) = \frac{577}{408}$$

as an extremely accurate approximate value. On converting this to sexagesimal notation we find the value (1;24, 51, 10 . . .), that would have been obtained if the procedure had been carried out in that number system. See Otto Neugebauer, *The Exact Sciences in Antiquity,* Providence, R.I.: Brown Univ. Press, 1957, pp. 35, 50, and plate 6a.

74. In this age of table books and calculating devices, it may be worth indicating the traditional "long method." It begins as shown in Fig. A8. This method is readily justified by reference to a diagram showing a square of area 15 square units, in which a square of side 3 is placed so as to leave a gnomon of area 6.

	3. 8 7 2
3	15.000000. . .
	9
68	6 00
	5 44
767	5600
	5369
7742	23100
	etc.

Fig. A8

From this gnomon a second gnomon of width 0.8 and area 5.44 is removed leaving on the outside a gnomon of area 0.56. From this is taken a gnomon of width 0.07 and area 0.5369, leaving a gnomon of area 0.0231, the process being repeated as far as is required. The side of a square of area 15 is thus found to be 3 + 0.8 + 0.07 + 0.002 + ... For a Greek example of such a calculation (in sexagesimal numbers) with a diagram and explanation, see Theon of Alexandria, *Commentary on Ptolemy's Syntaxis,* I, 10, reproduced with English translation in Ivor Thomas, *Selections Illustrating the History of Greek Mathematics,* vol. 1, pp. 56–61. Cf. Lam Lay Yong, "The Geometrical Basis of the Ancient Chinese Square-root Method," *Isis 61* (1970):92–102.

75. "Most of the finest products of an applied mathematician's fancy must be rejected, as soon as they have been created, for the brutal but sufficient reason that they do not fit the facts" (G. H. Hardy, *A Mathematician's Apology,* Cambridge, England: Cambridge Univ. Press, 1940, p. 75, or 1967, p. 135). "There is between reality and this mathematical construction the same relation as that between the tangent plane and a point of the embedded manifold which it touches; as soon as one moves a little way from the point of contact, the validity of the abstract model diminishes and disappears in general, because this 'tangent plane' strays away from reality" (René Thom, "Modern Mathematics: Does It Exist?" in A. G. Howson (ed.), *Developments in Mathematical Education, Proceedings of the Second International Congress on Mathematical Education,* Cambridge, England: Cambridge Univ. Press, 1973, pp. 205–206). See also Stephen Gaukroger, *Explanatory Structures: A Study of Concepts of Explanation in Early Physics and Philosophy,* Hassocks, Sussex, England: Harvester Press, 1978, chap. 6, and, for "Galileo's mature view of the necessarily approximate and progressive character of physical science," Stillman Drake, *Galileo at Work,* Chicago: Univ. of Chicago Press, 1978, pp. 376–381.

76. *Posterior Analytics,* 78 b 34 - 79 a15, translated by G. R. G. Mure.

77. That more thoroughgoing explanations of natural phenomena are needed in the substantive sciences than can be effected by reference to the abstract domain of mathematics alone was an inevitable consequence of the doctrine of the four causes and was an important tenet of Aristotelian philosophy of science. Perhaps the best known illustration of this tenet is the distinction between *astronomy,* understood as applied abstract geometry and kinematics, and *physics,* as a profound natural philosophy. A passage in Simplicius' commentary on Aristotle's *Physics,* II, 2, preserves the account of this by Geminus of Rhodes, 1st cent. B.C., or of his contemporary, Posidonius; it is clearly in the Aristotelian tradition:

> The things, then, of which alone [mathematical] astronomy claims to give an account it is able to establish by means of arithmetic and geometry. Now in many cases the astronomer and the physicist will propose to prove the same point, e.g. that the sun is of great size, or that the earth is spherical; but they will not proceed by the same road. The physicist will prove each fact by considerations of essence or substance, of force, of its being better that things should be as they are, or of coming into being and change; . . . but the astronomer [*qua* mathematician] when he proves facts from external conditions, is not qualified to judge the cause, . . . he invents, by way of hypothesis, and states certain expedients by the assumption of which the phenomena will be saved. . . . But he must go to the physicist for his first principles, namely that the movements of the stars are simple, uniform, and ordered, . . .

(Translated by Thomas L. Heath, *Greek Astronomy,* London: Dent, 1932, pp. 124–125.) See also Pierre Duhem, *To Save the Phenomena, An Essay on the Idea of Physical Theory from Plato to Galileo,* translated (from the French edition of 1908) by E. Doland and C. Maschler, Chicago: Univ. of Chicago Press, 1969.

78. *Posterior Analytics,* 100 a 11–13.

79. William P. D. Wightman referred to "the hinterland of thought and practice which made possible the emergence of this [or that] idea into explicit form" (*The Growth of Scientific Ideas,* Edinburgh: Oliver & Boyd 1950, p. 6). Lewis Mumford asked, "Would man have ever dreamed of flight in a world destitute of flying creatures?" (*The Myth of the Machine,* [vol. 1:] *Technics and Human Development,* New York: Harcourt Brace Jovanovich, 1967, p. 36).

80. Henri Poincaré, in the lecture already referred to in notes 18 and 44, "Mathematical Discovery," *Science and Method,* translated by F. Maitland, 1914, New York: Dover reprint, p. 46. Cf. Paul Halmos, "Innovation in Mathematics," *Scientific American* (September 1958):66–73: "Practical application, curiosity and history are the main wellsprings of innovation, but two others should also be mentioned. One is failure; the other is error. Everyone who has so far tried to prove Fermat's conjecture [his "last theorem"] has failed, but some of the efforts have produced the most fruitful concepts in modern algebra and in number theory. As for error, whole books full of brilliant mathematics have been inspired by it. An oversight or a misstatement by mathematician X is often just what mathematician Y needs to find the truth; if X and Y happen to be the same person, so much the better." (p. 71.) Cf., however, Morris Kline, *Why the Professor Can't Teach: Mathematics and the Dilemma of University Education,* New York: St. Martin's Press, 1977, for the dangers of triviality consequent upon too much "inbreeding" in pure mathematics conducted independently of stimulation from the sciences, technology, or any other activity of cultural significance. (See excerpts from this challenging book in the first issue of the *Mathematical Intelligencer 1* (1978):5–14, and P. J. Hilton's response, ibid., pp. 76–80.)

81. The following are typical elementary examples. A curve of reflected light from the distant metal dome of a new cathedral led W. H. Roever to determine "the locus of the actual brilliant points of the meridians of a sphere with respect to a source of light which is infinitely distant and an observer who is (practically) infinitely distant" (*American Mathematical Monthly 20* (1913):299–303). Cf. the example of the catacaustica of reflection on the surface of a liquid in a cup, which has attracted mathematicians for several centuries (see Arnold Emch, *Scripta Mathematica 18* (1952):31–35). G. H. Hardy and J. E. Littlewood proposed the problem of showing that "if we neglect the differences in the scoring values of different strokes, and psychological or other factors peculiar to the game of cricket, then a batsman who has a large average a should score $2a$ in about e^{-2} of his innings" (*Mathematical Gazette 18* (1934):195). M. Sholander set a problem concerning the motion of a hula hoop turning about "a vertical girl whose waist is circular, not smooth, and temporarily at rest" (*American Mathematical Monthly 66* (1959):917–918). Problems involving maxima and minima are legion, of course. The pseudopractical clothing of many of them need not be taken seriously. Yet mathematicians do have resources to be stimulated in otherwise tedious situations. For example, J. P. Ballantine was led to ask, for a person drinking from a conical glass and tipping it at a constant rate, "at what angle will the delivery be the maximum and at what angle will the surface be a maximum?" (*American Mathematical Monthly 30* (1923):339–341).

82. Johann Kepler, *Strena, seu de nive sexangula,* 1611, see the English edition, *The Six-Cornered Snowflake,* modernized Latin text with translation by Colin Hardie, Oxford: Clarendon Press, 1966; John Dalton, *A New System of Chemical Philosophy,* Part I, 1808, facsimile, London: Dawson, n.d., p. 133 ff; John G. Burke, *Origins of the Science of Crystals,* Berkeley: Univ. of California Press, 1966. The geometry of sphere packing utilized by Kepler in his *Strena* is now known to have been worked out some years earlier by Thomas Har[r]iot, who began his investigations in 1591 by actually stacking round bullets into pyramidal piles (see Muriel Rukeyser, *The Traces of Thomas Hariot,* New York: Random House, 1971, pp. 117–121). Brahmagupta, *c.* A.D. 628, in his *Brāhma-sphuṭa-siddhānta* ["The Revised System of Brahma"], XII, iii, 20, mentions that the summation rules for certain series can be illustrated by means of piles of equal balls, but there is no indication that anything more than the number of layers and the number of balls present was under consideration. The English translation is in H. T. Colebrooke, *Algebra with Arithmetic and Mensuration from the Sanscrit of Brahmegupta and Bhāscara,* London: Murray, 1817, pp. 293–294. The hexagonal symmetry of snowflakes had been written about by the Chinese from at least as early as the second century B.C., but the writers in that tradition remained "content to accept it as a fact of Nature and to explain it in accordance with the numerology of symbolic correlations ['six being the true number of water']" (pp. 103–104 in Joseph Needham, *Clerks and Craftsmen in China and the West,* Cambridge, England: Cambridge Univ. Press, 1970; see pp. 99–106, 134).

83. The problem was to prove the impossibility of crossing the seven bridges of the Prussian city of Königsberg (now Kaliningrad) in a continuous walk, starting and finishing at the same point, without crossing any bridge more than once. This problem was already well known to the citizens of Königsberg when Euler became interested in it. It took an Euler to respond to it mathematically, treating it abstractly, definitively, and in more general terms than those in which it had been proposed. For an English translation of Euler's paper of 1736, see James R. Newman (ed.), *The World of Mathematics,* New York: Simon and Schuster, 1956, vol. 1, pp. 573–580. An alternative translation of the first part of the paper is given in Dirk J. Struik (ed.), *A Source Book in Mathematics, 1200–1800,* Cambridge, Mass.: Harvard Univ. Press, 1969, pp. 184–187. Euler denoted the land regions by capital letters and the bridges by lower case letters, in effect reducing the former to points and the latter to connecting lines. (It was Euler who first designated the sides of a triangle *ABC* by the letters *a, b, c* placed respectively opposite vertices *A, B, C;* see solution to Problem 16b in the present book.) His paper is now considered the first ever on graph theory. (See Oystein Ore, *Graphs and Their Uses,* New York: Random House, 1963, pp. 23–24.)

84. The student was Francis Guthrie, later professor of mathematics at the South African University, Cape Town. Guthrie's conjecture (that *four* colors are sufficient to distinguish all the bordered regions on any map so that no two adjacent regions have the same color) was reported by his professor, Augustus De Morgan, in a letter dated 23 October 1852. Even before that, about 1840, the attention of A. F. Möbius was drawn to virtually the same problem, but he does not appear to have worked seriously upon it. See Kenneth O. May, "The Origin of the Four-Color Conjecture," *Isis 56* (1965):346–348 (reprinted in *Mathematics Teacher 60* (1967):516–519):

> There is no evidence that mapmakers were or are aware of the sufficiency of four colors. A sampling of atlases in the large collection of the Library of Congress indicates no tendency to minimize the number of colors used. Maps utilizing only four colors are rare, and those that do usually require only three. Books on cartography and the history of mapmaking do not mention the four-color property, though they often discuss various other problems relating to the coloring of maps. . . . the coloring of a geographical map is quite different from the formal problem posed by mathematicians because of such desiderata as coloring colonies the same as the mother country and the reservation of certain colors for terrain features, e.g. blue for water. The four-color conjecture cannot claim either origin or application in cartography.

(p. 346.) Cf. John Wilson, "New Light on the Origin of the Four-color Conjecture," *Historia Mathematica 3* (1976):329–330. The problem itself is clearly discussed by Philip Franklin, *The Four Color Problem,* New York: Scripta Mathematica, 1941; also by Oystein Ore, *The Four-Color Problem,* New York: Academic Press, 1967. See further, Kenneth Appel and Wolfgang Haken, "The Solution of the Four-Color-Map Problem," *Scientific American* (October 1977): 108–121; Thomas L. Saaty and Paul C. Kainen, *The Four-Color Problem, Assaults and Conquest,* New York: McGraw-Hill 1977; and W. T. Tutte, "Colouring Problems," *Mathematical Intelligencer 1* (1978):72–75.

85. For the beginnings of the mathematical theory of games, see M. Fréchet, "Emile Borel, Initiator of the Theory of Psychological Games and Its Application [with English translations of and comments on three pioneering notes by Borel dated 1921, 1924, 1927]," *Econometrica 21* (1953):95–127. See especially pp. 124–125 for an important statement by von Neumann. See also the special supplement to vol. 60 of the *Bulletin of the American Mathematical Society* (May 1958), titled "John von Neumann, 1903–1957," especially pp. 100–122, for H. W. Kuhn and A. W. Tucker's article, "John von Neumann's Work in the Theory of Games and Mathematical Economics"; and further: Brian Martin, "The Selective Usefulness of Game Theory," *Social Studies of Science 8* (1978): 85–110.

86. *Studies in the History of Statistics and Probability,* vol. 1, edited by E. S. Pearson and M. G. Kendall, 1970, and vol. 2, edited by M. G. Kendall and P. L. Plackett, 1977, London: Griffin; Oystein Ore, "Pascal and the Invention of Probability Theory," *American Mathematical Monthly 67* (1960):409–419; O. B. Sheynin, "On the Prehistory of the Theory of Probability," *Archive for History of Exact Sciences 12* (1974):97–141; J. van Brakel, "Some Remarks on the Prehistory of the Concept of Statistical Probability," ibid. *16* (1976):119–136; Ian Hacking, *The Emergence of Probability, A Philosophical Study of Early Ideas about Probability, Induction and Statistical Inference,* London: Cambridge Univ. Press, 1975 and the review of this book by Colin Howson, *British Journal for the Philosophy of Science 29* (1978):274–280. In Britain and on the Continent, not only elementary probability theory, but also "political arithmetic," statistical recording and actuarial studies, long preceded what could properly be called a theory of statistics. See lectures by Karl Pearson, 1921–1933, published as *A History of Statistics in the 17th and 18th Centuries against the Changing Background of Intellectual, Scientific and Religious Thought,* edited by E. S. Pearson, London: Griffin, 1978.

Not well known in the West is the combinatorial work in Indian mathematics as exemplified by examples such as the following: "In a pleasant, spacious and elegant edifice, with eight doors, constructed by a skillful architect, as a palace for the lord of the land, tell me the permutations of apertures taken one, two, three, etc." "Say, mathematician, how many are the combinations in one composition, with ingredients of six different tastes, sweet, pungent, astringent, sour, salt and bitter, taking them by ones, twos, or threes, etc." [Correct solutions are given: $8 + 28 + 56 + 70 + 56 + 28 + 8 + 1 = 255$ arrangements of the apertures; $6 + 15 + 20 + 15 + 6 + 1 = 63$ preparations.] (Bhāskara, *Līlāvatī* [A.D. 1150], IV, vi, 114; see translation by Henry Thomas Colebrooke, *Algebra with Arithmetic and Mensuration from the Sanscrit of Brahmegupta and Bhāscara,*

1817, p. 50.) For reference to similar problems dating back to several centuries B.C., see Bibhutibhusan Datta, "The Jaina School of Mathematics," *Bulletin of the Calcutta Mathematical Society 21* (1929):115–145.

87. Cf. Bertrand Russell, "The Study of Mathematics," *New Quarterly* (1907), reproduced in his *Mysticism and Logic,* Harmondsworth, England: Penguin Books, 1953, pp. 60–73, especially pp. 62, 72.

88. See David Eugene Smith, "On the Origin of Certain Typical Problems," *American Mathematical Monthly 24* (1917):64–71; Louis C. Karpinski, "The Origin of the Mathematics as Taught to Freshmen," *Scripta Mathematica 6* (1939):133–140; also the standard works on recreational mathematics, such as Maurice Kraitchik, *Mathematical Recreations,* 1942; reprinted, New York: Dover, 1953.

89. Boris Hessen's pioneering paper, "The Social and Economic Roots of Newton's 'Principia'," in *Science at the Cross Roads,* Papers Presented to the International Congress of the History of Science and Technology, London, 1931, by delegates of the USSR, 2nd ed., London: Cass, 1971, pp. 149–212, should be studied in conjunction with G. N. Clark, *Science and Social Welfare in the Age of Newton,* 1937, 2nd ed., Oxford: Clarendon Press, 1949, especially chap. 3. See further, Ruth Schwartz Cowan, *Isis 63* (1972):509–528; also *Social Studies of Science 8* (1978), part 1, special issue on the sociology of mathematics; Barry Barnes, *Scientific Knowledge and Sociological Theory,* London: Routledge and Kegan Paul, 1974, chap. 5, and Jerry Gaston, *The Reward System in British and American Science,* New York: Wiley, 1978.

90. After rejecting the idea that the perimeter of an ellipse is (like the area) the mean proportional between the corresponding measures of circles having as diameters the major axis and the minor axis of the ellipse, respectively, Kepler correctly concluded that the arithmetic mean of the circumferences of these two circles—our $p = \pi(a + b)$ formula—would give a very close estimate of the perimeter in the case of an ellipse with eccentricity as small as that of the orbit of Mars. But the general theory of the elliptical perimeter was quite beyond his reach and he was unable to suggest a more accurate formula than that just mentioned. Kepler's struggles with the orbit of Mars are the subject of part IV of his *Astronomia nova,* 1609, for which see *Johannes Kepler Gesammelte Werke,* edited by W. von Dyck, M. Caspar, and F. Hummer, vol. 3, Munich: Beck, 1937, pp. 271–381; refer especially to pp. 306–307, and to the secondary discussions mentioned in note 72 supra.

91. "The fact that trigonometry was cultivated not for its own sake, but to aid astronomical inquiry, explains the rather startling fact that spherical trigonometry came to exist in a developed state earlier than plane trigonometry" (Florian Cajori, *A History of Mathematics* [first published in 1893], New York: Macmillan, 1919, p. 47).

92. In about 1641, Galileo, aged and blind, had described a design for a pendulum clock. The details were taken down by Galileo's son, Vincenzio, who himself died in 1649 without having been able to produce a completed workable model. The details of Galileo's design became known to Huygens only in 1660, after Huygens had produced a pendulum clock of his own design in 1656. Over the years Huygens improved his design, introducing numerous refinements up to 1694. As early as 1657 Salomon Coster, clockmaker of the Hague, was constructing practical pendulum clocks to Huygens's specifications, the work of this early stage being reported by Huygens in his *Horologium* of 1658 (Christiaan Huygens, *Oeuvres Complètes,* vol. 17, The Hague: Nijhoff, 1932). Although it no longer seems likely that Galileo retained his early belief in the isochronism of the pendulum with changes of amplitude (see Stillman Drake, "New Light on a Galilean Claim about Pendulums," *Isis 66* (1975):92–95), it was Huygens who devised an ingenious means for continuously increasing the effective length of the pendulum as the amplitude decreased, a compensation that, in theory, would render the pendulum isochronous (or *tautochronous,* from *ταὐτό*, a contraction of *τὸ αὐτό*, 'the same'). Huygens suspended the rod of the pendulum from a short flexible silk strap, which by wrapping itself against small fixed "cheeks" caused the weight to deviate from the circular path over which it otherwise would have moved. He had not worked out the theoretical form of the requisite arc and of the controlling cheeks in time for his publication of 1658, but within a couple of years he had succeeded. It was in connection with his theorem on the tautochronism of the cycloid, viz., that "on a cycloid with a vertical axis whose vertex is below, the times of descent in which a mobile point, starting from rest at any arbitrary point of the curve, reaches the lowest point, are all equal . . . ," that Huygens was led to his tautochronous pendulum in which the bob moved in a *cycloidal* arc. The problem of the form of the cheeks led to his discovery that the *evolute* of an arch of a cycloid consists of two half-arches of a congruent cycloid meeting in a cusp, and from here he went on to develop a theory of evolutes and involutes, published in his magnum opus, *Horologium oscillatorium,* 1673 (*Oeuvres,* vol. 18). An English translation of several propositions is given in Dirk J. Struik (ed.), *A Source Book in Mathematics, 1200–1800,* Cambridge, Mass.: Harvard Univ. Press, 1969, pp. 263–269. Related to the foregoing tautochrone problem is the "brachistochrone" problem proposed by Johann Bernoulli in 1696, the problem of finding the equation of the path down which a particle will fall from one point to another in the shortest possible time (*βράχιστος*, superlative of *βραχύς*, 'short'). The solution to this problem also is a cycloid! See Struik's *Source Book,* pp. 391–399; also Piero Ariotti's detailed study, "Aspects of the Conception and Development of the Pendulum in the 17th Century," *Archive for History of Exact Sciences 8* (1972):329–410.

A certain anticipation of the theory of evolutes, in the case of conic sections, is to be found in Apollonius's *Conics,* Book V; for a concise statement, see Margaret Baron, *The Origins of the Infinitesimal Calculus,* Oxford: Pergamon Press, 1969, pp. 51–52. As for the motivation for the Greek study of the conic sections, William Whewell's oft-repeated sentiments are worth quoting: "The Greeks . . . had by no means confined themselves to those propositions which had a visible bearing on the phenomena of nature; but had followed out many beautiful trains of research concerning various kinds of figures, for the sake of their beauty alone; as for instance in their doctrine of Conic Sections. . . . But . . . if the properties of conic sections had not been demonstrated by the Greeks and thus rendered familiar to the mathematicians of succeeding ages, Kepler would probably not have been able to discover those laws respecting the orbits and motions of planets which were the occasion of the greatest revolution that ever happened in the history of science." (*The Philosophy of the Inductive Sciences,* 2nd ed., 1847, reprinted, London: Cass, 1967, vol. 1, pp. 156–157.) Cf., however, the conjecture of Philippe de la Hire, 1682, and of Otto Neugebauer, 1948, that the Greeks were led to study the conic sections via a consideration of the nature of the shadow curves generated on sundials; see W. W. Dolan, "Early Sundials and the Discovery of the Conic Sections," *Mathematics Magazine 45* (1972):8–12 (p. 8 for quotation from La Hire); also, O. Neugebauer, "The Astronomical Origin of the Theory of Conic Sections," *Proceedings of the American Philosophical Society, 92* (1948):136–138.

93. Capriciousness is of the essence of discovery, because we can only know where we are going, and whether it is worth going to, when we are there. Accordingly we cannot help following many false trails, and going astray in many ways. . . . Nevertheless, the development of mathematics is perhaps less capricious than that of other sciences, more completely determined (or less undetermined), if not by external factors, at least by internal ones, for each theory presses forward as it were, and the mathematicians who are playing with it must needs perceive some of its consequences. The desire to follow them to the limit is then likely to prove irresistible, whether these consequences be useful or not.

(George Sarton, *The Study of History of Mathematics,* 1936, Dover reprint, 1957, p. 19.)

HORA SEXTA

94. In *Laws,* VII, 818 a–b, Plato has his Athenian distinguish mathematical necessity from the pseudonecessities of popular discourse (or, as we should now say, *logical* necessity from technical, political, conventional, and suchlike "necessities"): "We simply cannot dispense with its character of *necessity;* in fact, it is this which the author of the proverb presumably had in view when he said that 'even God is never to be seen contending against necessity.' No doubt he meant the necessity which is *divine,* for if you understand the words of mere human necessities, like those to which men in general apply such sayings, they are far and away the silliest of speeches." Also *Laws,* V, 741 a: "with necessity, as the proverb says, not even a god can cope." (A. E. Taylor translation.) Cf. Descartes, Letter to Father Mesland, 2 May 1644:

> You raise the difficulty of conceiving how God could have chosen, freely and indifferently, that it should not be true that the three angles of a triangle are equal to two right angles, or in general that it should not be true that contradictories cannot be together. But this is easily removed by considering that God's power cannot have

> any limits; and also by considering that our mind is finite, and was created of such a nature that it can conceive the possibility of the things God chose should actually be possible, but not of things that God could have made possible, but in fact chose to make impossible. . . . although God has chosen that some truths should be necessary, that is not to say that he chose them necessarily.

Philosophical Writings, translated by Elizabeth Anscombe and Peter Thomas Geach, Edinburgh: Nelson, 1954, p. 291.)

95. It is true that Sagredo, though a wealthy man himself, was not corrupted on that account; yet it is proper to remark that the exchange here between Simplicio and Salviati constitutes a fictional extrapolation from the austerity with which Sagredo pursues truth in the dialogues to a more general austerity in the conduct of his life. The devotion to science shown by the historical Sagredo does not appear to have been associated with any notable acts of renunciation. This Giovanfrancesco Sagredo (1571–1620), whom Galileo immortalized in his greatest writings, was a man of wide interests, a great lover of art as well as of science.

96. Power, riches, wisdom, strength, honor, glory, blessing—and other listings derived from biblical sources. Cf. also the seven virtues, consisting of the four "cardinal virtues," prudence, temperance, fortitude, and justice (already distinguished by Plato and the Stoics) together with the "theological virtues," faith, hope, and love, added by Christian writers. For a concise statement of the widespread importance that has been attached to the number seven, see Arnold Whittick, *Symbols, Signs and Their Meaning,* 1960, p. 225. For the "Hippocratic" treatise on the number seven, Περὶ ῾Εβδομάδων (*'De Septimanis'*), see W. H. Roscher, *Die hippokratische Schrift von der Siebenzahl,* Paderborn: Schöningh, 1913. In this eclectic work with Pythagorean affinities, "the whole world, the winds, the seasons, the human soul, the human body, the functions of the head, each and all were . . . stamped with the hall-mark of seven" (Theodor Gomperz, *Greek Thinkers, A History of Ancient Philosophy,* translated by L. Magnus and G. C. Berry, vol. 1, London: Murray, 1901, p. 294). In the first century A.D., Philo Judaeus, steeped in both Greek and Hebrew philosophy, expanded upon the marvelous properties of the number seven and its supposedly auspicious role in the phenomenal world; see his *De Opificio Mundi,* xxx–xliii (*Philo,* Loeb Classical Library, vol. 1, pp. 72–101). See also Grace Murray Hopper, "The Ungenerated Seven as an Index to Pythagorean Number Theory," *American Mathematical Monthly 43* (1936):409–413; C. J. de Vogel, *Pythagoras and Early Pythagoreanism,* Assen: Van Gorcum, 1966, especially p. 168 ff.; Maurice H. Farbridge, *Studies in Biblical and Semitic Symbolism,* 1923, republished New York: Ktav, 1970, pp. 97, 119–139; Vincent F. Hopper, *Medieval Number Symbolism: Its Sources, Meaning, and Influence on Thought and Expression,* 1938, republished New York: Cooper Square, 1969; Morton W. Bloomfield, *The Seven Deadly Sins: An Introduction to the History of a Religious Concept . . . ,* Michigan State Univ. Press, 1952 and 1967; Bernard Orchard et al. (eds.), *A Catholic Commentary on Holy Scripture,* London: Nelson, 1953, refer index, s.v. "seven."

97. Euler, who wrote a paper on these "orbiform curves," was perhaps the first to consider them theoretically. It would be interesting to find out whether combinations of blending arcs forming "ovals" of constant diameter were ever used in practice before mathematicians became interested in them. Perhaps such forms did find occasional architectural or similar application, but prior practical use of particular examples, stumbled upon now and again, would not imply that the uniformity of their diameters was recognized, or recognized as a subject for theoretical study.

98. The seven equally spaced points (*A, B, C, D, E, F, G* in Figs. 31, 33) cannot be laid out by means of a strict straight-edge-and-compasses construction, though a construction involving a kind of *νεῦσις* for the regular heptagon (or heptagram) was devised by Archimedes, according to the Arabian tradition (see Thomas L. Heath, *A Manual of Greek Mathematics,* Oxford: Clarendon Press, 1931, or reprint, New York: Dover, 1963, pp. 340–342; also Johannes Tropfke, "Die Siebeneckabhandlung des Archimedes," *Osiris 1* (1936):636–651, and Crockett Johnson, "A Construction for a Regular Heptagon," *Mathematical Gazette 59* (1975):17–21). For a simple example of a *neusis* construction, see below, solution to Problem 21(b). For the regular polygons constructible by "Euclidean" methods, see Felix Klein, *Famous Problems of Elementary Geometry,* 2nd English ed., 1930 (New York: Dover reprint 1956, or *Famous Problems and Other Monographs,* New York: Chelsea, 1962), pp. 16–41, 81–85; also L. E. Dickson, "Constructions with Ruler and Compasses," in J. W. A. Young (ed.), *Monographs on Topics of Modern of Modern Mathematics,* New York: Longmans, Green, 1911, or Dover reprint (with introduction by Morris Kline) 1955, pp. 351–386.

99. The term *oval* is used here in the usual mathematical sense to refer to any closed curve that is everywhere outwardly convex (but not necessarily egglike to the extent of having one end more pointed than the other). Each of the ovals in Figs. 31, 33 has seven concurrent axes of symmetry, meeting at the centroid of the figure. These figures have "sevenfold symmetry," which is to say that if each is rotated in its plane about its centroid through one-seventh of 360° then it forms the same figure as before. But, as in the case of regular polygons with an odd number of sides, the centroid is not a center of symmetry; that is, not every chord through the centroid is bisected at that point. Only very exceptional central chords are so bisected. Consequently *wheels* (as opposed to rollers without axles) having peripheries in the form of any of these ovals would *not* function satisfactorily.

100. Faint memories of a Nietzschean dream may be supposed as the source of this allusion. Nietzsche, in his *Birth of Tragedy,* 1872 etc., distinguished two tendencies, the "Apollonian" and the "Dionysian," the former the classical genius of restraint, rational, critical, as delineated by Plato, the latter the longing for unrestrained expression, passionate exultation. Something powerful in Simplicio's unconscious has perhaps been touched by his encounter with these "circles that are not circles"; Sagredo, cautious and analytic as usual, is intent on removing the mystery from them. Whereas the circles of classical geometry (including one in a particular context named after "the Great Geometer," Apollonius) are perfectly restrained and completely symmetrical, these new ovals could suggest some pulsating or unfolding process in which a nucleus of centers, initially coincident and with undisclosed potentialities, develops kinematically or organically to yield a vital differentiated form. The infinite variations possible will be sufficiently evident from later diagrams. The continuous changes could best be displayed by means of a computer-produced film, a project with great aesthetic possibilities.

101. In isoperimetry, or the comparison of the areas (or volumes) of figures with equal perimeter (or surface area), important results were obtained by Zenodorus, now believed to have flourished in the early part of the second century B.C. (see *Dictionary of Scientific Biography*), and Pappus, first half of the fourth century A.D., but their theory was markedly incomplete. See note 224. Kepler (see note 90) in his *Astronomia Nova,* 1609, IV, xlviii (*Gesammelte Werke,* vol. 3, pp. 306-307) appealed to the "well-known result" that of (plane) figures that are equally capacious, the circle has the least perimeter, in order to justify his finding that the perimeter of an ellipse must be greater than the circumference of a circle of equal area (in our symbolism, $p_e > 2\pi\sqrt{(ab)}$). A fuller discussion is contained in Hora Quarta Decima.

102. It may be left to the reader to show that for the inner oval in Fig. 35, $BB':AA' = (\sqrt{14} - \sqrt{2}):2 \approx 1.164:1$.

103. Named after Franz Reuleaux (1829–1905). See his *Kinematics of Machinery,* translation of *Theoretische Kinematik,* 1875, by A. B. W. Kennedy, 1876; reprinted with an introduction by E. S. Ferguson, New York: Dover, 1963, pp. 129–160; refer also to chap. 9 and 10 for pumps, blowers, and rotary steam-engine designs—all items in the long prehistory of present-day rotary internal combustion engines. For the latter, see David H. Nash, "Rotary Engine Geometry," *Mathematics Magazine 50* (1977):87–89. A drill for cutting square holes, invented by H. J. Watts in 1914, consists of a shaft having a cross-section in the form of a Reuleaux triangle except for three cut-away sections to provide the flutes. For an illustration and brief description of this drill, see p. 150 of Martin Gardner, "Curves of Constant Width," *Scientific American* (February 1963):148–156.

104. Refer, Charles Mugler, *Dictionnaire historique de la terminologie géométrique des Grecs,* Paris: Klincksieck, 1958–59, pp. 127–128, 132–135. The genuineness of the exceptional appearances of διαγώνιος in Euclid (XI, 28, 38) may be doubted as the term is not used in the surviving texts of Aristarchus, Archimedes, and Apollonius, and Gandz was of the opinion that before the first century A.D. the Greeks "had no special term for the diagonal, but called it a 'line stretched from corner to the other corner.' The Sophists created the term διάμετρος 'diameter,' using it promiscuously for the diagonal and the diameter of the circle. Hero introduced the term διαγώνιος 'diagonal' " (Solomon Gandz, in his edition of "The Mishnat ha Middot, . . .," *Quellen und Studien zur Geschichte der Mathematik, Astronomie und Physik A, 2* (1932):1–96 & plates; p.

14, n. 16). Cf. Proclus, *Commentary on the First Book of Euclid's Elements,* 156, 6 ff. (p. 124 of translation by Glenn R. Morrow, Princeton, N.J.: Princeton Univ. Press, 1970).

105. Apollonius, *Conics,* Book I (see p. 5 of Apollonius of Perga, *Treatise on Conic Sections,* translated and edited by Thomas L. Heath, Cambridge, England: Cambridge Univ. Press, 1896, and cf. the introduction, p. xlix).

106. For an ellipse, all support lines are tangents and all tangents are support lines.

107. There is a certain lack of terminological uniformity in the literature. Derek G. Ball, "A Generalization of π," *Mathematical Gazette 57* (1973):298–303, actually gives the following definition: "The width of a shape is the maximum value of the distance between two points on its boundary" (p. 298). (Although builders sometimes refer to the length of a kitchen sink as its width, it would be very surprising to find the diagonal called the width. Cf., however, the *adspeak* used in T.V.-set promotion where *a* "23-inch screen" might only measure 23 inches along a diagonal!) J. H. Cadwell writes with reference to closed convex curves: "The breadth of the curve in a given direction is the distance between the unique pair of parallel support lines perpendicular to this direction. . . . As the chosen direction varies the breadth will change. We call the minimum value taken the width of the curve." (*Topics in Recreational Mathematics,* Cambridge, England: Cambridge Univ. Press, 1966, p. 144.)

108. There is an analogy that may be found not unilluminating between this situation in formal mathematics and a point made with respect to empirical science by operationalists intent upon stressing that alternative methods of measurement are not to be dismissed merely as a matter of laboratory convenience. Whereas it is, or was, usual simply to suppose that discrepancies arising between different methods of measuring what is customarily regarded as the same physical quantity are attributable to experimental error, operationalists urged that we should be prepared instead to recognize these discrepancies as probably being due, to some extent, to the presence of genuinely different properties (interactions). See, for example, Herbert Dingle: "The result of each operation should initially have its own name, and if we find that different operations yield approximately the same result, then we have made the empirical discovery that the quantities represented by the different names are approximately equal. . . . From this point of view the results yielded by our ten methods of measuring surface tension are not ten strivings after the same ideal but ten independent quantities which we have discovered to be approximately equal." (P. 7 of "A Theory of Measurement," *British Journal for the Philosophy of Science 1* (1950):5–26.) Not unnaturally, investigators have been more impressed by close agreements than by small discrepancies between different methods of measurement. Nevertheless the possibility cannot be ignored that a seeming identity between the indications of different measurement methods—or, in the concern of our discussion, between the denotations of alternative definitions—may break down in new circumstances to reveal significant distinctions.

109. See notes 58, 59, 94 supra.

110. Once a particular terminology or definition, or, more generally, a particular abstract characterization, has been adopted, one's expectations can hardly fail to be markedly influenced if not rigidly molded by it. Cf. Arthur Koestler: "Words are essential tools for formulating and communicating thoughts, and also for putting them into the storage of memory; but words can also become snares, decoys, or straight-jackets" (*The Act of Creation,* London: Hutchinson, 1964, or New York: Dell, 1967, p. 176); also: "Among all forms of mentation, verbal thinking is the most articulate, the most complex, and the most vulnerable to infectious diseases. It is liable to absorb whispered suggestions, and to incorporate them as hidden persuaders into the code" (p. 177). Daily examples provided by the news media as well as the extreme cases presented by George Orwell and Max Frisch have made us all too familiar with the use of deceptive vocabularies of "laundered" words: "They're whitewash words, cover-up words, corrupt words, words with a silver lining that are used to make ugly things look nicer and immoral things seem more moral. . . . And the distortions work the other way as well. . . . Those who criticize [the official policy or propaganda] do not have an 'opinion,' they are 'provocateurs.' They never convince others, they 'mislead' them." (Amos Elon, *Newsweek* (5 February 1979):56.)

HORA SEPTIMA

111. Do we say that a pentagram, the symbol of the Pythagorean Order, has 5, or 10 vertices? The polygons considered in the most elementary work are such that the sides meet only at their ends, and if the vertices are to be defined on the understanding that this restriction holds, it will be immaterial whether they are taken as the ends of the sides or as their points of intersection. But with a more general concept of "polygon," the possibility arises of some of the sides crossing others at points that are not endpoints; so the two criteria for a point to be a vertex, which previously were quite interchangeable, have now to be dissociated. (By defining a *plane rectilinear polygon* as a "plane figure consisting of n points, $p_1, p_2, p_3, \ldots, p_n$ (the vertices), $n \geq 3$, and of the line segments $p_1p_2, p_2p_3, \ldots, p_{n-1}p_n, p_np_1$ (the sides)," what we have been calling a *pentagram* becomes included in the class of pentagons.) It will be seen that, according to this definition, the pentagram is taken to have *five* vertices rather than ten. (This is in keeping with the understood ordering of the sides and the implicit requirement that the vertices be points common to adjacent sides.) Yet in some contexts it is the ten points rather than the five which are important: cf., a "complete quadrilateral," defined as "a figure consisting of four lines in a plane and their six points of intersection." The definitions quoted here are from G. James and R. C. James (eds.), *Mathematics Dictionary*, 4th ed., New York: Van Nostrand Reinhold, 1976. See also, Howard Eves, *An Introduction to the History of Mathematics*, on "Star-Polygons," New York: Holt, Rinehart, and Winston, 3rd ed., p. 229, or 4th ed., p. 231, and Imre Lakatos, *Proofs and Refutations*, 1976, pp. 17–18.

112. Cf. René Thom's statement that "each proof is, above all, the revelation of a new structure whose elements lie disconnected in man's intuition until reason joins them together. In this sense, each proof is a Socratic experience requiring the re-creation in the reader of the psychological processes necessary to elicit the implicit truth, all the elements of which he possessed but which had remained hidden in an unformulated state. . . . It is in the intuition that the *ultima ratio* of our faith in the truth of a theorem resides. And, according to a now-forgotten etymology, a theorem is above all the object of a vision." (P. 697 of " 'Modern' Mathematics: An Educational and Philosophic Error?" *American Scientist 59* (1971):695–699.)

113. "Sophistication is acquired by going through stages of being dissatisfied with different types of precision. . . . The whole point is to cultivate a logical approach by using 'local' logic (i.e., deducing interesting consequences of 'reasonable' assumptions) instead of being 'globally' logical [which belongs to a later stage] . . . so often the by-now-traditional approach to mathematics is to hammer the student into the ground, telling him what proofs are, asserting that one proof is 'rigorous' and another is not, so that he completes a course of mathematics feeling too afraid to question anything at the end." (Pp. 300–301 of Brian Griffiths, "The Languages of Mathematical Communication," *Mathematical Gazette 60* (1976):298–301.)

114. Cf. Richard Courant and Herbert Robbins, referring first to the mathematical innovators of the seventeenth and eighteenth centuries:

> In a veritable orgy of intuitive guesswork, of cogent reasoning interwoven with nonsensical mysticism, with a blind confidence in the superhuman power of formal procedure, they conquered a mathematical world of immense riches. Gradually the ecstasy of progress gave way to a spirit of critical self-control. In the nineteenth century the immanent need for consolidation and the desire for more security in the extension of higher learning that was prompted by the French revolution, inevitably led back to a revision of the foundations of the new mathematics, in particular of the differential and integral calculus and the underlying concept of limit. Thus the nineteenth century not only became a period of new advances, but was also characterized by a successful return to the classical ideal of precision and rigorous proof. In this respect it even surpassed the model of Greek science. Once more the pendulum swung toward the side of logical purity and abstraction.

(*What Is Mathematics*?, London: Oxford Univ. Press, 1941, p. xvi.)

115. As already mentioned in note 92, the evolute of an arch of a cycloid consists of two half-arches of a congruent cycloid meeting in a cusp. If a cycloid is laid out by taking various positions of a point (P, say) on a circle rolling along a fixed straight line, then any straight line through P and the point of contact (C, say) of the circle with the line along which it rolls is a normal to the cycloid. In this way, normals to the curve may be found at the same time as the points on it are obtained. The radius of curvature is not simply CP but may be shown to equal *twice* CP, a simple relation enabling the evolute to be plotted readily if required. See

Margaret Baron, *The Origins of the Infinitesimal Calculus*, Oxford: Pergamon Press, 1969, p. 164. Cf. F. R. Honey, "Determination of the Radius of Curvature of the Cycloid without the Aid of the Calculus," *American Mathematical Monthly 12* (1905):103–104.

HORA OCTAVA

116. "*Puncta inflexionum*" (Fermat, mid 1630s), "*punctum flectionis* [contrarii]" (Huygens, 1654), "*punctum flexus contrarii*" (Newton, 1671, and Leibniz, 1684)—see Thomas F. Mulcrone, "A Plea for the terminology 'Flex-point'," *Mathematics Teacher 61* (1968):475–478. An early use of part of a sine curve was made by Roberval, *c.* 1635, in connection with his ingenious determination of the area under an arch of the cycloid. This auxiliary curve, which we recognize as a half-cycle of a sine curve symmetrical about a point of inflexion, was referred to by Roberval simply as "the companion of the cycloid." In another determination, where he finds the area under a half-arch of a sine curve, he denotes the region only as the "curved figure equal to the [unit] square." Refer, Evelyn Walker, *A Study of the Traité des Indivisibles of Gilles Persone de Roberval,* New York: Teachers College, Columbia University, 1932, pp. 174–177 and 180. As Carl Boyer noted, after referring specifically to Roberval, "that the first consciously constructed graphs of a trigonometric function were motivated by the geometric transformation of sine lines rather than by the analytical function concept probably explains the surprising fact that the notion of periodicity did not enter into the theory of goniometric functions until the following century" (pp. 304–305 of "Historical Stages in the Definition of Curves," *National Mathematics Magazine 19* (1945):294–310). A different (and earlier?) origin of the sine curve is implicitly indicated in Fig. P19 for Problem 36.

117. For a proof, see H. Rademacher and O. Toeplitz, *The Enjoyment of Mathematics, Selections from Mathematics for the Amateur,* translated by H. Zuckerman, Princeton, N.J.: Princeton Univ. Press, 1957, pp. 174–177.

118. This proposition, Barbier's theorem, is named after Joseph-Émile Barbier (1839–1889). The following concise demonstration was given by Howard Eves: "By infinitesimal geometry, we can very readily show that if r_1 and r_2 are the radii of curvature at a pair of opposite points on the closed curve, then $r_1 + r_2 = b$. Let the intrinsic equation of the curve be $s = f(\phi)$, where $f(0) = 0$. Then, if p is the perimeter of the curve, $f(2\pi) = \text{p}$. Since $f'(\phi) = r$, the radius of curvature, we have: $f'(\phi + \pi) + f'(\phi) = b$. Integrating with respect to ϕ, we get: $f(\phi + \pi) + f(\phi) = \phi b + c$. To determine the constant of integration c, we observe that when $\phi = 0$, $f(\phi) = 0$. Hence $c = f(\pi)$, and we have $f(\phi + \pi) + f(\phi) = \phi b + f(\pi)$. Putting $\phi = \pi$, we get $p = f(2\pi) = \pi b$." ("Partial Solution to Problem E610," *American Mathematical Monthly 51* (1944):532–533.) For a more elementary (and longer) proof, see L. A. Lyusternik, *Convex Figures and*

Polyhedra, translated by T. Jefferson Smith, New York: Dover, 1963, pp. 31–34. (Ovals with double points and cusps, constructed as indicated in Figs. 39(ii) and 50, for example, satisfy Barbier's theorem provided that arcs that have to be traced in a reversed sense during one full circuit are counted as negative.) "Much less is known about solids of constant width than about curves of constant width. Though there is no direct analog of Barbier's theorem, Minkowski has pointed out that the shadows, formed by orthogonal projection, of a solid of constant width are of constant circumference." (Eves, *An Introduction to the History of Mathematics,* 3rd ed., p. 390, or 4th ed., p. 364.)

119. The two fallacious argument forms mentioned are as follows:

(i) *If p, then q*
not-p
———
∴ *not-q*

(ii) *If p, then q*
q
———
∴ *p*

120. A simple way of obtaining two dissimilar figures equal to each other in both area and perimeter is as follows. Let *ABCD* be any parallelogram, *G* the point of intersection of its diagonals, and *EF* any straight line through *G* meeting a pair of opposite sides internally at *E* and *F*. Then both the area and the perimeter of the parallelogram are bisected by *EF*. By turning half the parallelogram over on itself and reuniting it with the remaining half along *EF*, a new figure is obtained that is obviously equal to the original parallelogram in both area and perimeter. The new figure will be a hexagon with a reentrant angle except when *EF* is perpendicular to the sides it meets at *E, F*, in which case it will be a quadrilateral with one pair of opposite sides parallel. For a different example, see the regions marked A_1 and A_2 in Fig. S1, accompanying the solution to Problem 1.

121. To say that journals are not designed to carry accounts of wasted efforts would be to beg the question: Are we so intent on "efficiency" that every consideration not leading directly to "results" is to be regarded as *wasted?* Cf. note 16.

122. For the inaccuracy of this statement, see note 99. (No triangle has a center of symmetry.)

123. The term "centroid" was introduced as recently as the 1870s, and it was used at first in a sense quite different from that in which it came to be used very soon afterward. Acknowledging W. K. Clifford as its originator, the term was used by A. B. W. Kennedy in his English translation of Reuleaux's *Kinematics of Machinery*, 1876 (already referred to in note 103). There it was applied to designate the locus of an instantaneous center of rotation. But such loci very soon came to be termed "centrodes," by Clifford himself, in 1878. G. Minchin used the two terms in their present-day sense in 1882, but John Sturgeon Mackay referring specifically to the point of concurrency of the medians of a triangle, remarked that this point is called the centroid, "an expression due to T. S. Davies" (*Elements of Euclid*, London: Chambers, 1884, pp. 100–101).

124. Cf. Imre Lakatos who had one of his characters exclaim: "My quest is not only for *certainty* but also for *finality*. The theorem has to be certain—there must not be any counterexamples *within* its domain; but it has also to be *final*: there must not be any examples *outside* its domain. I want to draw a dividing line between examples and counterexamples, and not just between a safe domain of a few examples on the one hand and a mixed bag of examples and counterexamples on the other." (*Proofs and Refutations*, 1976, p. 63.)

HORA NONA

125. The best example from elementary geometry of a well-known theorem having a little-known but far more general counterpart is provided by the theorem of Pythagoras and its remarkable generalization given by Pappus concerning the sum of the areas of *any* two parallelograms on two sides of *any* triangle and a parallelogram on the third side (not necessarily the longest!), which shall be equal to the sum of the other two. For details, see Problem 2(a); a historical note follows the solution. In this case it is the restricted theorem that is the important one (some would say the most important in all mathematics), important because of the great number of connections it makes with items from various branches of mathematics.

126. F. L. G. Frege (1848–1925), however, pointed out how troublesome redefinition might be avoided:

> Instead of first defining a symbol for a limited domain and then using it for the purpose of defining itself in regard to a wider domain, we need only choose different signs, confining the reference of the first, once for all, to the narrower domain; in this way the first definition is now complete and draws sharp boundary-lines. This in no way prejudges the relation between the reference of one sign and that of the other; we can investigate this, without its being possible that the result of the investigation should make it questionable whether the definitions were justified. It really is worth the trouble to invent a new symbol if we can thus remove not a few logical difficulties and ensure the rigour of the proofs. But many mathematicians seem to have so little feeling for logical purity and accuracy that they will use a word to stand for three or four different things, sooner than make the frightful decision to invent a new word.

(Gottlob Frege, *Grundgesetze der Arithmetik*, vol. 2, 1903, § 60, translated by Peter Geach in Geach and Max Black (eds.), *Translations from the Philosophical Writings of Gottlob Frege*, 2nd ed., Oxford, Blackwell, 1960, p. 164.) It is understandable, however, that the mathematicians' desire for austerity and elegance has inclined them to avoid a proliferation of words and to seek terminological as well as conceptual economy. A compromise is often attainable by the use of extension-indicating adjectives, subscripts, etc.

127. As is well known, Euclid's *Elements* have been subjected both to high praise and severe disparagement. Thomas Heath, in the preface to his second edition of

The Thirteen Books of Euclid's Elements (Cambridge, England: Cambridge Univ. Press, 1926) gave it as his opinion that, "so long as mathematics is studied, mathematicians will find it necessary and worth while to come back again and again, for one purpose or another, to the twenty-two-centuries-old book which, notwithstanding its imperfections, remains the greatest elementary textbook in mathematics that the world is privileged to possess." It was, of course, inevitable that certain logical shortcomings, in particular the employment of unstated and presumably unconscious assumptions, would have been detected in such a pioneering work so continuously studied (over one thousand editions since the introduction of printing). Educators became dissatisfied with the *Elements* on different grounds. The work was written for a different milieu and "not written for schoolboys but for grown men," as Heath remarked in the preface to his first edition. It is to be seen as the great systematization of the mathematics that was known in that generation when Alexandria rather than Athens became the leading center of Greek learning, and to be seen also, perhaps, as a specific response to certain challenges implicit in the works of Plato and Aristotle. It is one thing to recognize the *Elements* as "one of the noblest monuments of antiquity" (Heath, preface to the first ed., 1908), it is another to attempt to defend it against the conclusion reached by Imre Lakatos:

> Growing concepts are the vehicles of progress, . . . the most exciting developments come from exploring the boundary regions of concepts, from stretching them, and from differentiating formerly undifferentiated concepts. In these growing theories intuition is inexperienced, it stumbles and errs. There is no theory which has not passed through such a period of growth; moreover, this period is the most exciting from the historical point of view and should be the most important from the teaching point of view. These periods cannot be properly understood without understanding the method of proofs and refutations, without adopting a fallibilist approach. This is why Euclid has been the evil genius particularly for the history of mathematics and for the teaching of mathematics, both on the introductory and the creative levels.

(*Proofs and Refutations*, 1976, p. 140.) Euclid's *Elements* ought to have been recognized for what it is—a great reference work for serious mathematicians. As an introductory text, it overpowered the student far more often than it inspired him. Many great books have been misused, and their authors are not to be held responsible on that account.

128. For a solid angle cannot be constructed with two triangles, or indeed planes. With three triangles the angle of the pyramid is constructed, with four the angle of the octahedron, and with five the angle of the icosahedron; but a solid angle cannot be formed by six equilateral and equiangular triangles placed together at one point, for, the angle of the equilateral triangle being two-thirds of a right angle, the six will equal four right angles: which is impossible, for any solid angle is contained by angles less than four right angles [XI, 21]. For the same reason, neither can a solid angle be constructed by more than six plane angles. By three squares the angle of the cube is contained, but by four it is impossible for a solid angle to be contained, for they will again be four right angles. By three equilateral and equiangular pentagons

> the angle of the dodecahedron is contained; but by four such it is impossible for any solid angle to be contained, for, the angle of an equilateral pentagon being a right angle and a fifth, the four angles will be greater than four right angles: which is impossible. Neither again will a solid angle be contained by other polygonal figures by reason of the same absurdity. Therefore, etc. Q.E.D.

(*The Thirteen Books of Euclid's Elements*, translated by Thomas L. Heath, 2nd ed., 1926, vol. 3, pp. 507–508.) The Greek text and an alternative translation are given in Ivor Thomas, *Selections Illustrating the History of Greek Mathematics*, Loeb Classical Library, vol. 1, pp. 476–479. This argument, usually referred to as Euclid's, was possibly inserted into the *Elements* later by some editor; on the other hand it might have been given by Theaetetus (flourished 380 B.C.), and perhaps even originated with the Pythagoreans of the previous century. Conceptually, it is very closely related to the simplest proof for establishing the result that among the regular polygons only three kinds (equilateral triangles, squares, regular hexagons) can be used alone to just fill up the region of a plane around a point in it. Proclus gives the simple proof of this two-dimensional theorem (the result he calls "paradoxical" for some reason!) and he explicitly states, probably on the authority of Eudemus, that the proposition is Pythagorean—Proclus, *Commentary on the First Book of Euclid's Elements*, 304, 11 - 305, 3 (Morrow ed., p. 238). The analogous three-dimensional space-filling problem is more difficult. Aristotle appears to have mistakenly believed that congruent regular tetrahedra will fill space on their own, as do cubes (*On the Heavens*, III, 8; 306 b 5–7). This is discussed by Heath in *Mathematics in Aristotle*, Oxford: Clarendon Press, 1949, p. 178.

129. Euclid, XI, def. 27. Definitions 25 to 28 are of a cube, octahedron, icosahedron, and dodecahedron, respectively. Euclid did not provide a corresponding definition of a tetrahedron. In Book XIII, Prop. 13, his regular tetrahedron is simply referred to as a pyramid (πυραμίς), though his definition of a pyramid is the general one: "A pyramid is a solid figure, contained by planes, which is constructed from one plane to one point" (Book XI, def. 12).

130. Kepler discovered two convex polyhedra, one with twelve and the other with thirty faces, which, although having the edges all equal, and the faces all congruent and equally inclined to each other at the edges, nevertheless have faces which are nonregular polygons; they are rhombuses. For illustrations, see *Harmonice mundi*, 1619, II, xxvii (J. Kepler, *Gesammelte Werke*, vol. 6, edited by M. Caspar, p. 83). These two polyhedra are sometimes described as "semi-regular", but this term is alternatively applied to a different species, the thirteen Archimedean solids soon to be referred to (see note 133). The first-mentioned solid of the present note, the rhombic dodecahedron, is well known to crystallographers; the angle between adjacent faces is 120°, and since this form is stackable in space without leaving any openings and is also circumscribable about a sphere, it has a place in the analysis of close packing of spheres. See L. G. Weld, *American Mathematical Monthly 23* (1916):346, and R. B. Hayward, "On Some Semi-Regular Solids," *Mathematical Gazette 1* (1897): 73–78.

131. Discussed and illustrated by M. S. Wahl, "The Orthotetrakaidekahedron—A Cell Model for Biology Classes," *Mathematics Teacher 70* (1977): 244–247.

132. This same solid may otherwise be obtained by removing eight triangle-based corner pyramids from a cube by planes passing through the midpoints of each set of conterminous edges. This solid, known as a cuboctahedron, is illustrated in diagram (iii) of Fig. P15 for Problem 28(c) and in the solution to that problem. A third solid with fourteen regular polygonal faces may be obtained by truncating a cube by the removal of small corner pyramids so as to leave the six inscribed regular octagons of the faces of the cube together with the eight equilateral triangles as the faces of the resulting solid.

133. The remarks just made concerning the two fourteen-faced polyhedra apply also to the third one, mentioned in note 132; it also is one of the Archimedean solids, each of which is inscribable within but not circumscribable about a sphere. An Archimedean solid is a polyhedron with all its faces regular polygons of at least two different kinds and with not less than four faces of any one kind, and having all its solid angles alike. See the note following the solution to Problem 28 and, for a systematic listing and concise description of these solids, refer to Pappus of Alexandria, *Collection*, V, xix (*La Collection mathématique*, translated by P. Ver Eecke, Paris: De Brouwer/Blanchard, 1933, vol. 1, pp. 272–276). Archimedes' own account of his thirteen polyhedra, being lost, was not available to European mathematicians, while Pappus' *Collection* remained almost completely unknown until the late sixteenth century. (For an apparent early exception, see S. Unguru, "Pappus in the Thirteen Century in the Latin West," *Archive for History of Exact Sciences 13* (1974):307–324.) Certainly, Pacioli (1509), Dürer (1525), Barbaro (1568), Jamnitzer (1568), and Stevin (1583), who rediscovered some of the Archimedean solids, were unacquainted with Archimedes' investigation, and, one and all, they fell short of matching it. See, for example, Stevin's *Problematum geometricorum*, Book 3, where the "nets" (surface developments) of ten of the thirteen solids are given, nine of these ten being defined in terms of truncations of the regular solids. (*The Principal Works of Simon Stevin*, Amsterdam: Swets & Zeitlinger, vol. 2, pp. 222–297, and D. J. Struik's introd., pp. 124–128.) With the posthumous publication in 1588 of Commandino's complete Latin translation of, and commentary on, the surviving books (III to VIII) of Pappus's *Collection*, mathematicians became informed of the scope of Archimedes' investigation and of many other important items of Greek geometry. It remained for Kepler, in the second book of his *Harmonice mundi*, 1619, to publish a systematic study of all thirteen of the Archimedean solids and to give them the names by which they are still known (J. Kepler, *Gesammelte Werke*, vol. 6, edited by M. Caspar). The solids are discussed and/or illustrated in Heath's *History of Greek Mathematics*, vol. 2, pp. 98–101; H. Martyn Cundy and A. P. Rollet, *Mathematical Models*, 2nd ed., Oxford: Clarendon Press, 1961, pp. 100–115; Magnus J. Wenninger, *Polyhedron Models*, Cambridge, England: Cambridge Univ. Press, 1971, pp. 20–32. A more technical treatment is included in H. S. M. Coxeter, M. S. Longuet-Higgins, and J. C. P. Miller, "Uniform Polyhedra," *Philosophical Transactions of the Royal Society of London, A246* (1954):401–450 and plates. ("A polyhedron is said to be *uniform* if its faces are regular while its vertices are all alike. By this we mean that one vertex can be transformed into any other by a symmetry operation. A uniform polyhedron whose faces are all alike is said to be *regular*."—p. 402.)

134. These are so-called from the use made of them in Plato's *Timaeus*, 53 c–57 d, where the four that have square or triangular faces are associated with the four elements of Empedocles—the cube with earth, the tetrahedron with fire, the octahedron with air, the icosahedron with water—the dodecahedron being assigned to the cosmos as a whole. The crude paraphrase of the *Timaeus* made in the first century A.D., entitled *On the Soul of the World and Nature*, forged in the name of the legendary Timaeus Locrus, indicated as the reason for this last correspondence, the fact that the dodecahedron "comes nearest" to the sphere. But it seems not to have been known at the time of Plato that the volume of a regular dodecahedron exceeds that of a regular icosahedron inscribable in the same sphere (though the brilliant Theaetetus might just possibly have discovered this somewhat counterintuitive relation)—see Problem 52. Cf., *Phaedo,* 110 b, where the Earth is likened to "one of those balls made of twelve pieces of skin." (Modern manufacturers of soccer balls prefer to use the Archimedean polyhedron with 32 faces of which 20 are regular hexagons and 12 are regular pentagons, the truncated icosahedron, thus achieving a very close approximation to a sphere on inflation.) Refer, *Plato's Cosmology, The Timaeus of Plato*, translation and commentary by Francis MacDonald Cornford, London: Kegan, Paul, Treach, Trubner, 1937, pp. 210–239.

135. Cf. the common definition: "A *regular polyhedron* is a polyhedron whose faces are congruent regular polygons and whose polyhedral angles are congruent" (James and James (eds.), *Mathematics Dictionary*). Leaving aside the question of the appropriate definition of "polygons" here (see note 111 supra), numerous variations in the definition appear in the literature, some being more "economical" than that just given. H. E. Webb, for example, proposed that a regular polyhedron be defined as "a convex polyhedron whose faces are congruent regular polygons, the same number at each vertex" (*American Mathematical Monthly 22* (1915):174–175). See also H. S. M. Coxeter, *Regular Polytopes* (1948) New York: Macmillan, 1963, pp. 15–16 (cf. pp. 100–104). ("A polytope is a geometrical figure bounded by portions of lines, planes, or hyperplanes"—ibid., preface.)

136. In a very muddled passage surviving from Hero's *Definitions* (Def. 104), it is asserted that Plato knew of a *τεσσαρεσκαιδεκάεδρον* (a fourteen-faced figure), probably the cuboctahedron. For a discussion of this passage, see William C. Waterhouse, "The Discovery of the Regular Solids," *Archive for History of Exact Sciences 9* (1972):212–221, on pp. 219–221, where the author conjectures that this early isolated example might have been introduced into discussions about the appropriate way to define a regular solid, perhaps as a counterexample to some proposed general definition. In support of this view Waterhouse points out that "we know specifically that the use of counter-examples was a common practice in the Academy. The most famous is doubtless Diogenes' plucked chicken, produced in response to Plato's definition of man as a featherless biped. (Diog. Laert. VI, 40. This of course may be apocryphal, but that does not affect its value as evidence that the practice was familiar.) We can observe the same process at work more seriously in the Socratic dialogues. And Aristotle in the *Topics* explains the use of counter-examples immediately after describing what a definition is (*Top.* A, 102, esp. 102 b 29–33)." (p. 221.)

137. For a discussion of Proclus' view that Euclid "set before himself, as the end of the whole *Elements*, the construction of the so-called Platonic figures" (*Commentary on the First Book of Euclid's Elements*, 68, 22–24), see Heath's introd. to his edition of Euclid's *Elements*, vol. 1, p. 2.

138. Heath, ibid. For Aristotle's criticism of the relevant section of the *Timaeus*, see *On the Heavens*, Book III, passim, and Heath, *Mathematics in Aristotle*, pp. 174–178.

139. H. E. Buchanan, "On Poristic Quadrilaterals," *American Mathematical Monthly 49* (1942):364–371.

140. *Topics*, VI, 14; 151 b 12–17, translated by W. A. Pickard-Cambridge.

141. M. Fréchet, *Les Espaces Abstraits*, 1928, p. 18, as quoted by Imre Lakatos, *Proofs and Refutations*, p. 89, n. 2. An equivalent statement by Bertrand Russell has been quoted above, note 63; see also the discussion toward the end of Hora Quarta.

142. Apart from recognizing right-angled triangles (very probably long thought of simply as half-rectangles) the early geometers, of the predemonstrative era, apparently ignored all but *isosceles* triangles. As Gandz noted, only much later were scalene triangles considered, "hence the name σκαληνός = 'limping'. The normal triangle was that with two even legs." (Solomon Gandz, *Quellen und Studien zur Geschichte der Mathematik, Astronomie und Physik A, 2* (1932):16, n. 2.)

143. See again the first part of note 43.

144. *Posterior Analytics*, I, 24; 85 b 37 – 86 b 28, translated by G. R. G. Mure.

145. In a delightful article, the geologist Thomas C. Chamberlin has explained the circumstances in which a theory becomes a *ruling theory:*

> The search for facts, the observation of phenomena and their interpretation, are all dominated by affection for the favored theory until it appears to its author or its advocate to have been overwhelmingly established. The theory then rapidly rises to the ruling position, and investigation, observation, and interpretation are controlled and directed by it. . . . So long as intellectual interest dealt chiefly with the

> intangible, so long it was possible for this habit of thought to survive, and to maintain its dominance, because the phenomena themselves, being largely subjective, were plastic in the hands of the ruling idea; but so soon as investigation turned itself earnestly to an inquiry into natural phenomena, whose manifestations are tangible, whose properties are rigid, whose laws are rigorous, the defects of the method became manifest, and an effort at reformation ensued.

("The Method of Multiple Working Hypotheses," *Science,* o.s., *15* (1890):92 ff., reprinted, *Science 148* (1965):754–759, quotation from p. 755.)

146. According to F. C. S. Schiller, traditional logic

> did not spring from interest in the exploration of nature, and did not aim at its prediction and control. Nor did it presuppose an incomplete system of knowledge which it was desired to extend and improve. It originated in a very special context, from the social need of regulating the practice of dialectical debate in the Greek schools, assemblies, and law-courts. It was necessary to draw up rules for determining which side had won, and which of the points that had been scored were good. . . . But for the purpose of apprehending scientific procedure the syllogism is a snare. . . . Aristotle had no sooner worked out the classic formulation of the rules of dialectical proof than he proceeded to extend their scope by applying them to the theory of science, in the *Posterior Analytics*. . . . [and when he sought to support this theory by mathematical examples] it easily escaped notice that the logical superiority of mathematics was an achievement, not a datum. Just because the mathematical sciences were very ancient, their origins had been forgotten, and with them the tentative gropings which had first selected, and subsequently confirmed, their principles. They had become immediately certain and 'self-evident', and no one was disposed to dispute them. . . . After its apparent success in analysing mathematical procedure there was no more disputing the supremacy of the theory of 'proof'. The facts that its field of application was soon found to be much narrower than that of science, and that it failed egregiously to apply to the procedures of the (openly) empirical sciences, and *a fortiori* could not justify them, if they were noticed at all, were held merely to show that these sciences stood on a low level of thought, which from the loftier standpoint of logic could be contemplated only with contempt.

(Pp. 237–240 of "Scientific Discovery and Logical Proof," in Charles Singer (ed.), *Studies in the History and Method of Science*, Oxford: Clarendon Press, 1917, pp. 235–289.) It now appears that the axioms as well as the postulates of Greek geometry were *not* regarded by all Greek philosophers as so self-evident that their truth was to be accepted without reservation, or without further justification. The Greek mathematicians, on the other hand, were probably not inclined to doubt the necessary truth of these first principles, for all that they were concerned to prove as much as possible—most notably Euclid's fifth postulate. Apollonius even went so far as to attempt to prove some of the Euclidean axioms themselves. Ian Mueller, "Greek Mathematics and Greek Logic," in John Corcoran (ed.), *Ancient Logic and Its Modern Interpretations*, Dordrecht: Reidel, 1974, pp. 35–70, argues that the principles of Greek logic were not derived from the forms of proof characteristic of Greek mathematics, and that Aristotle failed to analyze mathematical proof to find out that it was not syllogistic and could not fairly be pressed into syllogistic form. (See particularly p. 50 for Mueller's criticism of *Prior Analytics*, I, 23.) Mueller concludes also that "the codification of elementary mathematics by Euclid and the rich development of Greek mathematics in the third century are independent of logical theory." Others have argued for closer connections between Aristotle's logic and the contemporary mathematics, and likewise between Socratic dialectic and the efforts of fifth century mathematicians to discover the first principles from which deduction could proceed. See H. D. P. Lee, "Geometrical Method and Aristotle's Account of First Principles," *Classical Quarterly 29* (1935):113–124, and cf. Heath's *Mathematics in Aristotle*, especially pp. 55–56; also Benedict Einarson, "On Certain Mathematical Terms in Aristotle's Logic," *American Journal of Philology 57* (1936):33–54, 151–172, and cf. Robin Smith, "The Mathematical Origins of Aristotle's Syllogistic," *Archive for History of Exact Sciences 19* (1978):201–209; also Árpád Szabó, "Anfänge des euklidischen Axiomensystems," *Archive for History of Exact Sciences 1* (1960):37–106, and the same author's "Transformation of Mathematics into Deductive Science and the Beginnings of Its Foundation on Definitions and Axioms," *Scripta Mathematica 27* (1964):27–48A, 113–139.

147. Notwithstanding the differing opinions regarding the controversial matters touched on in note 146, it is clear that Aristotle saw in the mathematics of his day a model for the empirical sciences. Not only was it concerned with knowledge for its own sake and concerned with the universal rather than the particular, it was already largely systematized into a deductive form. But it required elaborate supplementing since it was a highly abstract model. For the mathematician,

> "before beginning his investigation . . . strips off all the sensible qualities, e.g. weight and lightness, hardness and its contrary, and also heat and cold and other sensible contrarieties, and leaves only the quantitative and continuous, sometimes in one, sometimes in two, sometimes in three dimensions, and the attributes of these *qua* quantitative and continuous, and does not consider them in any other respect, and examines the relative positions of some of the attributes of these, and the commensurabilities and incommensurabilities of others, and the ratios of others."

(*Metaphysics*, XI, 3; 1061 a 29–b 2; translated by W. D. Ross.)

148. See Simplicio's remark, and Sagredo's response, on p. 83.

149. Almost nothing survives of the works that Aristotle prepared for public reading. His extant treatises are edited versions of what we might now perhaps describe as draft reports of work in progress. Some of these may have survived only from students' notes of his lectures. The overall impression gained, at least from the works with which we are concerned, is that while Aristotle longed to produce an organized system for the whole of knowledge, he was honest enough to raise difficulties for which he had no solution. It seems safe to conclude that

Aristotle was less "aristotelian" than many of his followers, and certainly less dogmatic. Even today, the old confidence in the truth of "scientific insight," the faith at least in the intuition of the greatest scientists, is not entirely dead. Norman R. Campbell went so far as to write of Newton:

> As soon as it had occurred to him that the fall of the apple and the fall of the moon might be the same thing, he was utterly sure that they were the same thing; so beautiful an idea *must* be true. To him the confirmation of numerical agreement added nothing to the certainty; he examined whether the facts agreed with the object of convincing others, not himself. And when the facts as he knew them did not agree, we may be sure that his faith in the theory was in no way shaken; he *knew* that the facts must be wrong, but he had to wait many years before evidence of their falsity was found which would appeal to those who had not his genius.

(P. 103 of the 1952 Dover reprint of *What Is Science*?, 1921.) Cf. Lakatos, writing with acknowledgement to Popper: "Newton's mechanics and theory of gravitation was put forward as a daring guess which was ridiculed and called 'occult' by Leibniz and suspected even by Newton himself. But a few decades later, in the absence of refutations, his axioms came to be taken as indubitably true. The debate, from Kant to Poincaré, was no longer about the truth of Newtonian theory but about the nature of its certainty." (*Proofs and Refutations*, p. 49, n. 1.)

150. In the solution to Problem 1, the double segments E and V in Fig. S1 are right-angled biangles. A note on Hippocrates' quadratures of lunes is included in the same solution.

151. Occasionally, in generalizing a theorem, there might spring to mind a general method of proof that is actually simpler than the methods most likely to have been associated with more restricted versions of the theorem. For an example, see the discussion preceding the proof of Theorem β (pp. 144–145).

HORA DECIMA

152. Domenico Theotocopulos (1541–1614), called *El Greco* ('the Greek') for short, born in Crete and settled in Toledo, was noted for his characteristic dramatic paintings, many with elongated figures. Peter Medawar relates that "an opthalmologist, who shall remain nameless," surmised that this artist's figures were drawn elongated "because El Greco suffered a defect of vision that made him *see* people that way, and as he saw them so he would necessarily draw them." Medawar invites his readers to see that this explanation is nonsense. (*Advice to a Young Scientist*, New York: Harper & Row, 1980, p. 9.)

153. Very early examples are to be found in aboriginal cave art, in Europe, Africa, and Australia, where greatly attenuated figures represent agile fast-running hunters or, in other cases, spirit men believed (according to the descendants of the original artists in Arnhem Land) to be able to slip in and out of narrow crevices in rocks. Constructions involving various methodical distortions of squared or other backgrounds were used by Dürer for the comparison of different body and facial types (Albrecht Dürer, *Vier Bücher von menschlicher Proportion*, 1528, facsimile, Zürich: Stocker, 1969). D'Arcy Wentworth Thompson, who referred to Dürer, exploited the same ideas to exhibit affinities of form of various biological species. See his *On Growth and Form*, Cambridge, England: Cambridge Univ. Press, 1917 or 1942, or abridged edition 1961, final chapter. Compare further: G. D. Scott and M. R. Viner, "The Geometrical Appearance of Large Objects Moving at Relativistic Speeds," *American Journal of Physics 33* (1965):534–536.

154. The formula for the area of a triangle, which we now write in the form $\triangle = \sqrt{\{s(s-a)(s-b)(s-c)\}}$, was established by Hero of Alexandria (2nd half of 1st cent. or 1st half of 2nd cent. A.D.). Hero's proof involved the incenter of the general triangle and it began with the demonstration, equivalent to that given by Sagredo, showing that the perimeter times the incircle radius is equal to twice the area of the triangle (*Metrica*, I, 8; also *Dioptra*, chap. 30). For the Greek text and an English trans., see Ivor Thomas, *Selections Illustrating the History of Greek Mathematics*, vol. 2, pp. 470–477. According to al-Bīrūnī (973–after 1050) in his book on the determination of chords, a proposition specifying the area of a triangle of known side lengths had already been demonstrated by Archimedes. A suggested reconstruction of Archimedes' proof has been offered by C. M. Taisbak, "An Archimedean Proof of Heron's Formula for the Area of a Triangle, Reconstructed," *Centaurus 24* (1980):110–116.

155. Al-Khowārizmī stated the following rule in his short *Treatise on Mensuration, c.* A.D. 820: "The area in each circle is half of the diameter multiplied by half of the circumference. For, in every figure with equal sides and angles, such as a triangle, quadrangle, pentagon, or any polygon, you find the area by multiplying half of the perimeter with half of the diameter of the largest inscribed circle" (translated by Solomon Gandz in *Quellen und Studien zur Geschichte der Mathematik, Astronomie, und Physik A, 2* (1932), refer p. 70). As has just been shown in the present text, the phrase "with equal sides and angles" introduces a quite unnecessary restriction. But the reference to the circle reminds us of the traditional way of ensuring in this context that polygons will be circumscribable and inscribable with respect to a circle and that they will, as we now express it, approach, in area and in perimeter, the inscribed or circumscribed circle as a limit, as the number of sides increases without limit. The convenient, though not essential, way of ensuring this, is to specify that the polygons be regular. Classical examples are provided by Euclid XII, 2, and by the first proposition of Archimedes' "Measurement of a Circle." These "exhaustion-method" proofs follow a general pattern initiated by Eudoxus (flourished *c.* 360 B.C.). It is clear that the rule quoted from al-Khowārizmī was already known to the Greeks: see for example the quotation in note 224.

In Indian mathematics, Mahāvīra, *c.* A.D. 850, gave the rule: *Divide the area [of a triangle or quadrilateral] by one-quarter of the perimeter and the result is the diameter of the inscribed circle.* But it becomes apparent that the range of applicability of this formula was not properly understood by Mahāvīra when he proceeds to ask that the diameters of the inscribed circles be calculated "in relation to the already specified figures," since these include three clearly *noncircumscribable* quadrilaterals (Mahāvīra, *Gaṇita-sāra-saṅgraha* ["A Brief Exposition of the Compendium of Calculation"], chap. VII, stanzas numbered $223\frac{1}{2}$ and $224\frac{1}{2}$ (M. Raṅgācārya, *The Gaṇita-sāra-saṅgraha of Mahāvīrācārya*, with English translation and notes, Madras: Government Press, 1912, p. 254 of the English section)). Several centuries earlier than Mahāvīra, Āryabhaṭa I, *c.* A.D. 500, had given the rule, "Half of the circumference multiplied by half the diameter is the area of a circle" (*The Āryabhaṭīya of Āryabhaṭa*, translated with notes by Walter Eugene Clark, Chicago: Univ. of Chicago Press, 1930, p. 27). (This statement, the equivalent of the enunciation of the first proposition of Archimedes' "Measurement of a Circle," is immediately followed, in the second part of the same stanza, with the erroneous assertion that "this area multiplied by its own square root is the exact volume of a sphere." This incorrect second rule has been much remarked upon, sometimes with surprising ineptness. The most obvious suggestion dates back at least to Moritz Cantor, *Vorlesungen über Geschichte der Mathematik*, vol. 1, Leipzig: Teubner, 1880, p. 549. This not unlikely suggestion is simply that Āryabhaṭa relied upon a misleading analogy between a sphere and a cube. For the latter, $V = e^3 = A\sqrt{A}$, where A is the measure of the square cross-section.)

It remains to mention the unexpected discovery that one and one-half millenia before Archimedes, the Old Babylonians had arrived at a calculation procedure agreeing with the formula $A_0 = \frac{1}{4}Cd$. At least, tablet BM 85210 contains a calculation in which the area of a semicircle is found by obtaining one-quarter of the product arc times diameter (François Thureau-Dangin, *Textes mathématiques babyloniens*, London: Brill, 1938, pp. 50–51). The most commonly used Old Babylonian circle rules were those we should now express as $C = 3d$ and $A = \frac{1}{12}C^2$. Though individually very rough, these two rules are consistent: they accord with the true relation $A = \frac{1}{4}Cd$. Combining them, $A = \frac{1}{12}C(3d) = \frac{1}{4}Cd$, or $\frac{1}{2}Cr$. The rough circumference and area rules just quoted are so simple that we are not obliged to assume that the relation $A = \frac{1}{4}Cd$ was used as the path from one to the other. More likely, $C = 3d$ and $A = \frac{3}{4}d^2$ were arrived at independently by rough empirical estimation, and then, since the circumference of many circular objects is more easily measured than the diameter, the expression $\frac{3}{4}d^2$ could have been replaced by $\frac{3}{4}(\frac{1}{3}C)^2$, which gives $\frac{1}{12}C^2$. (It is not, of course, necessary to have algebraic symbolism for such simple transformations.) It may be further conjectured that, as a result of calculating and tabulating both the circumferences and areas of circles against their diameters, a methodical and attentive calculator could have arrived at the generalization, by simple induction, that $A = \frac{1}{4}Cd$. Alternatively, it is possible that this relation might have been intuited in connection with the Old Babylonian calculations of the areas of regular polygons. For these, see E. M. Bruins and M. Rutten, *Textes mathématiques de Suse*, Paris: Guethner, 1961, pp. 18–34.

156. For the difficulties associated with Pythagoras' supposed sacrifice of an ox (or a hecatomb of oxen) in celebration of his discovery of the theorem that bears his name (or possibly of some other theorem), see Heath's *History of Greek Mathematics*, vol. 1, pp. 144–145; W. K. C. Guthrie, *A History of Greek Philosophy*, vol. 1, Cambridge, England: Cambridge Univ. Press, 1971, pp. 188–191; Walter Burkert, *Lore and Science in Ancient Pythagoreanism*, translated by E. L. Minar, Cambridge, Mass.: Harvard Univ. Press, 1972, pp. 180–183.

157. 'Ενθουσιασμός, the state of being ἔνθεος, possessed by a god; supernatural inspiration; prophetic or poetic frenzy.

158. The term is A. H. Maslow's. See, for example, his last book, *The Farther Reaches of Human Nature*, New York: Viking, 1971, pp. 175–179:

> Mathematics can be just as beautiful, just as peak-producing as music; of course, there are mathematics teachers who have devoted themselves to preventing this. I had no glimpse of mathematics as a study in aesthetics until I was thirty years old, until I read some books on the subject. . . . the creative scientist lives by peak experiences. He lives for the moments of glory when a problem solves itself, when suddenly through a microscope he sees things in a very different way, the moments of revelation, of illumination, insight, understanding, ecstasy (pp. 177–178).

See, however, D. R. Weidman, "Emotional Perils of Mathematics," *Science 149* (1965):1048.

159. The tendency to fail to give adequate attention to cases that might *not* harmonize with one's favored beliefs is, of course, a very general characteristic of thought, and one that has never been more delightfully depicted than by Francis Bacon, *Novum Organum*, Book I, Aphorism 46. The best thinkers have deliberately sought out and investigated potential paradoxes and counterexamples. Thus Darwin was able to write of himself: "I had . . . , during many years, followed a golden rule, namely, that whenever a published fact, a new observation or thought came across me, which was opposed to my general results, to make a memorandum of it without fail and at once; for I had found by experience that such facts and thoughts were far more apt to escape from memory than favourable ones" (*The Autobiography of Charles Darwin, 1809–1882, with original omissions restored*, edited by Nora Barlow, London: Collins, 1958, p. 123).

160. "A striking, but by no means infrequent, fact observed in scientific research is that a result may be overlooked by a scientist although it is an obvious and immediate consequence of those he has himself obtained. Barrow, in his *Lectiones Geometricae*, mentions it, most appropriately calling it an ἀβλεψία —we should say, today, a psychical blindness." (P. 1 of Jacques Hadamard, "History of Science and Psychology of Invention," *Mathematika 1* (1954):1–3.)

161. Aristotle, *On the Soul*, II, 2; 413 a 19.

162. Contraposition: "the mode of inference which proceeds by transposing the subject and predicate, antecedent and consequent, or premise and conclusion, with negation of the transposed parts" (*Century Dictionary*, 1903 ed. C. S. Peirce was responsible for the definitions of the logical and mathematical and many of the scientific terms in this great work of scholarship).

HORA UNDECIMA

163. Not *entirely* at pleasure, of course! Bhāskara, in the mid-twelfth century, specified as follows the values between which the diagonals of a quadrilateral with given side lengths must lie: "Rule restricting the arbitrary assumption of a diagonal. The sum of the shortest pair of sides containing [an angle opposite] the diagonal being taken as a base, and the remaining two as the legs [of a triangle], the perpendicular is to be found: and, in like manner, with the other diagonal. The diagonal cannot by any means be longer than the corresponding base, nor shorter than the perpendicular to the other. Adverting to these limits an intelligent person may assume a diagonal" (*Līlāvatī*, VI, 185; p. 77 of the translation by H. T. Colebrooke, *Algebra with Arithmetic and Mensuration from the Sanscrit of Brahmegupta and Bhāscara*, 1817). Bhāskara's rule obviously gives the correct upper limits, but the lower limits are unsatisfactory from our point of view. In concave quadrilaterals (not under consideration by Bhāskara), much shorter diagonals than those provided by the rule are frequently possible (depending on the relative side lengths), whereas, if a restriction to convex quadrilaterals is accepted, the diagonals will usually have greater minima than those which the rule suggests. Perhaps Bhāskara did not intend that *any* diagonal within the specified range would be possible. The traditional Hindu mensuration of quadrilaterals involved the building up of complexes of matching right-angled triangles to produce quadrilaterals with diagonals perpendicular to each other, or to a side, or with sides parallel or perpendicular to one another. It may well have been that the quoted rule was intended to provide only a test that any proposed quadrilateral would have to satisfy.

164. For an alternative proof, see Problem 31(a).

165. Plutarch, *Isis and Osiris*, chap. 42, remarks incidentally that the Pythagoreans had noticed that the 4-by-4 square and the 6-by-3 rectangle are the only (rectangular) figures in which "the perimeters are equal to the areas." He adds that they abhorred the number 17, which separated the 16 from the 18, and he sees a connection between this and the Egyptian belief that Osiris died on the seventeenth of the month, at which time the waning of the moon first becomes distinctly noticeable! See Plutarch, *De Iside et Osiride*, (Greek and English) translated and edited by J. G. Griffiths, Cardiff: Univ. of Wales Press, 1970, pp. 184–185, 459–460; also Heath's *History of Greek Mathematics*, vol. 1, pp. 96–97.

166. Galileo, *Two New Sciences*, First Day, 73–76.

167. It follows—to return to the three-dimensional context—that the solid bounded by the cylindrical surface, its lower base, and the hemispherical surface, has every section parallel to the base equal to the coplanar section of the cone; consequently the *volume* of the hemispherically scooped out bowl is equal to that of the cone. So stated, this represents a direct application of (and, historically, could have been the inspiration for) Cavalieri's famous principle; see note 187. But neither Valerio nor Galileo justified the volume relation simply on the basis of the equality of the corresponding cross-sectional areas; they had too much respect for the tradition of Archimedean rigor to do so. Boyer was in error when he claimed that Galileo's demonstration here was one "involving the use of indivisibles and the so-called 'theorem of Cavalieri' " (Carl Boyer in Ernan McMullin (ed.), *Galileo, Man of Science*, New York: Basic Books, 1967, p. 250). Actually, Galileo refrained from giving any proof of the volume relation, referring his readers instead to the demonstration of "the new Archimedes of our age," Luca Valerio, *De centro gravitatis solidorum*, 1603–04, II, xii (*Two New Sciences*, First Day, 76). For Valerio's methods, see Margaret E. Baron, *The Origins of the Infinitesimal Calculus*, pp. 101–107, referring to pp. 105–106 for the particular proposition just mentioned. A comparable but not identical arrangement of a cone, sphere, and cylinder is to be found in Archimedes' *Method* (Proposition 2). But this work apparently remained quite unknown for some centuries prior to its rediscovery in 1906, in the original text of a palimpsest. For an English translation, see the supplement of 1912 to Heath's edition of *The Works of Archimedes*. Valerio's whole approach is clearly dependent upon the available orthodox works of Archimedes; we are not tempted to believe that he, or the other mathematicians of his age, had any knowledge of the brilliant, though nonrigorous, arguments set out in the *Method*.

168. *Two New Sciences*, First Day, 75.

169. *Two New Sciences*, First Day, 78; Drake translation, p. 40. Cf. Kepler, who in the fifth book of his *Harmonice mundi*, 1619, avowed that he had made manifest as much of the infinite glory of creation as the narrowness of his intellect could apprehend. (An English translation of Book V is included in the *Great Books of the Western World*, vol. 16; the passage referred to is on p. 1,080.) And Bacon, in his *Novum Organum*, 1620, Book I, Aph. 48, referred to the "weakness of our minds" that does not allow us to conceive of the infinitude of space or of time, or to understand infinite divisibility, etc. Descartes, in his *Principia Philosophiae*, 1644, Pars I, princ. xxvi, xxvii, even went so far as to assert that

> we must not try to dispute about the infinite . . . since it would be absurd that we who are finite should undertake to decide anything regarding it, and by this means in trying to comprehend it, so to speak regard it as finite. That is why we do not care to reply to those who demand whether the half of an infinite line is infinite, and whether an infinite number is even or odd and so on, because it is only those who imagine their mind to be infinite who appear to find it necessary to investigate such

> questions. As for our part, while we regard things in which, in a certain sense, we observe no limits, we shall not for all that state that they are infinite, but merely hold them to be indefinite . . . in order to reserve to God alone the name of infinite.

(*The Philosophical Works of Descartes*, translated by Elizabeth S. Haldane and G. R. T. Ross, Cambridge, England: Cambridge Univ. Press, 1911, vol. 1, pp. 229-230.) Cf. Pascal, *De l'esprit géométrique* (1658 or 1659?) in which it is asserted that geometrical minds are able to

> admire the greatness and power of nature in that double infinity which surrounds us on every side, and learn by the consideration of such marvels to know themselves, seeing themselves placed between an infinity and a nothing of extension, between an infinity and a nothing of number, between an infinity and a nothing of motion, between an infinity and a nothing of time. Whereupon we can learn to assess ourselves at our just value and to make reflections worth more than all the rest of geometry itself.

(Translated by R. Schofield under the title "On Geometrical Demonstration, Section I," *Great Books of the Western World*, vol. 33; passage quoted from p. 439. Cf. Pascal, *Pensées*, II, 72, 121, VII, 537, ibid. pp. 181-185, 195, 265.)

170. "Something surprising, a paradox, is a paradox only in a particular, as it were defective, surrounding" (Ludwig Wittgenstein, *Remarks on the Foundations of Mathematics*, translated by G. E. M. Anscombe, Oxford: Blackwell, 1956, V, 36, p. 186 e). Sagredo is endeavoring to provide a "surrounding" that will effectively dispel the paradox.

171. This means that an annulus similar to that considered here will have only $1/75^2$ of the area of a circle equal to it in perimeter; consequently the shape parameter for such an annulus is approximately 0.000014.

172. Before its appearance in the *Two New Sciences*, the paradox just considered had been sent by Galileo to Cavalieri in order, as Drake notes, "to caution him regarding the perils of the 'method of indivisibles' in geometry." In particular, it shows "the pitfalls of analogy in transferring the word 'equal' from entities of n dimensions to their supposed counterparts of $n - 1$ dimensions." (Stillman Drake in his edition of the *Two New Sciences*, p. 35, n. 18). See further, Drake, *Galileo at Work*, 1978, pp. 361, 365. Galileo followed this "soupdish" paradox with another, famous and much admired, involving an extension of the notion of "equality" based on well-defined one-to-one correspondences between members of an infinite set and of a proper infinite subset of itself, viz., between the natural numbers and the squares (or cubes, etc.) of these. See note 227.

173. The problem would be to try to find something special about these sets of numbers that might enable other sets to be found without the need (or, from a mathematician's point of view, without the indignity!) of continued trials involving empirical measurement. As Otto Neugebauer remarked, in connection with both Old Babylonian and ancient Egyptian mathematics, "problems concerning areas or volumes do not constitute an independent field of mathematical research but are only one of many applications of numerical methods to practical problems" (*The Exact Sciences in Antiquity*, 2nd ed., Provdence, R.I.: Brown Univ. Press, 1957, p. 79). For the surprisingly advanced knowledge of "pythagorean" triangles possessed by the Old Babylonian mathematicians, see note 178 below. But there is no satisfactory evidence from ancient Egypt to suggest that these triangles were known there. Numerical examples involving the areas of three squares, such that the sum of the two smaller is equal to the larger, could have been anciently devised simply as convenient examples for calculation before the connection with the sides of a right-angled triangle was ever suspected. An Egyptian problem from the nineteenth century B.C. (in the Berlin Papyrus 6619) requires the division of 100 square cubits into squares, the sides of one being three-quarters of the sides of the other. (See Heath's *Manual of Greek Mathematics*, pp. 467–468 and the corrigendum on p. viii; cf. pp. 422–423 of Thomas Eric Peet's "Mathematics in Ancient Egypt," *Bulletin of the John Rylands Library 15* (1931):409–441.) Neither in this problem nor in two or three other fragments of apparently similar calculations is there anything to suggest that it was realized that these squares, or the triad 3, 4, 5, was in any way connected with a right-angled triangle. R. S. Williamson presented what he took to be evidence of the use of the 3-4-5 triangle in the construction of Egyptian arch-forms ("The Saqqâra Graph: Its Geometrical and Architectural Significance," *Nature, 150* (Oct. 17 and Nov. 21, 1942): 460–461 and 607). But his kind of analysis, in which isolated surviving artefacts are matched against constructions that come to the mind of a modern investigator, must remain, generally speaking, inconclusive. It is hardly possible to produce satisfactory evidence on the basis of only a few individual items, be they pyramids, temples, statues, or vases, where the total number of these is large and no independent criterion is available for the selection of those exhibiting (or claimed to exhibit) the hypothetical proportions apart from the more numerous ones that do not. In the case of an arch form, an artisan, or even an architect preparing a drawing with dimensioned ordinates, might have used a flexed lath or palm midrib to determine one-half of the contour, or a hanging-rope form might be inverted, or the sweep of a freehand arc selected after several trials.

The *accuracy* of the right angles anciently achieved by some of the pyramid builders is not in question, but the 3-4-5 rope-stretching suggestion is not needed to account for it and in its crude form is not adequate to do so. (B. L. van der Waerden's caustic comment on the historiography of this suggestion is worth noting, *Science Awakening I,* translated by A. Dresden, 1954, Groningen: Nordhoff, 1961, p. 6.) Several primitive surveying techniques of remarkable accuracy have been suggested and tested in modern times. One of these, a symmetry method with sightings to distant pegs for laying out right angles, might easily have evolved in the field long before the presence of mathematicians who could have validated it theoretically. For this, see Somers Clarke and Reginald Engelbach, *Ancient Egyptian Masonry: The Building Craft*, Oxford: Oxford Univ. Press,

1930, pp. 67–68; see also pp. 46–59 for architects' plans including an example of the specification of a curve by dimensioned ordinates, and pp. 181–191 for arches in Egypt.

It should be mentioned that A. Seidenberg has written several challenging articles, of which the one most relevant to the present question is entitled "The Ritual Origin of Geometry," *Archive for History of Exact Sciences 1* (1962): 488–527. Largely on the ground of general plausibility, and in view of the contemporaneous Babylonian and (later?) Indian knowledge, Seidenberg, an ardent diffusionist, has little doubt but that the 3-4-5 prescription for obtaining right angles *would* have been known to the Egyptians, say of the Middle Kingdom period. He supports his case by reference to Plutarch, *Isis and Osiris*, chap. 56. In this passage Plutarch supposed that the 3-4-5 triangle *may have been* associated by the Egyptians with Osiris symbolized by the perpendicular, 3, Isis by the base, 4, and Horus by the hypotenuse, 5—but this idea seems to stem from what Plutarch takes to be a related use of the same triangle by Plato in the obscure passage of the *Republic*, VIII, 546 b–c. Plutarch's knowledge of Egyptian theology was demonstrably superficial and neither the context nor the constructive air of this reference encourages us to attach historical accuracy to it. (Seidenberg's application of the origins-in-ritual thesis appears much more reasonably applicable in connection with early Indian mathematics of the *Śulvasūtra* period.) Cf. Heath's *Thirteen Books of Euclid's Elements*, vol. 1, p. 417, and J. G. Griffiths' edition of Plutarch's *De Iside et Osiride*, pp. 206–209, 509–512.

174. Sometimes we can confidently glimpse an insight without which a documented calculation could not have arisen, but altogether more hazardous are conjectures as to how two or three such insights might have been combined to produce a proto-proof. A relevant ancient example is worth mentioning as suggestive of the type of restricted, incomplete, but ingenious and compelling proof that it is tempting to suppose was occasionally devised in predemonstrative eras. The example is from the *Chou Pei Suan Ching* ["Arithmetical Classic of the Gnomon and the Circular Paths of Heaven"]. This compilation is of uncertain date; the traditional dating to the twelfth century B.C. can no longer be accepted. Perhaps the first part, from which the following item is taken, may be dated to about the fourth century B.C. Presumably it is quite independent of Greek mathematics, though there might have been Indian influences. In the text, the sage prince, Chou Kung, is discoursing with his learned minister, Shang Kao. The latter indicates that geometrical knowledge is founded upon calculation [!], then asserts that the diagonal of a 3-by-4 rectangle has length 5 and immediately proceeds to the following argument to show that this is so: "After drawing a square on this diagonal, circumscribe it by half-rectangles like that which has been left outside, so as to form a [square] plate. Thus the [four] outer half-rectangles . . . together make two rectangles [of area 24]; then [when this is subtracted from the square plat of area 49], the remainder is of area 25." *Hence*, the side of this square, which is the diagonal of the original rectangle, has length 5. (See Joseph Needham, vol. 3, with the collaboration of Wang Ling, *Science and Civilization in China*, Cambridge, England: Cambridge Univ. Press, 1959, pp.

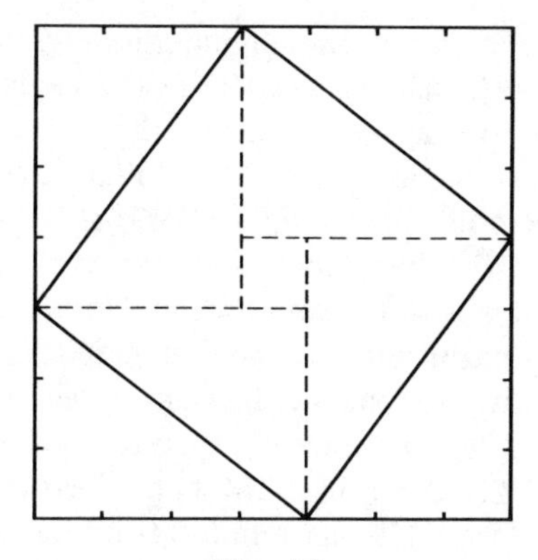

Fig. A9

19–24, 95–96.) In Fig. A9 I have dotted the lines that are not needed in connection with the foregoing calculation. But it was this inner part of the figure that was emphasized by the commentators. Chao Chün-Chhing, 3rd century A.D., obtained from it the *generalization* that we symbolize as $c^2 = a^2 + b^2$. In India, nine centuries later, Bhāskara gave the inner portion of the same diagram, showed how the parts could be rearranged to give a step figure congruent with that obtained by adjoining the squares on "the upright and side," and stated the general rule: "Twice the product of the upright and side, being added to the square of their difference, is equal to the sum of their squares, just as with [any] two unknown quantities. Hence, for facility, it is rightly said 'The square-root of the sum of the squares of upright and side, is the hypotenuse'." (*Bījagaṇita* ["The Science of Calculation with Elements"], V, 147, translated by H. T. Colebrooke, *Algebra with Arithmetic and Mensuration from the Sanscrit of Brahmegupta and Bhāscara*, 1817, p. 222.) We might go back to an earlier level and surmise as to the reasoning associated with a diagram on Old Babylonian tablet BM 15285. This is shown in Fig. A10, in which the dimensions and shading, not on the orig-

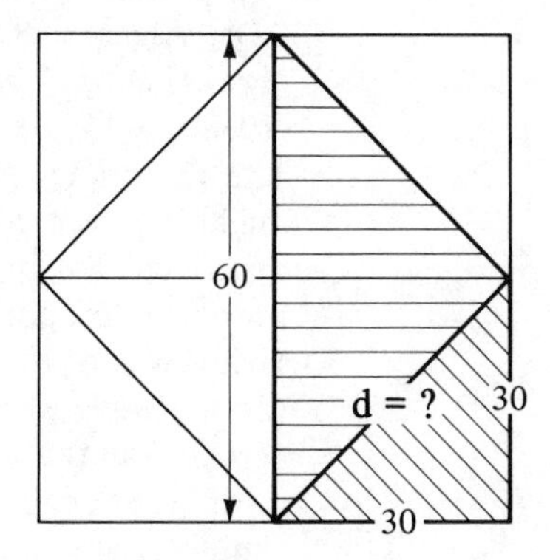

Fig. A10

inal, have been added. The text accompanying this diagram reads: "A square, the side is 1 [1 *ǓS* = 60 *GAR*]. Inside it I have drawn a square. The square which I have drawn touches the [outer] square. Inside the square I have drawn four triangles." (P. 138 of H. W. F. Saggs, "A Babylonian Geometrical Text," *Revue d'assyriologie et d'archéologie orientale 54* (1960):131–146.) Unfortunately, no calculations appear on the tablet, and so attempts to reconstruct insights that might have been associated with this and the many other diagrams from the same document are especially hazardous. Perhaps suggested in the first place by a traditional all-over decorative design, the diagram under consideration could possibly have been used to show that the square on the diagonal of a square is equal to twice the square itself; Socrates used the same diagram to teach the slave boy (or to allow him to recollect!) this very relation (Plato, *Meno*, 84 d–85 b). But the following suggestion for a first step toward the general Pythagorean theorem also fits in well with other indications from Old Babylonian mathematics. Suppose that $d = m$ times 30, where the value of the multiplier m is to be found. From the implied construction, it would have been evident that $60 = m$ times d. Hence $60 = m \times m \times 30$, and so $m \times m = 2$, where m clearly lies between $1\frac{1}{3}$ and $1\frac{1}{2}$. Another Old Babylonian tablet, YBC 7289, already mentioned in note 73, shows a square with the side actually labeled to indicate a side length of 30 units, while along the diagonal appear the sexagesimal equivalents of the multiplier 1.41421 and 30 times this multiplier. The most likely procedure for the determination of this accurate estimate has been indicated in the same note. It is to be expected that, eventually, a corresponding calculation would have been undertaken with respect to a *non*square rectangle. For the case of a 2-by-1 rectangle, illustrated in Fig. A11, the natural procedure may be outlined as follows:

$$a = 1, \quad b = 2, \quad c = ?, \quad d = 2b = 4, \quad e = d + a = 5.$$

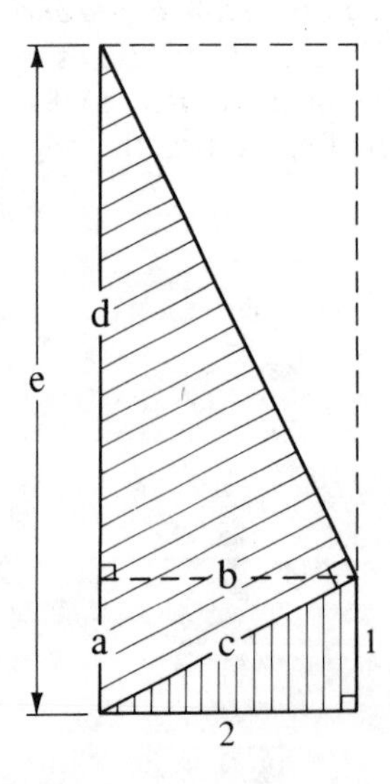

Fig. A11

But e is the same multiple of c as c is of 1. Hence, $e\ (= 5) = c \times c$, so that $c = \sqrt{5}$. (For Old Babylonian calculations involving similar triangles, see H. Goetsch, "Die Algebra der Babylonier," *Archive for History of Exact Sciences 5* (1968):79–153, especially pp. 105–108.) The reasoning involved in this example might be seen as an early stage of an approach that in Greek times led to the proof by proportion of Euclid VI, 31. Heath was of the opinion that this proof (or a simple version of it) was known earlier than that of Euclid I, 47, and that it was the proof involving the adjacent similar triangles that suggested to Euclid the famous demonstration of Book I. On this view, the transformation of the earlier method into one which could be given so near to the beginning of the *Elements* was "extraordinarily ingenious . . . [and] a veritable *tour de force* which compels admiration" (Heath, *The Thirteen Books of Euclid's Elements*, vol. 1, pp. 353–354).

Individual problems, initially regarded as isolated, frequently come to be recognized as representative of a whole class of problems solvable by the one procedure. When understood from the more general point of view, the particular solution may be transposable into a demonstration of a relation covering the whole class. The problems of the present note are clearly of this kind. It may be remarked also that they exemplify the so-called macroreductive or molar approach described in note 43. See further, Problem 29.

175. Salviati proceeds to answer this question without difficulty, but there is a further matter which ought to be acknowledged. It is assumed that the pioneer mathematician would not have been misled, through imperfect empirical measurement, into accepting such triads as 10, 15, 18 and 9, 19, 21 as positive instances. This may appear to be a not inconsiderable assumption since the largest angles of the triangles corresponding to the two triads mentioned are each less than a fifth of a degree away from a right angle. However, it is not at all necessary to suppose that the Old Babylonian investigators had to rely upon practical measurements of the diagonals of carefully-drawn rectangles. A methodical series of calculations of the type illustrated in connection with the last diagram of the previous note (e.g., with a = 1, b = 2; a = 1, b = 3; . . . a = 2, b = 3; a = 2, b = 5; . . . a = 3, b = 4; a = 3, b = 5; . . . , etc.) could naturally have led to the tabulating of the results, enabling a general relation between a, b, c (i.e., a short-cut recipe for obtaining the diagonal given the length and the width) to have been discovered by simple induction.

176. Cf. Proclus, *Commentary on the First Book of Euclid's Elements*, 428, 7 ff.

> Certain methods have been handed down for finding such triangles, one of them attributed to Plato, the other to Pythagoras. The method of Pythagoras begins with odd numbers, positing a given odd number as being the lesser of the two sides containing the [right] angle, taking its square, subtracting one from it, and positing half of the remainder as the greater of the sides about the right angle; then adding one to this, it gets the remaining side, the one subtending the [right] angle. . . . The Platonic method proceeds from even numbers. It takes a given even number as one of the sides about the right angle, divides it into two and squares the half, then adding one to the square gets the subtending side, and subtracting one from the square gets the other side about the right angle.

(Morrow translation, p. 340.) For a way in which these formulae could have been arrived at (or, more likely, might have been *justified*) in the Greek context, see Heath's *History of Greek Mathematics*, vol. 1, pp. 80–81.

177. In the first place, we have the following lemma: *If any two of (the measures of) the sides, a, b, c, have a common factor, then the same factor is contained in the remaining side measure.* This is easily shown since, if $a = ka'$ and $b = kb'$, then $c = k\sqrt{(a'^2 + b'^2)}$, $= kc'$ say, all letters denoting integers in the cases under consideration. Obviously it may be shown in the same way that for k a factor common to the hypotenuse and a side, the remaining side also contains k. If m and n have a common factor, then their product and their sum (and also their difference) have the same factor, and so $2mn$ ($= a$) and $m^2 - n^2$ ($= b$) each contain this factor (squared) whence, by the preceding lemma, so also does $m^2 + n^2$ ($= c$). Again, if m, n were both odd, then $m^2 - n^2$ as well as $2mn$ would be *even*, and consequently *all* sides would have a common factor of 2. Thus the necessity is shown, and the sufficiency may be proved as follows: Let $m = \alpha \cdot \beta \cdot \gamma \ldots$, $n = \rho \cdot \sigma \cdot \tau \ldots$, with no factor common to m, n. Then,

$$c + b = (m^2 + n^2) + (m^2 - n^2) = 2m^2 = 2(\alpha^2\beta^2\gamma^2 \ldots),$$
$$c - b = (m^2 + n^2) - (m^2 - n^2) = 2n^2 = 2(\rho^2\sigma^2\tau^2 \ldots).$$

These relations show that c and b could have no common factor other than 2, since if any two expressions have a common factor, that factor must appear in their sum and also in their difference. But 2 must also be ruled out as a possible factor common to $(m^2 + n^2)$ and $(m^2 - n^2)$, since m^2 and n^2 are neither both odd nor both even, given that m, n are of different parity; hence neither the sum nor the difference of m^2, n^2 can be even. Thus c ($= m^2 + n^2$) and b ($= m^2 - n^2$) have no common factor, and neither, by the initial lemma, have a and b, nor a and c.

178. This formula is implicit in Euclid X, Lemma following Prop. 28, and its equivalent was stated in verbal form by Brahmagupta, *c.* A.D. 628, in his *Brāhma-sphuṭa-siddhanta*, XII, 33 (Colebrooke translation, p. 306). On the evidence of a remarkable Old Babylonian table text, No. 322 of the Plimpton Collection at Columbia University, New York, it is now generally concluded that the Old Babylonian mathematicians also discovered the $2mn \mid m^2 - n^2 \mid m^2 + n^2$ procedure in some form, possibly in a one-parameter version corresponding to $a:b:c = 1:\frac{1}{2}(\lambda - 1/\lambda):\frac{1}{2}(\lambda + 1/\lambda)$. The damaged Plimpton 322 gives an incomplete list of what was perhaps intended as a sequence of 38 "Pythagorean" triads running from 2,0 | 1,59 | 2,49 to 20,0 | 49 | 20,1. Expressed decimally, these sexagesimal values are 120 | 119 | 169 (almost isosceles) and 1200 | 49 | 1201 (very acute). An extrapolation from what survives of the first fifteen lines has been set out by Derek J. de Solla Price, "The Babylonian 'Pythagorean-Triad' Tablet," *Centaurus 10* (1964):1–13. See further, Olaf Schmidt, "On Plimpton 322: Pythagorean Numbers in Babylonian Mathematics," *Centaurus 24* (1980):4–13, and Jöran Friberg, "Methods and Traditions of Babylonian Mathematics: Plimpton 322, Pythagorean Triples, and the Babylonian Triangle Parametric Equations," *Historia Mathematica 8* (1981):277–318.

Promising leads for uncovering possible paths to the discovery, and/or to the early justification, of such relations as (i) $a^2 + b^2 = c^2$ and (ii) $a = 2mn$, $b = m^2 - n^2$, $c = m^2 + n^2$ are provided by typical problems involving the determination of unknown distances. For (i), see note 174 supra; for (ii), see J. Pottage, "The Mensuration of Quadrilaterals and the Generation of Pythagorean Triads," *Archive for History of Exact Sciences 12* (1974):299–354, especially pp. 322–332. In this article it is shown how certain calculations may do far more than merely deliver the particular answer initially sought. To the extent to which they induce the recognition of generalizable relations and provide some insight into the necessity of these, such calculations would have been likely to have commanded attention and to have become appreciated as paradigm illustrations of indubitable truths long before the idea had occurred to anyone that they might with advantage be reformulated into explicit general demonstrations. Perhaps it was just because certain particular calculations and diagrams appeared to be such effective indicators of the appropriateness of this or that procedure, in all cases "of the same kind," that the notion of a deliberately set-out general justification possibly never arose before Greek times. In any case, attempts—if they had been made—to transpose the especially illuminating individual examples into generally applicable demonstrative proofs could be expected to bring into view the need for a tedious succession (if not an infinite regress) of subsidiary propositions or assumptions. It would be understandable if such a program were judged to be unrewarding in comparison with the simple revelation of key insights in the context of calculations, or of diagrams, composed of familiar, unquestioned elements. In spite of the sophisticated levels of mathematical knowledge now known to have been reached by the Old Babylonian mathematicians, the really challenging question is *not*: Why was the step to explicit deductive proof not taken in the Age of Hammurabi? Rather we should ask: How did it come about that the Greek mathematicians, by the time of Plato, recognized that this was a goal they were obliged to seek with respect to each prospective item of mathematical knowledge? Doubtless an answer along the same sort of sociological lines would be applicable here as might reasonably be offered to explain Greek ethical and political philosophy in relation to Egyptian, Babylonian, and Hebrew writings in the same area. But to attempt to give an answer of this kind here would almost certainly be to indulge in an unpardonable exercise in superficiality. One is reminded of Collingwood's remark: "In Galileo's time something happened to natural science (only a very ignorant or a very learned man would undertake to say briefly what it was) which suddenly and enormously increased the velocity of its progress and the width of its outlook" (R. G. Collingwood, *An Autobiography*, London: Oxford Univ. Press, 1939, p. 79).

HORA DUODECIMA

179. The five regular solids have respectively 4, 6, 8, 12 and 20 faces; refer to pp. 144, 148 and note 128.

180. For the area of the curved surface of a cone, see p. 5 and note 7.

181. The following editions of Ozanam's well-known work have been compared: Jacques Ozanam, *Recréations mathématiques et physiques*, 1696; the same, nouvelle éd., 1770; another edition "refondue & augmentée" by M. de C. G. F. [pseudonym of J. E. Montucla], 1778; English edition, entitled *Recreations in*

Science and Natural Philosophy, being C. Hutton's translation of Montucla's edition, revised by E. Riddle, 1851. The passage quoted is reproduced from pp. 165–166 of this English edition; the corresponding passage in the French edition of 1778 is on pp. 380–382 of vol. 1. Nothing of relevance appeared in the earlier editions examined.

182. This example of local priority may be considered in relation to Aristotle's global doctrine of the prior and the posterior. In connection with the third of five senses of "prior to" distinguished by Aristotle in his *Categories*, chap. 12, 14 a 25 - 14 b 22, we read that "in sciences which use demonstration there is that which is prior and that which is posterior in order; in geometry, the elements are prior to the propositions; in reading and writing, the letters of the alphabet are prior to the syllables . . ." For the meaning of "elements," see Proclus, *Commentary on the First Book of Euclid's Elements*, 72–73 (Morrow translation, pp. 59–60). Cf. *Posterior Analytics*, I, 2: " 'prior' and 'better known' are ambiguous terms, for there is a difference between what is prior and better known in the order of being and what is prior and better known to man. I mean that objects nearer to sense are prior and better known to man; objects without qualification prior and better known are those further from sense" (71 b 33 - 72 a 3). Also *Topics*, VI, 4: "absolutely the prior is more intelligible than the posterior, a point, for instance, than a line, a line than a plane, and a plane than a solid; just as also a unit is more intelligible than a number; for it is the prius and starting-point of all number. Likewise, also, a letter is more intelligible than a syllable. Whereas to us it sometimes happens that the converse is the case: for the solid falls under perception most of all—more than a plane—and a plane more than a line, and a line more than a point; for most people learn things like the former earlier than the latter . . ." (141 b 5–13). But, according to Aristotle, scientific knowledge must proceed from what is better known "absolutely," "in nature," or "in the order of being." This doctrine of priority—along with that of demonstrative knowledge derived from the basic principles and of definitions that capture "essential natures"—apart from its importance in epistemology, was doubtless influential also in the development of Western educational attitudes. See also pp. 154–155 in the present book.

183. In a formal proof, this assertion would require justification. Cf. the two-dimensional case discussed in note 19. It was perhaps because of the difficulty of formalizing such an approach that Archimedes avoided the use of such pyramidal elements of volume in his work on the sphere. Compare, however, the related problem, already solved in Euclid XII, 17. There it is shown how, given two concentric spheres, no matter how close together, it is always possible to inscribe within the outer sphere a certain polyhedral solid that nowhere meets the inner sphere.

184. In a precious heuristic avowal, so rare in the history of mathematics, Archimedes tells us how, after obtaining the result for the *volume* of a sphere (by his "conceptual weighing" method), he intuited the result for the surface area:

"From this theorem, to the effect that the sphere is four times as great as the cone with a great circle of the sphere as base and with height equal to the radius of the sphere, I conceived the notion that the surface of any sphere is four times as great as a great circle in it; for, judging from the fact that any circle is equal to a triangle with base equal to the circumference and height equal to the radius of the circle, I apprehended that, in like manner, any sphere is equal to a cone with base equal to the surface of the sphere and height equal to the radius." ("The Method," statement following Proposition 2, Heath edition, *Works of Archimedes*, Supplement of 1912, pp. 20–21.)

185. *The Thirteen Books of Euclid's Elements*, and Ptolemy (Claudius Ptolemaeus), *The Almagest* also in thirteen books. A translation of the latter work by R. C. Taliaferro is included in vol. 16 of *Great Books of the Western World*, 1952.

186. Archimedes, "On the Sphere and the Cylinder," I, 42–44; Heath (ed.), *Works*, pp. 52–55.

187. According to Cavalieri's well-known theorem, the contents of two figures (whether plane or solid) are equal if they have their corresponding sections (line or plane, respectively) everywhere equal to each other, where by *corresponding sections* is meant sections, one in each figure, equally distant from the (equal) bases (*Geometria indivisibilibus continuorum*, 1635, VII, 1; English translations in Dirk J. Struik (ed.), *A Source Book in Mathematics, 1200–1800*, Cambridge, Mass.: Harvard Univ. Press, 1969, pp. 210–214, and David Eugene Smith, *A Source Book in Mathematics*, (1929) New York: Dover, 1959, pp. 605–609). The detailed argument offered here by Cavalieri in proof of this fundamental proposition was *not* an argument in terms of indivisibles; it certainly did not simply amount to the assertion that the contents are equal because each is equal to the same sum of equal (infinite) sets of "indivisibles" (lines or planes) of which the figures (plane or solid) might be supposed to be composed. Instead we find that Cavalieri proceeds by superposition: The first figure is imagined to be brought into partial coincidence with the second, then it is immediately evident that the parts not coinciding, called the "residua," must have their correspondingly placed sections equal; and so, if one of these first residua is (partially) superimposed on the other, the residua then remaining must likewise have equal sections at equal "heights." (Typically, each residuum becomes fragmented into more and more, smaller and smaller, "islands.") The case of two equally high parallelograms on equal bases exemplifies the exceptional situation where complete coverage can be achieved with a small number of superpositions. But, except in the case of specially related figures, there will still be residua remaining after any finite number of superpositions. Cavalieri avoided mentioning this in his demonstration; he showed only that *if* superpositions are carried out until the whole of one figure has been covered, so that it has no residuum, then the other figure can have no residuum either. It was on this basis that he concluded that the given figures are equivalent. "The argument is ingenious and intuitive, but it contains a weak point in that it is not proved that the residuals, in the described operations, became exhausted" (Ettore Carruccio, "Cavalieri," *Dictionary of Scientific Biography*). But Cavalieri then proceeded to give an extended demonstration involving the ancient method of exhaustion, and this is rigorous for certain classes of figure; see *Geometria*, VII, lemmas following Proposition 1, or *Exercitationes geometricae sex*, 1647, Exercitatio Secunda (much of which consists of a reproduction of key theorems from the earlier treatise). And other arguments given by Cavalieri in support of his principle are included in the same works.

Cavalieri has not been justly treated by secondary writers. Especially unfortunate was Felix Klein's highly misleading "historical" remark on the principle: "Its establishment by Cavalieri amounts precisely to this, that he thinks of both solids as built up of layers of thin leaves which, according to the hypothesis, are congruent in pairs, i.e., one of the bodies could be transformed into the other by translating its individual leaves; but this could not alter the volume, since this consists of the same summands before and after the translation" (*Elementary Mathematics from an Advanced Standpoint*, vol. 1: *Arithmetic, Algebra, Analysis*, translated by E. R. Hedrick and C. A. Noble, New York: Dover, n.d., p. 210). Cavalieri did *incidentally* mention that plane figures are to be imagined covered by parallel chords "like a warp composed of parallel threads, and solids like books made up of parallel leaves," adding that "while in a warp the threads, and in books the leaves, are always limited in number, for there is some thickness, we must . . . suppose in plane figures the lines and in solids the planes to be unlimited in number, as if they were without any thickness" (Exercitatio Prima, notes iv, v, pp. 3–4. And in Exercitatio Tertia (which is Cavalieri's specific rejoinder to criticisms appearing in Guldin's *Centrobaryca*, vol. IV, 1641), in alluding to the charge that he had appropriated his method from Kepler's *Stereometria doliorum* of 1615, Cavalieri asserted that Kepler had regarded his bodies as made up of minute solids (e.g., laminae) packed together, "whereas I say only that planes are like [*ut*] aggregates of all equidistant lines, and bodies like aggregates of all equidistant planes," and then added that "no one can fail to see how different these things are from each other" (chap. 1, p. 180). Cavalieri deliberately avoided any categorical assertion that a continuum could actually be *composed of* elements of the next lower dimension; see *Geometrica,* II, Scholium to Prop. 1. The following concise discussions will be found helpful: Carl B. Boyer, "Cavalieri, Limits and Discarded Infinitesimals," *Scripta Mathematica 8* (1941):79–91, and a section in D. T. Whiteside's "Mathematical Thought in the Later 17th Century," *Archive for History of Exact Sciences 1* (1961):179–388, on pp. 311–318. Whiteside remarks: "Cavalieri's thought in detail is unbelievably rich—he had read widely in Archimedes, Stevin, Kepler and others (and had absorbed the medieval theory of latitude of forms, especially the geometrical aspects developed by Oresme), and his ideas are an amalgam of what he has read and of the thoughts that reading inspired" (p. 312, n. 2). In addition, Cavalieri was in frequent communication with Galileo; over one hundred letters from

Cavalieri to Galileo appear in the latter's *Opere*, though only two of Galileo's to Cavalieri survived.

It remains to remark that Cavalieri's principle is frequently stated and used, as indeed it was by Cavalieri himself, in a generalized form to apply to figures in which the corresponding sections, while not equal to each other, are in a constant ratio at all levels. (That equally-high pyramids or cones have their volumes in the ratio of their base areas, a relation already demonstrated in Euclid, XII, 5, 6, and 11, may be taken as illustrating this extended principle.) Further, Cavalieri used two methods, which he referred to as the "earlier" and the "later," in which the indivisibles were compared "collectively" and "distributively," respectively: "To obtain the ratio and the measure of two given figures, plane as well as solid, according to the earlier method there must be extricated the ratio between the aggregates of all the indivisibles conceptually describable in them with respect to a given regula. And according to the later method it must be sought out whether between individual aligned [correspondingly placed] indivisibles, be they straight lines or planes, there is to be found some common ratio. For [the contents of] the figures too will have the same ratio." (Exercitatio Prima, note x, p. 6.)

A remarkable determination of the volume of a sphere by Tsu Ch'ung-chih (or perhaps by his son, Tsu Keng-chih) in the second half of the fifth century A.D. has been described by T. Kiang, "An Old Chinese Way of Finding the Volume of a Sphere," *Mathematical Gazette 56* (1972):88–91. This involved two applications of what amounts to Cavalieri's principle, though there appears to have been no attempt to justify this fundamental theorem. (In his account, Kiang shows himself to have been unaware that Archimedes had studied the geometry of the region common to two equal cylinders intersecting at right angles; see Archimedes' "Method," prefatory letter to Eratosthenes, and Zeuthen's reconstruction of Proposition 15, Heath (ed.), *Works of Archimedes*, Supplement, pp. 12, 48–51, and cf. pp. 18–20.) A fuller account of the Chinese volume determination has now been published by D. B. Wagner, who attributes it to the son, Tsu Keng-chih, late 5th century. See D. B. Wagner, "Liu Hui and Tsu Keng-chih on the Volume of a Sphere," *Chinese Science 3* (1978):59–79. The basic assumption to which appeal was made has been translated thus: "If volumes are constructed of piled up blocks,/And corresponding areas are equal,/Then the volumes cannot be unequal" (p. 61).

HORA TERTIA DECIMA

188. See Heath (ed.) *The Thirteen Books of Euclid's Elements,* vol. 3, pp. 512–520 for the so-called Book XIV (by Hypsicles, *c.* 180 B.C.) and Book XV (by, or partly by, a pupil of Isidorus of Miletus, 6th cent. A.D.).

189. A verbal rule equivalent to formula (1) was obtained—or rather, was justified—by Leonardo Fibonacci in his *Practica Geometriae,* A.D. 1220, by a somewhat complex geometrical derivation based on the difference of two pyramid volumes (*Scritti di Leonardo Pisano,* edited by Baldassare Boncompagni, vol. 2, Rome, 1862, pp. 174–176). According to Solomon Gandz (*Quellen und Studien zur Geschiche der Mathematik, Astronomie und Physik A, 2* (1932):1–96, on p. 35) the same result is to be found in Savasorda's *Geometry,* over one century earlier, but this claim is unconfirmed. What appears in Curtze's edition of Plato of Tivoli's translation of Savasorda's *Geometry,* entitled *Liber Embadorum,* is a calculation of the volume of a conical frustum by a method that appears to be simply of the following form: $V = (a^2 + b^2 + ab) \times (11/14) \times (h/3)$, where a, b, h are the measures of the base diameters and height (see p. 172 of M. Curtze's edition cited in note 12 supra).

Gandz certainly was mistaken in accepting (loc. cit.) Touraeff's assumption that the ancient Egyptian mathematicians were able to lay out their calculations in accordance with our formula (1). The occasion for this erroneous interpretation was Touraeff's encounter with Problem 14 of the *Moscow Mathematical Papyrus* (*MMP* 14). All who have since made a study of this problem, in which the volume of a frustum of a square-based pyramid is correctly obtained, agree that the calculation would have been seen by the Egyptian responsible for it as an application of a method which we should symbolize by $V = (a^2 + ab + b^2) \times \frac{1}{3}h$, and that is remarkable enough! Yet Touraeff, who was the very first writer on the problem ("The Volume of the Truncated Pyramid in Egyptian Mathematics," *Ancient Egypt* (1917):100–102), referred this calculation from the nineteenth century B.C. to the more general formula, $V = \{A_B + \sqrt{(A_B A_T)} + A_T\}(h/3)$. Thus possessed of an inappropriate hindsight, he was at a loss to explain what he was led to describe as the Egyptian use of the "mean proportional" between the base areas! While no calculation of the volume of a *complete* pyramid is known to have survived from ancient (pre-Greek) Egyptian mathematics, this Middle Kingdom example, *MMP* 14, is taken as strong evidence that the calculators of that era would have known the rule we write as $V = \frac{1}{3}A_B h$. (The *Moscow Mathematical Papyrus,* edited with a commentary by W. W. Struve, was published as "Mathematischer Papyrus des Staatlichen Museums der Schönen Künste in Moskau," *Quellen und Studien zur Geschichte der Mathematik, A1* (1930):1–197 and plates.)

190. Indian mensuration from Brahmagupta to Bhāskara, for the most part highly conservative, did undergo some development. In calculating the volumes of prismoid-shaped excavations, the Hindu mathematicians started by taking a correctly weighted mean of an underestimate and an overestimate (denoted below by V_u and V_o, respectively); several modifications of this brought them to a rule practically equivalent to our formula (2). The domain of validity of the rules was not stated or discussed in the texts but only imprecisely "suggested" by examples that were to be regarded as typical. The excavations mentioned in the examples were mostly in the form of inverted frusta of square-based or rectangular-based pyramids. Brahmagupta's rules were as follows:

> The area, deduced from the moieties of the sums of the sides at top and bottom, being multiplied by the depth, is the practical measure of the content. Half the sum

> of the areas at top and bottom, multiplied by the depth, gives the gross content. Subtracting the practical content from the other, divide the difference by three, and add the quotient to the practical content, the sum is the neat content.

(Brahmagupta, *Brāhma-sphuṭa-siddhānta,* c. A.D. 628, XII, v, 45–46; H. T. Colebrooke translation, pp. 312–313.) Symbolizing these rules, for a frustum or other prismoid with parallel rectangular faces $a_1 \times a_2$ and $b_1 \times b_2$ distance h apart, we have $V_u = \frac{1}{2}(a_1 + b_1) \cdot \frac{1}{2}(a_2 + b_2) \cdot h$, for the "practical measure," $V_o = \frac{1}{2}(a_1a_2 + b_1b_2)h$, for the "gross content," and $V_n = \frac{1}{3}(V_o - V_u) + V_u$, for the "neat content." These rules were doubtless tested by comparing answers obtained from them with that yielded by a difference-of-two-pyramids calculation for a few particular cases.

The Jain, Mahāvīra (mid-9th century A.D.), in his chapter on excavations, likewise not explicitly defined, gave the same three methods as Brahmagupta with the only modification that as well as the bases, "sections" (presumably parallel to the bases and, one would hope, equally spaced!) are allowed for in the volume calculations. Unfortunately, Mahāvīra apparently failed to realize that the weights to be applied to the underestimate and the overestimate depend upon the number of sections upon which these values are based, and so his "neat" formula is not generally exact. For example, in the case of any prismoid (to which class Mahāvīra did not always restrict himself) and utilizing three sections, understood as the upper and lower bases along with the midsection parallel to them, his rules amount to the following: $V_u = A_{MS}h$, $V_o = \frac{1}{3}(A_T + A_B + A_{MS}) \cdot h$, and, rigidly retaining the traditional weighting, $V_n = \frac{2}{3}V_u + \frac{1}{3}V_o$; but the exact value in *this* case would actually be given by $V = \frac{1}{2}V_u + \frac{1}{2}V_o$. Refer, *Gaṇita-sāra-saṅgraha,* VIII, 9–18$\frac{1}{2}$; Raṅgācārya, trans., pp. 259–261.

Śrīdhara, who wrote not more than about one century after Mahāvīra, gave the following as Rule 54 of his *Triśatikā*:

> To find the contents of a well add together the square of the diameter at the top, the square of the same at the bottom, and the square of the sum of the two diameters. Take ten times the square of this total and extract its root. Multiply this by the depth and divide the product by twenty-four.

(Translation from p. 215 of N. Ramanujacharia and G. R. Kaye, "The *Triśatikā* of Śrīdharācārya," *Bibliotheca Mathematica,* 3rd series, *13* (1912–13):203–217.) We recognize here the traditional Indian use of $\sqrt{10}$ for π (used consistently in both circumference and area rules). Expressing Śrīdhara's rule symbolically and taking out the common factor $(\sqrt{10})/4$ in order to obtain the rule in its simplest form, for the square-based case, we find $V = \{a^2 + b^2 + (a + b)^2\}(h/6)$. Bhāskara, writing in A.D. 1150, gave the rule as follows:

> The aggregate of the areas at the top and at the bottom, and of that resulting from the sum [of the sides of the summit and base], being divided by six, the quotient is the mean area: that multiplied by the depth is the neat content. A third part of the content of the regular equal solid is the content of the acute one.

(*Līlāvatī,* VII, 221; Colebrooke translation, p. 98.) In the next stanza, VII, 222, the use of the rule is illustrated by an example of the determination of the volume of an excavation in the form of a pyramid frustum having rectangular bases. We cannot hope now to discover Bhāskara's own opinion as to the range of solid forms for which the volume rule provides exact measures. The concluding sentence is to be understood to assert that one-third of a prism gives the volume of the equally based inscribed pyramid. The two individually plausible but mutually inconsistent rules described by Brahmagupta as giving the "practical" and the "gross" values, respectively, are now omitted since the new rule for the "neat" (exact) volume is itself directly and simply applicable. The sixteenth-century commentator, Gaṇeśa, remarked in connection with the formation of the mean (cross-sectional) area in Bhāskara's rule:

> Half the sum of the breadth at the mouth and bottom is the mean breadth; and half the sum of the length at the mouth and bottom is the mean length; their product is the area at the middle of the parallelipipedon. (Four times that is the product of the sums of the length and breadth.) This, added to once the area at the mouth and once the area at the bottom, is six times the mean area.

(Colebrooke's translation, loc. cit.; the sentence in parentheses is presumably Colebrooke's.) The term "parallelipipedon" (Colebrooke's spelling) does not, of course, do any sort of justice even to the most typical forms of excavation to which the rule was apparently to be applied. Possibly it was not until Gaṇeśa's formulation that the rule first came to be thought of in a general form corresponding to our prismoidal formula,

$$V = \{A_T + A_B + 4A_{MS}\} \cdot \frac{h}{6},$$

and not merely as

$$V = \{a_1a_2 + b_1b_2 + (a_1 + b_1)(a_2 + b_2)\} \cdot \frac{h}{6},$$

for rectangular bases, or as

$$V = \{d_T^2 + d_B^2 + (d_T + d_B)^2\}\frac{\sqrt{10}}{4} \cdot \frac{h}{6},$$

for circular ones.

It seems that in the West the rule had to be discovered anew—and then in connection not with excavations, but with wine barrels! For the volume of a barrel having the form of a segment of a spheroid of two equal bases, Cavalieri obtained the rule, which, with reference to his figure reproduced here, we may write as $V = (1/3)IM\{2A_{CG} + A_{BH}\}$, or in our general notation, $V = (h/3)\{2A_{MS} + A_B\}$. There is no evidence here that Cavalieri himself was conscious of other applications of this rule, or that he was aware of the form that we call the general prismoidal formula. But, as suggested in the following note, Cavalieri's barrel rule may well have been the source of an important sequence of mathematical development:

> In Vino Veritas:—"Si adunque moltiplicaremo la terza parte di *IM* lunghezza della Botte *BDFH* in due cerchi maggiore *CG* e uno de minore *BH, DF* come in *BH,* ci

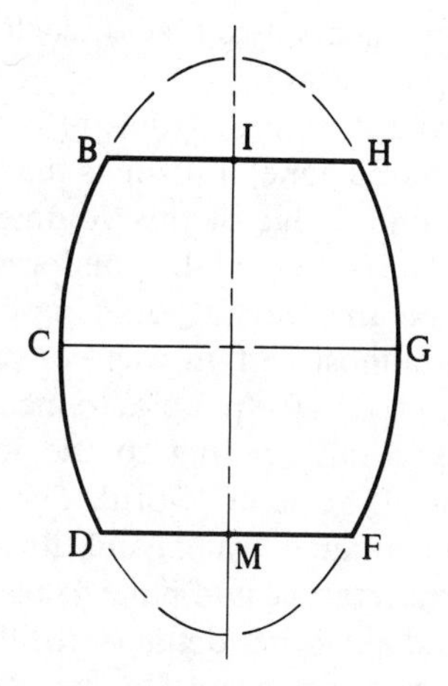

Fig. A12

verrà la capacità di detta Botte." F. B. Cavalieri, *Una centuria di varii Problemi nella Prattica Astrologica,* (Bologna, 1639), Problema 80. (Per Dr. G. N. Watson. It was probably this formula for the capacity of a cask of wine which suggested to James Gregory (who spent the years 1665–8 in Padua) the more general formula for approximate integration, often written in the form

$$\int_0^{2h} y\, dx = h(y_0 + 4y_1 + y_2)/3,$$

which, in its turn, led to the outstanding generalisations due to Newton and Cotes, and subsequently to the minor generalisation known as 'Simpson's rule'; about this rule (published in 1743) there is not much to be said except that it is more accurate than the 'trapezoidal rule.' (Per Dr. G. N. Watson).

(*Mathematical Gazette, 44* (1960):268.) In translation the rule is as follows: "If therefore we multiply the third part of *IM,* the length of the barrel *BDFH* by two major circles *CG* and one of the minor ones *BH* (*DF* is like *BH*), it will give the capacity of the above-mentioned barrel."

191. Unthinking statements are not infrequently encountered concerning the supposed necessity of certain items of mathematical knowledge in connection with this or that technical achievement. Following Reginald Engelbach, whose *Problem of the Obelisks* appeared in 1923, Battiscombe Gunn and Eric Peet stated that for the Egyptians it was "very necessary to know the exact weight" of an obelisk in order to prepare for the difficulties of transportation and erection (p. 184 of "Four Geometrical Problems from the Moscow Mathematical Papyrus," *Journal of Egyptian Archaeology* 15 (1929):167–185). It is surprising that such claims are so confidently made. We may be permitted to inquire just how the "exact" weight could have amounted to a significant datum in plans made to overcome handling problems. Was it necessity also for the Egyptians to know the force in each rope, the mechanical advantage of each lever and the frictional resistance between the sliding surfaces?

Bernard P. Grenfell and Arthur S. Hunt, referring to another frustum example, this time the frustum of a cone, inform us that "if the instrument in question were a water-clock, a knowledge of this volume would be of great importance" (*The Oxyrhychus Papyri* Part III, London: Egypt Exploration Fund, 1903, p. 145). This claim, like the previous one, implies some kind of theoretical framework that we can be almost certain was *not* possessed by the Egyptians. Egyptian water clocks were used in the temples to measure the night hours for the purpose of regulating the rituals relating to the journey of the sun-god Rē through the twelve domains of the nether world. The size of the orifice needed to just deliver the bulk of the contained water, and the relative lengths of the scales for the different times of the year, would have to be determined empirically, by trial and error, as also would any other detail of the design that was not arbitrarily assumed. The search for an exact calculation procedure for a frustum of a pyramid or cone could have been *suggested* by ancient technological practice, but it is not required to explain such practice. The fact that the Egyptians discovered an exact calculating procedure for the volumes of frusta, and presumably also for complete pyramids (see note 189), may be taken as evidence that they were responsive to challenges over and above the merely practical.

192. The conventionalizations of decorative artists, simplifying the forms of familiar objects and facilitating their detachment from the customary physical and emotional contexts, could eventually have induced in suitably motivated and experienced observers levels of abstraction of the kind properly to be counted as mathematical. And, as Herbert Read has written, "do not let us underestimate the faculty of abstraction: it was the foundation, not only of fine art, but of logic, of science, of all scientific method. If a distinction is to be made between *homo faber* and *homo sapiens,* it lies in this faculty of abstraction." (*Icon and Idea, The Function of Art in the Development of Human Consciousness,* London: Faber & Faber, 1955, p. 39.)

193. On one of the Late Babylonian tablets providing evidence for a continuity in the mathematical tradition from Old Babylonian times to the Greek era (AO 6484 in the Louvre), there appears a calculation of the sum of the squares of the integers from 1 to 10 by a short method, viz., (55)($\frac{1}{3}$ of 1 + $\frac{2}{3}$ of 10). This will be recognized as being of the correct form,

$$\sum_{r=1}^{n} r^2 = \left(\sum_{r=1}^{n} r\right)\left(\frac{1 + 2n}{3}\right) = \frac{n}{6}(n + 1)(2n + 1).$$

A possible Babylonian derivation was given by Otto Neugebauer, *Vorlesungen über Geschichte der antiken mathematischen Wissenschaften,* vol. 1: *Vorgriechischer Mathematik,* Berlin: Springer, 1934, pp. 172–174. For a better suggestion (based on a simple induction from tabulated values), see pp. 112–113 of Solomon Gandz, "Studies in Babylonian Mathematics III," *Isis 32* (1940):

103–115. There is no evidence that this summation was connected with the problem of the pyramid volume in Babylonian mathematics.

Archimedes proved that if $L_1, L_2, L_3, \ldots L_n$ be n lines forming an ascending arithmetical progression in which the common difference is equal to least term L_1, then

$$(n + 1)L_n^2 + L_1(L_1 + L_2 + L_3 + \ldots + L_n)$$
$$= 3(L_1^2 + L_2^2 + L_2^3 + \ldots + L_n^2).$$

See "On Conoids and Spheroids," lemma to Proposition 2; cf. "On Spirals," Propositions 10, 11; Heath, ed. *Works of Archimedes,* pp. 107–109, 162–165. (With $L_1 = 1$, $L_2 = 2$, $L_3 = 3, \ldots L_n = n$, this proposition gives

$$\sum_{r=1}^{n} r^2 = \frac{n}{6}(n + 1)(2n + 1).)$$

Archimedes uses his proposition for mensurational demonstrations, and the same interest is evident in the work of Ibn al-Haytham, *c.* A.D. 1000, who established, in his *Treatise on the Measuring of the Paraboloid,* that "if one has a series of numbers beginning with unity and always increasing by [a common difference of] one . . . the sum of the squares of the numbers is equal to the third of the cube of the greatest number plus half of the square of the greatest number plus a sixth of this greatest number itself." In the same work, Ibn al-Haytham finds also the sums of the cubes, and of the fourth powers, of the first n integers. The demonstrations are very prolix, but the ideas are both simple and ingenious. For a German translation and commentary, see Heinrich Suter, *Bibliotheca Mathematica,* 3rd series, *12* (1911–12):289–332. For a concise account in modern symbolism, refer to Margaret Baron, *The Origins of the Infinitesimal Calculus,* pp. 66–70. Some related seventeenth century work was discussed by Carl Boyer, "Pascal's Formula for the Sums of Powers of the Integers," *Scripta Mathematica 9* (1943):237–244.

In Indian mathematics, the sum-of-squares rule, and the more easily found sum-of-cubes rule, were given in A.D. 499 by Āryabhaṭa, *Āryabhaṭīya,* II, 22: "The sixth part of the product of three quantities consisting of the number of terms, the number of terms plus one, and twice the number of terms plus one is the sum of the squares. The square of the sum of the (original) series is the sum of the cubes." (*The Āryabhaṭīya of Āryabhaṭa, An Ancient Indian Work on Mathematical and Astronomy,* translated by Walter Eugene Clarke, Chicago: Univ. of Chicago Press, 1930, p. 37.) Equivalent rules were given by Brahmagupta, *c.* A.D. 628, in his *Brāhma-sphuṭa-siddhānta,* XII, 20 (Colebrooke translation, pp. 293–294).

194. For another example of this kind, in which inconvenient empirical measurement was replaced by calculation, see p. 25. Cf. also Henry E. Roscoe, *John Dalton and the Rise of Modern Chemistry,* London: Cassell, 1901, p. 66, where Angus Smith is reported to have remarked that Dalton seemed to have begun his experiments with his hands and finished them off with his head.

195. Undue emphasis has been placed on the proportions of the Great Pyramid of Cheops, the angle of slope of which (slant face to horizontal) is, or was, within about one minute of 51°51′; consequently this angle was practically indistinguishable from (i) arcot $\pi/4$, or from (ii) arcot$\sqrt{\{\frac{1}{2}(\sqrt{5}-1)\}}$. These close agreements have been exploited by certain writers who have claimed (i) that the Egyptians of the Pyramid Age had a knowledge of a far more accurate value of π than is anywhere else suggested in what is known of the history of mathematics prior to Archimedes, or (ii) that the Egyptian architect of the Great Pyramid based his design on the golden section, or on some (unlikely!) relation leading to golden-section proportions, or (iii) that though the Egyptians themselves were innocent of such sophistication, they were instruments of a divine plan and unknowingly built in accordance with a design of the Deity Himself, to dimensions having prophetic, as well as mathematical, metrological and physical significance! The two best-known (or most notorious) of the older pyramidist works were: Piazzi Smyth, *Our Inheritance of the Great Pyramid* (1st ed., 1864, 5th ed., 1890), and David Davidson and Herbert Aldersmith, *The Great Pyramid, Its Divine Message* (eight editions to 1940). The Cheops Pyramid originally had a height of 280 cubits and a base edge length of 440 cubits. The horizontal departure of the side faces from the vertical was therefore $5\frac{1}{2}$ hands for every cubit (of 7 hands) measured vertically, and *this* is how the slope would have been planned and specified in the first place, measurement in degrees being then unknown. In the *Rhind Mathematical Papyrus* several problems involving pyramids make use of precisely this system of slope determination. Refer, Noel Wheeler, "Pyramids and Their Purpose," *Antiquity 9* (1935):5–21, 161–189, 292–304, where extensive empirical data on the differing proportions of the various surviving pyramids is given. (Mention might also be made of the recent spate of books with such titles as *Pyramid Power* (1973), *The Secret Power of Pyramids* (1975), *The Secret Forces of the Great Pyramid* (1975), *The Psychic Power of Pyramids* (1976), *The Pyramid Energy Handbook* (1977), *Life Force in the Great Pyramids* (1978). Most of these books, dealing with the putative "energy-concentrating" properties of hollow model pyramids, have unfortunately called upon the older, discredited pyramid literature for support and assumed uncritically that it is the precise form of the Cheops Pyramid that is crucial in bringing about the numerous surprising effects claimed.)

196. It is not possible to stack together 27 pyramids congruent with *Oabcd* so as to form *OABCD:* they do not themselves fill space. But 19 of them together with 16 tetrahedra (each having half the volume of *Oabcd*), could be packed to fill up *OABCD*.

197. An Egyptian would have said, "since 9 − 1 = 8" —his exemplar for our $x - \frac{1}{9}x =$ a constant.

198. Some indication of the very considerable incidental mastery of an informal geometry of joints, inlays, catches, and so on, achieved by ancient Egyptian joiners, stone masons and architects is given in *A History of Technology,* edited by

Charles Singer et al., vol. 1, Oxford: Clarendon Press, 1954, chap. 17, pp. 475–484, and especially chap. 25, "Fine Woodwork," by Cyril Aldred. Although it might be claimed that, in the absence of more theoretical studies, some such practical work would have provided an almost indispensable background of experience in the manipulation and understanding of spatial relations, needless to say it cannot be supposed that in itself this would have sufficed to ensure the occurrence of a hypothetical experiment for finding the pyramid volume as sophisticated as that described in the dialogue. (Of course, if it were only a matter of explaining the discovery of a rule for the volume of a pyramid, we might be satisfied to assume that this had been deduced from the weighing of model pyramids and prisms. But the documented procedure in the *Moscow Mathematical Papyrus* corresponding to our formula $V = (a^2 + ab + b^2)h/3$—see note 189—strongly suggests more rational heuristic or justificatory procedures.)

As craftsmanship became more complex and where numerous workers were to be involved in a single project, it must have become apparent how advantageous it would be to organize resources in ways that were more effective technologically, or more economical, and not only in ways followed for their supposedly magical significance or their aesthetically pleasing results. The challenge to plan in advance in ever greater detail, to *preconstruct* in one's head, with or without the assistance of material tokens, would bring with it further challenges to *explain* and to *justify* the initial conceptions, as well as to actualize them. Yet retrospective attempts on our part to correlate supposed patterns of ancient understanding, or even individual insights, with the manifest practical accomplishment must remain hazardous in the absence of specific evidence.

199. This thought leads to the invention (conceptualization) of whole new classes of solid forms, including some, perhaps, never before imagined. For example, for the solid represented in Fig. A13 in orthographic projection it may be supposed that $x = c_1 z^n$, $y = c_2 z^m$, where $m + n = 2$, the cross-section parallel to the base being everywhere rectangular, but with shape, as well as area, a function of the distance z from the "highest" point. These cross-sectional areas are given by $A(z) = c_1 c_2 z^n z^{2-n} = cz^2$, exactly as for the solids considered in the discussion. Here, then, for each pair of constants, c_1 and c_2, the members of an infinite set of generalized "pyramids"—or "pyramidoids"—are definable via the parameter n. For all such solids,

$$V = \int_0^h cz^2 dz = \frac{1}{3}ch^3,$$

and since $A_B = ch^2$, the volume formula may be expressed as $V = \frac{1}{3}A_B h$, exactly as for traditionally recognized pyramids and cones. And further, for each value of n, infinite variations are possible from the right pyramidoids with rectangular sections assumed here so far, by allowing oblique, twisted, and various versions of arbitrary cross-sectional form (for example, elliptical with axes x, y). There is no need to restrict n to the range 0 to 2. With n outside this range, the "apex" at

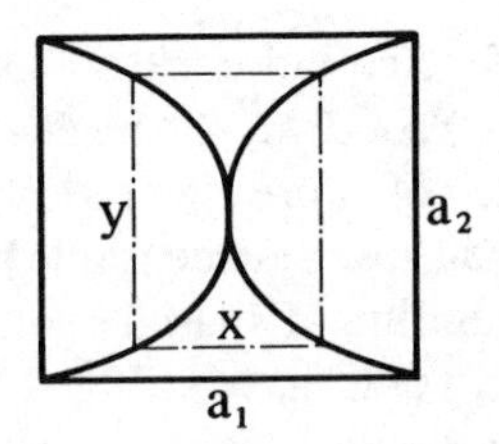

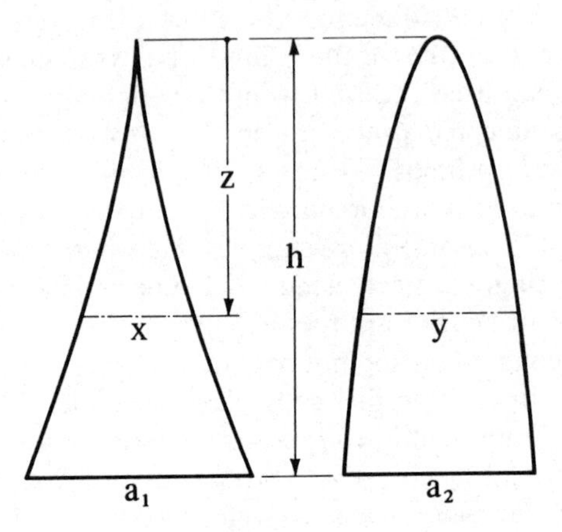

Fig. A13

$z = 0$ is replaced by an infinitely long razor-edge. Frusta, or other segments of such examples as the following might be of interest to abstract sculptors:

$$x = z^3, \qquad y = z^{-1}; \qquad \text{or} \qquad x = c_1 z^n \cdot f(z), \qquad y = c_2 z^{2-n}/f(z),$$

where, for example, $f(z)$ might be taken as a^z, or as $p + \sin(z/q)$, $p > 1$. Frusta of these solids have their volumes correctly measured by both formulae (1) and (2). (In the seventeenth century the term "pyramidoid" was used in a different sense. For this now defunct use, see *The Mathematical Papers of Isaac Newton,* vol. 1: 1664–1666, edited by D. T. Whiteside, Cambridge, England: Cambridge Univ. Press, p. 92, n. 8. Cf., James and James, *Mathematics Dictionary,* s. v. "Conoid," no entry being given for pyramidoid.)

200. Cf. the following works of Karl Popper: *The Logic of Scientific Discovery* (translated by the author and J. and L. Freed, London: Hutchinson 1959), *Conjectures and Refutations, The Growth of Scientific Knowledge* (London: Routledge & Kegan Paul, 1963), *Objective Knowledge, An Evolutionary Approach,* (Oxford: Clarendon Press, 1972), referring to the indexes especially under "Corroboration." Several critical discussions of Popper's views on corroboration are to be found in *The Philosophy of Karl Popper,* edited by Paul Arthur Schilpp, Library of Living Philosophers, 2 vols, La Salle, Ill.: Open Court, 1974.

201. The condition for the equality of V_1 and V_2 having been found *not* to be satisfied in the case of the prismoid of Fig. 97, it is illuminating to investigate the general case represented in Fig. A14. All the faces are assumed to be plane, the top and bottom faces being parallel rectangles, not necessarily similar; that is, we begin with no assumption as to the relative values of a_1, a_2, b_1, b_2, except that they are nonnegative.

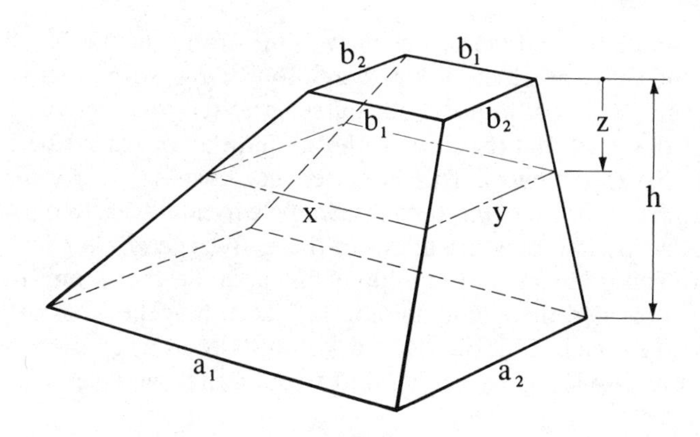

Fig. A14

$$x = b_1 + \frac{z}{h}(a_1 - b_1), \qquad y = b_2 + \frac{z}{h}(a_2 - b_2),$$

$$\begin{aligned} xy &= \{b_1h + z(a_1 - b_1)\}\{b_2h + z(a_2 - b_2)\}/h^2 \\ &= \{b_1b_2h^2 + (a_1b_2 + a_2b_1 - 2b_1b_2)hz + (a_1 - b_1)(a_2 - b_2)z^2\}/h^2. \end{aligned}$$

$V = \int_0^h xy\,dz,$ from which is obtained

$$\begin{aligned} V &= \frac{h}{6}\{2a_1a_2 + 2b_1b_2 + a_1b_2 + a_2b_1\} \\ &= \frac{h}{6}\{a_1a_2 + b_1b_2 + (a_1 + b_1)(a_2 + b_2)\}, \end{aligned}$$

regardless of the relation between the magnitudes of a_1, b_1, a_2, b_2. Hence, for all solids of this kind,

$$V = \frac{h}{6}\{A_T + A_B + 4A_{MS}\}. \tag{2}$$

An alternative simplification is

$$V = \frac{h}{3}\{a_1a_2 + b_1b_2 + \frac{1}{2}(a_1b_2 + a_2b_1)\}.$$

Now provided only that $a_1b_2 = a_2b_1$,

$$\frac{1}{2}(a_1b_2 + a_2b_1) = a_1b_2 = a_2b_1 = \sqrt{\{(a_1b_2)(a_2b_1)\}} = \sqrt{\{(a_1a_2)(b_1b_2)\}}.$$

Hence, in these circumstances, the last volume expression may be written,

$$V = \frac{h}{3}\{A_T + A_B + \sqrt{(A_TA_B)}\} \qquad (1)$$

Thus this formula correctly measures the volume of rectangular-based prismoids for which, and only for which, in our symbolism, $a_1b_2 = a_2b_1$, that is, for which $b_2 : a_2 = b_1 : a_1$. This condition is, of course, precisely that associated with the similarity of the bases and (because of the assumption of plane side faces) of all sections parallel to the bases. This indicates that formula (1) is generally applicable only to frusta of *pyramids* (and to complete pyramids) and to (the infinitely numerous) Cavalierian transformations of these. By *generally* applicable is meant applicable to frusta having base and top lying in parallel planes any arbitrary distance apart, provided that these are not so placed that the cross-sectional area vanishes between them. This proviso will be understood by considering a double pyramid (see Fig. A15) for which it will be found that the volume is given by

$$V = \frac{h}{3}\{A_T + A_B \textit{ minus } \sqrt{(A_TA_B)}\}.$$

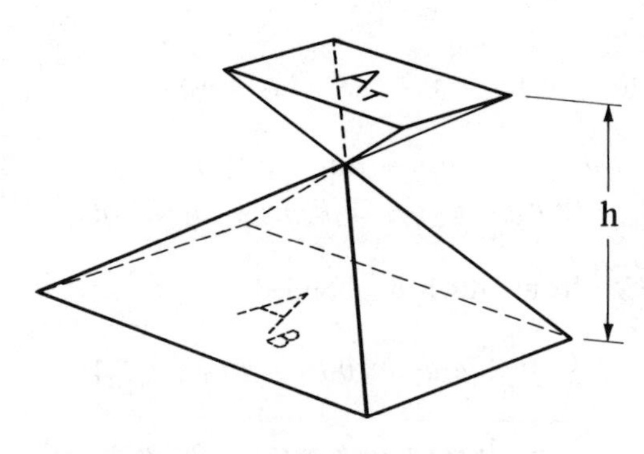

Fig. A15

Formula (2), on the other hand, applies unchanged to double pyramids and cones.

202. Carl Boyer, *The History of the Calculus and Its Conceptual Development*, New York: Dover, 1959, pp. 119–128, 142–161, 171–173. The most detailed early treatment of these generalizations was set out in Wallis's *Arithmetica Infinitorum*, 1656; for this, see J. F. Scott, *The Mathematical Work of John Wallis, D.D., F.R.S. (1616–1703)*, London: Taylor and Francis, 1938, chap. 4.

203. Cf. Popper, *The Logic of Scientific Discovery,* pp. 121–123, where increase of universality (widening the subject or antecedent class) is given as one of the two outstanding ways of increasing the empirical content of natural laws. (The other of these ways, by an increase in the *degree of precision* of the predicate, is not applicable in its most obvious sense in our example since the relationship predicated is one of exact equality. But Popper's notion of degree of precision is sometimes relevant within mathematics, for example in connection with inequalities, stronger claims being preferred to weaker ones for most purposes. It might be noted that narrowing the range of a characteristic predicated of a subject is only one way of increasing the degree of precision of a proposition: any restriction of the predicate suffices, and this may be achieved by supplementing claims already made by further claims of a different kind. "All *A*'s are both *B* and *C*" would have a higher precision in Popper's sense than "All *A*'s are *B,*" if any *B* is not *C*.)

204. "The aim of proof is, in fact, not merely to place the truth of a proposition beyond all doubt, but also to afford an insight into the dependence of truths upon one another. After we have convinced ourselves that a boulder is immovable, by trying unsuccessfully to move it, there remains the further question, what is it that supports it so securely?"—Gottlob Frege, *Die Grundlagen der Arithmetik,* German text of 1884 published with translation by J. L. Austin as *The Foundations of Arithmetic,* Oxford: Blackwell, 1950 and 1953, p. 2. See also note 228.

205. If alternative antecedents can be found that will support the same conclusion, it is natural to seek a single specification for the original and the new antecedents, if only for the sake of conceptual economy. Bacon has an especially pertinent discussion on generalization by way of alternative antecedents in the eleventh chapter of his uncompleted treatise, "Of the Interpretation of Nature," which he wrote under the name "Valerius Terminus" (*The Works of Francis Bacon,* edited by James Spedding, R. L. Ellis, and D. D. Heath, vol. 3, 1857, pp. 235–241). Bacon's concern was with empirical, not mathematical, examples; his antecedents are causes, or "directions," as he calls them, for bringing about effects, and he takes it for granted that the several directions corresponding to any given effect are ultimately reducible to a single one.

206. The following definition probably gives the earliest meaning conveyed by the term "prismoid": "a solid Figure, contained under several Planes whose Bases are rectangular Parallelograms, parallel and alike situate" (John Harris, *Lexicon Technicum,* I, 1704 (according to the *Oxford English Dictionary,* s.v. "prismoid"). The same definition is given in the 4th edition of Harris's *Lexicon,* vol. 1, 1725, and is repeated in Edmund Stone's *New Mathematical Dictionary,* 1726. In Hero of Alexandria's *Metrica,* II, 8, such a solid is referred to as a βωμίσκος, "little altar," and its volume is found by dissecting it into a parallelepiped, two triangular prisms, and a pyramid. For details, see Heath's *History of Greek Mathematics,* vol. 2, pp. 332–334.

Just which boundaries are to be drawn in a definition is a matter liable to be determined by numerous unexpressed considerations. No one has ever seriously suggested that one ought to define a prismoid as any solid for which the volume is measured by the expression $(h/6)\{A_T + A_B + 4A_{MS}\}$. The volume is not the only property to be taken into account. Even if we suppose it to be the chief focus of interest we should still want an independent definition in which the defining characteristics were such that they would *entail* the volume formula and not only for the whole solid but for any frustum of it with base and top parallel to the base and top of the whole solid.

Consideration of the first diagram of Fig. 96 may suggest a further generalization to solids having parallel but *non*similar bases with *curvilinear* boundaries. A general extension along these lines that has been adopted (or, at least, proposed) allows of indefinitely many prismoids for any pair of parallel bases in a given alignment, the different prismoids resulting from differences in the motions of the ends of the straight line that generates the "side" surface: "A *prismoid,* in the general sense, is a solid figure bounded by two closed curves, lying in parallel planes, and a surface which is generated by a straight line, the *generator,* moving continuously with its ends on the plane curves, these ends moving always in the same direction on each of the curves and describing each curve once only" (J. H. Michell and M. H. Belz, *The Elements of Mathematical Analysis,* London: Macmillan, 1937, vol. 2, p. 814). The authors supplement their definition with the following remarks: "This definition does not exclude the possibility of one end of the line remaining stationary while the other end describes a finite arc. For example, a polyhedral prismoid is formed when the parallel plane curves are polygons, not necessarily of the same number of sides, and the remaining faces are triangles, each of which has as vertex an angular point of one of the plane polygons and as opposite side one of the sides of the other polygon."

The tradition according to which prismoids are to be understood (like prisms) to be bounded entirely by *plane* surfaces is one that few authors have been willing to relinquish. James and James in their well-known *Mathematics Dictionary,* 4th ed., 1976, make no mention of any prismoid not so bounded. Yet foreshadowings of Michell and Belz's general definition are to be found at least back to the late eighteenth century. Charles Hutton, in the first edition of his *Mathematical and Philosophical Dictionary,* 1795, though defining a prismoid as "a solid, or body, somewhat resembling a prism, but that its ends are any dissimilar parallel plane figures of the same number of sides, the upright sides being trapezoids," nevertheless proceeds to remark that "if the ends of the prismoid be bounded by dissimilar curves, it is sometimes called a cylindroid." This appears to be an anticipation of Michell and Belz's "prismoid in the general sense," yet, on turning to Hutton's own definitions of "cylinder" and "cylindroid," we find nothing more exciting than the following:

> CYLINDER, a solid having two equal circular ends, and every plane section parallel to the ends a circle equal to them also.
>
> CYLINDROID, a solid resembling the figure of a cylinder; but differing from it as having ellipses for its ends or bases, instead of circles, [as] in the cylinder.

These are, of course, poor definitions. Hutton himself apparently wished to exclude cases in which the centers of the parallel cross-sections do not lie on a straight line, since he notes that "the right cylinder may be conceived to be generated by the rotation of a rectangle about one of its sides." Presumably, for Hutton himself, cylindroids were simply elliptical cylinders. Already Archimedes had made use of oblique elliptical cylinders and had shown that while sections of such cylinders can always be found that are circles, only in special cases will such sections be perpendicular to the axis of the cylinder. See his "On Conoids and Spheroids," especially Propositions 9, 19, 20, *Works* Heath ed., pp. 119–121, 129–130.

It is to be noticed that, in the case of Michell and Belz's generalized prismoids, the area of cross-section (parallel to the ends) turns out to be a *quadratic* function of the distance from an end (as the authors show on pp. 815–816 of their second volume); hence even their very general definition excludes solids with cross-sections following a cubic law, although, as has been shown (p. 232 of this book), the "prismoidal formula"—our formula (2)—is applicable to them.

It remains to remark that the entry on the *prismoidal formula* in James and James's *Mathematics Dictionary* (all editions) contains an inaccuracy. After giving the usual volume formula as $V = \frac{1}{6}h(B_1 + 4B_m + B_2)$, and correctly noting that it applies when the cross-sectional area is a linear, a quadratic, or a cubic function of the distance from a base, it is then asserted: "The prismoidal formula is sometimes given as $V = \frac{1}{4}h(B_1 + 3S)$, where S is the area of a section parallel to the base and $\frac{2}{3}$ the distance from B_1 to B_2. This is equivalent to the preceding form." Actually, however, the exact equivalence breaks down if the law of cross-sectional area goes beyond the quadratic form. For a cubic law, $A(z) = a + bz + cz^2 + dz^3$, we have

$$B_1 = A(0) = a, \qquad S = A\left(\frac{2}{3}h\right) = a + \frac{2}{3}bh + \frac{4}{9}ch^2 + \frac{8}{27}dh^3.$$

So $\frac{1}{4}h(B_1 + 3S)$ comes to

$$\frac{1}{4}h\left(a + 3a + 2bh + \frac{4}{3}ch^2 + \frac{8}{9}dh^3\right), \text{ that is } ah + \frac{1}{2}bh^2 + \frac{1}{3}ch^3 + \frac{2}{9}dh^4,$$

whereas the true volume measure is given by

$$\int_0^h (a + bz + cz^2 + dz^3)\,dz = ah + \frac{1}{2}bh^2 + \frac{1}{3}ch^3 + \frac{1}{4}dh^4.$$

The particular two-term prismoidal formula given by James and James was first published by G. D. Hermann Kinkelin, "Zur Theorie des Prismoides," *Archiv der Mathematik und Physik 39* (1862):181–186, an English translation of which is included in George Bruce Halsted, "The Criterion for Two-Term Prismoidal Formulae," *Transactions of the Texas Academy of Science 1,* no. 5, (1896): 19–32.

HORA QUARTA DECIMA

207. For the equivalent of this in early Hindu mathematics, in the Śulvasūtras of Baudháyana, Apastamba, and Kátyáyana, see p. 245 of G. Thibaut, "On the Śulvasūtras," *Journal of the Asiatic Society of Bengal 44*, Part 1 (1875):227–275; or, more conveniently, C. N. Srinivasiengar, *The History of Ancient Indian Mathematics,* Calcutta: World Press Private, 1967, p. 12.

208. For the significance of Euclid II, 5, as an important theorem of Greek "geometrical algebra," see Heath's commentary in his edition of *The Thirteen Books of Euclid's Elements,* vol. 1, pp. 383–385. The appropriateness of the orthodox modern interpretations was challenged by Sabetai Unguru, "On the Need to Rewrite the History of Greek Mathematics," *Archive for History of Exact Sciences, 15* (1975):67–114. Three severe rejoinders were published: B. L. van der Waerden, "Defence of a 'Shocking' Point of View," ibid. *15* (1976):199–210; Hans Freudenthal, "What Is Algebra and What Has It Been in History?," ibid. *16* (1977):189–200; and André Weil, "Who Betrayed Euclid?," ibid. *19* (1978): 91-93. See also Unguru's further paper, "History of Ancient Mathematics: Some Reflections on the State of the Art," *Isis, 70* (1979):555–565.

209. *Nova stereometria doliorum vinariorum,* 1615, as quoted in D. J. Struik (ed.), *A Source Book in Mathematics, 1200–1800,* p. 222. Cf. Carl Boyer, *The History of the Calculus and Its Conceptual Development,* 1959, pp. 110–111. A somewhat similar remark had been made by Oresme two and one-half centuries earlier. This is quoted and discussed in Boyer, pp. 85–86.

210. This table was given by Solomon Gandz, "Studies in Babylonian Mathematics III," *Isis, 32* (1940):103–115, on pp. 108–109, in connection with his conjectural reconstruction of the heuristic path that could have led the Old Babylonian mathematicians to their method of finding two numbers when their sum and product are given. The same suggestion was referred to, not as a hypothesis but as though it were a known fact, and without acknowledgement to Gandz, in J. F. Scott, *A History of Mathematics, From Antiquity to the Beginning of the Nineteenth Century,* London: Taylor and Francis, 1958 (and 1960), p. 11.

211. For an English translation of Fermat's own paper containing this method (received by Descartes in January 1638, via Mersenne), see Struik's *Source Book,* pp. 223–224; cf. Boyer's *History of the Calculus,* pp. 155–156.

212. For a general comment, see note 114. Of particular relevance is the following item given by Newton in all editions of his *Principia,* as the first illustration of Lemma II of Book II. It is reproduced here as it appears on p. 250 in Florian Cajori's edition (Cambridge Univ. Press, 1934 and subsequent printings) of Andrew Motte's 1729 translation of Newton's third edition:

> Case 1. Any rectangle, as AB, augmented by a continual flux, when, as yet, there wanted of the sides A and B half their moments $\frac{1}{2}a$ and $\frac{1}{2}b$, was $A - \frac{1}{2}a$ into $B - \frac{1}{2}b$, or $AB - \frac{1}{2}aB - \frac{1}{2}bA + \frac{1}{4}ab$; but as soon as the sides A and B are aug-

mented by the other half-moments, the rectangle becomes $A + \frac{1}{2}a$ into $B + \frac{1}{2}b$, or $AB + \frac{1}{2}aB + \frac{1}{2}bA + \frac{1}{4}ab$. From this rectangle subtract the former rectangle, and there will remain the excess $aB + bA$. Therefore with the whole increments a and b of the sides, the increment $aB + bA$ of the rectangle is generated. Q.E.D.

It follows that if the "moments" (increments) are such that $b = -a$, as required for the case of a varying rectangle of constant perimeter, the increment of the area will be $a(B - A)$, and this $\gtreqless 0$ according as $B \gtreqless A$, provided $a > 0$. With this proviso, A is the measure of the sides being *increased* and B that of the sides being *decreased;* therefore the conclusion to be drawn is that so long as it is the shorter sides of an isoperimetric rectangle that are increasing and the longer sides that are decreasing, the area will be increasing, whereas if the longer sides are increasing at the expense of the shorter sides, the area will be decreasing.

In such a work as the *Principia,* however, Newton did not stop to consider our simple maximization problem. Rather he was establishing the equivalent of what we now refer to as the rule for differentiating products of two (and then of three or more) functions of the same variable. So he obtains, as a special case, naA^{n-1} as the moment of any power A^n, while in his final illustration of Lemma II, he reaches $maA^{m-1}B^n + nbB^{n-1}A^m$ as "the moment of any generated quantity A^mB^n . . . whether the indices m and n of the powers be whole numbers or fractions, affirmative or negative." Yet in this, his greatest work, Newton can hardly be said to have been concerned to promote his calculus as a new branch of mathematics. As Cajori remarked, "At this place [Book II, Lemma II] also in Book I, Lemmas I, II, XI (Scholium), Newton makes a brief general statement of the principles of fluxions and fluents (that is, of the differential and integral calculus), but he does not give his notation. No one is known to have obtained from the *Principia* alone a working knowledge of fluxions." (Ibid., p. 654, n. 31.)

When Berkeley criticized the faulty foundations of the calculus of his day, the item quoted at the beginning of this note came in for special attention. After giving Newton's argument, Berkeley continued:

> But it is plain that the direct and true method to obtain the moment or increment of the rectangle AB, is to take the sides as increased by their whole increments, and so multiply them together, $A + a$ by $B + b$, the product whereof $AB + aB + bA + ab$ is the augmented rectangle; whence, if we subduct AB the remainder $aB + bA + ab$ will be the true increment of the rectangle, exceeding that which was obtained by the former illegitimate and indirect method by the quantity ab. And this holds universally be the quantities a and b what they will, big or little, finite or infinitesimal, increments, moments, or velocities. Nor will it avail to say that ab is a quantity exceeding small: since we are told [by Newton himself, Introd. ad *Quadraturam curvarum*] that *in rebus mathematicis errores quam minimi non sunt contemnendi* [in matters mathematical even the smallest errors are not to be disregarded].

(*The Analyst,* 1734, §9 in *The Works of George Berkeley,* edited by A. A. Luce and T. E. Jessop, London, Nelson, vol. 4, pp. 53–102.)

Most eighteenth-century British mathematicians found it difficult to accept that Newton could be so vulnerable to criticism, and excessive deference to the master resulted in some incompetent defences offered on his behalf. In the particular item under discussion, Newton's presentation really was arbitrary; it amounted to a fiddle, a sleight of hand. As William Rowan Hamilton was to write:

> His mode of getting rid of *ab* appeared to me long ago (I must confess it) to involve so much *artifice*, as to deserve to be called *sophistical;* although I should not like to say so publicly. He subtracts, you know, $(A - \frac{1}{2}a)(B - \frac{1}{2}b)$ from $(A + \frac{1}{2}a)(B + \frac{1}{2}b)$; whereby, of course, *ab* disappears in the result. But by *what right*, or *what reason* other than to given an unreal air of *simplicity* to the calculation does he *prepare* the *products* thus? Might it not be argued similarly that the difference
>
> $$\left(A + \frac{1}{2}a\right)^3 - \left(A - \frac{1}{2}a\right)^3 = 3aA^2 + \frac{1}{4}a^3$$
>
> was the moment of A^3; and is it not a sufficient *indication* that the mode of procedure adopted is not the fit one for the subject, that it quite *masks* the notion of *limit;* or rather has the appearance of treating that notion as foreign and irrelevant, notwithstanding all that has been said so well before, in the First Section of the First Book [of the *Principia*]?

(From a letter to Augustus De Morgan written in 1862, quoted in Florian Cajori, *A History of the Conceptions of Limits and Fluxions in Great Britain from Newton to Woodhouse,* Chicago: Open Court, 1919, pp. 91–92.)

In sections 13 to 16 of the *Analyst,* Berkeley proceeded to examine the alternative method of finding the fluxion of x^n, which Newton used in the introduction to the *Tractatus de quadratura curvarum,* published as an appendix to the first edition of his *Opticks,* 1704. Newton's determination was as follows:

> In the same time that the quantity x, by flowing, becomes $x + o$, the quantity x^n will become $(x + o)^n$, that is, by the method of infinite series, $x^n + nox^{n-1} + \frac{n^2 - n}{2}oox^{n-2} + \&c.$ And the augments o and $nox^{n-1} + \frac{n^2 - n}{2}oox^{n-2} + \&c.$ *are to one another* as 1 and $nx^{n-1} + \frac{n^2 - n}{2}ox^{n-2} + \&c.$ Now let these augments vanish, and their ultimate *ratio* will be 1 to nx^{n-1}. [Emphasis added.]

(Translation of J. Stewart, 1745; D. J. Struik's *Source Book,* p. 306.) To this Berkeley objected:

> when it is said, let the increments vanish, i.e. let the increments be nothing, or let there be no increments, the former supposition that the increments were something, or that there were increments, is destroyed, and yet a consequence of that supposition, i.e. an expression got by virtue thereof, is retained.

(*Analyst,* §13, *Works,* vol. 4, p. 72.) Berkeley overlooked (or consciously ignored) the all-important reference to *ratios* by Newton, who was considering not the increments for their own sake but the effect of the increments on certain ratios. Yet, having said that, we must acknowledge that Berkeley's criticisms help us refrain from simply reading the calculus of Newton's time through modern eyes. It will not do to adopt the facile assumption that the procedures were in fact based upon an intuitive, implied notion of *limiting values.* Although it is possible to find isolated statements that could appear to support that assumption—for instance in *Principia,* Book I, Scholium following Lemma XI—a coherent treatment along these lines was lacking. Progress in this direction was made by Benjamin Robins in his *Discourse Concerning the Nature and Certainty of Sir Isaac Newton's Methods of Fluxions and of Prime and Ultimate Ratios,* 1735. Robins pointed out, for example, that in the expression $nox^{n-1} + \frac{n^2 - n}{2}oox^{n-1} + \&c.$, appearing in the passage quoted above from Newton's *Quadratura curvarum,*

> all the terms after the first taken together may be made less than any part whatever of the first, that shall be assigned. Consequently the proportion of the first term $nx^{n-1}o$ to the whole augment may be made to approach within any degree whatever of the proportion of equality; and therefore the ultimate proportion [being sought] . . . is that of $nx^{n-1}o$ only to o, or the proportion of nx^{n-1} to 1.

(*Discourse,* §147.) And Newton's "ultimate ratio of vanishing quantities," Robins explicated as a

> fixed quantity, which some varying quantity, by a continual augmentation or diminution, shall perpetually approach, but never pass . . . provided the varying quantity can be made in its approach to the other to differ from it by less than any quantity how minute soever, that can be assigned.

(Ibid., §95, and cf. §110.)

Unfortunately, there was an untoward delay in following up this approach, considering that it was already to some extent anticipated in the writings of Stevin (1586), Valerio (1604), and a number of their seventeenth-century successors, such as Wallis (1656). For the state of the analytic art around 1700, see Carl B. Boyer, *The History of the Calculus and Its Conceptual Development,* chap. 5 (and chap. 4 for the anticipations just mentioned); also H. J. M. Bos, "Differentials . . . and the Derivative in the Leibnizian Calculus," *Archive for History of Exact Sciences 14* (1974):1–90. For some subsequent developments, see Ivor Grattan-Guinness, *The Development of the Foundations of Mathematical Analysis from Euler to Riemann,* Cambridge, Mass.: MIT Press, 1970; and Judith V. Grabiner, "The Origins of Cauchy's Theory of the Derivative," *Historia Mathematica 5* (1978):379–409.

213. The method is explicitly set out in Al-Khowārizmī's *Algebra, c.* A.D. 830. For this, see *Robert of Chester's Latin Translation of the Algebra of Al-Khowārizmī,* with an introduction, notes, and an English version by L.C. Karpinski, 1915 (reprinted as the first part of Louis C. Karpinski and John G. Winter, *Contributions to the History of Science,* Ann Arbor: Univ. of Michigan Studies, Humanistic Series vol. 11, 1930); an extract is reproduced in D. J. Struik's *Source Book,* pp. 56–60. For the long historical background, see Solomon Gandz, "The Origin and Development of the Quadratic Equations in Babylonian, Greek, and Early Arabic Algebra," *Osiris 3* (1938):405–556; also A. Seidenberg, "The Origin of Mathematics," *Archive for History of Exact Sciences 18* (1978):301–342.

214. See note 154.

215. In general, $\sqrt[n]{(f_1 f_2 f_3 \ldots f_n)} \leq (1/n)(f_1 + f_2 + f_3 + \ldots + f_n)$. For a proof of this "theorem of the geometric and arithmetic means," see Hans Rademacher and Otto Toeplitz, *The Enjoyment of Mathematics,* translated by H. Zuckerman, Princeton, N.J.: Princeton Univ. Press, 1957, pp. 101–103, or Edwin Beckenbach and Richard Bellman, *An Introduction to Inequalities,* New York: Random House, 1961, pp. 54–59.

216. Galileo, *Two New Sciences,* First Day, 102; Drake, translation, p. 62. The numerals I to IV are inserted here for convenience. The first part has been called "Galileo's Theorem," for example in the *Century Dictionary,* s.v., "theorem." Galileo himself acknowledges that this whole theorem was suggested by a passage he encountered while studying the thirteenth-century *Tractatus de sphaera* of Johannes de Sacrobosco in conjunction with a learned commentary upon it. This commentary, by Clavius, 1581, though perhaps the most thoroughgoing of the numerous ones on this much-published work, is unfortunately not included in Lynn Thorndike's edition of *The* Sphere *of Sacrobosco and Its Commentators,* Chicago: Univ. of Chicago Press, 1949. For related work in Greek mathematics, see note 219 and especially note 224.

217. Bryson of Heraclea, who lived at the end of the fifth century B.C., is said to have put forward a suggestion for finding the area of a circle in this way. See Heath's *History of Greek Mathematics,* 1921, vol. 1, pp. 223–224; also A. Wasserstein, "Some Early Greek Attempts to Square the Circle," *Phronesis 4*(1959): 92–100.

218. "Measurement of a Circle," Prop. 1. This reduction of the problem of quadrature to the problem of the rectification of the circumference led Archimedes (in Prop. 3 of the same work) to establish close rational approximations between which the ratio $C:d$ (and hence, in view of Prop. 1, the ratio $A:r^2$ also) must lie (Heath (ed.) *Works,* pp. 91–98).

219. *Componendo et permutando,* by composition $[a:b = c:d \Rightarrow (a+b):b = (c+d):d]$ and by permutation $[a:b = c:d \Rightarrow a:c = b:d]$. Cf. Euclid, Book V, especially Prop. 18. The initial steps of this part of Galileo's proof are characteristically Greek; the same comparisons between sectors and triangles related in the same way appear already in Prop. 8 of Euclid's *Optics.* This proposition is given with English translation in Ivor Thomas, *Selections Illustrating the History of Greek Mathematics,* vol. 1, pp. 502–505; for the first English translation of the whole work, see H. E. Burton, "The Optics of Euclid," *Journal of the Optical Society of America 35* (1945):357–372. The most apt moves of the great pioneer investigators and their most valuable insights, insofar as they enter the common pool and become widely disseminated, typically become known to later mathematicians in the form of standardized procedures, theorems, or formulae, tending in time to become regarded as mere commonplaces.

220. Nowadays, of course, such a treatment would be quite routine. It is natural to consider two regular polygons of equal perimeter, p say, having m, n sides, respectively. Then the sides are of length p/m, p/n, and they subtend angles at the centers of $2\pi/m$ and $2\pi/n$ radians. Then the area ratio

$$\frac{(p^2/4n)\cot(\pi/n)}{(p^2/4m)\cot(\pi/m)} = \frac{(1/n)\cot(\pi/n)}{(1/m)\cot(\pi/m)} = \frac{(\pi/n)\tan(\pi/m)}{(\pi/m)\tan(\pi/n)} = \frac{\beta \cdot \tan\alpha}{\alpha \cdot \tan\beta},$$

where $\alpha = \pi/m$, $\beta = \pi/n$. Suppose $n > m$, then $\beta < \alpha < \pi/2$, in which case $\tan\alpha/\tan\beta > \alpha/\beta > 1$. (The geometric equivalent of this was demonstrated in Greek mathematics, for example in the proposition from Euclid's *Optics* referred to in the preceding note. The result, again in entirely geometric form, was assumed as known, a generation after Euclid, by Aristarchus in Prop. 7 of his "On the Sizes and Distances of the Sun and Moon" (Thomas Heath, *Aristarchus of Samos, The Ancient Copernicus,* Oxford: Clarendon Press, 1913, pp. 376–381.) Continuing, we may now write,

$$\frac{\beta \cdot \tan\alpha}{\alpha \cdot \tan\beta} > \frac{\beta}{\alpha} \times \frac{\alpha}{\beta} = 1; \quad \text{consequently,} \quad \frac{\text{area of } n\text{-gon}}{\text{area of } m\text{-gon}} > 1.$$

Thus, for regular polygons of equal perimeter, the greater the number of sides, the greater the area. In the limit, as $n \to \infty$,

$$\begin{aligned}\text{area of regular } n\text{-gon} &= \lim_{n\to\infty} \{(p^2/4n)\cot(\pi/n)\} \\ &= \lim_{n\to\infty} \left\{ (p^2/4\pi)(\pi/n) \cdot \frac{\cos(\pi/n)}{\sin(\pi/n)} \right\}.\end{aligned}$$

That is,

$$\begin{aligned}\text{area} = \lim_{\theta\to 0}\{(p^2/4\pi)(\theta/\sin\theta) \cdot \cos\theta\} &= p^2/4\pi \\ &= \text{area of circle of circumference } p.\end{aligned}$$

221. Compare note 19, especially toward the end.

222. A quadrilateral having given side lengths has its greatest area when its opposite angles are supplementary, that is, when it is inscribable within a circle. For a quadrilateral $ABCD$ (in which the lengths of the sides AB, BC, CD, DA are denoted by a, b, c, d units, respectively), this may be shown by beginning with the area expression $\frac{1}{2}ab \sin B + \frac{1}{2}cd \sin D$, and, with the assistance of the equivalent expressions for AC^2 (viz., $a^2 + b^2 - 2ab\cos B$, and $c^2 + d^2 - 2cd\cos D$), obtaining, after appropriate algebraic and trigonometric manipulation, the area formula $Q = \sqrt{\{(s-a)(s-b)(s-c)(s-d) - T\}}$. Here $s = \frac{1}{2}(a+b+c+d)$, and $T = abcd\cos^2\{\frac{1}{2}(B+D)\}$. Clearly, T takes its minimum value of zero when the angles B and D (and hence A and C also) are supplementary. The maximum possible area, for given values of a, b, c, and d, is therefore

$$Q_{\max} = \sqrt{\{(s-a)(s-b)(s-c)(s-d)\}},$$

the well-known result for a cyclic quadrilateral. (If O is the circumcenter of $ABCD$, the isosceles triangles OAB, OBC, OCD, ODA may be arranged in different orders around their common vertex to produce three different cyclic quadrilaterals. These differ in shape, but, being composed of the same four isosceles triangles, obviously they cannot differ in area or in circumradius.)

223. Trigonometry is certainly not needed! For the most elegant proof of (i), make use of a parallelogram $B_1B_2A_2C$ and right-angled triangle A_1B_1C. For (ii), let A_2B_2 cut A_1B_1 at X and draw A_1Y parallel to OB_1 to meet A_2B_2 at Y, then compare Δs A_1A_2X and B_1B_2X via ΔA_1YX.

224. Zenodorus, who flourished early in the second century B.C., wrote a lost treatise on isoperimetry that is known through versions in Pappus's *Collection,* in the commentary by Theon of Alexandria on Ptolemy's *Almagest,* and in the anonymous *Introduction to the Almagest.* A concise account is given in Heath's *History of Greek Mathematics,* vol. 2, pp. 207–213. Of the four propositions from the *Two New Sciences* given in our dialogue (pp. 247–250), Zenodorus actually found proofs for those we have numbered II and IV, the first of these being of particular interest on account of its simplicity:

> The proof that a circle is greater than any regular polygon with the same perimeter is deduced immediately [by Zenodorus] from Archimedes's proposition that the area of a circle is equal to the right-angled triangle with perpendicular side equal to the radius and base equal to the perimeter of the circle. . . . The perpendicular from the centre of the circle circumscribing the polygon is easily proved to be less than the radius of the given circle with perimeter equal to that of the polygon; whence the proposition follows.

(Heath, p. 209.) An analogous proof applied to a sphere in relation to any of the five regular solids will immediately suggest itself, and must surely have occurred to Zenodorus. In any case, we find it given (five centuries later) by Pappus in his *Collection,* V, xix, Prop. 18 (*La Collection mathématique,* translated by P. Ver Eecke, vol. 1, pp. 276–277). The relation of these propositions to Theorems α and β of our dialogue should be noticed.

225. Jacob Steiner, "Über Maximum und Minimum bei den Figuren in der Ebene, auf der Kugelfläche und im Raume überhaupt," 1841 (*Gesammelte Werke,* edited by K. Weierstrauss, Berlin: Reimer, 1882, vol. 2, pp. 177–308 and plates 9–14). H. A. Schwarz, "Beweis des Satzes, dass die Kugel kleinere Oberfläche besitzt, als jeder andere Körper gleichen Volumens," 1884 (*Gesmmelte mathematische Abhandlungen von H. A. Schwarz,* vol. 2, pp. 327–340; reprinted, Bronx, N.Y.: Chelsea, 1972). Steiner's elegant methods are featured in the following popular discussions: George Polya, *Mathematics and Plausible Reasoning,* vol. 1: *Induction and Analogy in Mathematics,* Princeton, N.J.: Princeton Univ. Press, 1954, chap. 10; Rademacher and Toeplitz, *Enjoyment of Mathematics,* chaps. 21, 22; Nicholas D. Kazarinoff, *Geometric Inequalities,* New York: Random House, New Mathematical Library 4, 1961, pp. 58–63,

68–70; Heinrich Dörrie, *100 Great Problems of Elementary Mathematics,* translated from the German edition of 1948 by David Antin, New York: Dover, 1965, especially Problems 99 and 100; H. G. Eggleston, "The Isoperimetric Problem," in N. J. Hardiman (ed.), *Exploring University Mathematics I,* Oxford: Pergamon Press, 1967, chap. 7.

The brilliant Steiner had a severely disadvantaged childhood and did not learn to write until he was fourteen, yet he came to be considered as the greatest geometer since Apollonius. He established in five different ways that for any oval, with the exception of a circle, there always exists another oval having a greater area for the same perimeter. The following outline of one of his proofs is included for the way in which it so elegantly complements the introductory discussion of the dialogue as well as in acknowledgement of the priority due to its author. It is enough to consider *convex* ovals, since (as has been indicated in our Fig. 109) any outward concavity may be removed so as to obtain a convex oval of greater area without any increase in the perimeter. If the convex oval is not a circle, then, observes Steiner, four *non*concyclic points must lie on it. Denoting these points by A, B, C, D, and joining them in order, a five-part subdivision of the enclosed region is obtained: viz., the inner quadrilateral $ABCD$ and four segments exterior to it. Let

$$A_o = Q + S_1 + S_2 + S_3 + S_4$$

denote the area of the oval as the sum of the areas of these component parts. The inner noncyclic quadrilateral may now be replaced by a cyclic quadrilateral having the same side lengths and with the same coplanar segments exterior to it. That this transformation increases the area has been shown in note 222. We may now write

$$A_i = Q_{\max} + S_1 + S_2 + S_3 + S_4,$$

where A_i (the improved area) $> A_o$ (the original area). Clearly, it is only the circular form that is not subject to such improvement.

226. The use of this example to indicate the incompleteness of demonstrations of the isoperimetric theorem devised by Steiner is due to O. Perron, according to Eggleston (p. 100 of the chapter referred to in the preceding note).

227. Galileo's paradox is rather different as will be seen from the following extract (*Two New Sciences,* First Day 82–83):

> In our discussion a little while ago, we concluded that in the infinite number, there must be as many squares or cubes as all the numbers, because both [squares and cubes] are as numerous as their roots, and all numbers are roots. Next we saw that the larger the numbers taken, the scarcer became the squares to be found among them, and still rarer, the cubes. Hence it is manifest that to the extent that we go to greater numbers, by that much and more do we depart from the infinite number. From this it follows that turning back (since our direction took us always farther from our desired goal), if any number may be called infinite, it is unity. And truly,

> in unity are those conditions and necessary requisites of the infinite number. I refer to those [conditions] of containing in itself as many squares as cubes, and as many as all the numbers [contained].

(Drake translation, p. 45.)

228. Proofs help us refine our imperfect intuitions, and intuition helps us refine our imperfect proofs. Mathematics progresses via ongoing interactions between the proof-sketches offered by mathematicians and the intuitions stirred to consciousness in the course of working out, or working through, such partial demonstrations. Complete formal demonstrations would be impossibly long. In practice, a proof is regarded as good if, without unnecessarily multiplying the necessary connections between the proposition proposed for proof and other propositions (whose truth is not in question), it renders a sufficient and appropriately ordered selection of those connections so transparent that the proposed proposition is able to be placed with those others as equally indubitable. But shortcomings in proofs are sometimes overlooked by authors and referees alike. And published theorems by competent mathematicians are occasionally disallowed, and not infrequently doubted, by other competent mathematicians. For example:

> In 1879, Kempe published a proof of the four-color conjecture that stood for eleven years before Heawood uncovered a fatal flaw in the reasoning. The first collaboration between Hardy and Littlewood resulted in a paper they delivered at the June 1911 meeting of the London Mathematical Society; the paper was never published because they subsequently discovered that their proof was wrong. Cauchy, Lamé, and Kummer all thought at one time or another that they had proved Fermat's Last Theorem. In 1945, Rademacher thought he had solved the Riemann Hypothesis; his results not only circulated in the mathematical world but were announced in *Time* magazine.

(P. 272 of R. A. De Millo, R. J. Lipton, and A. J. Perlis, "Social Processes of Theorems and Programs," *Communications of the Association for Computing Machinery 22* (1979):271–280; see p. 280 for documentation not included here.)

229. A. S. Besicovitch, "Another Proof of the Isoperimetric Theorem," *Mathematical Gazette 47* (1963):43.

230. "Ordinary" (or "simple") points are nonisolated points of a curve, at which the curve does not cross itself and at which the curve has a smoothly turning tangent. If, for example, the Reuleaux triangle of Fig. 38 were under consideration, the initial part of the argument that follows would not apply for a chord such as B_1B_3, but it would hold for a chord $B_1'B_3$, where B_1' is any neighbouring point taken as close as desired to B_1; and so the overall conclusion follows.

XVI

The best effect of any book is that it excites the reader to self activity.

—Thomas Carlyle

He [Thomas Hobbes] was forty years old before he looked on geometry; which happened accidentally. Being in a gentleman's library, Euclid's Elements *lay open, and 'twas the 47* El. libri *I [Pythagoras' Theorem]. He read the proposition. . . . 'By God,' sayd he, 'this is impossible.' So he reads the demonstration of it, which referred him back to such a proposition; which proposition he read. That referred him back to another, which he also read.* Et sic deinceps, *that at last he was demonstratively convinced of that truth. This made him in love with geometry.*

—John Aubrey

At the age of eleven, I began Euclid, with my brother as my tutor. This was one of the great events of my life, as dazzling as first love. I had not imagined that there was anything so delicious in the world.

—Bertrand Russell

The Greeks . . . , as Littlewood said to me once, are not clever schoolboys or 'scholarship candidates,' but 'Fellows of another college.' So Greek mathematics is 'permanent,' more permanent even than Greek literature. Archimedes will be remembered when Aeschylus is forgotten, because languages die and mathematical ideas do not.

—G. H. Hardy

Whenever, and to the extent that, ideas pre-requisite for understanding have not been made available to the learner, then whatever is communicated can only be in the form of assertions; and these will not provide nourishment for a growing intelligence. (The food metaphor is a close one. Genuine nourishment becomes part of the bodily self of the person who eats it: indigestible material is internalized, but not assimilated, and efforts to retain it indefinitely are contrary to our natural functions.)

—R. R. Skemp

Without the concepts, methods and results found and developed by previous generations right down to Greek antiquity one cannot understand either the aims or the achievements of mathematics in the last fifty years.

—Hermann Weyl

PROBLEMATA

The problems of this section are intended to complement those investigated in the dialogue by providing the reader with opportunities for discovering important geometrical facts or rediscovering forgotten ones in contexts designed to facilitate an appreciation of their interconnectedness. Though some of the problems are new, many belong in some form to the common stock accumulated by textbook writers over the centuries. Indeed, perhaps half of the examples would have caused no great surprise to a student of mathematics at Alexandria two thousand years ago. But in many modern mathematics curricula much of this kind of work has been excluded to make way for other topics. What was for so long regarded as of fundamental importance in the training of mathematicians and others has recently been discarded in some places. Consequently it is possible that some students at present majoring in mathematics will find many of the problems unfamiliar and even quite demanding. It may be that the interested amateur will be as well prepared to deal with this material as many of these specialist students.

The aim has been to try to provide the reader with a carefully graded sequence of fundamental, significant, and beautiful problems, rather than to exhibit novelty for novelty's sake. Merely curious problems and those having only their difficulty to recommend them have been omitted. The purpose of including geometrical constructions is primarily *theoretical,* as it so clearly was in the case of the traditional interest in such problems. The Greek mathematicians were not concerned to train draughtsmen! The devising of a valid construction was correctly seen as an excellent exercise leading to—and, with its formal demonstration, constituting part of—the theoretical structure of the discipline. For the most elegant solutions to the present set of problems, the student should refrain from the use of trigonometry. Where arithmetic or algebraic manipulations are required, they are of a very simple kind. Exact answers to numerical exercises should be given; thus simplest surd forms, π, etc., are to be retained in preference to decimal approximations.

Serious students will not, of course, prematurely consult the solutions, hints, and discussions given in the section following the problems. Where it is necessary to refer to these in order to be able to proceed, the problem should be marked down for a new attempt at a later date.

1. Find the area of the shaded region in each of the diagrams of Fig. P1. The arcs are all circles, semicircles, or quadrants of circles. In the first five cases (to be assumed symmetrical), answers are to be given in terms of the area of the square (of side a, say) upon which the figure is based. The asterisks indicate midpoints of sides. In the last diagram, the two crescents together are to be proved equal in area to the right-angled triangle on the three sides of which the semicircles are described.

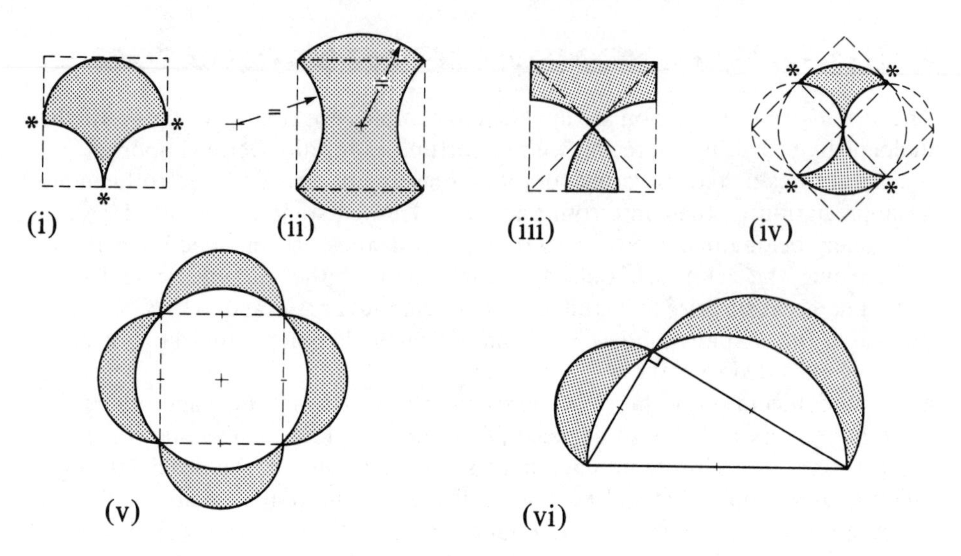

Fig. P1

2. (a) Parallelograms $ABED$, $BCFG$ are drawn exterior to a triangle ABC. DE, FG produced meet in H, and HB produced meets AC in K. AL, CM drawn parallel to KH meet DE, FG in L, M, respectively. Prove that $ALMC$ is a parallelogram equal in area to the sum of the parallelograms $ABED$, $BCFG$.

   (b) Prove that in any right-angled triangle, the perpendicular to the hypotenuse from the vertex of the right angle divides the hypotenuse into two segments that are to one another as the squares on the sides about the right angle.

   (c) Two equal squares $ABCD$ and $PQRS$ are so placed that A, P, B, Q, C, R, D, S, A can be joined to form a regular octagon. Find the ratio of the area of this octagon to the area of the region common to the two squares.

3. (a) Prove that the region shaded in the left-hand diagram of Fig. P2 (bounded by four semicircular arcs) is equal in area to the circle of diameter EF for any r less than R.

   (b) In the left half of the right-hand diagram of Fig. P2, prove that the semicircular arc of radius r divides the quadrant into two equal areas. Prove that in the right half of the diagram the shaded areas are equal.

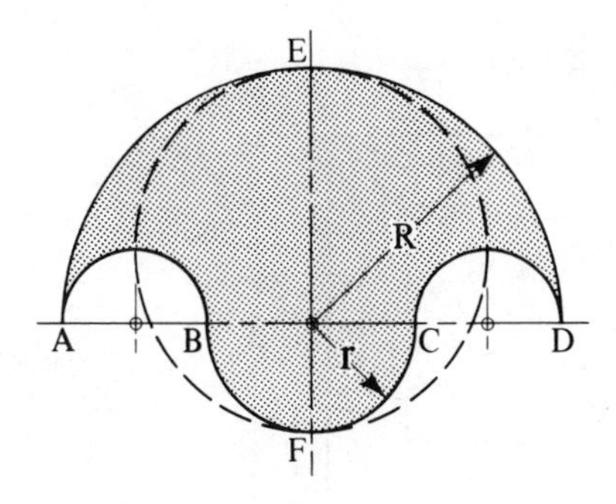

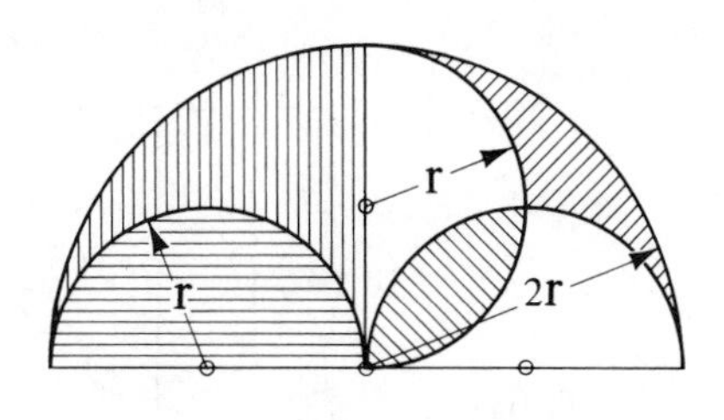

Fig. P2

4. (a) Describe two circles of radii 2 cm, 3 cm, with their centers 4 cm apart. By further construction, obtain a circle of radius 5 cm making external contact with each of the first two circles. Prove that (in spite of appearances) the triangle having its vertices at the centers of the three circles is *not* right-angled.

(b) Starting with the same two circles as in part (a), obtain a circle of radius 5 cm that touches these two circles and encloses both of them. Prove that the triangle having the center of this circle as a vertex and the common chord of the two smaller circles as a side *is* right-angled.

(c) Three spheres, centers A, B, C, rest on a horizontal plane and touch each other. Find their radii given the distances between their centers: $AB = 6$, $AC = 8$, $BC = 9$ units. If the points of contact of the spheres with the plane are A', B', C', show that $A'B' \cdot B'C' \cdot C'A'$ is equal to the product of the diameters of the three spheres (i) in this particular case, (ii) in general for any three spheres in contact with each other and with a plane.

5. (a) P, Q are points in the equal sides, AB, AC, respectively, of a triangle ABC such that $AP = PQ = QB = BC$. Prove that the angles of triangle ABC are as $1:3:3$.

(b) A straight line through vertex A of parallelogram $ABCD$ cuts CD at X and BC produced at Y; prove that triangles BXC and DXY are of equal area.

(c) P is a variable point on a given circle, center C; chord PQ is perpendicular to a fixed diameter BD. Where does the bisector of angle CPQ cut the circle as P moves around it?

(d) Explain how Fig. P3, may be used to show that

$$\frac{1}{2}(a^2 + b^2) > \left\{\frac{1}{2}(a + b)\right\}^2 > ab,$$

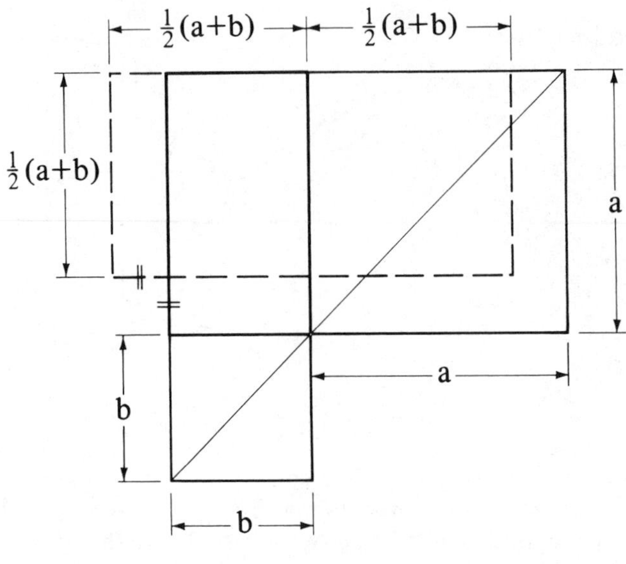

Fig. P3

provided $b \neq a$. (Hence: for two different positive magnitudes, the r.m.s. value exceeds the arithmetic mean, and this in turn exceeds the geometric mean.)

6. Plato, after having referred to geometric means (intermediate terms of a sequence of terms in continued proportion—*Timaeus,* 31c–32a), speaks of the two other kinds of mean that had been distinguished by the early Pythagoreans: "the one [harmonic] exceeding the one extreme and being exceeded by the other by the same fraction of the extremes, the other [arithmetic] exceeding the one extreme by the same number whereby it was exceeded by the other" (ibid., 36a). Express Plato's definition of the harmonic mean symbolically, obtaining convenient expressions for the harmonic mean, h say, of two magnitudes a and b. Show (i) that if three quantities are in harmonic progression then their reciprocals are in arithmetic progression; (ii) that the construction indicated in the left-hand diagram of Fig. P4 provides a valid method for obtaining the harmonic mean, h, between two given magnitudes represented by a, b; (iii) that the geometric mean of two quantities is also the geometric mean of their arithmetic mean and their harmonic mean; (iv) that in the right-hand diagram (in which APB is a right triangle inscribed in the semicircle, center C) p, q, r are, respectively, the geometric, harmonic, and arithmetic means of a and b.

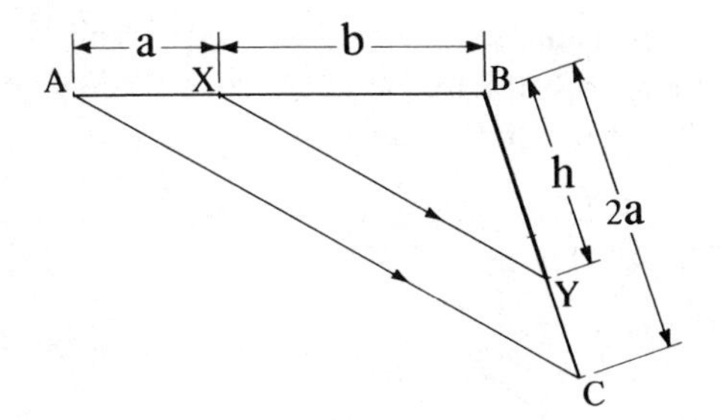

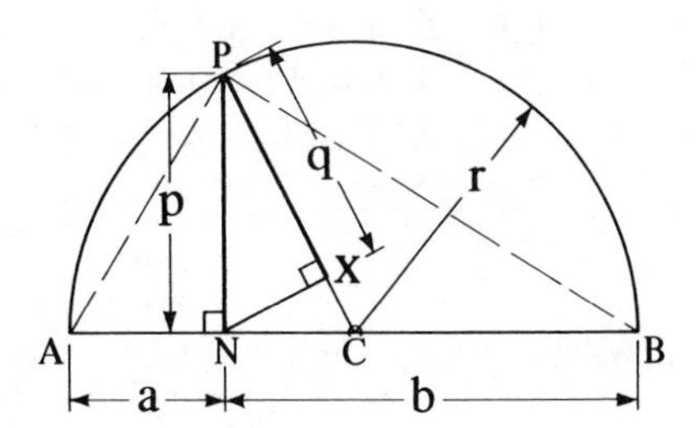

Fig. P4

7. (a) Two spheres, radii R, r, touch each other at point X and they touch a common plane at points P, Q. Prove (i) that PQ is twice the geometric mean of the radii and (ii) that the distance of X from the line PQ is equal to the harmonic mean of the radii.

(b) Altitudes PW, QX of a triangle PQR intersect at Z; prove that (i) $PZ \cdot ZW = QZ \cdot ZX$, and (ii) $PZ \cdot PW = PX \cdot PR$.

(c) P, Q, R, S are the respective midpoints of sides AB, BC, CD, DA of a square $ABCD$; AQ, BR, CS, DP divide the square into nine regions. Express the area of the largest of these as a fraction of the area of $ABCD$.

(d) P is a point in the plane of a square $ABCD$ such that $\angle BPC = \frac{1}{2}\angle CPD = \angle DPA$. Devise a construction for locating P. (P must, of course, lie on the perpendicular through the midpoint, M say, of CD. If CN meets PB perpendicularly at N, prove that $CN = CM$ and hence identify the shape of ΔBCN.) Can you find more than one position for P?

(e) Within a triangle with sides of length 2, 3, 4 units an equilateral hexagon is inscribed by cutting off the corners of the triangle with lines parallel to the opposite sides. Show that the hexagon has a perimeter 8/13 that of the triangle and an area equal to 108/169 of the area of the triangle.

8. (a) Prove that the area between two concentric circles is equal to the area of a third circle having its diameter equal to a "chord-tangent" of the annulus (that is, equal to a chord of the outer circle drawn to touch the inner circle).

(b) A cube $ABCD,EFGH$ is divided into two parts by a plane passing through vertices B, D, G. Prove that the volumes of the two portions are in the ratio 5 : 1. (*Note:* The lettering $ABCD$, $EFGH$ is to be taken as indicating that AE, BF, CG, DH are parallel edges.)

(c) $BFCGA$ is a semicircle in which F and G are determined by the perpendicular bisectors of the sides of the inscribed triangle in the manner indicated in Fig. P5. Prove that $AE - EG = BD - DF$.

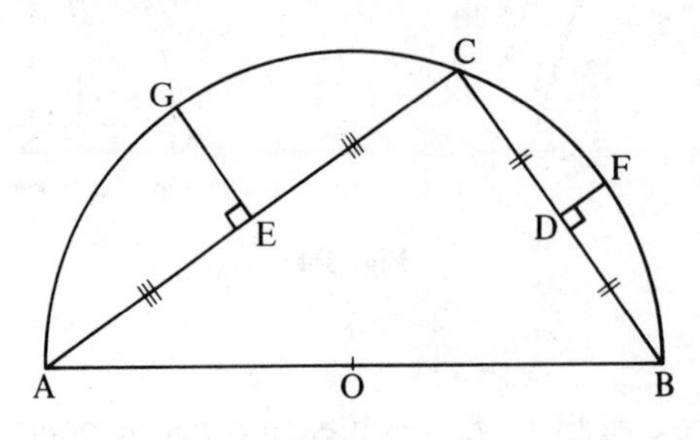

Fig. P5

(d) OX, OY are two perpendicular lines; points A, B move along OX and OY, respectively, in such a way that the distance from A to B remains constant. What is the locus of the point of intersection of the parallels to OX and OY through B and A, respectively?

(e) Construct a triangle with sides $\sqrt{2}$, $\sqrt{3}$, $\sqrt{5}$ units. What is the area of its circumcircle?

9. (a) A quadrilateral with sides of length 2, 3, 7, and 4 units has its longest side parallel to its shortest side. Calculate the area and show that this is only 27/40 of the area of a square of equal perimeter.

(b) $ABCD$ is a quadrilateral having DC parallel to AB; X is the point of intersection of the diagonals. Prove that if $CD = 3AB$, the areas of triangles ABX, BCX, CDX, DAX are as $1:3:9:3$.

(c) In diagram (c) of Fig. P6, $AB//EF//CD$; prove that if $ABFE$ and $EFDC$ are of equal area, then $EF = \sqrt{\{\frac{1}{2}(AB^2 + CD^2)\}}$.

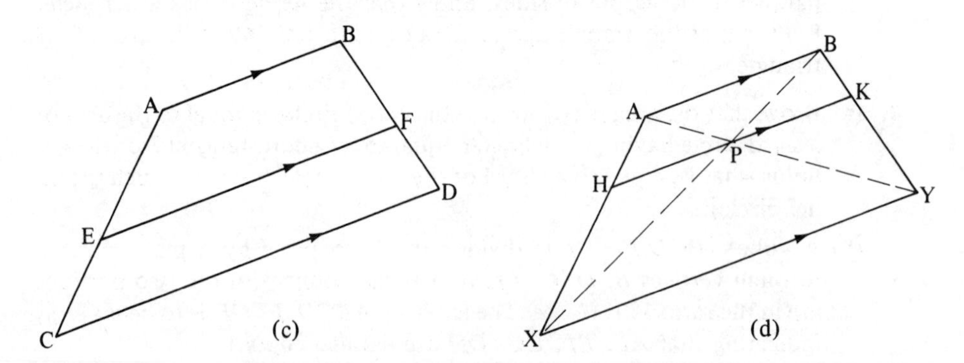

Fig. P6

(d) In diagram (d) of Fig. P6, $AB // HK // XY$, and HK passes through the point of intersection of the diagonals of $ABYX$. Prove that HK is the harmonic mean between AB and XY.

Note: The quadrilateral in each of the four parts of Problem 9 has one pair of parallel sides and one pair of nonparallel sides. In Britain, such quadrilaterals are termed *trapezia,* quadrilaterals with no side parallel to another being called *trapezoids.* The meanings attached to these terms were confusingly interchanged early in the nineteenth century, and the reversed usage has remained customary in the United States. There appears to have been a lack of uniformity in the Greek usage, even on the part of Euclid himself; see Thomas Heath, *The Thirteen Books of Euclid's Elements,* vol. 1, pp. 188–190.

10. (a) M is the midpoint of side PQ of square $PQRS$; PR intersects SM at X. Prove that quadrilateral $MQRX$ has $2\frac{1}{2}$ times the area of triangle PSX.

(b) Points X, Y, Z divide sides AB, BC, CA of a triangle in the ratios $AX:XB = 1:2$, $BY:YC = 3:4$, $CZ:ZA = 5:6$. Prove that the areas of triangles XYZ, ABC are as $3:11$.

(c) Two spheres with radii as $1:2$ touch externally and are circumscribed symmetrically by a right cone. Prove that the volume of the cone is exactly twice that of the larger sphere.

(d) Four spheres, each of radius r, are arranged with their centers A, B, C, D, at the vertices of a square of side $2r$. Find the distance from the plane of $ABCD$ of the center E of a fifth equal sphere that is in contact with each of the first four. Hence, find how many such spheres can be placed within a cubical container of internal edge length $12r$, given that they are packed with alternate layers of 36, 25, 36, 25 . . . spheres.

11. (a) E is the midpoint of side AB of square $ABCD$; CN meets DE perpendicularly at N. Express the area of triangle CDN as a fraction of the area of $ABCD$.

(b) Find a construction for obtaining points B', C' in sides AB, AC, respectively, of a triangle ABC such that $B'C' // BC$ and $BB' + CC' = B'C'$.

(c) T is a point inside a square $PQRS$; $TP = 2$ units, $TQ = TS = 3$ units. Find TR, (i) by construction, (ii) by calculation. Repeat for the case where T is a (coplanar) point exterior to the square, the data being otherwise unchanged.

(d) XYZ is an isosceles triangle right-angled at Z; the bisectors of angles X, Y, Z meet the opposite sides at P, Q, R, respectively. Prove that $\triangle PQR : \triangle XYZ = (3\sqrt{2} - 4) : 1$.

(e) A, B, C, D are four collinear points, and ADE is a right angle. Given that $AB = BC = CD = DE$, prove that triangles ACE and ECB are similar and *hence* show that $\angle EAD + \angle EBD = \angle ECD$.

12. (a) ABC is any triangle, M is the midpoint of BC and T is the point of trisection of AC nearer to C. BT intersects AM at X; prove that $AX:XM = 4:1$.

(b) Figure P7 shows the construction of a regular three-pointed star within an equilateral triangle using the joins from the points of trisection of the sides to the vertices of the triangle. Show that the star covers exactly two-fifths of the area of the triangle.

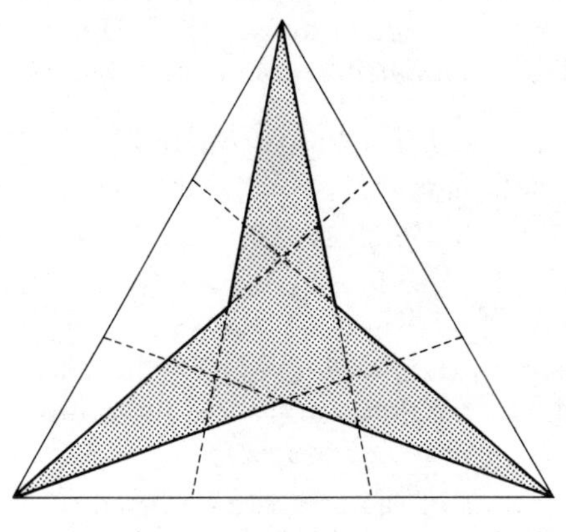

Fig. P7

(c) Find the ratios $r:R$, $r':R$ for the symmetrical arrangements of circles and semicircles shown in Fig. P8.

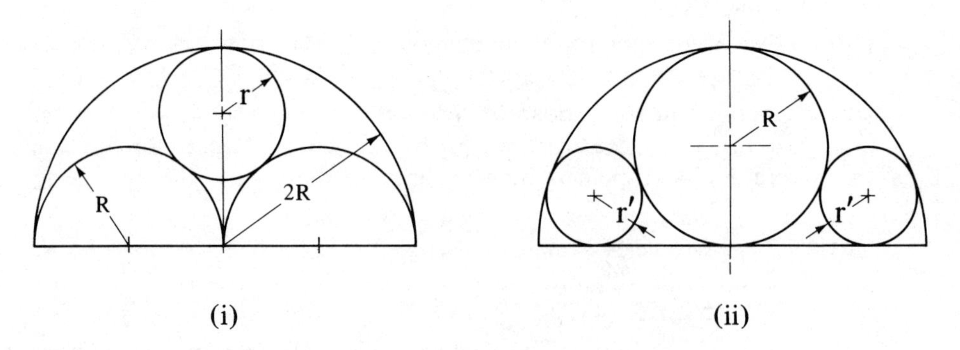

Fig. P8

13. (a) Six equal circles are inscribed within a given circle so that they touch each other externally as indicated in Fig. P9(a). Find the ratio of radius of the given circle to that of one of the six equal cirles and hence determine the size of an inner circle that will make external contact with each of the six equal circles.

(b) Determine the altitude of a regular tetrahedron of edge length e units and show that the volume of the tetrahedron is between one-eighth and one-ninth of the volume of a cube of equal edge length.

(c) Six spheres of unit radius are held in contact with a seventh equal sphere on a horizontal plane. Three more unit spheres rest in stable equilibrium upon these seven and an eleventh equal sphere rests centrally upon the three in the second layer. (See Fig. P9(c).) Calculate the overall height of the stack.

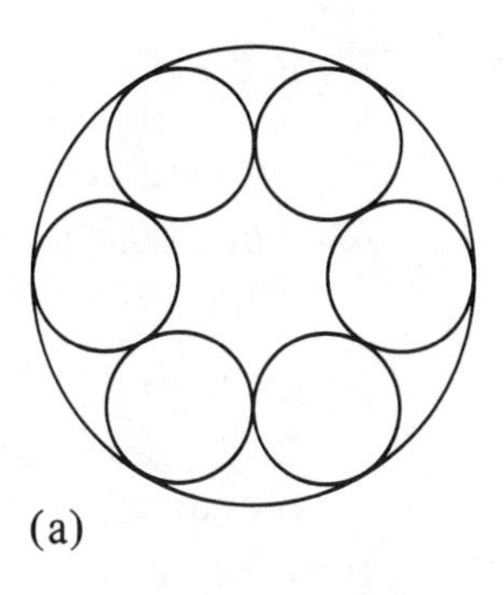

(a)

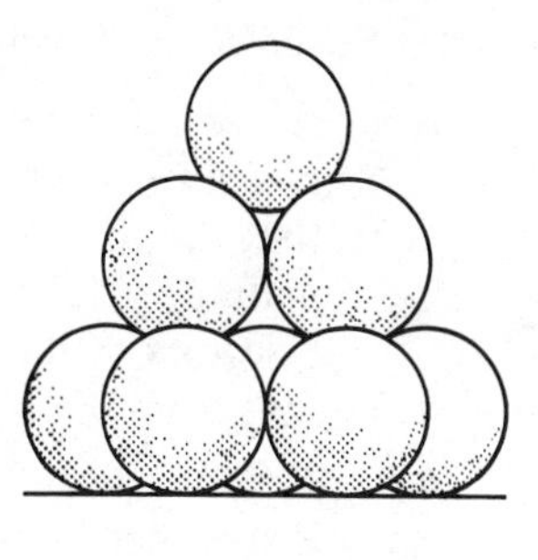

(c)

Fig. P9

14. (a) Prove that the angle that an arc of a circle subtends at the center is double that which it subtends at any point on the remaining part of the circumference.

(b) P is any point on the arc of a quadrant of a circle bounded by radii OA, OB. What is the sum of angles OAP, OBP?

(c) Prove that in any triangle circles described on two sides as diameters intersect on the third side, produced if necessary.

(d) Construct a triangle ABC in which $AB = 5$ cm, $BC = 7$ cm, $CA = 6$ cm, and by further construction obtain a point P at which sides AB and BC subtend angles of 90° and 135°, respectively.

(e) HK, KL are two coterminus chords of a circle of diameter 3 units. If $HK = KL = 2$ units, calculate the length of HL.

15. (a) Prove that (i) the opposite angles of a cyclic quadrilateral are supplementary, (ii) a quadrilateral with its opposite angles supplementary is cyclic, (iii) the angles made by a tangent to a circle with a chord terminating at the point of contact are respectively equal to the angles in the alternate segments of the circle.

(b) Prove that the tangents at the extremities of two perpendicular chords of a circle intersect in four concyclic points, provided that the chords do not meet at the circumference.

(c) AB, CD are two perpendicular chords of a circle that intersect inside the circle; prove that

$$\text{arc } AC + \text{arc } BD = \text{arc } CB + \text{arc } DA = \text{half-circumference.}$$

Investigate also the case where the perpendicular chords intersect only if produced.

(d) If two chords AB, CD of a circle intersect at right angles at X, prove that $AX^2 + BX^2 + CX^2 + DX^2 = (\text{diameter})^2$.

16. (a) Prove that the point of intersection of the perpendicular bisectors of two sides of a triangle is equidistant from the vertices, and prove further that the join from this point of intersection to the midpoint of the remaining side meets that side at right angles. (Hence: The perpendicular bisectors of the three sides of a triangle meet in a point that is equidistant from the vertices. This point of concurrency, usually denoted by O, is called the *circumcenter*.)

(b) Prove that the area of any triangle is equal to the product of the three sides divided by four times the radius of the circumcircle. Show how this result works out for right-angled triangles.

(c) The altitudes AD, BE of a triangle ABC intersect at H; DE, CH are joined. Prove that $\angle ECH = \angle EDH = \angle EBA$, and that CH (produced if necessary) meets AB at right angles. (Hence: The three altitudes of a triangle meet at a point. This point of concurrency, H, is called the *orthocenter*.)

(d) Through the vertices of a triangle ABC, lines are drawn parallel to the opposite sides so as to form a triangle XYZ. Show that the circumcenter of one of the triangles is the orthocenter of the other, and hence that in any triangle the distance of each vertex from the orthocenter is twice the distance from the circumcenter to the side opposite that vertex.

17. (a) Prove (i)–(iii). (i) The bisectors of the (interior) angles of a triangle are concurrent and this point of concurrency (the "incenter," I) is equidistant from the sides of the triangle. (ii) If U, V, W are the respective points of contact of the incircle with the sides BC, CA, AB of the triangle ABC, then $AV = s - a$, $BW = s - b$, $CU = s - c$. (In the standard notation, $BC = a$, $CA = b$, $AB = c$, and $s = \frac{1}{2}(a + b + c)$.) (iii) If the escribed circle remote from A touches BC at X, AC produced at Y, and AB produced at Z, then $AY = AZ = s$, $CX = CY = s - b$, $BX = BZ = s - c$. (Each of the three escribed circles of a triangle—sometimes called the *excircles*—touch one side and the other two sides produced.)

(b) (i) Prove that the orthocenter of an acute-angled triangle ABC with altitudes AD, BE, CF is the incenter of triangle DEF. Investigate also the obtuse-angled case. (ii) Prove that the incenter of a triangle ABC coincides with the orthocenter of the triangle having its vertices at the excenters—that is, at the centers, E_a, E_b, E_c, of the escribed circles.

(c) Medians AX, BY of a triangle ABC intersect at point G. Prove that if CG is produced to meet AB at Z, then Z is the midpoint of AB. (In your proof, produce CG to P so that $GP = CG$ and show that $AGBP$ is a parallelogram.) Prove also that $AG = 2GX$, $BG = 2GY$, $CG = 2GZ$. (Hence: The three medians of a triangle meet in a point that is a point of trisection of each median. This point of concurrency is called the centroid G of the triangle.)

(d) Prove that in any triangle other than an equilateral one the circumcenter O, the centroid G, and the orthocenter H, are collinear and that $GH = 2 \cdot OG$. (Join OG and produce to P such that $GP = 2 \cdot OG$ and prove, using similar triangles, that P is in fact the orthocenter.)

18. (a) For a triangle with sides of lengths 8, 4, 8 units, calculate the distances from the shortest side to (i) the orthocenter H, (ii) the incenter I, (iii) the centroid G, (iv) the circumcenter O, and hence show that these distances are as $1:3:5:7$. Calculate the corresponding distances from the longest side of a triangle with sides 5, 8, 5 units, and show that in this case $OH:IG = 39:2$.

(b) AD, BE, CF are the altitudes of a triangle of an acute-angled triangle ABC with orthocenter H. AD is produced to H' such that $DH' = HD$; BE is produced to H'' and CF to H''' where $EH'' = HE$ and $FH''' = HF$. (H', H'', H''' are thus the "images" or "reflections" of H in BC, CA, AB, respectively.) Prove (i) $\angle CHD = \angle ABC$, (ii) H', H'', H''' lie on the circumcircle of $\triangle ABC$, (iii) the tangent at A to the circumcircle of ABC is parallel to EF. Consider also the case where $\triangle ABC$ is obtuse-angled.

(c) Calculate the area of a triangle having medians of lengths 3, 4, 5 units. Find also the lengths of its sides and determine whether or not it is right angled.

19. (a) P is a variable point in the plane of a parallelogram $ABCD$. What can be said about the sum of the distances from P to the sides of $ABCD$ (produced where necessary)? Find also the minimum value of the sum of the distances from P to the *vertices* of $ABCD$.

(b) A triangle has sides of length 4, 5, 6 units. Calculate (i) the length of the altitude to the 5-unit side, and (ii) the radius of the inscribed circle. Prove that for any triangle with sides in arithmetic progression, the incircle radius is one-third of one of the altitudes.

(c) Two altitudes of a triangle have lengths 12 units and 20 units, respectively; prove that the third altitude must be longer than $7\frac{1}{2}$ units and shorter than 30 units.

(d) Prove that for any point within an equilateral triangle, the sum of the perpendicular distances to the sides is equal to the altitude of the triangle. (Viviani, 1622–1703, appears to have been the first to publish this theorem.) What difference does it make if the point is exterior to the triangle (but still coplanar with it)?

20. (a) Chords AB, CD of a circle intersect at E, inside the circle; prove that $AE \cdot EB = CE \cdot ED$. Prove also the corresponding theorem for the case in which the chords intersect only when produced outside the circle. Examine also the case where D coincides with C to give a tangent CE meeting a "secant" AE.

(b) Describe a semicircle of radius 4 units, and erect ordinates to the arc from the bounding diameter at distances of 1, 2, 3 units from the maximum ordinate. Write down the length of each ordinate.

(c) Show how to obtain graphically—by straightedge-and-compasses construction—line intervals of length $\sqrt{10}$, $\frac{1}{2}(\sqrt{5} + 1)$, $\sqrt[4]{2}$, $\sqrt[8]{7}$ units.

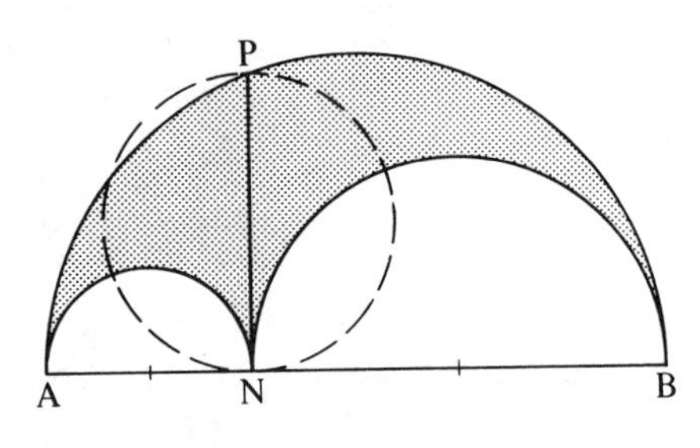

Fig. P10

(d) Prove that the shaded area of Fig. P10 (bounded by three semicircular arcs) is equal to the circle of diameter PN (where N is any point in AB and $PN \perp AB$).

21. (a) Explain how the constructions indicated in the first two diagrams of Fig. P11 provide graphical determinations of the width, w units, of a rectangle of given area, here 9 square units, when the length of the rectangle exceeds the width by an arbitrary amount, 4 units in this example.

(b) In the symmetrical pentagon $PQRST$ of the third diagram of Fig. P11, P, T, R are to be collinear. Show how to construct the figure, preferably by a straightedge-and-compasses method.

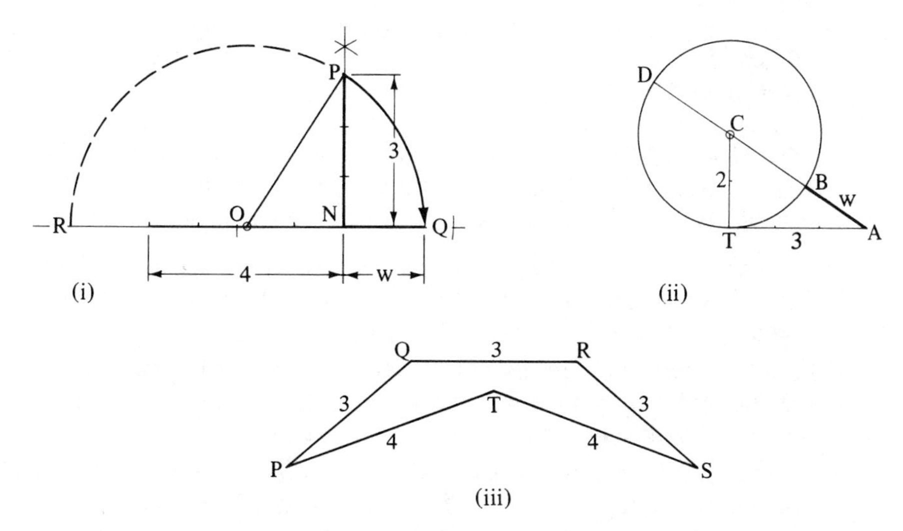

Fig. P11

22. (a) A fixed point P lies outside a given circle with center O. What is the locus of the midpoint M of a variable chord CD of the circle, given that C, D, P, are always collinear?

(b) Construct a rectangle $ABCD$ in which $AB = 3$, $BC = 2$ units, and find points P, Q in side CD at which AB subtends angles of $60°$. Also determine the distance between P and Q by calculation.

(c) A, B, C are collinear points such that $AB = 4$ units, $BC = 5$ units; P is a variable point whose locus is a line through A perpendicular to line ABC. Find, (i) by construction and (ii) by calculation, the position of P for which angle BPC is a maximum.

(d) Construct a parallelogram having its adjacent angles in the ratio $2:1$ and its diagonals in the ratio $2:3$. Also show that if the diagonals have lengths 4, 6 units, the area of the parallelogram will be $5\sqrt{3}$ square units.

23. (a) Prove that the bisectors of the interior angles of a quadrilateral intersect in four concyclic points. (Except where two or more of these points of intersection coincide, there will be *six* distinct points of intersection obtainable by sufficiently prolonging the bisectors, at least if they are nonparallel, but only four of these points will in general be concyclic.) Is the same true for the bisectors of the *exterior* angles? For this general case, are the interior and exterior quadrilaterals formed by the angle bisectors necessarily similar? What are the forms of these interior and exterior quadrilaterals when the original quadrilateral is (i) a rectangle, (ii) a parallelogram?

(b) P, Q are points in sides WX, XY, respectively, of square $WXYZ$, such that $WP = XQ$. WQ and ZP intersect at T. Prove that triangle WTZ and quadrilateral $PXQT$ are equal in area. Show that the angles of quadrilateral $PXQT$ are respectively equal to the angles of quadrilateral $QYZT$ but that these quadrilaterals cannot be similar. For the special case of P, Q the midpoints of WX, XY, prove that the areas of $PXQT$ and $QYZT$ are as $4:11$, while the areas of their circumcircles are as $2:5$.

24. (a) Construct a regular hexagon and inscribe a square within it (two sides of the square being parallel to a pair of sides of the hexagon). Calculate the ratio, side-of-square to side-of-hexagon.

(b) Find a construction for inscribing a semicircle within a given square (the semicircle being symmetrical about a diagonal of the square). Calculate the ratio, radius-of-semicircle to side-of-square.

(c) A square is inscribed in a semicircle (one side of the square lying along the bounding diameter). Find, by construction and by calculation, the ratio of the side of the square to the radius of the circle. Solve the corresponding problem for a cube within a hemisphere.

25. (a) Three circles with the same exterior common tangents touch each other as indicated in Fig. P12. Prove that the radius of the intermediate circle is the mean proportional (geometric mean) between the radii of the other two circles.

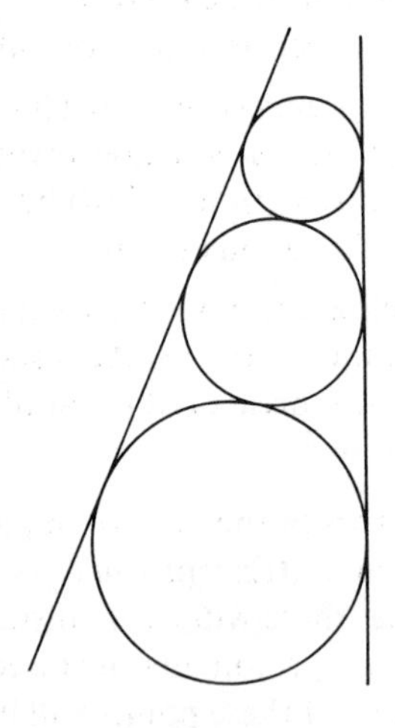

Fig. P12

(b) Two parallel tangents to a given circle are cut at points A and B by a third tangent, which touches the same circle at C. Prove that the radius of the circle is the geometric mean of AC and BC.

(c) P is a point in side AB of a triangle ABC; Q, R are the feet of the perpendiculars from P to sides AC, BC, respectively. Find a construction for determining the position of P in order that QR shall be parallel to AB.

(d) P is a given point lying between the arms of an angle ABC; a variable straight line through P meets BA, BC at X, Y, respectively. Show how to locate the critical positions of X, Y by construction in order that XY will be bisected at P. Prove that it is for this position of XY that $\triangle XBY$ has its least area.

26. (a) Find points P, Q in sides BC, CD, respectively, of a parallelogram $ABCD$ such that $3\triangle AQD = 2\triangle ABP = \text{quad.}APCQ$.

(b) Find how to divide side BC of a triangle ABC at points P and Q in order that the triangle will be divided into three equal areas by lines through P and Q parallel to the median through A.

(c) Devise a straightedge-and-compasses construction for dividing a given circular region into three parts of equal area by means of two concentric circles.

(d) A pyramid of perpendicular height a units is to be divided into two parts by a plane parallel to the base. Find the distance of the cutting plane from the apex of the pyramid, (i) if the division is to be made so that the base of the pyramid cut off will have half the area of the base of the original pyramid, (ii) if the original solid is to be divided into two parts of equal volume.

27. (a) P is a point in diagonal BD of rectangle $ABCD$ such that the distances of P from AB, BC, CD are as $1:2:4$. Prove (i) that $PA:PB = \sqrt{13}:1$, and (ii) that triangle PAC: rectangle $ABCD = 3:10$.

(b) (i) Prove that for any triangle APB in which PN is an altitude, $BN^2 - AN^2 = PB^2 - PA^2$.

(ii) $ABCD$ is a square and P is a coplanar point within it such that $PA:PB:PC:PD = 2:3:4:x$. Find x.

(c) Obtain by construction a rectangle twice as long as it is wide and with perimeter equal to that of a given square. What is the ratio of the areas, rectangle to square?

(d) Given two line intervals a, c (Fig. P13), find by construction two other intervals, b and d, such that $a:b = b:c = c:d$. Prove (i) that a rectangle with sides b, c is equal in area to a rectangle with sides a, d; (ii) that a right square-based prism with adjacent edges a, a, d is equal in volume to a cube of edge b.

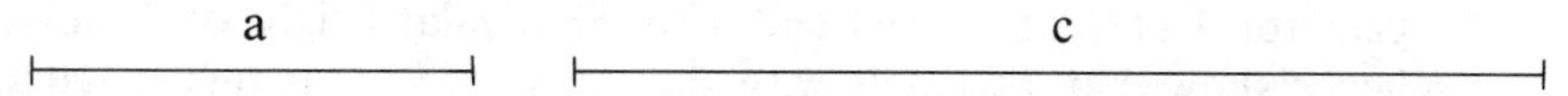

Fig. P13

28. (a) $ABCD,EFGH$ is a cuboid (a rectangular prism) in which AB, AD, AE are of lengths 1, 2, 3 units, respectively. Calculate the area of triangle ACF by two different methods.

(b) Cube $ABCD,EFGH$ is divided into three parts by a pair of parallel planes, one of which passes through vertices A, F, H, and the other through the midpoints P, Q, R, S, T, U of edges AB, BF, FG, GH, HD, DA. What fraction of the cube lies between these two planes? *Prove* that $PQRSTU$ is a regular hexagon.

(c) The diagrams of Fig. P14 show (i) a regular tetrahedron inscribed in a cube with each of its edges coinciding with a face diagonal of the cube, (ii) a regular octahedron with its vertices at the centers of the faces of a cube, (iii) a "cuboctahedron" with its vertices at the midpoints of the edges of a cube. In each case calculate both the ratio of the surface areas and the ratio of the volumes of the inner solid and the enclosing cube.

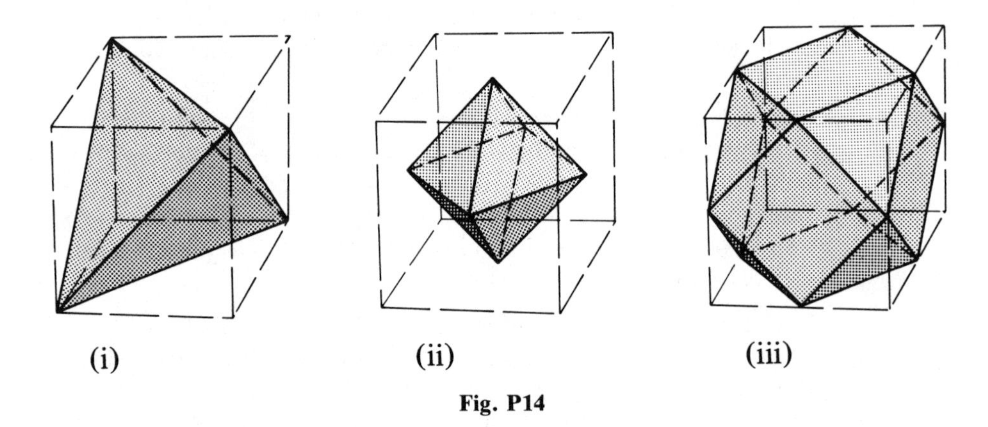

Fig. P14

29. (a) Find the diameter of a circle in which a chord of length 10 units cuts off a segment of height 2 units. Hence show that the right-angled triangle shown in the first diagram of Fig. P15 has its sides as $20:21:29$, and, consequently, that any triangle having its sides as $20:21:29$ is right-angled.

(b) P is a point inside of BC of a rectangle $ABCD$ such that $PC = PA$. Given that $AB = 2$, $AD = 5$ units, find the sides of triangle APB. Repeat for $AB = 4$, $AD = 7$ units, and hence show that a triangle with sides as $33:56:65$ is right-angled.

(c) Find the radius of the circumcircle of an isosceles triangle having its base of length 4 units and the altitude to the base of length 5 units. Repeat for a base of 6 units and a perpendicular height of 8 units, and hence show that a triangle with sides as $48:55:73$ is right-angled.

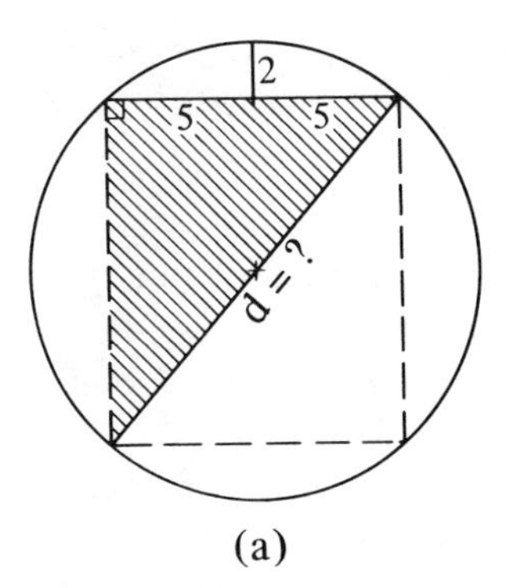

(a)

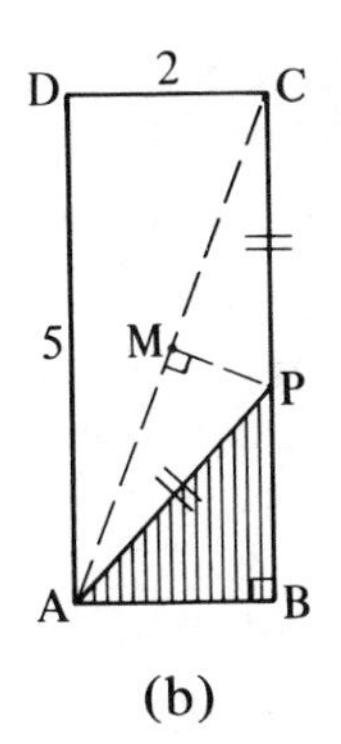

(b)

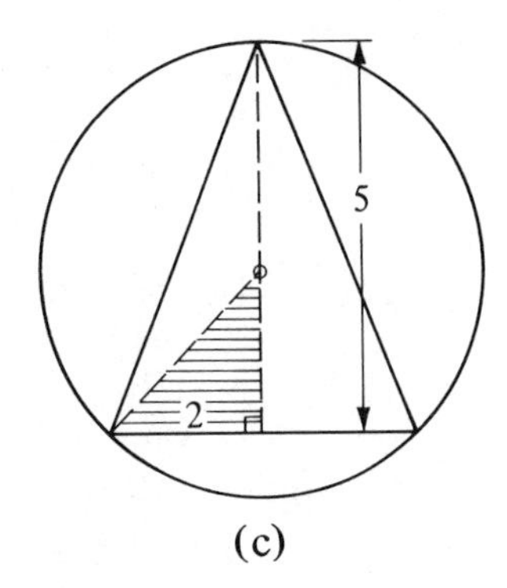

(c)

Fig. P15

(d) Repeat parts (a)–(c) of this problem with m, n respectively substituted in place of the 5 and the 2. (Hence, for m, greater than n, but m, n otherwise arbitrary, a triangle with sides as $2mn$, $m^2 - n^2$, $m^2 + n^2$ is right-angled.)

30. (a) Prove that the bisector of any (interior) angle of a triangle divides the opposite side into two parts that are to one another as the lengths of the sides containing the bisected angle. Extend this theorem to cover the case of any exterior angle bisector. (Let the bisector of angle A of a triangle ABC intersect BC at D, say. One method of proving that $BD:CD = AB:AC$ makes use of the perpendiculars from D meeting the sides AC, AB, respectively, at P and Q, say, and works with the ratio of the areas of triangles ABD, ACD. Another method makes use of lines drawn from B and C to meet the bisector of angle A perpendicularly, at X and Y say.)

(b) Find the locus of a point that moves in a plane so that its distances from two fixed points in the plane are in a constant ratio.

31. (a) $ABCD$ is a quadrilateral (not having any pair of adjacent sides equal); it is lettered so that $AB > DA$. If $AB + CD = BC + DA$, and E, F are points in AB, BC, respectively, such that $AE = AD$ and $CF = CD$, prove that the circumcenter of triangle DEF is equidistant from the sides of quadrilateral $ABCD$.

(b) Sides AD and BC of a cyclic quadrilateral $ABCD$ are produced to meet at E. AB, BC, CE, ED are, respectively, 4, 5, 3, 4 units long. (i) Find CD and DA. Hence (ii) prove that BD bisects angle ABC. (iii) If AC cuts BD at X, prove that the areas of triangles AXD, CXD, CDE, AXB, BXC are as $4:5:18:20:25$. (iv) If BA and CD produced meet at F, find FA, FD and the length of the tangent from F to the circle circumscribing $ABCD$.

(c) AB is the chord common to two coplanar circles and C is any point in AB produced. Prove that the circle centered at C and which cuts the circles through A, B orthogonally has a radius that is the mean proportional between AC and BC.

32. (a) Devise a construction for inscribing a circle within a given quadrant of a circle (or two equal circles within a given semicircle, etc.). Also determine by calculation the ratio of the incircle radius, r say, to the quadrant radius, R.

(b) The circle inscribed in a quadrant of a circle touches the arc of the quadrant at C and the bounding radii OA, OB at D and E, respectively. Prove (i) that the angles DCO, ECO, OAE, OBD are each one-quarter of a right angle, (ii) if CN meets OB perpendicularly at N, then NA intersects OC at the incenter, and (iii) if the square $AOBF$ is completed, then $CF =$ the incircle radius.

(c) Find by calculation the ratio of the radius of a sphere to the radius of the sphere inscribed in one of its octants. (A sphere is divided into eight congruent octants by three mutually perpendicular planes having their common point at the center of the sphere.)

33. (a) Given four coplanar points A, B, X, Y, find a construction for locating a point P in the line XY in order that $AP + PB$ shall be a minimum. (See Fig. P16(a) and make use of the fact that $AP + PB = AP + PB'$, where B' is the image of B in XY—that is, where BB' is perpendicular to XY and is bisected by XY.)

(b) $ABCD$ is a rectangle in which $AB = 5$, $BC = 10$ units; S, F are points in BC, DA such that $SC = FA = 1$ unit (Fig. P16(b)). Find by calculation

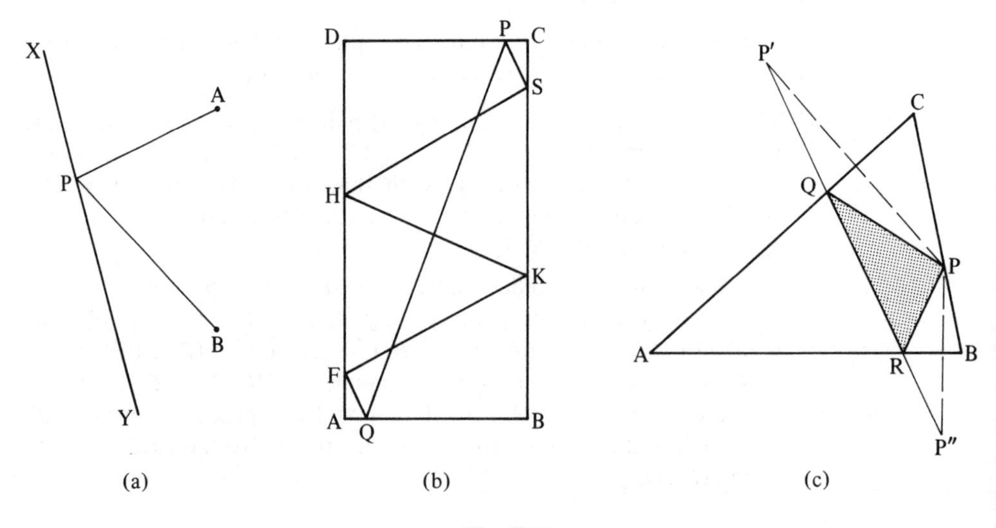

Fig. P16

the length of the minimum path from S to a point P in CD, thence to a point Q in AB and finally from Q to F. Find also the minimum value of $SH + HK + KF$, with H, K in AD, BC.

(c) P is a point in the side BC of an acute-angled triangle ABC; P', P'' are the images of P in CA, AB, respectively (Fig. P16(c)). Prove (i) that for P a fixed point in BC, the inscribed triangle PQR has the least perimeter when Q, R are collinear with P', P''; (ii) that $\angle P'AP'' = 2\angle CAB$, and hence, if P is free to move along BC, then $P'P''$ will be least when AP is least; (iii) that for P, Q, R all variable points, one in each side of an acute-angled triangle ABC, the perimeter of triangle PQR is a minimum when P, Q, R are the feet of the altitudes of triangle ABC.

34. AA' and BB' are direct (or "exterior") common tangents to two circles, centers C, C', of radii 11 cm, 4 cm, respectively; $CC' = 25$ cm. AA' and BB' (produced) meet at V''. FF' is a transverse (or "interior") common tangent intersecting AA', BB' at V', V, as indicated in Fig. P17. (The diagram is not to scale.) Determine: (i) AA' (or BB'); (ii) FF'; (iii) VF' (or VB'), VF (or VB); (iv) $V'F$ (or $V'A$), $V'F'$ (or $V'A'$); (v) $A'V''$ (or $B'V''$); (vi) the perimeter and area of triangle $VV'V''$. (The solution to this problem should be preserved as it will be needed in connection with Problem 42.)

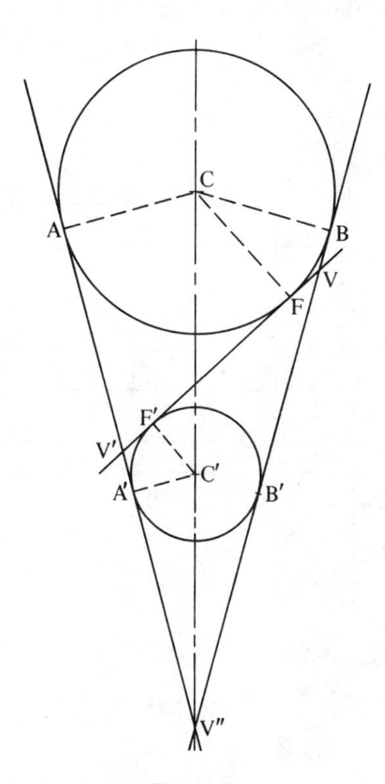

Fig. P17

35. (a) AB and CD are two perpendicular diameters of a circle. A circular arc centered at C and passing through A and B divides the region bounded by the circle into two parts equal in perimeter but with areas in a ratio exceeding $2:1$. Find an exact expression for this ratio.

(b) Find a straightedge-and-compasses construction for dividing a given interval (of length ℓ, say) into two parts such that the sum of their squares shall be equal to a given square (of side a, say, where $\frac{1}{2}\ell^2 < a^2 < \ell^2$).

(c) A circle of radius 1 unit touches a circle of radius 2 units internally. Two circles, each of radius r units, touch the unit circle externally and the two-unit circle internally. Calculate the value of r in order that the two equal circles will just touch each other.

(d) Describe two circles, radii 1 cm and 2 cm, centers 3 cm apart, and obtain a third circle of radius 4 cm that is touched internally by each of the first two circles. Calculate the distances between the three points at which the circles make contact.

36. Figure P18 shows a plan and elevation of an obliquely truncated cylinder with projections of the true shape of the elliptical slant face and of the development of the curved surface. Calculate (i) the area of the elliptical face; (ii) the dimensions "v", "w", "x", "y", "z"; (iii) the area of the curved surface and the volume of the solid.

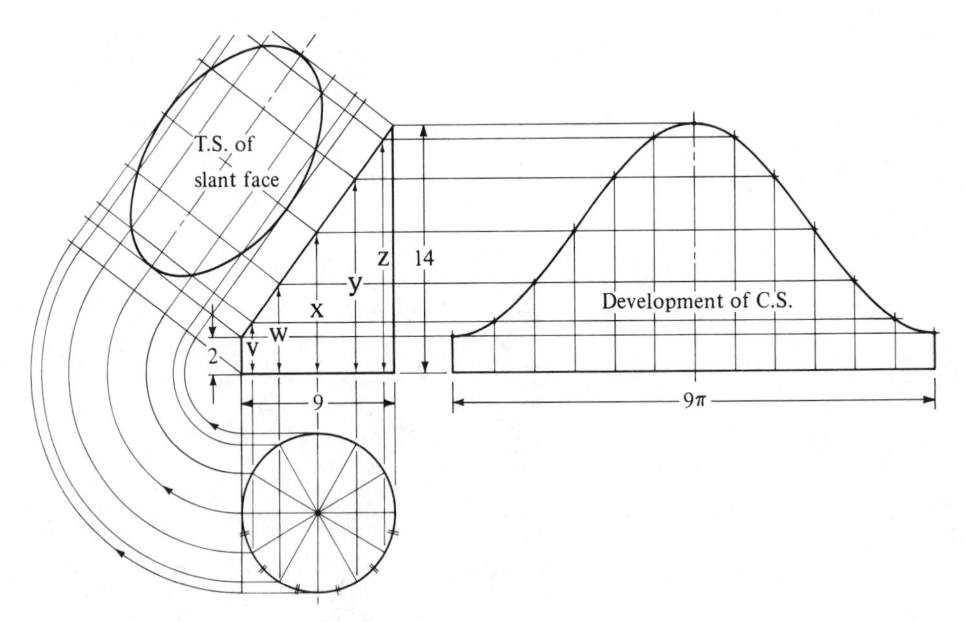

Fig. P18

37. (a) E is a point on the median AD of a triangle ABC such that $AE:ED = 2:3$. BE produced meets AC in F. Find $AF:FC$. (*Hint*: Draw EK parallel to BC to meet AC in K and use similar triangles. Alternatively: Join DF and work in terms of area of component triangles.)

(b) Triangle PQR has PQ equal to PR; a straight line cuts PQ, QR, and PR produced in points X, Y, Z, respectively. If $QX = RZ$, prove (i) that Y is the midpoint of XZ, and (ii) that $QY:YR = PZ:PX$.

(c) H, K are points in sides DE, DF of a triangle DEF such that $HE = \frac{1}{5}DE$, $DK = \frac{1}{5}DF$. Into what ratios does HK cut medians DR, FT? What fraction is quadrilateral $EFKH$ of triangle DEF?

38. The plans and elevations shown in Fig. P19 represent two different decapitations of two identical right square pyramids with bases 20×20 and heights 25 units. In each case, calculate (i) the area of the quadrilateral face $bcde$, (ii) the volume of the *removed* pyramid $Abcde$.

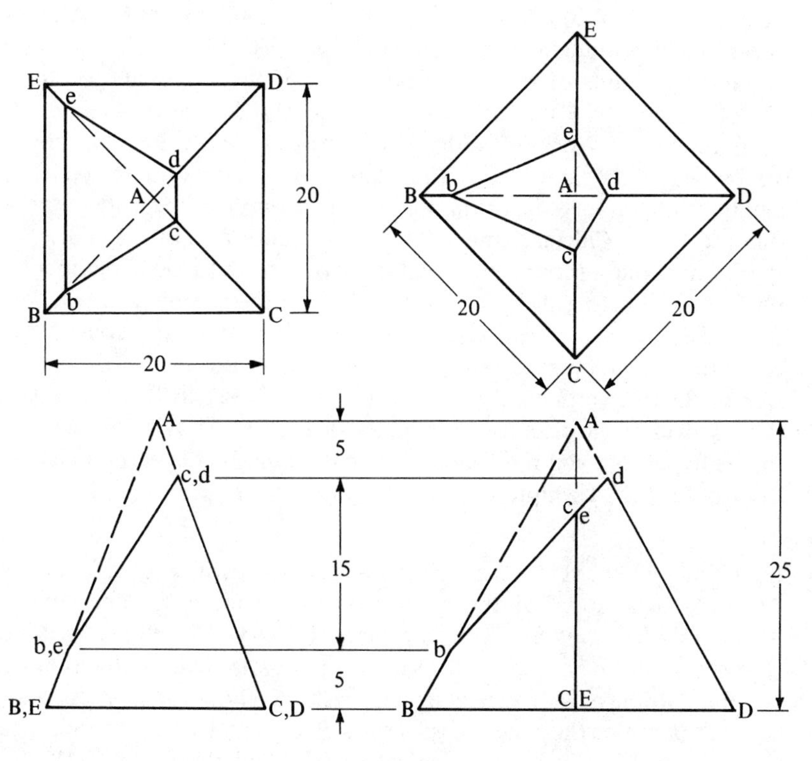

Fig. P19

39. Plot the loci of points P, E, H that move in a plane containing a fixed point F (the "focus") and a fixed line DD' (the "directrix"), such that (i) the distance of P from F is always equal to the distance of P from DD'; (ii) the distance of E from F is always a constant fraction (e.g., two-thirds) of its distance from DD'; (iii) the distance of H from F is always a constant multiple (e.g., three-halves) of its distance from DD'. It will be most convenient to work with compasses on squared paper. Plot all loci on the one sheet and notice that the locus of H has two distinct branches. In order to obtain checks on calculations to follow, take F at a distance of 5 units from DD' and use the suggested ratios $2:3$ and $3:2$. (The curve obtained is a *parabola*, or an *ellipse*, or a *hyperbola*, according as the constant ratio of the distances of the tracing point from the focus and from the directrix is equal to, less than, or greater than unity. These curves are called conic sections, or simply *conics*, as they are obtainable as the intersections of variously inclined planes with conical surfaces. The ratio of distance from focus to distance from directrix is called the "eccentricity", ϵ, of the conic.)

What are the greatest and least distances of E from DD'? Using the symmetry of the locus of E (about the midpoint of the join of the extreme positions of E), find a second focus F' and a second directrix $\Delta\Delta'$, with respect to which the same ellipse might be specified. Let E', E'' be the (orthogonal) projections of E on the major axis of the ellipse and on directrix DD', respectively. Take EE'' equal to ℓ units and dimension EF, FE', $E'F'$, in terms of ℓ. Working with right triangles EFE', $EF'E'$, show that $EF' = \frac{2}{3}(18 - \ell)$, and hence that $EF + EF' =$ a constant, namely, the length of the major axis, or (in this case) two-thirds of the distance between the directrices. Calculate the length of the minor axis of the ellipse.

Carry out a similar investigation with respect to the hyperbola traced by H, first locating the second focus F'' and the second directrix for this curve. Establish that the *difference* between HF'' and HF is constant and is equal to the minimum distance between the two branches of the curve (the join of the two vertices, also called the "transverse axis"), or (in this case) three-halves of the distance between the directrices. Draw in the asymptotes of the hyperbola you have plotted. What would be the eccentricity of the hyperbola if the asymptotes were at right angles?

40. (a) Figure P20(a) shows a parabolic line of intersection of a plane with a cone, axis $X'OX$ of the parabola—and hence any plane containing $X'OX$—being parallel to the generator $VU'V$. (An ellipse or a hyperbola would be obtained instead of a parabola if the inclination of the plane to the axis of the cone were, respectively, greater or less than the inclination of the generators.) P and P' are any two points on the parabolic intersection of the conical surface and the plane XOY. In the parallel circular sections of the cone through P and P', half-chords PA and $P'A'$ are ordinates perpendicular to $X'OX$ and parallel to $Y'OY$.

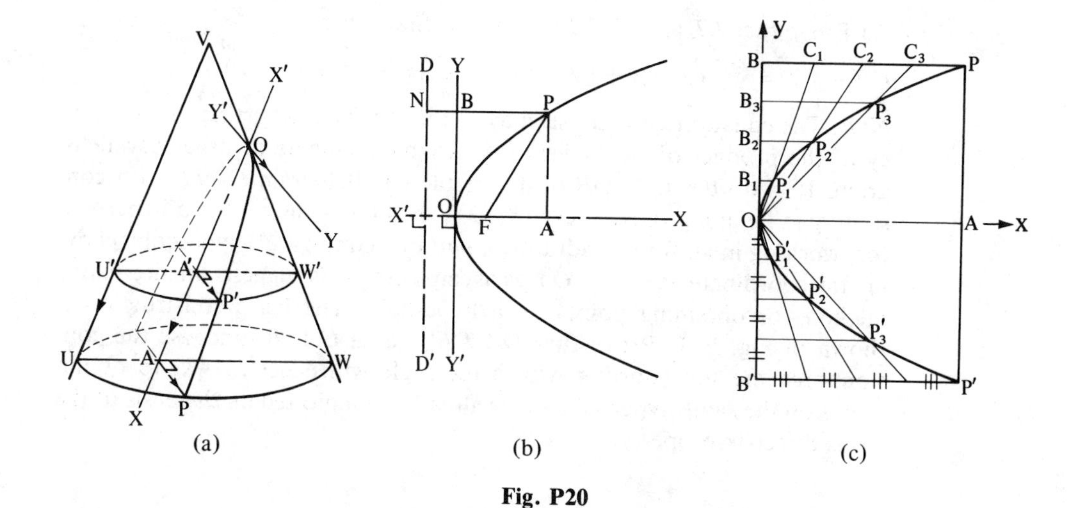

Fig. P20

Diameters UAW and $U'A'W'$ are drawn perpendicular to PA and $P'A'$, respectively, as shown. Prove that $A'P'^2 : AP^2 = OA' : OA$ (or $y_{P'}^2 : y_P^2 = x_{P'} : x_P$; in general $y^2/x = k$).

(b) The locus of a point P moving in a plane defined by a line DD' and point F is shown in Fig. P20(b). As indicated, axes are taken perpendicular and parallel to DD' through O, midway between F and DD'. AP, equal to OB, and BP, equal to OA, are the coordinates of P. If the distance of P from directrix DD' is always equal to its distance from focus F, prove that $OB^2 : BP$ is constant for all positions of P. (Thus the characteristic property of the curve—a parabola—may be expressed by the equation $y^2/x = k$; if F is at distance $2a$ from DD', $OF = a$, and $y^2 = 4ax$.)

(c) In the construction indicated in Fig. P20(c), sides OB, BP of rectangle $OAPB$ are divided into the same number of equal parts, and likewise sides OB', $B'P'$ of $OAP'B'$. Prove that the points P_1, P_2, P_3, etc. are such that $B_1P_1 : B_2P_2 : B_3P_3 : \ldots = OB_1^2 : OB_2^2 : OB_3^2 : \ldots$ (That is, P_1, P_2, P_3, . . . , and likewise P_1', P_2', P_3', . . . , have their coordinates related according to $y^2 = kx$, and are thus points on a parabola.)

41. Plot the hyperbola having (i) the coordinate axes as asymptotes; (ii) points (k, k) and $(-k, -k)$, as the foci F and F', respectively; (iii) directrices DD', through $(k, 0)$ and $(0, k)$, and $\Delta\Delta'$ through $(-k, 0)$, $(0, -k)$, and (iv) eccentricity $\epsilon = FT : NT = F'T : N'T = \sqrt{2} : 1$, where N, N' are the feet of the perpendiculars from the tracing point T to DD' and to $\Delta\Delta'$, respectively, T being on either branch of the hyperbola.

(a) Prove that the equation of the hyperbola is $xy = \frac{1}{2}k^2$. (Start with $FT = NT\sqrt{2} = QT$, where Q is the point on DD' with the same x-coordinate as T.)

(b) Prove that FT and $F'T$ differ by $2k$. Start with

$$F'T - FT = \sqrt{\{(x + k)^2 + (y + k)^2\}} - \sqrt{\{(x - k)^2 + (y - k)^2\}}.$$

Note: The equation xy = const. may be thought of as asserting the constancy of the product of two sides of a variable rectangle or other parallelogram. If two sides OA, OB of a variable parallelogram $OATB$ with constant area lie along fixed lines OX, OY, then the locus of T is a hyperbola (or, working in all four quadrants, a *pair* of "conjugate" hyperbolas) having the coordinate axes OX, OY as asymptotes. This suggests an alternative method for obtaining points on a hyperbola. The basic construction is shown in Fig. P21. Prove that $OA'T'B'$ = area $OATB$, and use the constant-area method (starting with a rectangle with sides k, $\frac{1}{2}k$) to obtain points on the same hyperbola as has already been plotted on the basis of the focus-directrix property.

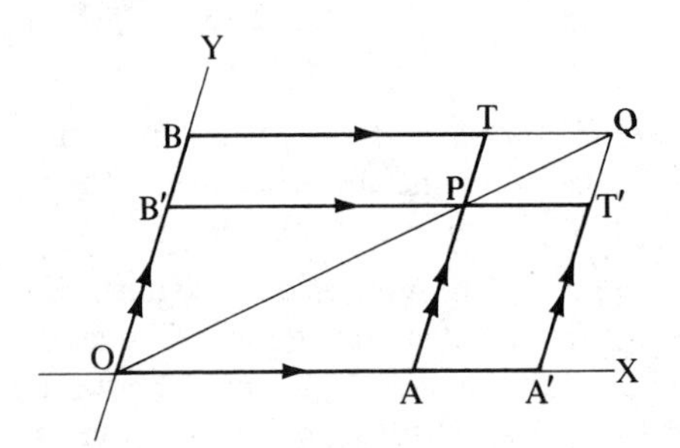

Fig. P21

42. Interpret Fig. P17, given for Problem 34, as the axial cross-section of a right circular cone cut by a plane through VV' that touches the sphere of radius 11 cm, center C, at F and the sphere of radius 4 cm, center C', at F' (see Fig. P22). As before, take CC' equal to 25 cm, and refer to previously calculated values as required. P is any point on the line of intersection of the cone with the plane that touches the spheres, and the line through V'', P (a generator of the cone) intersects the circles of contact, of diameters AB, $A'B'$, at Q and Q'. (i) Prove that $PF = PQ$ and $PF' = PQ'$, and hence that $PF + PF'$ is constant. (ii) Give the value of $PF + PF'$ and hence (knowing FF') determine the length of the minor axis of the elliptical section. (iii) Calculate the diameters AB, $A'B'$. (iv) If M is the midpoint of VV' and MN is perpendicular to the axis of the cone, meeting it at N, prove that N is halfway between AB and $A'B'$ and hence show that the radius of the circular section of the cone through N is $7\frac{1}{5}$ cm. (v) Calculate the radii of the circular sections of the cone through V and V' and hence determine NM. (vi) Calculate the length of the chord at distance NM from the center N of the circle of radius $7\frac{1}{5}$ cm, and hence obtain an independent evaluation of the length of the minor axis of the elliptical section.

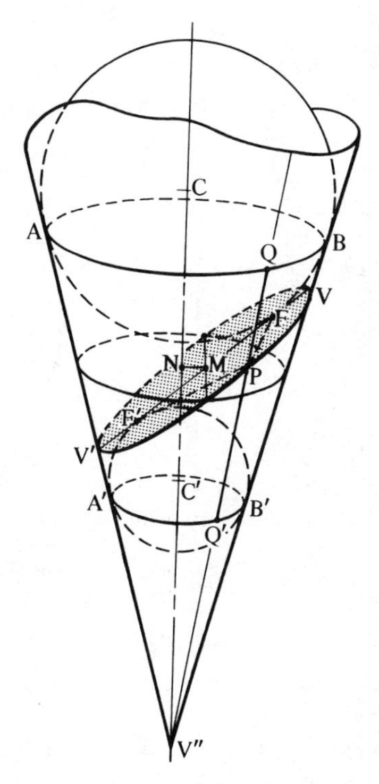

Fig. P22

43. (a) ABC is a triangle in which $AB = AC$ and D is a point in AC such that $BD = BC$. A line parallel to BC through M, the midpoint of CD, intersects side AB, the join BD, and median AX, in points M', O, N, respectively. DE, parallel to MM', meets AX in E. Given $BC = 9$ cm, $DE = 2$ cm, find MM', ON, AB.

(b) Interpreting triangle ABC in the figure of part (a) as an axial section of a right circular cone (base diameter 9 cm, slant height $\frac{27}{5}\sqrt{5}$ cm), calculate the length of the minor axis of the elliptical section (perpendicular to ABC) of which BD is the major axis. (Note that O is the midpoint of BD, and find the length of the appropriate chord in the circle of diameter MM'.)

(c) Calculate the distance between the foci and the eccentricity of an ellipse with axes of lengths 6 cm, 9 cm.

(d) Figure P23 shows the elevation of portion of a right circular cone cut by a plane ZZ, seen here edge on. Calculate the lengths of the axes and the distance between the foci of the elliptical intersection. Show how a congruent ellipse may be obtained by taking an appropriately inclined section of a cylinder of diameter 36 mm. Verify (i) that the points of contact of the sectioning plane with spheres inscribed in the cylinder co-

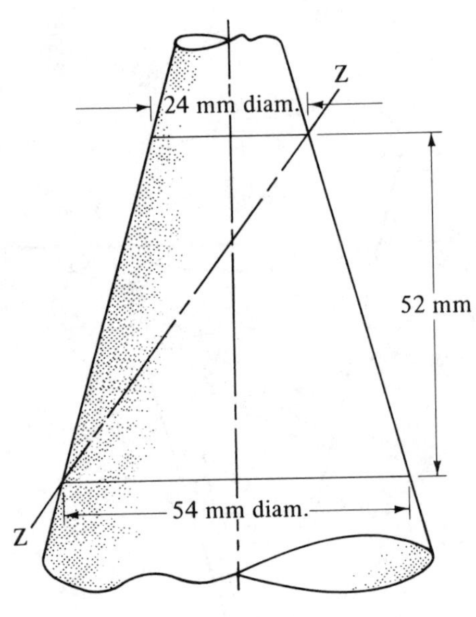

Fig. P23

incide with the foci, and (ii) that the distance between the circles of contact of these spheres with the cylinder is equal to the major axis of the ellipse.

44. Figure P24 shows the side elevation (with projected midsection) of a vessel having the form of a frustum of a right circular cone. Show that the ratio of the lower (filled) volume V to the upper (unfilled) volume U is given by $V:U = r\sqrt{r}:R\sqrt{R}$—hence, for example, if $r:R = 4:9$, the vessel would be $\frac{8}{35}$ full. (Include the volume C of the complementary cone in your working, showing that $\mathrm{V} + \mathrm{C} = \frac{1}{3}\pi abp$, where a, b measure the semiaxes of the elliptical section separating the filled and unfilled parts of the vessel, and $p = Rrh/a(R - r)$ by similar triangles.)

45. (a) Show how, by the appropriate choice of the lengths, ℓ units, of the opposite edges of the tetrahedron shown in Fig. P25, corresponding cross-sections (at any height z units) of the sphere and the tetrahedron may be made equal to each other. Hence, using Cavalieri's theorem, establish the formula for the volume of a sphere of radius R.

Show also how, by a simple extension of Cavalieri's theorem (relating one figure to the sum of two or more others), the volume of a sphere may be found via the following relationship:

sphere of radius R	+	two cones having radii R, heights R	=	cylinder of radius R, height $2R$.

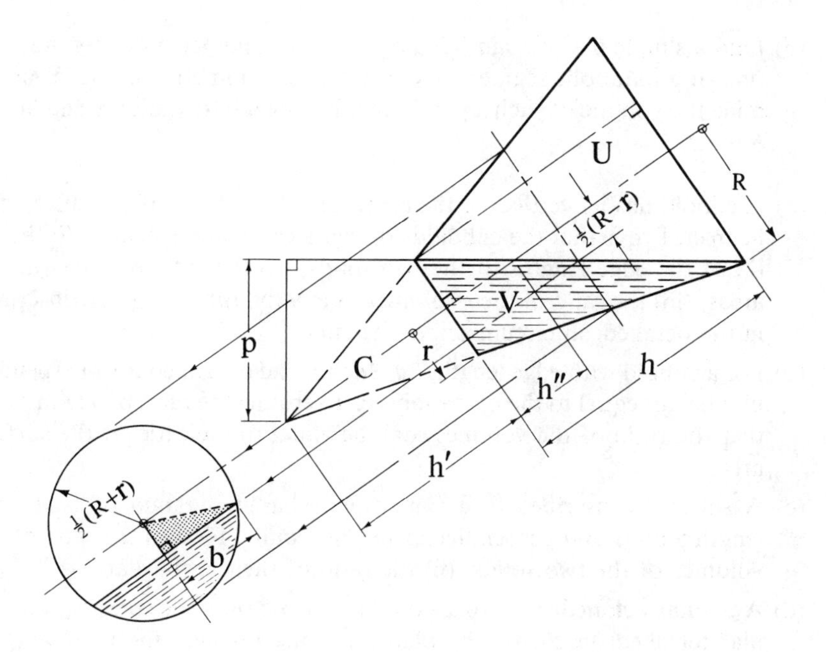

Fig. P24

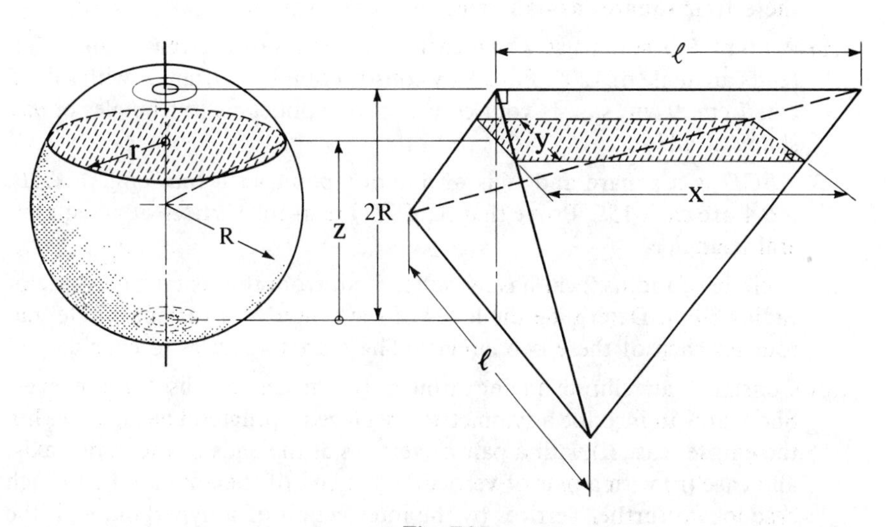

Fig. P25

(b) Find a simple Cavalierian "analogue" of a solid generated by the rotation of a parabolic segment about it axis of symmetry and hence determine the volume of such a paraboloid having a base radius R and height h units.

46. (a) A cuboid has its vertices at the centroids of the faces of a regular octahedron. Prove that the cuboid is in fact a cube, and calculate (i) the ratio of the edge lengths of the two solids, (ii) the ratio of their surface areas, (iii) the ratio of their volumes. Show (iv) that the sphere inscribed in the octahedron circumscribes the cube.

(b) For a cuboid with edge lengths $2a$, $3a$, $4a$, and a right cone of perpendicular height equal to the base radius r, determine the ratio of $a : r$ in order that the ratio of the volumes shall be equal to the ratio of the surface areas.

(c) A sphere is inscribed in a right square-based pyramid of base edge length 5 units and perpendicular height 6 units. Find (i) the ratio of the volumes of the two solids, (ii) the ratio of their surface areas.

(d) A regular octahedron is formed by removing the four corners of a regular tetrahedron cut off by planes passing through the midpoints of each set of three adjacent edges of the tetrahedron. Show that the octahedron has half the volume of the tetrahedron and also that it has half the surface area. Independently calculate the radius of the inscribed sphere of each of these solids.

47. (a) Four coplanar squares $ABST$, $BCUV$, $CDWX$, $DAYZ$ lie exterior to a parallelogram $ABCD$. Prove that the quadrilateral having the centers of these four squares as its vertices is itself a square.

(b) A point P is sought at which each of the sides of a given triangle subtends an angle of 120°. Find P by construction for a triangle with sides 5 cm, 7 cm, 9 cm, say. Is your construction applicable to triangles of *any* shape at all?

(c) $ABCD$ is a square and X is an interior point such that angles XAB, XBA are each 15°. Prove that X, C, D are at the vertices of an equilateral triangle.

(d) A circle of radius 2 cm has its center 5 cm from the center of a circle of radius 8 cm. Determine the locus of the center of a variable circle that touches each of these two circles. (There are two cases to consider.)

(e) Construct an ellipse having arbitrarily chosen lengths for the axes. Show how to inscribe a symmetrically placed equilateral hexagon within the ellipse: case (i) with a pair of vertices at the ends of the major axis, and case (ii) with a pair of vertices at the end of the minor axis. In each case locate further vertices by the intersection of a hyperbola with the ellipse. (For convenience a set-square or graph paper may be used to obtain parallels and perpendiculars, but otherwise straightedge-and-compasses methods should be used to obtain points through which the ellipse and the hyperbolas may be drawn freehand.)

48. (a) A quadrilateral has three equal sides, each of length a units, one of which is parallel to the fourth side, of length b units. Prove that the circular arc determined by its vertices as shown in the diagram (a) of Fig. P26 is greater or less than a semicircle according as $b \gtrless 2a$.

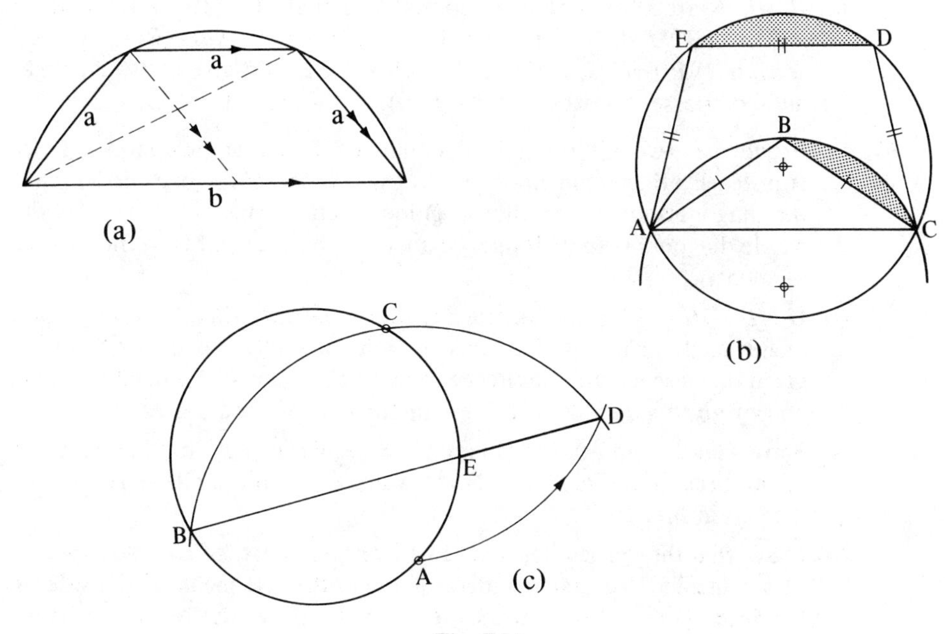

Fig. P26

(b) Within a quadrilateral $ACDE$ in which $AE = ED = DC$, and $ED // AC$, a triangle ABC is drawn with BC equal to BA. A circle is described through A, C, D, E, and a circular arc passes through A, B, C, as indicated in diagram (b). Prove that the shaded segments are similar if and only if A, B, D are collinear.

(c) Diagram (c) shows a circle of unknown center. Arc BCD, of any convenient radius, is centered at any point A of the circumference of the given circle. This arc is cut at D by the arc of radius CA centered at C. The join DB cuts the given circle at E. Prove that DE is equal to the radius of the given circle.

(d) Prove that for any triangle ABC—whether acute-angled or obtuse-angled—the circles AHB, BHC, CHA, ABC are all of equal radii, H being the orthocenter.

49. (a) Prove that the area of a regular dodecagon is exactly three times the area of the square on the radius of the circumscribing circle. Find also an expression for the area in terms of the side length, and, finally, express the area in terms of a diagonal subtending four sides of the dodecagon.

(b) Show that the area of a regular *decagon* is given by $A = \frac{1}{2}pR$, where p is the (measure of the) perimeter of a regular *pentagon* inscribed in the circle, of radius R, circumscribing the decagon.

(c) A, P, B are three collinear points such that $AB : AP = AP : PB$, or $AP^2 = AB \cdot PB$. (P is thus said to divide the interval AB in "golden section.") Prove that $AP : AB = (\sqrt{5} - 1) : 2$, and give a straightedge-and-compasses construction for locating P, given A and B.

50. (a) Prove that two intersecting diagonals of a regular pentagon cut off from each other segments equal to the side length. Prove further that the diagonals cut each other in golden section (that is, prove that the whole diagonal is to its longer segment as this segment is to the shorter segment.)

(b) If $ABCDE$ is a regular pentagon, show that the triangle ADB has its angles as $2:1:2$. Devise a construction for such an isosceles triangle, given the base length, and hence show how to construct a regular pentagon of given side length (using straightedge-and-compasses methods).

(c) Show that the radius of the incircle of a regular pentagon is to the radius of the circumcircle as $(\sqrt{5} + 1) : 4$. (Make use of the similar triangles indicated in Fig. P27(i).)

(d) Prove that the perpendicular drawn from the center of a circle to a side of an inscribed regular pentagon is the arithmetic mean of the side of the decagon inscribed in the same circle and the radius of this common

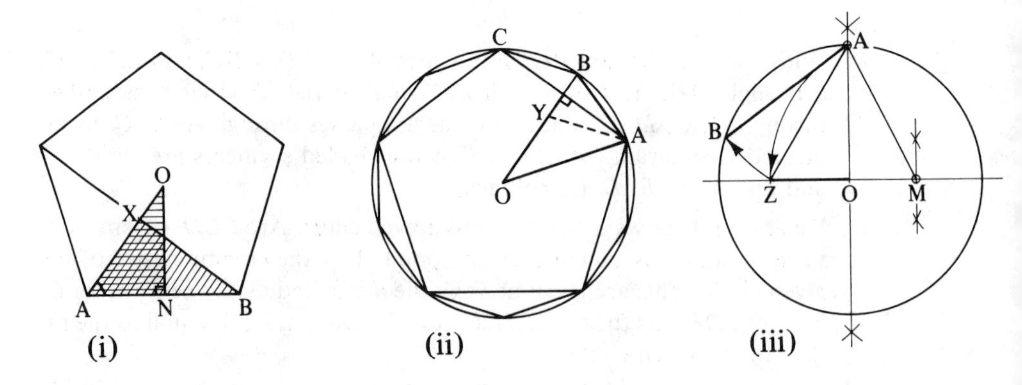

Fig. P27

circumcircle. (Make use of the line AY equal to AB as shown in Fig. P27(ii).)

(e) Express the length of the side of a regular pentagon in terms of the radius of the circumcircle. Repeat for the side of a regular decagon, using the results obtained in parts (c) and (d). Hence prove that the straight-edge-and compasses construction indicated in Fig. P27(iii) is correct for obtaining the side ($AB = AZ$) of a regular pentagon, and the side (equal to OZ) of a regular decagon, which may be inscribed in a circle of given radius OA.

51. (a) Show by a sketch how the diagonals of a regular pentagon divide the included area into eleven nonoverlapping regions. Determine the ratio of the area of the innermost region to the area of the complete figure.

(b) A pyramid on a regular pentagonal base has its slant faces all equilateral triangles. Find the angle between any two nonadjacent slant edges, and prove that the angle between the plane through such edges and the base is twice the angle between a slant edge and the base.

(c) Prove that the perpendicular distance, p units, from the center of a regular pentagon to a side, of length a units, is given by $p = \frac{1}{10}a\sqrt{\{5(5 + 2\sqrt{5})\}}$, and hence obtain a formula for the volume of a pyramid of the kind described in part (b).

52. (a) Prove that the area of a regular pentagon inscribed in a circle of radius ρ is given by the formula $P = \frac{5}{8}\{\sqrt{(10 + 2\sqrt{5})}\}\rho^2$. Find also a formula for the area, T square units, of an equilateral triangle inscribed in a circle of radius ρ units. Hence prove that $12P$ exceeds $20T$ (by almost 10 percent).

(b) Prove that the ratio of the radii of spheres inscribed in and circumscribed about a regular dodecahedron is equal to the ratio of the corresponding spheres in the case of a regular icosahedron. (Show that $r_d^2 : R_d^2$ and $r_i^2 : R_i^2$ each have the same value, viz., $(5 + 2\sqrt{5}) : 15$.) (Figure P28(i) indicates that the same sphere would circumscribe the dodecahedron and the centrally placed cube having its edges equal to the face diagonals of the dodecahedron: $R_d = R_c = \frac{1}{2}e_c\sqrt{3}$, where $e_c = \frac{1}{2}(\sqrt{5} + 1)e_d$. And, as is clear from Fig. P28(ii) the distance between a pair of parallel edges of the icosahedron is equal to a diagonal of a regular pentagon of side e_i; thus R_i = half the diagonal of a rectangle of length $\frac{1}{2}(\sqrt{5} + 1)e_i$ and width e_i).

(c) Using the results established in parts (a) and (b), prove that if it is given either that a regular dodecahedron and a regular icosahedron have equal inscribed spheres or that they have equal circumscribed spheres, then the *volume* of the dodecahedron is greater (by almost 10 percent) than the volume of the icosahedron.

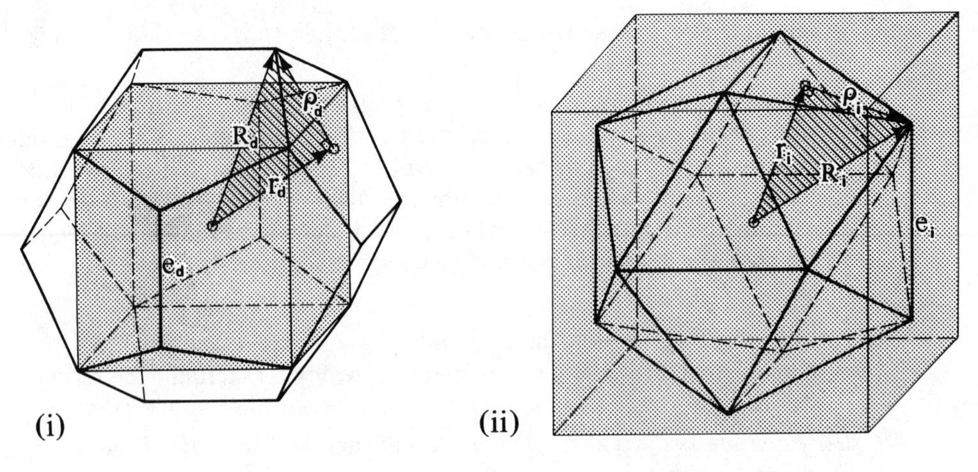

Fig. P28

(d) Notwithstanding (more accurately: in keeping with) the result established in part (c), prove that for the case of a regular icosahedron and a regular dodecahedron *with equal surface areas*, the volume of the icosahedron *exceeds* that of the dodecahedron (by about 5 percent).

XVII

Obscurity, indeed, is painful to the mind as well as to the eye; but to bring light from obscurity, by whatever labour, must needs be delightful and rejoicing.

—David Hume

How extremely stupid not to have thought of that!

—T. H. Huxley
[on reading Darwin's *Origin of Species*]

It is a common and tempting fallacy to view the later steps in a mathematical evolution as much more obvious and cogent after the fact than they were beforehand.

—John von Neumann

Ignorance is a tissue of positive, tenacious, interdependent errors . . . [which] are not easily destroyed one by one [because] they are coordinated.

—Gaston Bachelard

We used to think that if we knew one, we knew two, because one and one are two. We are finding that we must learn a great deal more about 'and'.

—A. S. Eddington

The whole question of imagination in science is often misunderstood by people in other disciplines. . . . They overlook the fact that whatever we are allowed to imagine in science must be consistent with everything else we know. . . . *The problem of creating something which is new, but which is consistent with everything which has been seen before, is one of extreme difficulty.*

—R. P. Feynman

Every mathematician knows that a proof is not really 'understood' as long as one has done nothing more than verify step by step the correctness of the deductions involved, and has not tried to gain a clear insight into the ideas which led to the construction of this chain of deductions in preference to all others. . . . What axiomatics sets as its essential goal is precisely what logical formalism alone cannot supply: the deep-lying intelligibility of mathematics.

—Nicolas Bourbaki

SOLUTIONES

To become adept at solving significant mathematical problems, not only is appropriate mathematical knowledge required but also the ability to draw upon a relevant selection of that knowledge. The mind has to be brought to bear on each problem in such a way that one becomes aware of promising implications of the data and of promising relationships that would imply what has to be established or discovered. In the case of geometrical problems, this process is greatly assisted by taking very seriously the drawing of clear, well-labeled diagrams. Thus depicting the problem is very often an essential part of that preparation of the mind that makes possible both the illumination and the verification stages of discovery (see again, note 18). Students, concerned to develop their strategic and tactical skills and to deepen their understanding generally, will be prepared to invest a considerable independent effort in order to give themselves the opportunity of discovering and comparing alternative methods and also to ensure that they will obtain optimum benefit from the explanations provided here when these are finally consulted.

Most of the solutions that follow are given in outline only, and parts of solutions, or even whole solutions, have usually been omitted in cases where the reader may reasonably be supposed to be able to master the problem with little or no assistance. Sometimes, however, solutions to simple problems are given in fair detail for the purpose of exhibiting some particular method of proof or use of symbolism that might otherwise go unnoticed. Alternative methods are to be cherished, since the development of a repertoire of different approaches to simple problems best prepares the mind to deal effectively with a wide range of more difficult material.

No simple answer can be given to the question of just what may be implicitly assumed in the composition of a particular proof. Mathematical practitioners, teachers, and students well know that, quite apart from considerations of pure logic, the most appropriate forms of proof are very much dependent upon the needs of the person composing the proofs and of those who are potential readers (cf. note 228). Advanced students of geometry will be aware of the occasional applicability of more sophisticated methods than are adopted here. The purpose of such advanced methods is, of course, to enable shorter, or more general, proofs to be devised, if not in the case of the particular problem under consideration, then for related problem situations that are not our present concern. Readers wishing to proceed to higher levels will find themselves well-served by such books as H. S. M. Coxeter's *Introduction to Geometry,* and Daniel

Pedoe's *A Course of Geometry for Colleges and Universities.* The present book may be regarded as introductory to these more advanced texts, as well as to the classical writings of Euclid, Archimedes, Apollonius, Galileo, and Newton. Any professional mathematician who should chance to inspect these pages is reminded that "it is not everyone that has a Genius fitted for the most exalted Speculations, or that is capable of reading the Works of the most sublime and celebrated Authors; and, therefore, tho' I should go no farther than to bring down some of the best and most useful Things already known to the Level of ordinary Capacities, I should think this might, in some measure, exempt me from Censure" (Thomas Simpson, *The Nature and Laws of Chance,* 1740, preface).

1. This first example was inspired by accounts of the investigations into the areas of certain "lunes," or crescents, by Hippocrates of Chios in the second half of the fifth century B.C. It is not unlikely that these investigations were suggested by such well-known decorative forms as are illustrated in Fig. S1. The basic intersecting-circles design is over 6,000 years old and predates the use of compasses. For very early freehand examples from Mesopotamia, see Max von Oppenheim, *Tell Halaf*, vol. 1, 1943, plate LI. An instrumentally produced version of the same pattern appears on an Old Babylonian mathematical tablet (BM 15285), illustrated in H. W. F. Saggs, *The Greatness That Was Babylon*, New York: Hawthorn, 1962, plate 24. The leaflike double segments, V and E, which could also have symbolized vulvae, eyes, seeds, fishes, or boats, were widely used in ancient decoration, as was the pelecoid, the axe-shaped form, of area A_2. Since each of A_1 and $A_2 = A + 2S - 2S = A$, we have here simple curvilinear figures that may

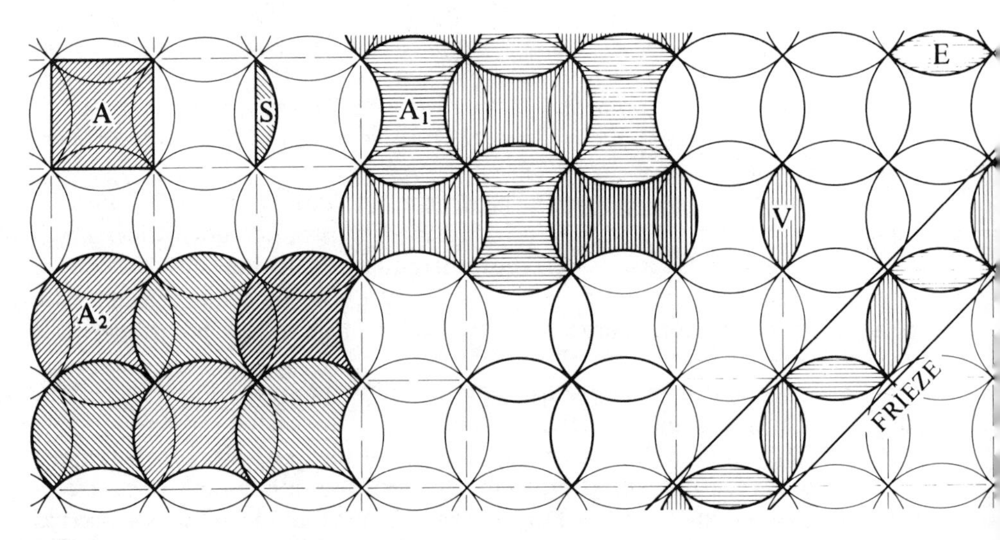

Fig. S1

be squared. This "give-and-take principle," together with the Pythagorean relation between the sides of a right-angled triangle, and the proposition that the areas of similar segments are to one another as the squares of their chords (or as the squares of their radii), was exploited by Hippocrates in his brilliant quadratures of several forms of lune. For this work, see Thomas L. Heath, *A History of Greek Mathematics,* Oxford: Clarendon Press, 1921, 1960, 1965, vol. 1, pp. 183–199; Ivor Thomas, *Selections Illustrating the History of Greek Mathematics,* London: Heinemann, Loeb Classical Library, 1939, vol. 1, pp. 234–253.

Of the diagrams given in Fig. P1 for the present problem, only (v) contains lunes actually recorded as having been among Hippocrates' own examples, but his general approach may be illustrated, somewhat symbolized for convenience, by our example (iv). Let

$$C = T + S - 2S' \tag{1}$$

denote the area equation: curvilinear figure $WXY = \triangle WXY +$ segment on chord $WY -$ segments on chords WX, XY. (See Fig. S2.) These segments, on WY, WX, and XY, are similar since their central angles are equal, each being a right angle. Hence,

$$S : S' = WY^2 : WY^2 = 2 : 1 \quad \Rightarrow S = 2S' \tag{2}$$

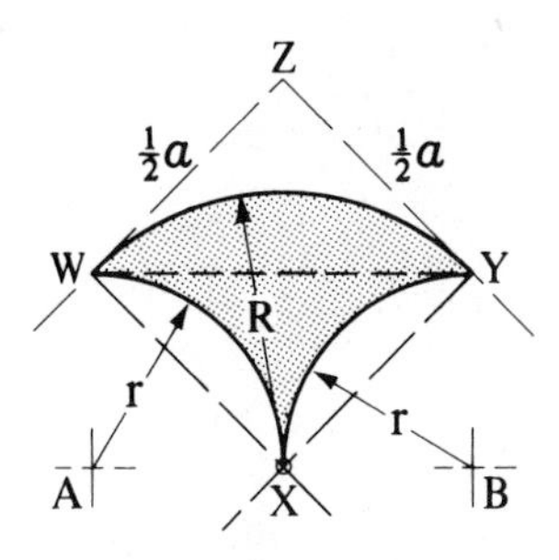

Fig. S2

From (1) and (2), $C = T$; consequently the required area, shaded in the given diagram (iv) of Fig. P1 is equal to twice T, that is, twice $\triangle WXY$; so the required region is equal to square $WXYZ$, area $\frac{1}{4}a^2$.

Example (vi) The two lunes = right triangle + semicircles on the two shorter sides – semicircle on the hypotenuse. As semicircles are to one another as the squares on their diameters, in this case as the squares on the sides of a right triangle, it follows that the two lunes are together equal in area to the right triangle. Strangely, this generalization of Hippocrates' simplest case is not mentioned in any surviving Greek source. It was given, however, by Ibn al-Haytham (called Alhazen in the West) who flourished around A.D. 1000, doing important work in optics, astronomy and mathematics. The great *Dictionary of Scientific Biography* devotes twice as many pages to Ibn al-Haytham as to Galileo!

2. (a) Since $ABHL$ and $BCMH$ are parallelograms, AL and CM are each equal and parallel to BH, and hence are equal and parallel to each other. Therefore $ALMC$ is a parallelogram, and

$$\begin{aligned} ALMC &= AKXL + KCMX \\ &= ABHL + BCMH \\ &= ABED + BCFG. \end{aligned}$$

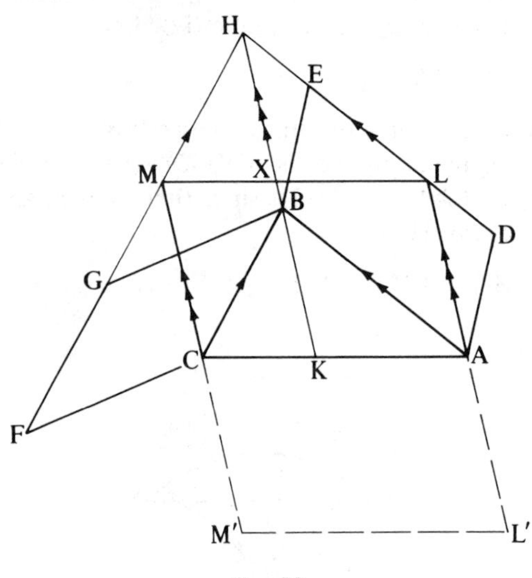

Fig. S3

(Hence we have a construction for finding a parallelogram $ALMC$, or $AL'M'C$, equal in area to the sum of the two arbitrary parallelograms $ABED$, $BCFG$—Fig. S3.)

This impressive generalization of the Pythagorean theorem involving the addition of any parallelograms (whether squares or not) on two sides of any triangle (whether right-angled or not) was given by Pappus of Alexandria (fl. early 4th cent. A.D.) in his work known as the *Synagoge* or the *Collection*, IV, 1, 1. This proposition, with an English translation, is given by Thomas, *Selections Illustrating the History of Greek Mathematics*, vol. 2, pp. 574–579. Pappus' demonstration occasionally makes its reappearance, without acknowledgment, in modern school texts in the form of Fig. S4, a diagrammatic "proof without words" for the special case of the Theorem of Pythagoras.

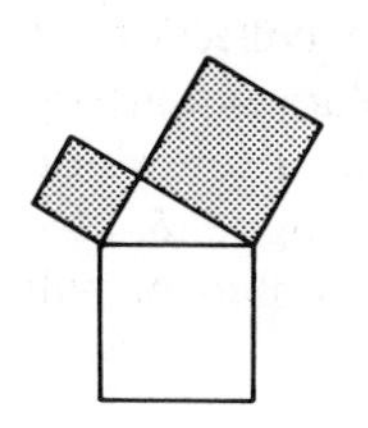
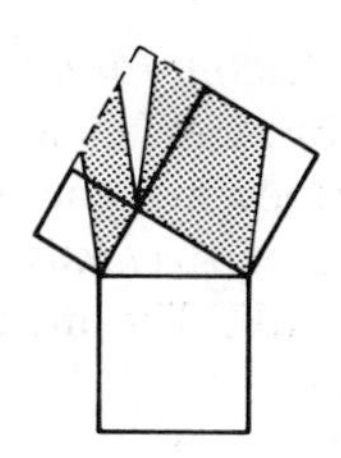
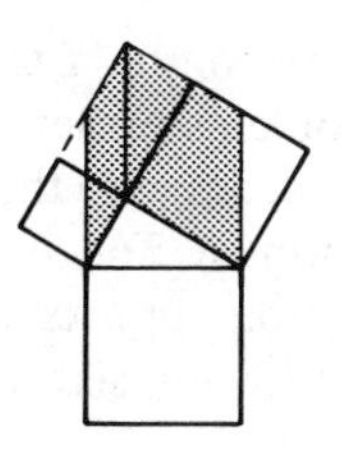
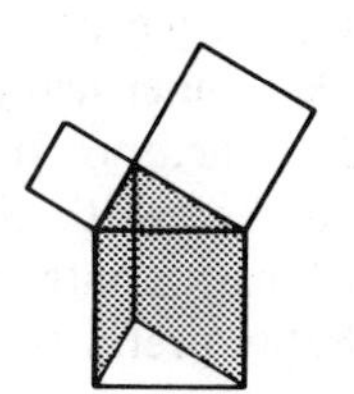
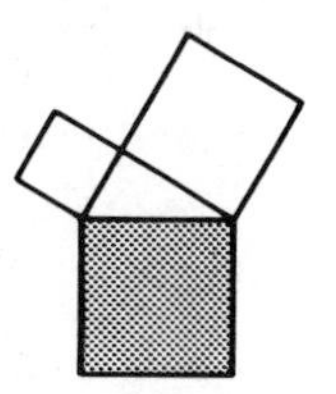

Fig. S4

(b) By similar triangles *BNC, ABC* (Fig. S5, upper diagram),

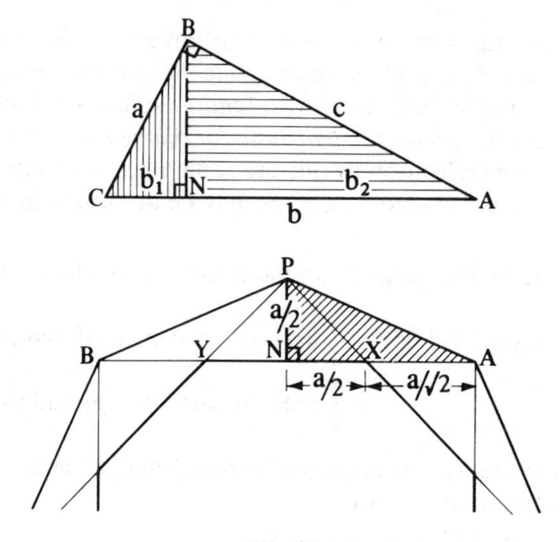

Fig. S5

$$b_1 : a = a : b \Rightarrow b_1 = a^2/b.$$

By similar triangles *ANB, ABC,*

$$b_2 : c = c : b \Rightarrow b_2 = c^2/b.$$

Hence, $b_1 : b_2 = a^2 : c^2$.

(c) Area ratio $= AP^2 : XY^2$ (Fig. S5, lower diagram). Let $XY = a$, then

$$AP^2 = (a/2 + a/\sqrt{2})^2 + (a/2)^2,$$

from which

$$AP^2 : XY^2 = (2 + \sqrt{2}) : 2.$$

("The areas of similar figures are to one another as the squares of corresponding linear dimensions.")

3. (a) This example comes from the *Liber Assumptorum*, a collection of Greek geometrical problems preserved by Arabian scholars and attributed by them to Archimedes. See the "Book of Lemmas," Proposition 14, in the Heath edition of *The Works of Archimedes*, pp. 315–316, also Heath's introduction, pp. xxxii–xxxiii. The modern student will probably prefer the following setting out:

$$\text{Shaded area} = \frac{1}{2}\pi R^2 + \frac{1}{2}\pi r^2 - \pi\left\{\frac{1}{2}(R - r)\right\}^2,$$

$$\text{which simplifies to } \frac{\pi}{4}(R + r)^2$$

$$= \text{area of circle of diameter } EF.$$

This use of the symbol π is less than three centuries old (see dialogue note 6); the Greek solution is made to depend on the proposition (following Euclid XII, 2): *Circles (and therefore semicircles) are to one another as the squares on their radii (or diameters)*. Solutions appealing to such a generalization as this are sometimes more elegant than those utilizing formulae involving π, as may be illustrated in part (b) following.

(b) First part. By the proposition italicized in the solution to part (a):

$$\text{Semicircle of radius } r = \frac{1}{4} \text{ area of semicircle of radius } 2r$$

$$= \frac{1}{2} \text{ area of quadrant of radius } 2r.$$

Hence the two regions indicated in the left half of the second diagram of Fig. P2 are of equal area.

Second part. Referring to Fig. S6:

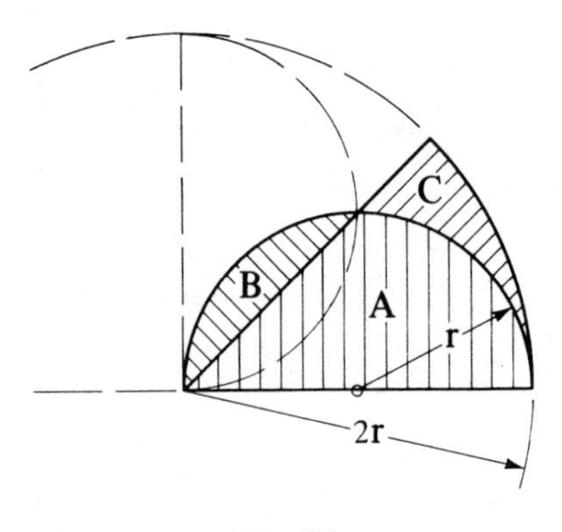

Fig. S6

$$A + B = \text{semicircle of radius } r$$
$$= \frac{1}{4} \text{ semicircle of radius } 2r$$
$$= A + C$$
$$\therefore B = C.$$

Consequently, in the right half of the second diagram of Fig. P2, the shaded areas are equal.

For Leonardo da Vinci's fascination with examples of this kind, see Julian Lowell Coolidge, *The Mathematics of Great Amateurs*, Oxford: Clarendon Press, 1949, or New York: Dover reprint, 1963, pp. 45–48.

4. (a) $4^2 + 7^2 \neq 8^2$, so the triangle is not right-angled. Since $8^2 < 4^2 + 7^2$, the largest angle of the triangle is acute. (Euclid, II, 13, Converse, as given by Hero of Alexandria; see Heath, ed., *The Thirteen Books of Euclid's Elements,* vol. 1, pp. 406–409, and cf. p. 405.)

(b) Make use of congruent triangles ABC, BAE (diagram (b) of Fig. S7); $CN = EX = DX$; hence $CD \,//\, AB$, whereas $DE \perp AB$, ...

(c) $r_A = 2\frac{1}{2}$, $r_B = 3\frac{1}{2}$, $r_C = 5\frac{1}{2}$; then $A'B' \cdot B'C' \cdot C'A' = \sqrt{35} \cdot \sqrt{77} \cdot \sqrt{55} = 5 \cdot 7 \cdot 11$. In general, $A'B' = \sqrt{\{(r_B + r_A)^2 - (r_B - r_A)^2\}} = \sqrt{\{(2r_B)(2r_A)\}} = 2\sqrt{(r_A r_B)}$. Likewise, $B'C' = 2\sqrt{(r_B r_C)}$; $C'A' = 2\sqrt{(r_C r_A)}$: hence $A'B' \cdot B'C' \cdot C'A' =$ etc.

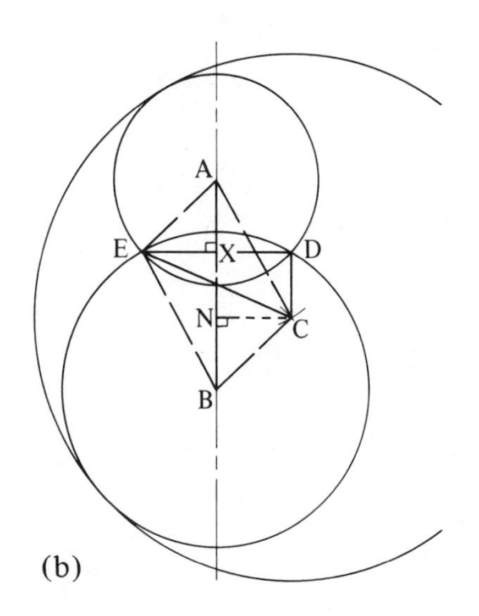

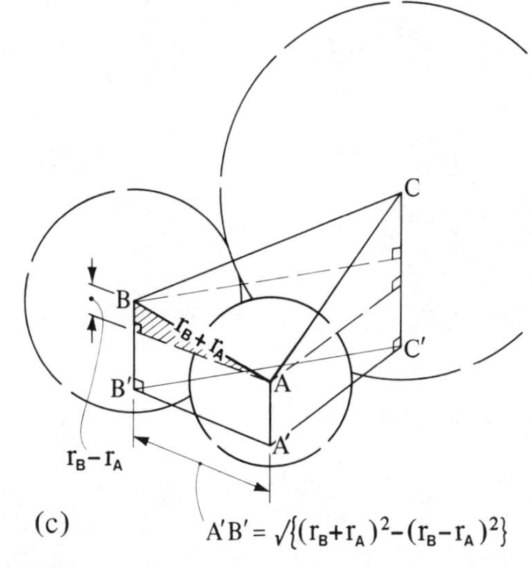

Fig. S7

5. (a) Let $\angle BAC = \alpha$, say, then $\angle AQP = \alpha$, $\angle BPQ = 2\alpha$, $\angle PBQ = 2\alpha$; $\angle BQC = 3\alpha$, $\angle BCQ = 3\alpha$; $\angle ABC = \angle ACB = 3\alpha = 3\angle BAC$.

(c) In diagram (c) of Fig. S8, $\alpha_2 = \alpha_1$, from data; $\alpha_3 = \alpha_1$, since ΔPCN is isosceles; $\therefore \alpha_3 = \alpha_2$, and consequently $CN \,//\, PQ$, so $CN \perp BD$. Thus the bisector of $\angle CPQ$ meets the circle at N, if P is on the semicircle $BN'D$ (or at N', if P is on the semicircle BND), where NN' is the diameter perpendicular to BD.

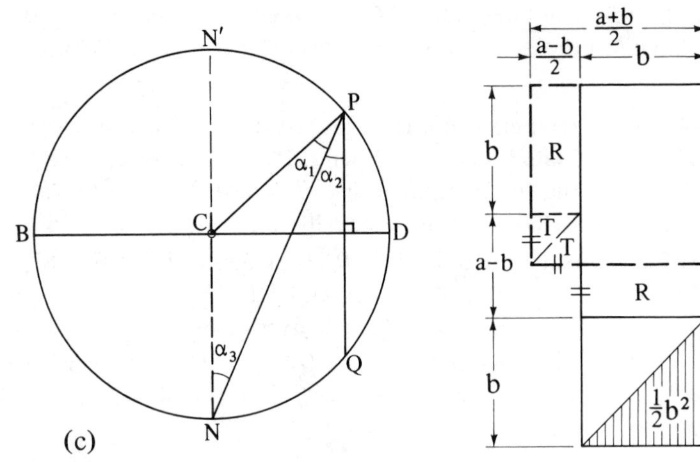

Fig. S8

(d) From the two upper shaded parts of diagram (d) of Fig. S8,

$$\frac{1}{2}a^2 - 2T + \frac{1}{2}b^2 = \left\{\frac{1}{2}(a+b)\right\}^2 \Rightarrow \frac{1}{2}(a^2+b^2) > \left\{\frac{1}{2}(a+b)\right\}^2;$$

and, comparing the upper *left* square of side $\frac{1}{2}(a+b)$ with the rectangle $a \times b$,

$$\left\{\frac{1}{2}(a+b)\right\}^2 + R - R - 2T = ab \quad \Rightarrow \left\{\frac{1}{2}(a+b)\right\}^2 > ab.$$

$$\text{Thus, } \frac{1}{2}(a^2+b^2) > \left\{\frac{1}{2}(a+b)\right\}^2 > ab;$$

hence, for *a, b unequal positive* quantities,

$$\sqrt{\left\{\frac{1}{2}(a^2+b^2)\right\}} > \frac{1}{2}(a+b) > \sqrt{(ab)}.$$

The same result may be shown more simply via the construction illustrated in Fig. S9.

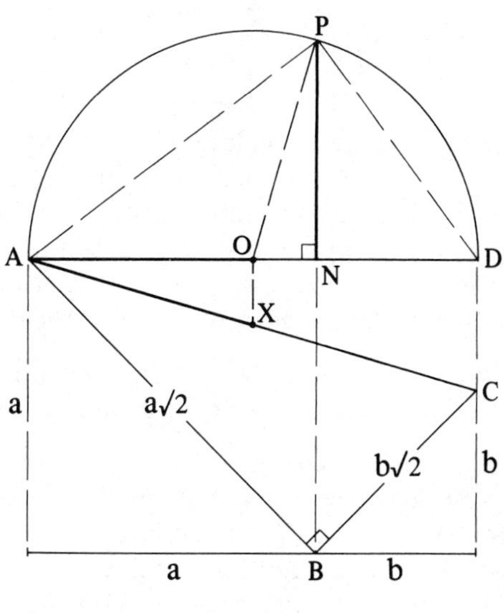

Fig. S9

$$AX = \frac{1}{2}AC = \frac{1}{2}\sqrt{(2a^2 + 2b^2)} = \sqrt{\left\{\frac{1}{2}(a^2 + b^2)\right\}}.$$

$$AX > AO = \frac{1}{2}(a + b). \quad AO = OP > NP = \sqrt{(ab)}.$$

6. $(h - a)/a = (b - h)/b$, from which $h(a + b) = 2ab$ or $h = 2ab/(a + b)$. (For example, 8 is the harmonic mean of 6 and 12 because $(8 - 6):6 = (12 - 8):12$. Archytas (early 4th cent. B.C.), who first convinced Plato of the importance of mathematics, is said to have adopted the name *harmonic* mean for what had been referred to as the *subcontrary* by earlier Pythagoreans. But a couple of generations before Archytas, Philolaus had regarded the cube, with its 6 faces, 8 vertices, and 12 edges, as constituting a "geometric harmony.")

(i) $$h = \frac{2ab}{a + b} \Rightarrow h = 2/\left(\frac{1}{a} + \frac{1}{b}\right) \Rightarrow \frac{1}{h} = \frac{1}{2}\left(\frac{1}{a} + \frac{1}{b}\right),$$

so $1/h$ is the arithmetic mean of $1/a$ and $1/b$, that is $1/a$, $1/h$, $1/b$ are in arithmetic progression.

(ii) In the first diagram, by similar triangles, $h/2a = b/(a + b)$, which implies $h = 2ab/(a + b)$.

(iii) $h = \dfrac{2ab}{a + b} \Rightarrow \dfrac{h}{\sqrt{(ab)}} = \dfrac{\sqrt{(ab)}}{(1/2)(a + b)}$,

that is,

$$\frac{\text{harmonic mean}}{\text{geometric mean}} = \frac{\text{geometric mean}}{\text{arithmetic mean}}.$$

(iv) $r = \frac{1}{2}(a + b)$. From similar triangles *ANP, PNB*, $a:p = p:b$, or $p = \sqrt{(ab)}$. From similar triangles *NXP, CNP*, $q:p = p:r$, or $q = p^2/r$, that is, $q = ab/\frac{1}{2}(a + b)$ = harmonic mean. This construction for "exhibiting the three means in a semicircle" was given, as the work of "another geometer," by Pappus (first half of the fourth century A.D.), in his *Collection*, III, xi (Ivor Thomas, *Selections Illustrating the History of Greek Mathematics*, vol. 2, pp. 568–571).

7. (b) In (i) use similar triangles *PZX, QZW*, and in (ii) *PZX, PRW*.

(c) The inner square has one-fifth the area of *ABCD*.

(d) The procedure indicated in parenthesis, and indeed the general approach typically used in devising proofs and constructions, is called *analytic*, for all that the final presentation is customarily in *synthetic* form. "The synthetic [method] says, 'Since A is true, it follows that B is true'; the analytic says, 'To prove that B is true, it is sufficient to prove that A is true.' The synthetic 'puts together' known truths, and by the combination perceives a truth theretofore unknown [or unproved]; the analytic 'pulls apart' the statement under question into simpler statements whose truth or falsity is more easily determined." (J. W. A. Young, *The Teaching of Mathematics*, New York: Longmans, Green, 1907, p. 54.) Historically, attempts to understand mathematical heuristic go back to the Greek method of mathematical analysis and are related also to the general Socratic method, in which the teacher (sometimes pretending ignorance and acting as "a midwife at the birth of ideas"!) places before the learner an appropriate sequence of ancillary problems more amenable to solution than the main problem. Just what the Greeks themselves understood by geometrical analysis is the subject of an ongoing controversy. See Norman Gulley, "Greek Geometrical Analysis," *Phronesis 33* (1958):1–14; Michael S. Mahoney, "Another Look at Greek Geometrical Analysis," *Archive for History of Exact Sciences 5* (1968):318–348; Jaakko Hintikka and Unto Remes, *The Method of Analysis, Its Geometrical Origin and Its General Significance*, Dordrecht: Reidel, 1974, and Erkka Maula's review of this book, "An End of Invention," *Annals of Science 38* (1981):109–122.

Consider now the particular problem: a sketch, indicating how the figure will appear when the construction has been carried out, shows that triangles *CPN* and *CPM* will have to be congruent, and consequently that $\triangle BCN$, right-angled at N, will have CN equal to $\frac{1}{2}BN$; thus BCN is half an equilateral triangle; so P will lie on a line through B inclined at 30° to BC. But two such lines may be drawn, and two positions are possible for P, one inside the square and the other outside it. (It is easy to show that the distance between these two positions is $\sqrt{3}$ times the length of the side of the square.) The same construction applies for any quadarilateral $ABCD$ for which $BC = CD = DA$ and $AB // DC$.

(e) Let the side length of the hexagon be s units, then, since the corner triangles are similar to the given triangle, their side lengths are readily dimensional in terms of s. If, for example, d units is the distance from the vertex of the smallest angle of the given triangle to the nearest vertex of the hexagon, then $d = 1\frac{1}{2}s$. Further, by similar triangles, $s/4 = (3 - s - d)/3$, where $3 - s - d$ (equal to $3 - 2\frac{1}{2}s$) units is the distance from the vertex of the largest angle of the given triangle to the nearer vertex of the hexagon in the 3-unit side of the triangle. From the preceding equations, find $s = 12/13$, and hence the perimeter ratio. The areas of the corner triangles are to the area of the given triangle as $s^2 : 2^2$, $s^2 : 3^2$, $s^2 : 4^2$. Hence, required area ratio,

$$\begin{aligned} H : T &= (T - Ts^2/2^2 - Ts^2/3^2 - Ts^2/4^2) : T \\ &= (1 - 36/169 - 16/169 - 9/169) : 1 \\ &= 108 : 169. \end{aligned}$$

8. (b) Vol. of pyramid $BCGD = \frac{1}{3}(\triangle BCG)(CD) = \ldots.$

(c) Join OD, OE; then $ODCE$ is a rectangle; radius $= OD + DF = OE + EG$, $\therefore\ EC + DF = DC + EG \Rightarrow AE + DF = BD + EG \Rightarrow AE - EG = BD - DF$.

(d) Let the parallels through A, B meet at P: then $OAPB$ is a variable rectangle having diagonal AB of constant length. Hence, diagonal OP also . . . .

(e) Assume a convenient unit and use straight-edge-and-compasses methods. Since $(\sqrt{2})^2 + (\sqrt{3})^2 = (\sqrt{5})^2$, the triangle is right-angled. Hence the hypotenuse is the diameter of the circumcircle. Area $= 5\pi/4$ square units.

9. (a) Divide the figure into a parallelogram and a triangle by a line through one end of the shortest side. Then determine the distance between the parallel sides (2.4 units) . . . .

(b) The main theorems involved here are as follows: (i) *Triangles on the same base and between the same parallels are equal in area* (Euclid, I, 37). (ii) *[The areas of] triangles of equal height are to one another as their bases* (Euclid, VI, 1). (iii) *Similar triangles are to one another as the duplicate ratio of [i.e., as the squares on] corresponding sides* (Euclid, VI, 19).

(c) Referring to diagram (c) of Fig. S10, we have from the data,

$$\frac{1}{2}(a + x)p = \frac{1}{2}\left\{\frac{1}{2}(a + b)q\right\},$$

which implies,

$$p/q = \frac{1}{2}(a + b)/(a + x) \quad (1)$$

By similar $\triangle s$,

$$p/q = (x - a)/(b - a) \quad (2)$$

From (1) and (2),

$$x^2 = \frac{1}{2}(a^2 + b^2), \text{ thus } EF = \sqrt{\left\{\frac{1}{2}(a^2 + b^2)\right\}}.$$

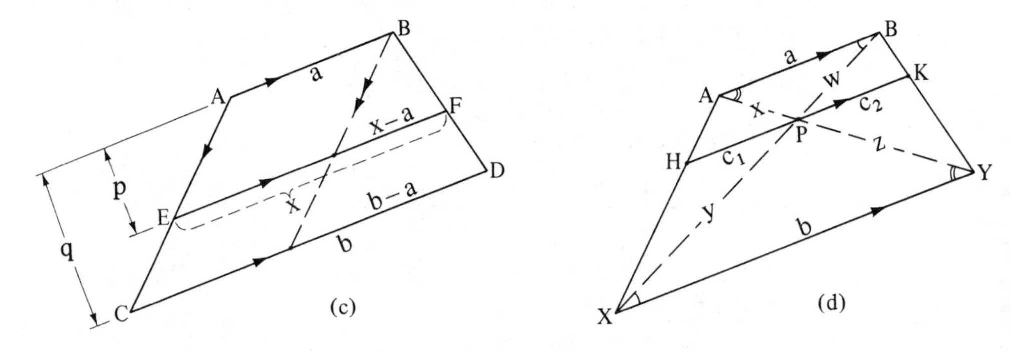

Fig. S10

This "root mean square" (r.m.s.) value is sometimes referred to as the "quadratic mean" (of a and b). Old Babylonian tablets bearing calculations involving such quadrilaterals, with given numerical dimensions, show that this relation was known nearly 4,000 years ago. It was doubtless discovered in the first place in connection with particular, simple examples, probably examples in which the parallel sides were taken perpendicular to one of the remaining sides.

(d) By similar triangles (diagram (d) of Fig. S10),

$$c_1/a = y/(y + w) \quad (1)$$
$$c_2/a = z/(z + x) \quad (2)$$
$$w/y = x/z = a/b. \quad (3)$$
$$(3) \Rightarrow (y + w)/y = (z + x)/z = (a + b)/b$$
$$\Rightarrow y/(y + w) = z/(z + x) = b/(a + b). \quad (4)$$

Substituting from (4) into (1) and (2):

$$c_1/a = b/(a + b),\ c_2/a = b/(a + b),$$
$$\Rightarrow c_1 = c_2 = ab/(a + b);$$
$$\therefore HK = 2ab/(a + b), \text{ the required harmonic mean.}$$

10. (a) The areas of similar triangles PMX, RSX are as $1:4$. Since triangles of equal height are to one another as their bases (Euclid, VI, 1), $\triangle PMX : \triangle PXS = 1:2$, and $\triangle PXS : \triangle XRS = \ldots.$ Show that $\triangle PXS = \frac{1}{6}$ of $PQRS$, $MQRX = 5/12$ of $PQRS \ldots.$

(b)
$$\triangle AXZ = \frac{1}{3} \times \frac{6}{11} \text{ area } \triangle ABC; \quad \triangle BYX = \frac{2}{3} \times \frac{3}{7} \text{ area } \triangle ABC;$$
$$\triangle CZY = \frac{4}{7} \times \frac{5}{11} \text{ area } \triangle ABC.$$
$$\therefore \triangle XYZ = \triangle ABC - \left(\frac{2}{11} + \frac{2}{7} + \frac{20}{77}\right)\triangle ABC = \ldots \text{ etc.}$$

This method of calculation may be justified by means of the theorem mentioned in the solution to part (a). In the present case, joining BZ,

$$\triangle AXZ : \triangle ABZ = AX : AB = 1:3 \qquad (1)$$
$$\triangle ABZ : \triangle ABC = AZ : AC = 6:11 \qquad (2)$$

From (1) and (2):

$$\triangle AXZ : \triangle ABC = \frac{1}{3} \times \frac{6}{11}.$$

Likewise, the two other initial statements of the calculation may be obtained.

(c) Taking the radius of the smaller sphere as the unit, the height, h say, of the cone is given by $(h - 5)/(h - 2) = \frac{1}{2}$, from which $h = 8$, and hence base radius $= \sqrt{8}$ units. Volume ratio, cone : larger sphere $= ((1/3)\pi \cdot 8 \cdot 8) : ((4/3)\pi \cdot 8)$, which comes to $2:1$.

(d) Required distance = perpendicular height of a right square pyramid having *all* its edges of length $2r$. Answer: $r\sqrt{2}$ units.
Last part: With alternate layers of 36 and 25 spheres, the number of layers is given by the maximum integral value of n for which, $r + (n - 1)r\sqrt{2} + r \not> 12r$, which comes to $n \not> 5\sqrt{2} + 1$. Hence 8 layers may be packed, and the required number of spheres is $4 \times 36 + 4 \times 25 = 244$.

11. (a) The triangle is *one-fifth* of the square.

(b) Start by analyzing a sketch of the figure as it will appear when completed and consider the bisectors of angles B and C.

(c) $TR = \sqrt{14}$ units (in each case!).

(d) Let QN meet XY perpendicularly at N. Then, since triangles YQN, YQZ are congruent, $QN = QZ$, $= a$ say. Further, QNX, QZP are congruent isosceles right triangles; hence $QX = QP = a\sqrt{2}$, and area of $\triangle PQR = (\frac{1}{2}\sqrt{2})a^2$. Let $XZ = b$; then area of $\triangle XYZ = \frac{1}{2}b^2$. $XZ = XQ + QZ$, so $b = a(\sqrt{2} + 1)$. Hence, $\triangle PQR : \triangle XYZ = (\frac{1}{2}\sqrt{2})a^2 : \frac{1}{2}a^2(\sqrt{2} + 1)^2, \ldots$ etc.

(e) In triangles ACE, ECB, $\angle ACE = \angle ECB$, and $AC:CE = EC:CB$, since each ratio $= \sqrt{2}:1$. Therefore triangles ACE, ECB are similar. (Euclid VI, 6: "If two triangles have one angle [of the one] equal to one angle [of the other] and the sides about the equal angles proportional, the triangles will be [mutually] equiangular and will have those angles equal which the corresponding sides subtend.") Further, $\angle EAC + \angle AEC = \angle ECD$, since $\angle ECD$ is exterior to $\triangle ACE$. But $\angle AEC = \angle EBC$, since these are corresponding angles in the similar triangles ACE, ECB. Hence, $\angle EAC + \angle EBC = \angle ECD$, that is $\angle EAD + \angle EBD = \angle ECD$.

12. (a) Draw TY parallel to CB, meeting AM at Y. From similar triangles XMB, XYT,

$$XM:XY = MB:YT = MC:YT \qquad (1)$$

From similar triangles AMC, AYT,

$$MC:YT = AC:AT = 3:2 \qquad (2)$$

From (1) and (2),

$$XM:XY = 3:2, \therefore XM = \frac{3}{5}YM = \frac{3}{5} \text{ of } \frac{1}{3}AM = \frac{1}{5}AM;$$

$$\text{hence } AX = \frac{4}{5}AM, \text{ so } AX:XM = 4:1.$$

(b) Applying the result of part (a), show that each isosceles triangle exterior to the star has one-fifth of the area of the outer equilateral triangle.

(c) Diagram (i) of Fig. P8: Let A be the center of the semicircle of radius $2R$, let B be the center of either one of the semicircles of radius R, and let C be the center of the circle of radius r. Then, in right triangle ABC, $AB = R$, $AC = 2R - r$, and hypotenuse $BC = R + r$. Hence,

$$R^2 + (2R - r)^2 = (R + r)^2,$$

from which $r:R = 2:3$.
Diagram (ii) of Fig. P8: Let A be the center of the semicircle of radius $2R$, let B be the center of the circle of radius R, and C the center of either of the circles of radius r', and let CN meet AB perpendicularly at N. Then, from right triangle BNC, $NC^2 = (R + r')^2 - (R - r')^2$, from which $NC^2 = 4Rr'$. All sides of right triangle ANC may now be dimensioned in terms of R and r' and the equation $r'^2 + 4Rr' = (2R - r')^2$ obtained, from which $r':R = 1:2$.

13. (b) The required altitude is the altitude of a right triangle having hypotenuse e and base $e/\sqrt{3}$, which comes to $e\sqrt{2}/\sqrt{3}$ (or $e\sqrt{6}/3$).

$$\text{Volume} = \frac{1}{3}\left(\frac{e}{2} \cdot \frac{e\sqrt{3}}{2}\right)\frac{e\sqrt{2}}{\sqrt{3}} = \frac{e^3 \cdot \sqrt{2}}{12} \quad (\approx 0.118e^3).$$

From $\frac{1}{64} > \frac{1}{72} > \frac{1}{81}$, it follows that $\frac{1}{8} > \frac{1}{6\sqrt{2}} > \frac{1}{9}$;

hence $\frac{e^3}{8} > \frac{e^3 \cdot \sqrt{2}}{12} > \frac{e^3}{9}$.

(c) Overall height $= 2\left(r + \frac{(2r)\sqrt{6}}{3}\right) = \frac{2}{3}(3 + 2\sqrt{6})r$ units.

14. (a) Join from the vertex of the angle at the circumference to the center and produce. Consider different cases. The theorem is Euclid, III, 20, but the Greeks did not recognize any angle greater than (or equal to) two right angles as being an angle at all, and so Euclid did not deal with the case where the angle subtended at the center is reflex. For Heath's extended commentary, see his *Thirteen Books of Euclid's Elements*, vol. 2, pp. 46–49.

(b) 135°.

(d) P is the point of intersection of two circles (or circular arcs), one on AB as diameter, the other passing through B, C and having its center, D say, outside the triangle, remote from A, such that the reflex angle $BDC = 270°$. (Angles BCD, CBD are, of course, each constructed equal to 45°.)

(e) Join from the center, O say, to K; let X denote the intersection of lines HL, KO, and let ON meet KL perpendicularly at N. Then, in right triangle ONK, $ON = \frac{1}{2}\sqrt{5}$ units. From similar triangles HXK, ONK, find HX and hence obtain $HL = (4/3)\sqrt{5}$ units.

15. (a) (i) Euclid, III, 22. (ii) Converse, not given by Euclid. (iii) Euclid, III, 32.

(ii) Let $ABCD$ be a quadrilateral having angles DAB, BCD supplementary. It will be sufficient to show that the circle through A, B, D passes through C. If this circle does *not* pass through C then it either cuts BC, at C' say, or BC produced, at C'' say. In the former case $BC'D$, as an exterior angle of triangle CDC', would exceed angle BCD. ("If one side of a triangle is produced, the exterior angle is greater than either of the interior opposite angles" —Euclid, I, 16.) But $\angle BC'D + \angle DAB = 2$ right angles (Euclid III, 22), and $\angle BCD + \angle DAB = 2$ right angles (datum); hence angles $BC'D$, BCD could not be unequal and the supposition (that the $\odot ABD$ cuts BC at C') having this implication must be rejected. In the latter case, in which the circle through A, B, D is supposed to cut BC

produced at C'', angle BCD, exterior to triangle CDC'', would have to exceed angle $BC''D$, but as each of these angles must be the supplement of angle DAB, the supposition requiring that they differ must likewise be rejected. Hence C cannot be other than on the circle through A, B, D, and so the quadrilateral $ABCD$ must be cyclic, given that its opposite angles are supplementary.

(b) By the alternate segment theorem, the angles marked α (see Fig. S11) are equal, and likewise the angles marked β, and $\alpha + \beta = 1$ right angle.

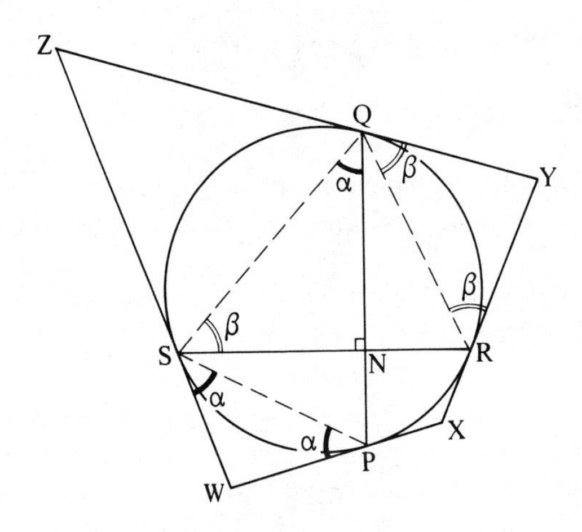

Fig. S11

Angles W and Y of quadrilateral $WXYZ$ are now easily shown to be supplementary. (Of course, if sufficiently prolonged, the tangents will be seen to meet in *six* points.) The same proof, with a slight modification, is applicable in the case where the perpendicular chords intersect only if produced outside the circle.

(c) See Fig. S12. Arcs AC, BD subtend complementary angles, θ and ϕ say, at the circumference, and therefore supplementary angles at the center. Hence

$$\text{arc } AC + \text{arc } BD = \text{half-circumference.}$$

Likewise, or by subtraction,

$$\text{arc } CB + \text{arc } AD = \text{half-circumference.}$$

If the perpendicular chords intersect outside the circle exactly the same relations hold, provided the *minor* arcs continue to be consistently taken: arc AC + arc BD (overlapping along arc AD) = (minor) arc BC + arc AD = half-circumference. (Or *major* arc BC *minus* minor arc

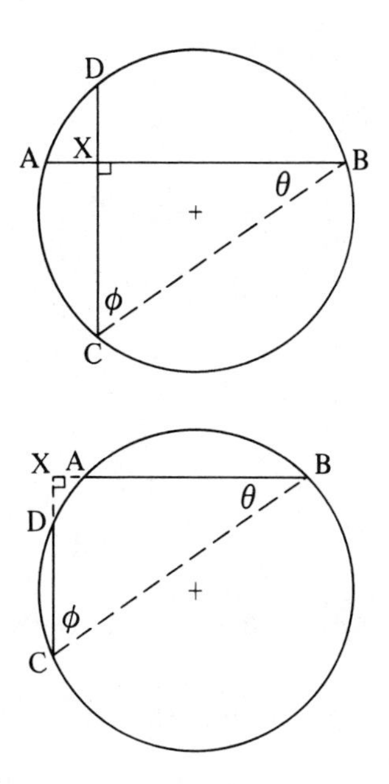

Fig. S12

AD = half-circumference.) Problems (c) and (d) are taken from the Archimedean (or pseudo-Archimedian?) *Book of Lemmas*, Propositions 9 and 11. The former proposition is worth consulting for a different solution to problem (c); see Heath (ed), *Works of Archimedes*, pp. 310–311.

(d) Complete the inscribed rectangle $CDD'C'$. The diameter parallel to CD bisects both arcs AB, DD' . . . . $(\text{diameter})^2 = CD'^2 = BD'^2 + BC^2$ $= \ldots .$

16. (a) The nearest proposition to this in Euclid is IV, 5: "About a given triangle to circumscribe a circle," in which it is proved that the point of intersection of the perpendicular bisectors of *two* sides is equidistant from the three vertices. If, then, this point is joined to the midpoint of the third side, it is easily proved (via congruent triangles) that the join meets the third side at right angles and so is the perpendicular bisector of this side. For an interesting commentary on this proof, see Hans Freudenthal, *Mathematics as an Educational Task*, Dordrecht: Reidel, 1973, p. 459.

(b) Referring to the upper diagram of Fig. S13, in which ABC represents any triangle with circumcenter O and OD perpendicular to BC: $\angle DOC = \frac{1}{2}\angle BOC = \angle BAC$. Hence, if $CN \perp AB$, $\triangle NAC$ is similar to $\triangle DOC$.

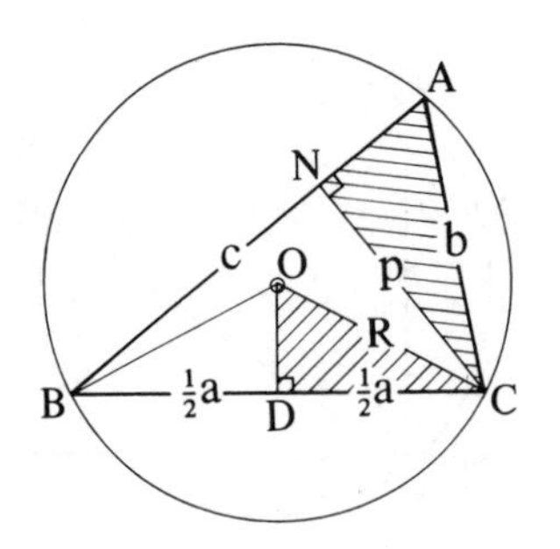

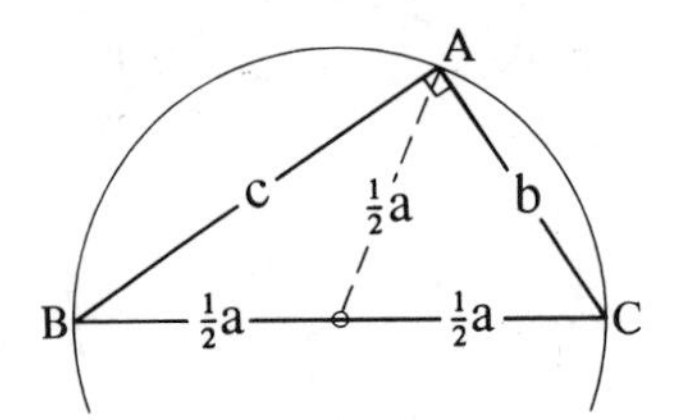

Fig. S13

$$\text{Area of } \triangle ABC = \frac{1}{2}cp \tag{1}$$

$$p : \frac{1}{2}a = b : R, \text{ giving } p = ab/2R \tag{2}$$

From (1) and (2): area $= abc/4R$. In the case of a right-angled triangle, the circumcenter is at the midpoint of the hypotenuse BC, equal to a in the second diagram. Thus $R = \frac{1}{2}a$, and so the expression $abc/4R$ immediately reduces to $\frac{1}{2}bc$.

As Boyer remarked, present-day notations owe more to Euler (1707–1783) than to any other mathematician: "The use of the small letters *a, b, c* for the sides of a triangle and of the corresponding capitals *A, B, C* for the opposite angles stems from Euler, as does the application of the letters *r, R*, and *s* for the radius of the inscribed and the circumscribed circles and the semiperimeter of the triangle respectively. The beautiful formula $4rRs = abc$ relating the six lengths also is one of the many elementary results attributed to him, although equivalents of this result are implied in ancient geometry. The designation lx for logarithm of x, the use of the now-familiar Σ to indicate a summation, and, perhaps most important of all, the notation $f(x)$ for a function of x . . . are other Eulerian notations related to ours." (Carl Boyer, *A History of Mathematics*, New York: Wiley, 1968, p. 485.)

(c) (i) Show that $CDHE$ is a cyclic quadrilateral, and hence that $\angle ECH = \angle EDH$.
(ii) Show that $ABDE$ is a cyclic, whence $\angle EBA = \angle EDA$ $(= \angle EDH)$.
(iii) From (i) and (ii), $\angle ECH = \angle EBA$ that is, $\angle ACF = \angle EBA$, where F is

the point of intersection of CH (produced if angles BAC and ABC are both acute) and AB.

(iv) Show that triangles AFC and AEB are similar, and hence that $\angle AFC = \angle AEB$ (= 1 right angle). . . .

(d) Divide $\triangle XYZ$ into six right triangles having a common vertex at its circumcenter and do the same for $\triangle ABC$ with respect to its circumcenter. Prove that the component triangles of $\triangle XYZ$ are, respectively, similar to and twice the linear scale of the (oppositely aligned) component triangles of $\triangle ABC$.

17. (b) (i) In the acute-angled case, make use of angles in the same segments of circles on AH, BH, CH as diameters. In the obtuse-angled case, supposing, for example, that angle A is obtuse, make use of angles in the same segments of circles on AH, AB, AC as diameters. In the obtuse-angled case, the orthocenter lies outside both triangles ABC and DEF, and so it cannot possibly coincide with the incenter of $\triangle DEF$. Nevertheless, it may be shown in the latter case that one of the altitudes of $\triangle ABC$ bisects one of the angles of $\triangle DEF$ internally while the remaining two altitudes bisect the other angles of $\triangle DEF$ externally. (The bisector of an exterior angle is called an external bisector.) The triangle formed by joining the feet of the altitudes of a given triangle is variously called the *pedal* triangle, the *altitude* triangle, and the *orthic* triangle of the given triangle. As this example shows, the altitudes of any acute-angled triangle bisect the interior angles of its pedal triangle, while those of any obtuse-angled triangle bisect one interior and two pairs of vertically opposite exterior angles of the pedal triangle. (An exterior angle of a triangle—or, in general, of a polygon—is formed by one side and the extension of an adjacent side; there are thus two vertically opposite, hence equal, exterior angles at each vertex.) Alternatively, we may say: the interior angles of the pedal triangle of any obtuse-angled triangle are bisected respectively by one altitude of the latter and two of its sides produced.

(ii) E_a (being equidistant from AB produced and BC) lies on the bisector of the exterior angles at B (as also does E_c, so E_aBE_c is a straight line); also E_a (being equidistant from AC produced and BC) lies on the bisector of the exterior angles at C (as also does E_b, so E_aCE_b is a straight line). Further, E_a (being equidistant from AB produced and AC produced) must lie on the bisector of the interior angle A of $\triangle ABC$. Hence AIE_a is a straight line. Likewise, BIE_b and CIE_c (and also E_bAE_c) may be shown to be straight lines. Consideration of the pairs of equal angles at A shows that $E_aA \perp E_bE_c$. So E_aA, and likewise E_bB, E_cC, are shown to be altitudes of $\triangle E_aE_bE_c$, and their point of concurrency is I, the incenter of $\triangle ABC$. (It is easy to show that $\triangle E_aE_bE_c$ is necessarily acute-angled no matter what the shape of its pedal triangle ABC.)

(c) A variation of the method indicated in the statement of the problem may be outlined as follows: Complete the quadrilateral $XYY'X'$, where Y', X' are the midpoints of AG, BG, respectively. Show that $XYY'X'$ is a parallelogram, and hence that AX, BY trisect each other at G. If G is defined as a point of trisection of one of the medians (remote from its vertex end), then, by the foregoing, it is also a point of trisection of a second median, and likewise of a third. (Make sure that you can prove the elementary propositions assumed here, especially (i) *The join of the midpoints of two sides of a triangle is parallel to the third side and half its length*, and (ii) *A quadrilateral having one pair of opposite sides both equal and parallel is a parallelogram*.)

(d) The line OGH is known as the Euler line of a triangle. The result is evident in the special case of a right-angled triangle. In general, consider a triangle ABC, with medians AX, BY, CZ, centroid G, circumcenter O. Produce OG to P such that $GP = 2 \cdot OG$. Then (i) triangles OGX, PGA are similar; consequently $PA /\!/ XO$ and so the straight line through P and A includes the altitude through A. Likewise (ii) it follows from the similarity of triangles OGY, PGB that $PB /\!/ YO$ and hence that the line through P, B includes the altitude through B. Finally (iii) by similar triangles OGZ, PGC, the line through P, C includes the altitude. Thus P is the orthocenter of $\triangle ABC$.

18. (a) $\frac{2}{15}\sqrt{15} : \frac{2}{5}\sqrt{15} : \frac{2}{3}\sqrt{15} : \frac{14}{15}\sqrt{15} = \frac{2}{15} : \frac{6}{15} : \frac{10}{15} : \frac{14}{15} = 1:3:5:7.$

Second part: $OH : IG = (1\frac{1}{6} + 5\frac{1}{3}) : (1\frac{1}{3} - 1) = 6\frac{1}{2} : \frac{1}{3} = 39 : 2.$

(b) (i) Angles CHD and ABC are each complements of $\angle HCD$ ($= \angle BCF$). . . . Otherwise use the fact that $\angle CHD$ is exterior to cyclic quadrilateral $BDHF$.
(ii) $\angle CH'A$ ($= \angle CH'D$) $= \angle CHD = \angle ABC$: hence H' and B lie on the same circle through A, C. . . .
(iii) The angle between the tangent at A and side AC = angle ABC in the alternate segment. Also $\angle ABC = \angle CHD$, from (i), and $\angle CHD = \angle AHF = \angle AEF$, since A, F, H, E are concyclic. The required parallelism may now be established via equal alternate angles. (Likewise, tgt at $B /\!/ FD$ and tgt at $C /\!/ DE$; thus the tangents to the circumcircle of a triangle at its vertices are respectively parallel to the sides of the pedal triangle.)

(c) Consider an auxiliary triangle formed by segments of (i) any median, (ii) a line through the vertex terminating this median drawn parallel to a second median, (iii) the remaining median produced. Show that the original triangle must have three times the area of any such auxiliary triangle. *Answer*: 8 square units.

Second part: sides $3\frac{1}{3}$, $(4/3)\sqrt{13}$, $\frac{2}{3}\sqrt{73}$ units. Since the square on the longest side is less than the sum of the squares on the other two sides, the triangle is acute-angled.

19. (a) For P inside the parallelogram, the sum of the distances to the sides is constant, being equal to the sum of the distances between each pair of opposite sides. (For P outside the parallelogram, two significantly different cases should be distinguished.) The minimum value of $PA + PB + PC + PD$ is equal to $AC + BD$, occurring for P at the point of intersection of the diagonals.

(b) (i) If the required altitude has length p units and divides the given triangle into two right triangles with sides x, p, 4 units and $5 - x$, p, 6 units, respectively, then $p^2 = 4^2 - x^2$, and $p^2 = 6^2 - (5 - x)^2$, from which $x = \frac{1}{2}$, $p = (3/2)\sqrt{7}$. Otherwise: area $= \frac{1}{2} \cdot 5 \cdot p = \sqrt{\{s(s-a)(s-b)(s-c)\}}$, etc.

(ii)

$$\frac{1}{2}\cdot 4 \cdot r + \frac{1}{2}\cdot 5 \cdot r + \frac{1}{2}\cdot 6 \cdot r = \text{area of } \Delta, \text{ that is } rs = \Delta.$$

$$r = \Delta/s = \frac{1}{2} \cdot 5 \cdot \frac{3}{2}\sqrt{7} \cdot \frac{2}{15} = \frac{1}{2}\sqrt{7}.$$

General case: Let the sides have lengths a, $a - d$, $a + d$ units, and the altitude to the side of length a be p units long, then the area is given by $\Delta = \frac{1}{2}ap$, and also by $\Delta = rs$ (where here $s = 1\frac{1}{2}a$); hence $1\frac{1}{2}ar = \frac{1}{2}ap$, giving $r : p = 1 : 3$.

(c) Let the side lengths be a, b, c units; altitudes 12, 20, x units;

$$\text{Area} = 6a = 10b = \frac{1}{2}cx \tag{1}$$

$$a + b > c > a - b \tag{2}$$

From (1), $b = \frac{3}{5}a$, hence (2) gives

$$\frac{8}{5}a > c > \frac{2}{5}a \tag{3}$$

Also from (1), $x = 20b/c$, while from (3),

$$\frac{5}{8a} < \frac{1}{c} < \frac{5}{2a} \tag{4}$$

Multiply (4) through by $20b$ (or its equivalent, $12a$) to obtain $7\frac{1}{2} < x < 30$.

(d) Let the side length of the equilateral triangle be a units, and its altitude h units, and let p, q, r denote the respective measures of the distances from the point to the sides. Working with alternative expressions for the area of the equilateral triangle, show that, when the point is anywhere inside it, $p + q + r = h$, whereas for the point in each of the six easily defined regions outside the triangle (lying between pairs of sides produced), $p - q - r = h$, $p + q - r = h$, $-p + q - r = h$, $-p + q + r = h$, $-p - q + r = h$, and $p - q + r = h$, respectively.

20. (a) In each case (see Fig. S14), triangles EAC, EDB are readily proved similar with $EA : ED = EC : EB$, implying $EA \cdot EB = EC \cdot ED$ [$= EC^2$ in case (iii)]. These three cases are the subject of Euclid III, 35, 36, 37, but Euclid avoided the use of similar triangles since the theorems on similar triangles came later in his sequence (Euclid VI, 4 ff.).

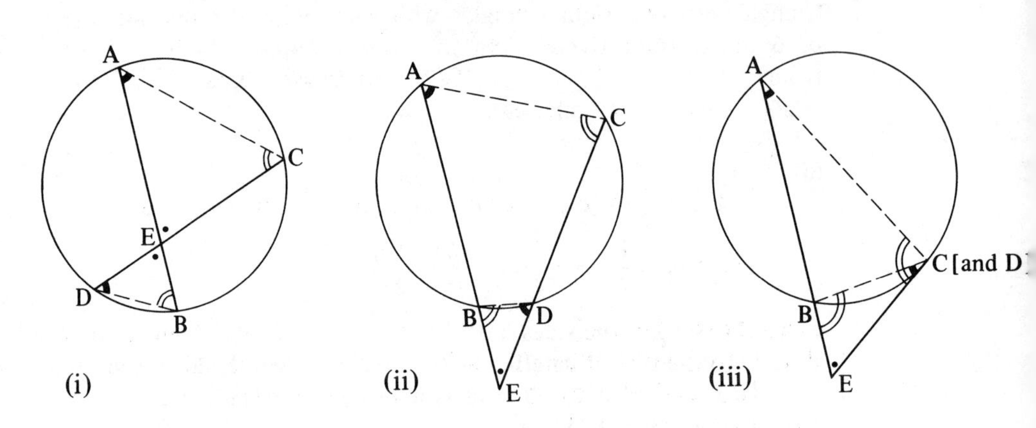

Fig. S14

(b) $A'A^2 = A'A \cdot AA'' = 5 \times 3$; $\therefore$ $AA' = \sqrt{15}$. Likewise, $BB' = \sqrt{12}$, $CC' = \sqrt{7}$ units. Otherwise, by Pythagoras' theorem applied to $\triangle OAA'$, $AA' = \sqrt{(4^2 - 1^2)} = \ldots$, etc.

The diagram on the right in Fig. S15, from Sawaguchi Kazuyuki's *Kokon Sampō-ki* ("Old and New Mathematical Methods"), *c.* A.D. 1670, illustrates an approximation to the area of a circle by the crude integration of rectangular strips. An analogous plan had already been used a few years earlier for estimating the volume of a sphere via the summation of cylindrical disks. In one such calculation, by Muramatsu, 1663, based on 100 equally spaced sections, a volume of 524 was obtained for a sphere of radius 5 units (correct value $(4/3)\pi \times 5^3 = 523.598\ldots$). See David Eugene Smith and Yushio Mikami, *A History of Japanese Mathematics*, Chicago: Open Court, 1914, pp. 84–87; our reproduction of Sawaguchi's illustration is taken from p. 87 of Smith and Mikami.

(c) A line segment of length $\sqrt{10}$ units may be obtained by the semicircle construction for finding the mean proportional between 1 and 10 or,

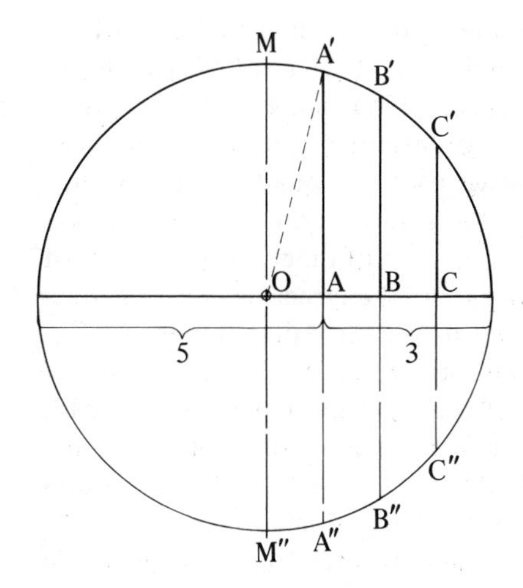

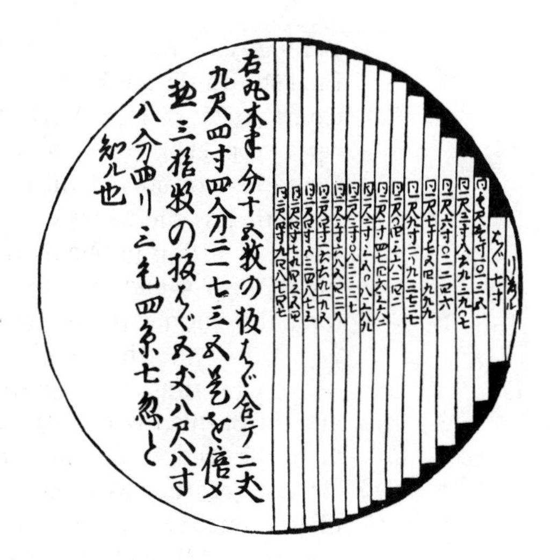

Fig. S15

better, between 2 and 5 units. Alternatively, of course, the hypotenuse of a right triangle with sides 1 and 3 has length $\sqrt{10}$ units. For $\sqrt[4]{2}$, obtain the mean proportional between 1 and $\sqrt{2}$. ($1 : \mu = \mu : \sqrt{2} \Rightarrow \mu^2 = \sqrt{2}$, so $\mu = \sqrt[4]{2}$.)

(d) This example is from the Archimedean "Book of Lemmas," Proposition 4. The proof attributed to Archimedes is as follows (Heath (ed.), *Works,* pp. 304–305):

$AB^2 = AN^2 + NB^2 + 2AN \cdot NB = AN^2 + NB^2 + 2PN^2$. But circles (or semicircles) are to one another as the squares of their diameters. Hence, (semicircle on AB) = (sum of semicircles on AN, NB) + 2(semicircle on PN). Thus, the circle on PN as diameter is equal to the difference between the semicircle on AB and the sum of the semicircles on AN, NB, i.e. is equal to the area of the ἄρβηλος [literally the "shoemaker's knife", indicating the region shown shaded in Fig. P10].

21. (a) The typically Greek construction method shown in diagram (i) of Fig. P12 might have been used as early as the fifth century B.C. By the intersecting-chords theorem (preceding problem, part (a)) or from similar triangles RPN, PQN, $w(4 + w) = 3^2$. Clearly, from the diagram, the radius of the circular arc is $\sqrt{13}$ units, and so the required value of w is $\sqrt{13} - 2$. The construction indicated in diagram (ii) of Fig. P12 provides an alternative determination of the same unknown. Since $AB \cdot AD = AT^2$, that is, $w(w + 4) = 3^2$, and $AB = AC - CB = \sqrt{13} - 2$, the same root is obtained as before. This construction and related constructions for other forms of quadratic equations appear in the first book of Descartes' celebrated *Géométrie* of 1637. See *The*

Geometry of René Descartes, facsimile of the original French with an English translation by D. E. Smith and M. L. Latham, Chicago: Open Court, 1925, or New York: Dover reprint, 1954, pp. 12–17. The classical Greek geometers had no notion of negative numbers, and even Descartes considered only the positive roots of quadratic equations. Nevertheless the negative as well as the positive root of the equation $w(w + 4) = 9$, is implicitly represented in each of diagrams (i) and (ii) of Fig. P12. If w and $w + 4$ are both negative, the larger of these two factors, the one nearer to zero, is given in the first diagram by $QN = 2 - \sqrt{13}$, while the smaller is 4 units less than this, namely, $NR = -2 - \sqrt{13}$; thus, the negative root of the equation is $w = -2 - \sqrt{13}$, and the product $w(w + 4)$ is easily verified as having the required value, 9. In the second diagram, in which $AB \cdot AD = AT^2$, if both AB and AD are regarded as negative, then AD, of magnitude $2 + \sqrt{13}$, will be dimensioned as $-2 - \sqrt{13}$ ($= w$), in which case AB, four units nearer to zero than AD, will be $+2 - \sqrt{13}$ ($= w + 4$).

(b) What is probably the most obvious method is sufficiently indicated in Fig. S16. This involves the use of an arc of the curve $ABCDE$ . . . . Termed a *conchoid* (alternatively *cochloid*), this "shell-like" curve was the subject of a lost work by Nicomedes in the 3rd century B.C. The curve may well have been used (or considered and rejected!) before Nicomedes' time for such "neusis" constructions as the one shown here. The term νεῦσις (= verging, inclination) came to be used by the Greeks to describe constructions in which a given line interval (PT, in the present example) has to be inserted between two curves (here the arc of radius 3, center Q, and the perpendicular bisector of QR) *and* to be

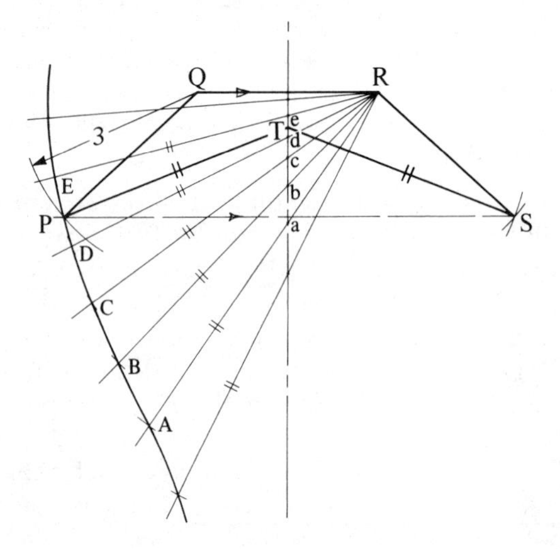

Fig. S16

placed in such a direction that it converges toward, or diverges away from, a given point (R). Without special apparatus enabling the curve to be generated mechanically, it would have to be drawn in freehand through previously plotted points, of which A, B, C, etc. are representative. Although such points could be located by straightedge-and-compasses construction, the curve itself could not be so produced. Departures from the preferred austerity symbolized by the straight line and circle were tolerated where they seemed unavoidable, and indeed the Greeks studied numerous higher curves. But in the present example, the following method would have been preferred.

Analysis: Complete the triangle RTQ; clearly this must be similar to triangle PQR. Then $RT:PQ = QR:RP$, or, denoting the length of RT by w units, $w:3 = 3:(w+4)$. Hence, $w(w+4) = 9$, where w can be found by straightedge-and-compasses constructions such as those given in part (a), or w can immediately be calculated from sketch diagrams corresponding to these constructions. Thus $w = \sqrt{13} - 2$, and $w + 4 = \sqrt{13} + 2$. Hence T may be located by the intersection of an arc of radius w, center R (or Q) with the perpendicular bisector of QR. RT produced 4 units beyond T locates P, or, better, locate P by the intersection of arcs of radii 3 and $w + 4$ units, respectively, centered at Q and R. S, of course, will be at the intersection of arcs equal to these two, but with their centers interchanged.

In connection with his quadrature of lunes, Hippocrates of Chios (second half of the 5th century B.C.) was faced with the construction of a symmetrical pentagon of the same general type as $PQRST$. Some details of his solution have survived in the account of Simplicius in his commentary on Aristotle's *Physics*. Simplicius (6th cent. A.D.) reproduced passages from Eudemus (4th cent. B.C.), and these in turn were claimed to contain direct quotations from Hippocrates. For references, see Solution 1; Heath, vol. 1, pp. 193–196; Thomas, vol. 1, pp. 242–247.

22. (a) OMP is a right angle for all allowable positions of CD. Consequently the locus is that part of a circle on diameter OP lying within the given circle.

(b) Let N, M be the midpoints of sides AB, CD, and let O, in the line MN, be the center of the major arc of the circle at all points on which AB subtends angles of 60°; then $\angle AON = 60°$, $\angle BAO = 30°$. P, Q are the points of intersection of the arc, center O and radius OA, with side CD. $AN = 1\frac{1}{2}$, $ON = \frac{1}{2}\sqrt{3}$, $OA = \sqrt{3}$, $OP = OQ = \sqrt{3}$, $OM = 2 - \frac{1}{2}\sqrt{3}$; hence

$$PQ = 2\sqrt{\left\{(\sqrt{3})^2 - \left(2 - \frac{1}{2}\sqrt{3}\right)^2\right\}},$$

simplifying to $\sqrt{(8\sqrt{3} - 7)}$ units.

(c) The critical positions of P are indicated in Fig. S17. $BO = CO = AM = 6\frac{1}{2}$ units, hence $AP = MO = \sqrt{\{(6\frac{1}{2})^2 - (2\frac{1}{2})^2\}} = 6$ units. Likewise, of course, $AP' = 6$ units. Alternatively, $AP^2 = AP'^2 = AB \cdot AC$ . . . . For any point, Q say, other than P or P' on the perpendicular to ABC, $\angle BPC$ (or $\angle BP'C$) $= \angle BXC > \angle BQC$, where X is the point of intersection of BQ and the circle.

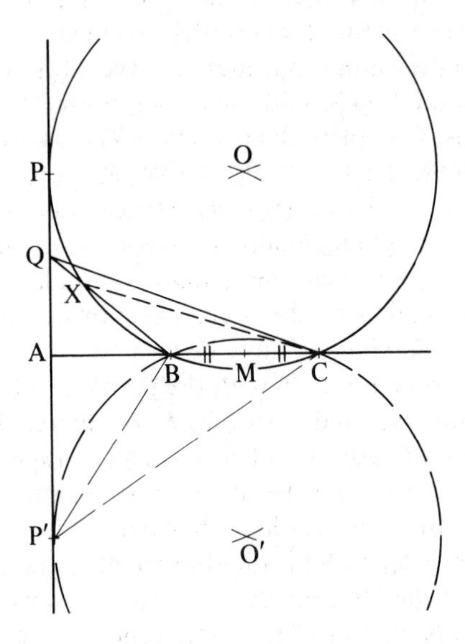

Fig. S17

(d) The diagonals of a parallelogram bisect each other; they do *not* bisect the angles of the parallelogram (except in the special case when it is a rhombus). Since the angles are as 2 : 1, the larger are each two-thirds of two right angles (120°), and the smaller are each one-third of two right angles (60°). Start by laying down one of the diagonals first; if this is the shorter one, then the longer diagonal will have its ends on the circle centered at the midpoint of the shorter one, and of radius equal to half the longer diagonal (or three-quarters of the shorter one). Also, since the shorter diagonal has to subtend angles of 60° at the ends of the longer diagonal, these endpoints must lie on the arcs of (major) segments of circles containing angles of 60°, the common chord of these two segments being the shorter diagonal itself.

In planning the final calculation, it is necessary to resist being led astray by the erroneous assumption that the shorter diagonal meets the shorter sides at right angles, for all that this might be suggested by a carelessly viewed scale drawing (see Fig. S18).

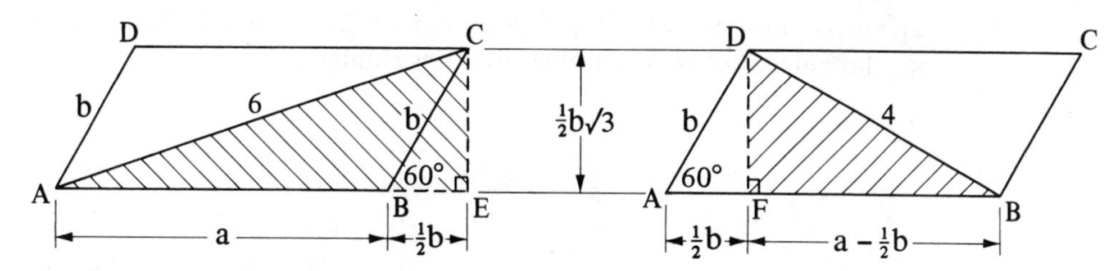

Fig. S18

From rt. $\triangle AEC$, $\left(a + \frac{1}{2}b\right)^2 + \left(\frac{1}{2}b\sqrt{3}\right)^2 = 6^2$,

giving $a^2 + ab + b^2 = 36$ (1)

From rt. $\triangle BFD$, $\left(a - \frac{1}{2}b\right)^2 + \left(\frac{1}{2}b\sqrt{3}\right)^2 = 4^2$,

giving $a^2 - ab + b^2 = 16$ (2)

From (1) and (2), $ab = 10$.

Area of parallelogram $= AB \cdot DF = a\left(\frac{1}{2}b\sqrt{3}\right) = 5\sqrt{3}\ \text{cm}^2$.

23. (a) Denote the interior angles of the general quadrilateral, $ABCD$ say, by 2α, 2β, 2γ, 2δ, respectively, and let the bisectors of these angles meet at P, Q, R, S, as shown. Then the interior angle P of $PQRS = \angle APB$ of $\triangle APB$ (first diagram of Fig. S19) *or* is vertically opposite to $\angle APB$ (second diagram). In either case, $\angle SPQ = 180° - \alpha - \beta$. Likewise, $\angle PQR = 180° - \beta - \gamma$, $\angle QRS = 180° - \gamma - \delta$, $\angle RSP = 180° - \delta - \alpha$. So $\angle SPQ + \angle QRS = 360° - \alpha - \beta - \gamma - \delta = 180°$. Thus $PQRS$ has its opposite angles supplementary and is therefore cyclic.

Let the bisectors of the exterior angles of $ABCD$ meet at P', Q', R', S', as indicated. Then, in quadrilateral $P'Q'R'S'$, $\angle P' = \alpha + \beta$, $\angle Q' = \beta + \gamma$, $\angle R' = \gamma + \delta$, $\angle S' = \delta + \alpha$. Hence the sum of each pair of opposite angles is $\alpha + \beta + \gamma + \delta$, or 2 right angles. Thus $P'Q'R'S'$ is cyclic.

It is to be noticed that even though the sides of $P'Q'R'S'$ are respectively perpendicular to the sides of $PQRS$, these two quadrilaterals are not mutually equiangular—the corresponding angles being supplementary, not equal (except where they might happen to be right angles). Hence, $PQRS$ and $P'Q'R'S'$ cannot in general be similar. But an exceptional case arises if $ABCD$ is a parallelogram, as is shown in Fig. S20. (Of course, mutual equiangularity is only a necessary condition and not a sufficient one for the similarity of quadrilaterals. The two quadrilaterals $ABCD$ in Fig. S19 have been drawn so as to be mutually equiangular, but they are obviously not similar.)

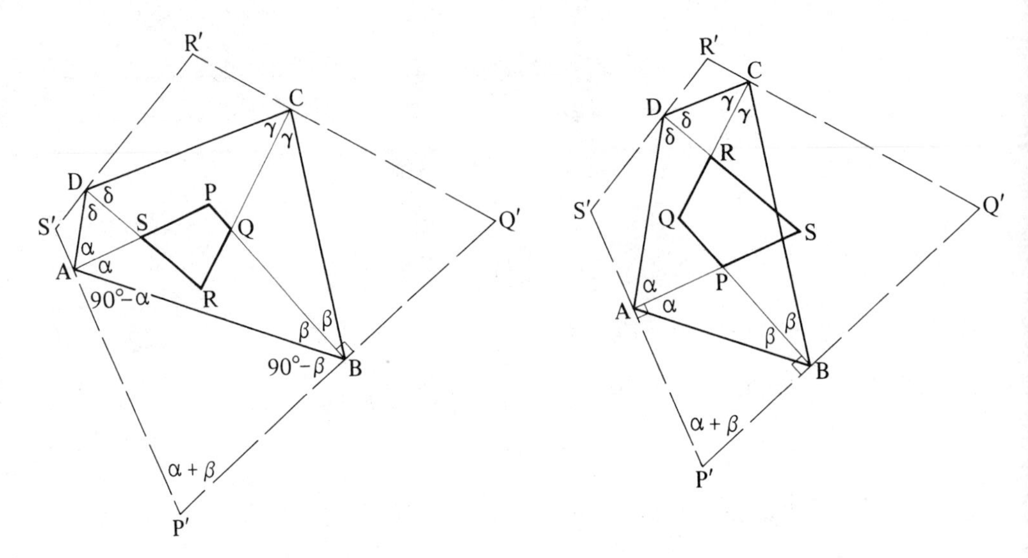

Fig. S19

Last parts, (i) squares, (ii) similar rectangles.
Working from Fig. S20, each angle of $PQRS$ and $P'Q'R'S' = \alpha + \beta = 1$ right angle; hence $PQRS$ and $P'Q'R'S'$ are rectangles. From the congruency of triangles ASD, ASZ, BQC, BQY, the lines marked a are equal, and likewise the lines marked b. Denoting the lengths of the adjacent sides of rectangle $PQRS$ by c, d, as indicated, the lengths of the sides of $P'Q'R'S'$ are given by $P'Q' = 2a + c$ and $Q'R' = 2b + d$. Further, since ASZ and YQB are congruent triangles, S, Q are equidistant from AB, that is $SQ \parallel AB$, and hence triangles QRS, ASZ are similar. Consequently,

$$\frac{d}{b} = \frac{c}{a}, \therefore \frac{2b + d}{b} = \frac{2a + c}{a}, \text{ so } \frac{Q'R'}{PQ} = \frac{P'Q'}{SP}.$$

Hence rectangle $P'Q'R'S'$ is similar to rectangle $QRSP$.

(b) $\triangle WPZ \equiv \triangle XQW \Rightarrow \triangle WPZ - \triangle WPT = \triangle XQW - \triangle WPT$, i.e., $\triangle WTZ =$ quad. $PXQT$. Quadrilaterals $PXQT$, $QYZT$ are easily shown to be mutually equiangular, with PX, XQ corresponding to QY, YZ, respectively, but, whereas $PX = QY$, $XQ < YZ$; thus the corresponding sides are not proportional and so the quadrilaterals cannot be similar.

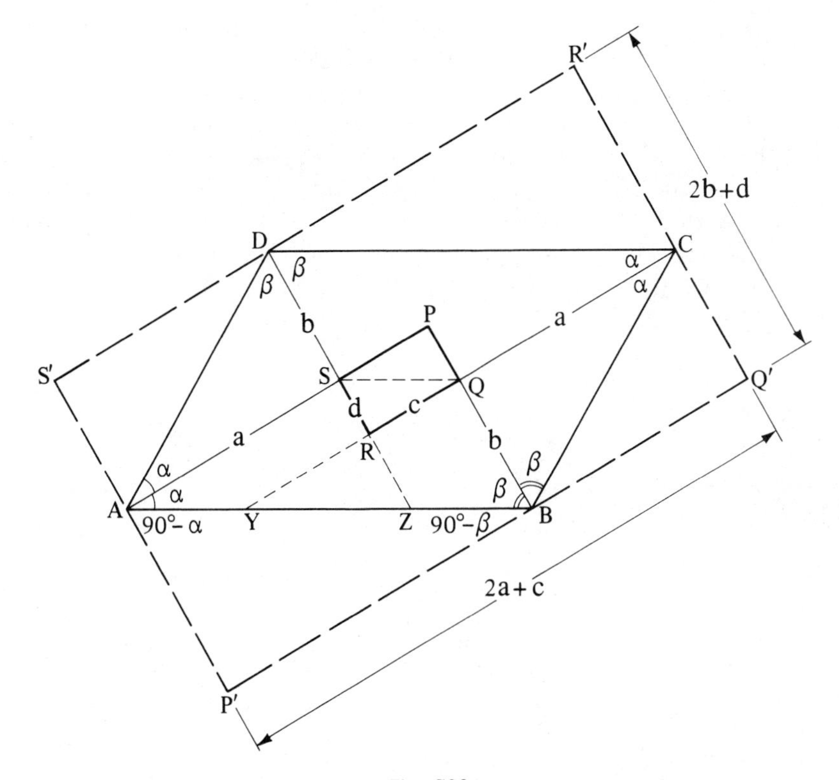

Fig. S20

In the last part, if the side length of $WXYZ$ be denoted by $2a$ units, then $WQ = ZP = a\sqrt{5}$, $WT = 2a/\sqrt{5}$, $PT = a/\sqrt{5}$, $ZT = 4a/\sqrt{5}$, $TQ = 3a/\sqrt{5}$, from which area $PXQT = 4a^2/5$, $QXZT = 11a^2/5$, ... Note that PQ, QZ are the diameters of circles circumscribing $PXQT$, $QYZT$. Hence the areas of these circumcircles are as $PQ^2 : QZ^2$, equal to $(a\sqrt{2})^2 : (a\sqrt{5})^2$, i.e., $2:5$.

24. (a) Let the square and the hexagon have side lengths a, b units, respectively; then $a/2 + a/2\sqrt{3} = b$, giving $a : b = (3 - \sqrt{3}) : 1$.

(b) A method different from the reduction construction indicated in diagram (b) of Fig. S21 can be devised by noting that $\angle OAN = \angle ONA = \angle BAN =$ one-quarter of a right angle. Further, $OC = ON \cdot \sqrt{2} = AO \cdot \sqrt{2}$, and $AC = AO + OC = r + r\sqrt{2}$. Alternatively, diagonal $= a\sqrt{2} = (r/\sqrt{2} + r) \cdot \sqrt{2} = r + r\sqrt{2}$. A third method, therefore, is to determine O by dividing the diagonal about which the semicircle is to be symmetrical internally in the ratio $1 : \sqrt{2}$. A neat way of achieving this is to describe an arc, center A, radius AB, cutting AC, at X

say. Then an arc, center X, radius XC, will cut AC again at the required center O (for $AO:OX:XC = r:r/\sqrt{2}:r/\sqrt{2} \Rightarrow AO:OC = 1:\sqrt{2}$). Required ratio, $r:a = (2-\sqrt{2}):1$. (Note: That the semicircle in this example is the *largest* that may be enclosed within the given square may be proved by comparing its radius with that of any semicircle *not* symmetrical about a diagonal but having the ends of its diameter on adjacent sides of the square and its arc touching one of the remaining sides. For, if the radius of this latter semicircle is denoted by r', it is easy to show that $r':a < (2-\sqrt{2}):1$.)

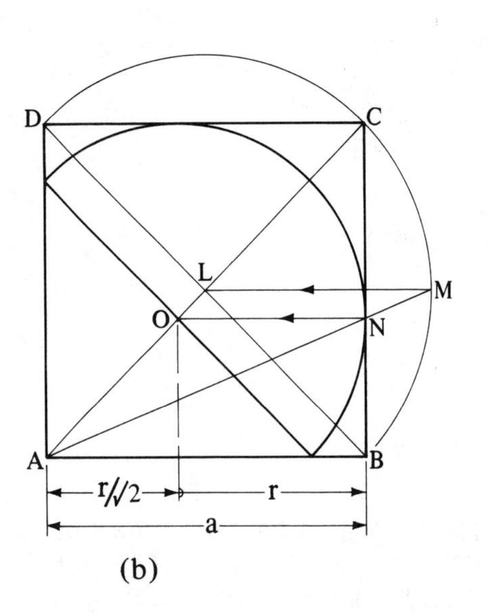

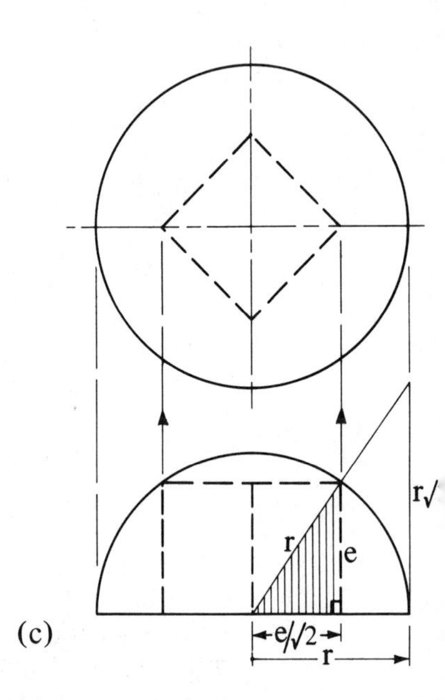

Fig. S21

(c) First part (not illustrated), $s:r = 2:\sqrt{5}$.
Second part (see diagram (c) of Fig. S21), $e:r = \sqrt{6}:3$.

25. (a) If the radii be denoted by r, ρ, R, in ascending order, the equation $(\rho - r)/(R - \rho) = (\rho + r)/(R + \rho)$, readily obtained by similar triangles, yields the required result.

(c) Analysis via a preliminary sketch (Fig. S22(i)) shows that when the figure is completed $PQCR$ will be a cyclic quadrilateral, and so $\angle BCP$ will be equal to $\angle PQR$ (which $= \angle QPA = 90° - \alpha$). Thus if $\angle A$ is acute P may be located by laying out $\angle BCP$ equal to the complement of $\angle A$. (The same construction is correct whether or not either $\angle B$ or $\angle C$ is obtuse.) The slight modification to the analysis appropriate for the case where $\angle A$ is obtuse will be evident from the diagram (ii).

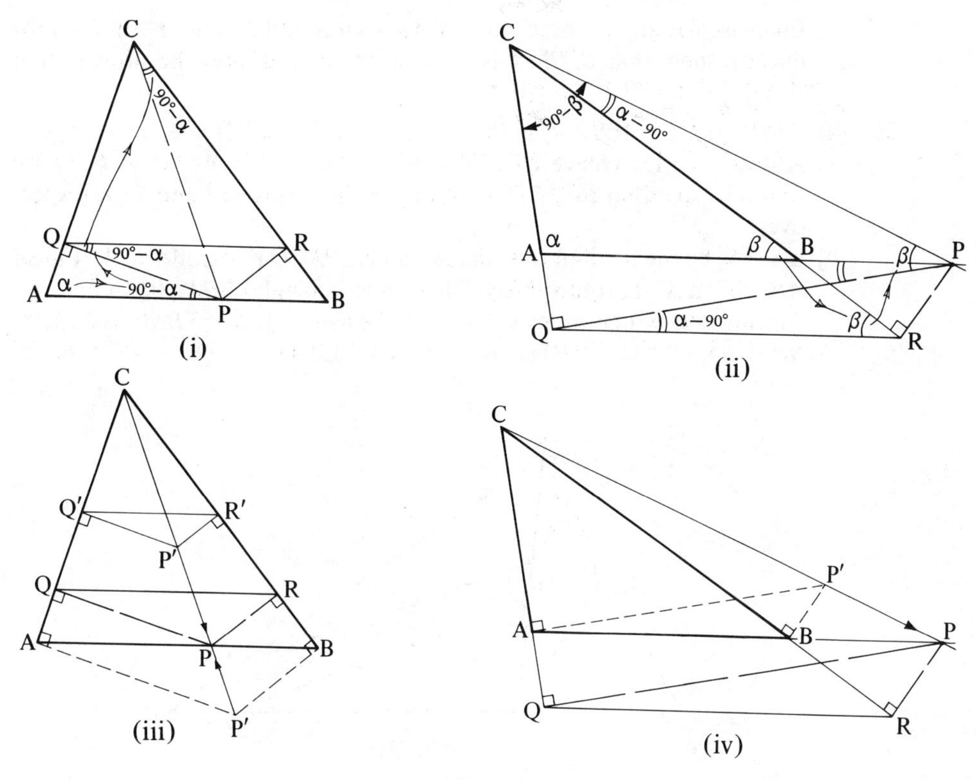

Fig. S22

Another method: If $Q'R'$ is any line drawn parallel to AB as shown in the diagram (iii), and P' is found by the intersection of the perpendiculars to AC, BC through Q', R', respectively, then $CQ'P'R'$ will be similar to $CQPR$ having P on AB and QR parallel to AB as required. Join CP', and produce if necessary to obtain P on AB. The construction is simplified if Q' and R' are taken to coincide with A and B, respectively. The same enlarging or reducing construction is applicable whether or not the given triangle is obtuse-angled (see diagram (iv)).

(d) Analysis indicates that X and Y can be located via parallelogram $XBYZ$ constructed by joining BP and producing to Z so that $PZ = BP$. . . . Let X' be any point on BA further from B than X; join $X'P$ and produce to meet BC in Y'. Then,

$$\begin{aligned}\triangle X'BY' &= \triangle XBY + \triangle XPX' - \triangle YPY'\\ &= \triangle XBY + \triangle XPX' - \triangle XPQ\end{aligned}$$

(where Q is the point of intersection of $X'P$ and ZX; $\triangle XPQ \equiv \triangle YPY'$),

$$\therefore\ \triangle X'BY' = \triangle XBY + \triangle XQX', \quad \text{and so} \quad \triangle XBY < \triangle X'BY'.$$

Likewise for any point X'' on BX between B and X, and Y'' on BY produced, such that $X''BY''$ is a straight line, it may be shown that $\triangle XBY < \triangle X''BY''$.

26. (a) $ABP:APCQ:AQD = 3:6:2 \Rightarrow ABP:ABCD = 3:11$; $AQD:ABCD = 2:11$. Hence $BP:BC = 6:11$; $QD:CD = 4:11$; so P, Q are found by dividing BC, CD internally in the ratios $6:5$ and $7:4$, respectively.

(b) Let AM be the median through A and let PX, QY, parallel to MA meet AB, AC in X, Y, respectively. Then, since triangles BPX, BMA are similar, and since their areas will have to be as $1:1\frac{1}{2}$, $BP^2:BM^2 = 2:3$, $\therefore$ $BP = BM\sqrt{2}/\sqrt{3} = 2BM/\sqrt{6} = BC/\sqrt{6}$. Likewise, $QC = BC/\sqrt{6}$. A

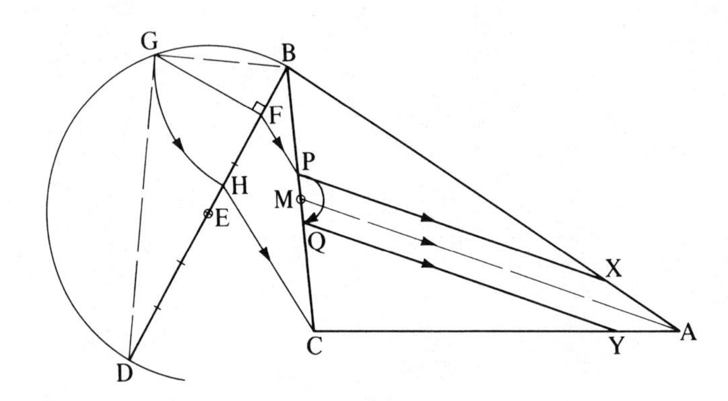

Fig. S23

convenient construction for dividing BC at P and Q is shown in Fig. S23. From similar triangles BFG, BGD, $BF:BG = BG:BD$, and the product of these ratios, $BF:BD$, $= 1:6$. $\therefore$ $BF:BG = 1:\sqrt{6}$. $BP:BC = BF:BH = BF:BG = 1:\sqrt{6}$. . . .

(c) First show that the radii are to be as $1:\sqrt{2}:\sqrt{3}$.

(d) In obvious notation,

(i) $A':A\ (= 1:2) = a'^2:a^2 \Rightarrow a':a = 1:\sqrt{2}$.

(ii) $V':V\ (= 1:2) = a'^3:a^3 \Rightarrow a':a = 1:\sqrt[3]{2}$.

(No straightedge-and-compasses construction is possible for obtaining lines in the ratio $1:\sqrt[3]{2}$.)

27. (a) (ii) $\triangle PAC = \triangle BAC - \triangle PAB - \triangle PBC$

$$= \frac{1}{2}ABCD - \frac{1}{10}ABCD - \frac{1}{10}ABCD$$

$$= \frac{3}{10}ABCD.$$ Thus, required ratio $= 3:10$.

(b) (ii) $PB^2 - PA^2 = PC^2 - PD^2$, from which $x = \sqrt{11}$.

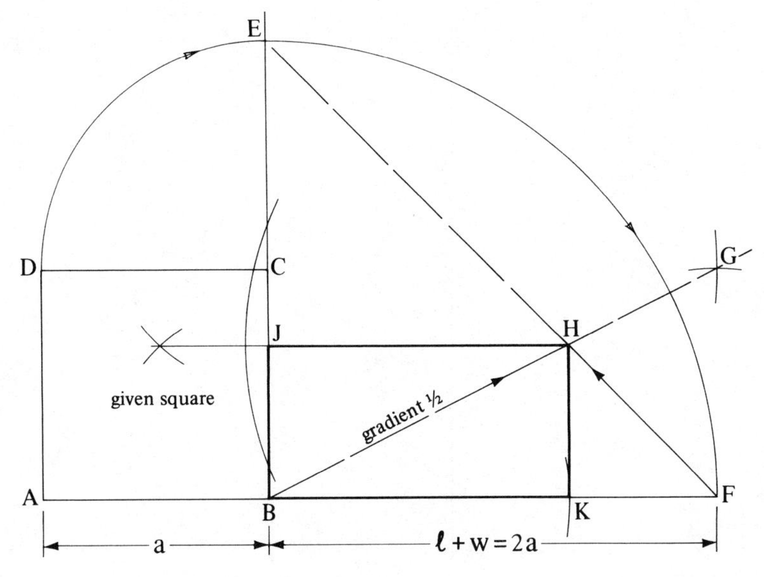

Fig. S24

(c) See Fig. S24. Rectangle: square = $\ell w/a^2 = 2w^2/a^2$, where $3w = 2a$, so $w/a = 2/3$; hence, required area ratio = 8 : 9.

(d) The construction (shown in Fig. S25 to a reduced scale) sets out three juxtaposed similar right-angled triangles, such that $a:b = b:c$, from the first pair, and $b:c = c:d$ from the second pair. Note how the dashed line and arc determine the right angle at R.

(i) $\dfrac{a}{b}\left(=\dfrac{b}{c}\right) = \dfrac{c}{d} \Rightarrow ad = bc.$

(ii) $\dfrac{a}{b} = \dfrac{b}{c} = \dfrac{c}{d}, = k$ say,

$$\begin{aligned} \Rightarrow a \cdot a \cdot d &= (kb)(kb)(c/k) \\ &= (b)(b)(kc) \\ &= b^3 \end{aligned}$$

This construction is related to the famous problem of the duplication of the cube in Greek mathematics. That problem was shown by Hippocrates of Chios to be reducible to the problem of inserting two mean proportionals (geometric means) between two given line segments. But this turned out to be impossible by straightedge-and-compasses methods. Notice in particular that the construction given here cannot be used to determine b and c if only a and d are given. See Heath's *History of Greek Mathematics*, vol. 1, pp. 200–201, 255–258.

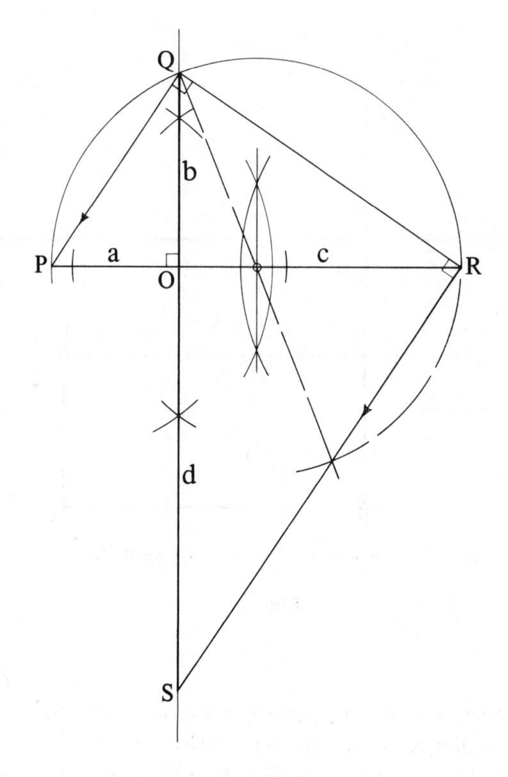

Fig. S25

28. (a) First method: $AC = \sqrt{5}$ (= a, say); $AF = \sqrt{10}$ (= b, say); $CF = \sqrt{13}$ (= c, say).

$$\text{Area of } \triangle ACF = \sqrt{\{s(s-a)(s-b)(s-c)\}}, \text{ where } s = \frac{1}{2}(a+b+c)$$

$$= \sqrt{\left[\frac{1}{2}(\sqrt{5}+\sqrt{10}+\sqrt{13}) \cdot \frac{1}{2}(-\sqrt{5}+\sqrt{10}+\sqrt{13}) \cdot \frac{1}{2}(\sqrt{5}-\sqrt{10}+\sqrt{13}) \cdot \frac{1}{2}(\sqrt{5}+\sqrt{10}-\sqrt{13})\right]}$$

$$= \frac{1}{2} \times \frac{1}{2}\sqrt{[\{(\sqrt{10}+\sqrt{13})^2 - (\sqrt{5})^2\} \cdot \{(\sqrt{5})^2 - (\sqrt{10}-\sqrt{13})^2\}]},$$

which simplifies to $3\frac{1}{2}$ (square units) exactly. The formula used here is given in verbal form by Hero of Alexandria, and is proved in two of his surviving texts: *Metrica*, I, 8 and *Dioptra*, chap. 30. But Hero appears to have been anticipated by Archimedes—see dialogue note 154.)

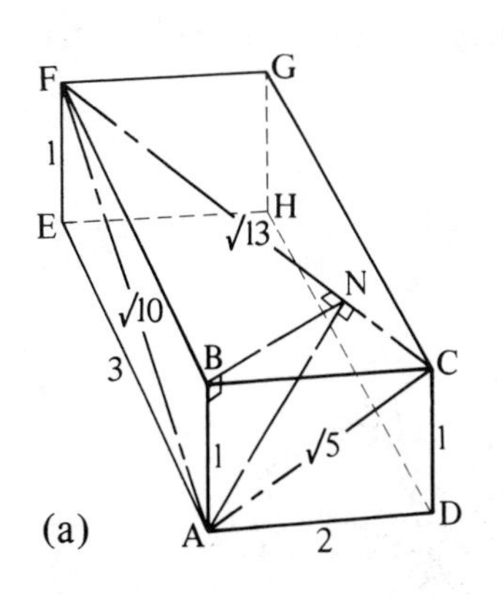

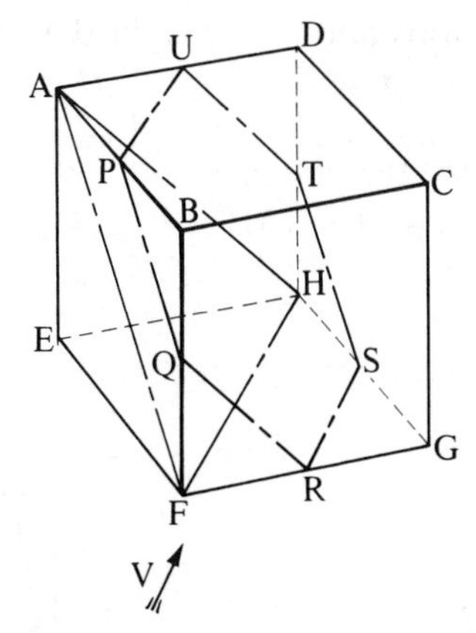

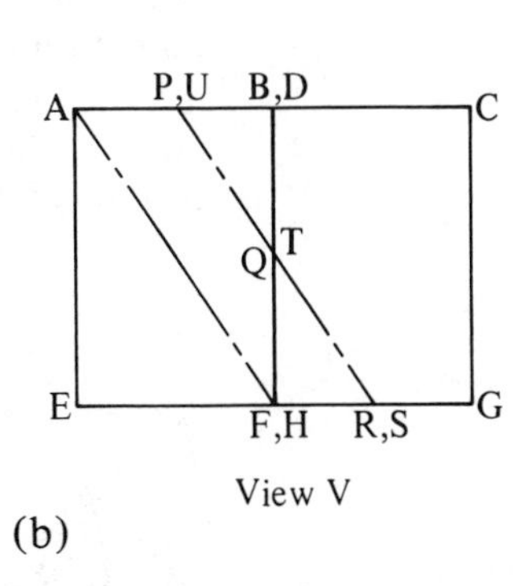

Fig. S26

Second method: See the first of the diagrams in Fig. S26. Area of $\triangle ACF = \frac{1}{2}FC \cdot AN$, where $AN = \sqrt{(AB^2 + BN^2)}$. From similar triangles BNC, FBC, $BN:3 = 2:\sqrt{13} \Rightarrow BN = 6/\sqrt{13}$. $AN = \sqrt{\{1^2 + (6/\sqrt{13})^2\}} \Rightarrow AN = 7/\sqrt{13}$. $\therefore$ Area of $\triangle ACF = \frac{1}{2}\sqrt{13} \cdot 7/\sqrt{13} = 3\frac{1}{2}$ square units.

(b) Volume on either side of plane $PQRSTU$ = half-volume of the cube; volume of pyramid $A,EFH = \frac{1}{3}(\frac{1}{2}EF \cdot EH) \cdot AE$ = one-sixth volume of cube. Hence, *one-third* volume of the whole cube lies between the two sectioning planes.

PQ is parallel to and half the length of AF; TS is parallel to and half the length of DG; DG is equal and parallel to AF; $\therefore$ TS is parallel and equal to PQ. Likewise it may be shown that each pair of opposite sides of $PQRSTU$ are parallel and equal, *and* that each side of the hexagon is parallel to a side of equilateral triangle AFH. Hence the hexagon is plane as well as equilateral, and its adjacent sides are inclined to each other at angles (exterior to itself) equal to the interior angles of an equilateral triangle. Hence the hexagon is equiangular as well as equilateral; therefore it is regular. (In more detail: the plane determined by PQ, QR is parallel to the plane determined by AF, AH; the plane determined by QR, RS is parallel to the plane determined by AH, FH, and so on around the hexagon, for: "If two straight lines meeting one another be parallel to two straight lines meeting one another, not being in the same plane, the planes through them are parallel"—Euclid, XI, 15.)

(c) (i) Surface area ratio, $A_{tet} : A_{cube} = 1 : \sqrt{3}$ ($= \sqrt{3} : 3$);
volume ratio, $V_{tet} : V_{cube} = 1 : 3$.
(ii) $A_{oct} : A_{cube} = \sqrt{3} : 6$; $V_{oct} : V_{cube} = 1 : 6$. Note that the octahedron in (ii) is related to the tetrahedron in (i) in the manner indicated in the first diagram of Fig. S27. Clearly, $A_{oct} : A_{tet} = 1 : 2$, and $V_{oct} : V_{tet} = 1 : 2$

also. These ratios allow checks on the foregoing independently determined ratios, since

$(A_{\text{oct}}/A_{\text{tet}})(A_{\text{tet}}/A_{\text{cube}}) = A_{\text{oct}}/A_{\text{cube}}$,

and likewise for the volumes.

(iii) $A_{\text{c-o}} : A_{\text{cube}} = (3 + \sqrt{3}) : 6$; $V_{\text{c-o}} : V_{\text{cube}} = 5 : 6$.

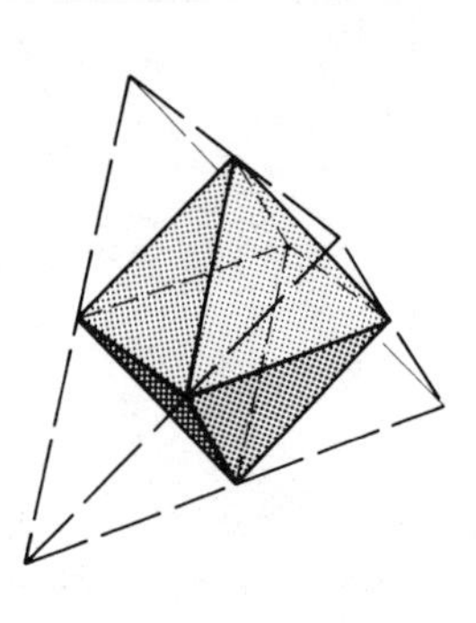

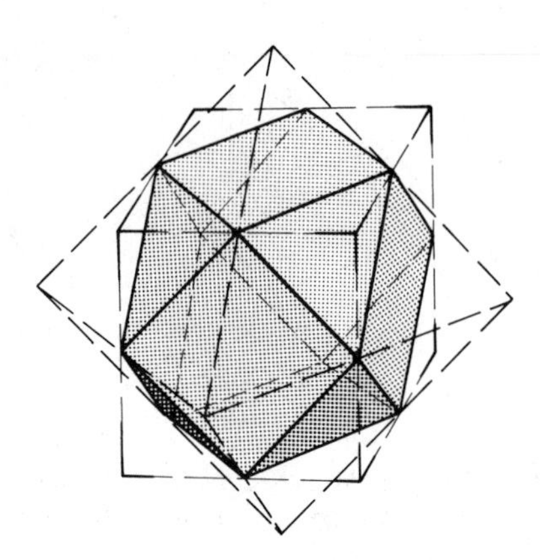

Fig. S27

The cuboctahedron can also be obtained by the removal of the six pyramids from a regular octahedron by sectioning planes through the midpoints of each set of four adjacent edges. Thus a cuboctahedron of edge length e may be regarded as the region common to a cube of edge $e\sqrt{2}$ and an octahedron of edge $2e$ arranged as indicated in the second diagram of Fig. S27.

Note: The cuboctahedron is one of the thirteen "semiregular" polyhedra investigated by Archimedes, viz., one of the solids having all their faces regular polygons of at least two different kinds, all their solid angles alike, and not less than four faces of any one kind. The last requirement might have been stated as "not fewer than *three* faces of any one kind" and equally well excluded the prisms with regular polygonal ends and square side faces, and the "antiprisms" with regular polygonal bases and equilateral side faces, but as a matter of fact no additional solids would be admitted by a change from the "not less than four" to the apparently less severe "not less than three" stipulation. Even apart from the last requirement, the condition that the solid angles be alike excludes the square-based and the regular pentagonal-based pyramids with equilateral triangular side faces. Refer to Hora Nona and especially to note 133. (An antiprism may be obtained from a prism by rotating one end face in its own plane so that each of its vertices corresponds to a side of the other end, and then, instead of having a twisted side surface, connecting up the end faces or "bases" by triangular side faces. The side edges zigzag back and forth alternately ending in a vertex first of one base and then of the other. An illustration is included in the historical note following Solution 52.)

29. (a) $d^2 - (d-4)^2 = 10^2$, from which $d = 14\frac{1}{2}$, $d - 4 = 10\frac{1}{2}$; hence sides of shaded triangle have lengths 10, $10\frac{1}{2}$, $14\frac{1}{2}$ units, and so are as 20 : 21 : 29. The "consequence" referred to in the question comes under Euclid VI, 5, viz.: "If two triangles have their sides proportional, the triangles will be [mutually] equiangular and will have those angles equal which the corresponding sides subtend."

(b) Triangles APM, ACD are similar ($\angle MAP = \angle DAC$, since each $= \angle MCP$); hence $AP : AC = AM : AD$, where $AM = \frac{1}{2}AC$, and $AC = \sqrt{29}$, from which $AP = 2.9$, and $AB = 2.0$, so $BP = 2.1$. For $AB = 4$, $AD = 7$, $AP/\sqrt{65} = (\frac{1}{2}\sqrt{65})/7 \Rightarrow AP = 65/14$; $AB = 56/14$, hence $BP = 33/14$. Consequently any triangle with sides as 33 : 56 : 65 is right-angled.

(c) Let R units be the radius sought; sides of shaded triangle have lengths R, $5 - R$, 2 units. $R^2 - (R-5)^2 = 2^2$, from which $R = 2.9$, $5 - R = 2.1$, . . . base = 2.0. . . .

The Old Babylonian mathematicians of around the 18th century B.C. were already solving problems like these, as were the Indian mathematicians at the time of Brahmagupta, 7th century A.D. The former probably and the latter certainly were in possession of the rule equivalent to the one given in parenthesis in part (d) of the present problem. Brahmagupta gave one formulation of it as follows: "The sum of the squares of two unlike quantities [is the measure of each of] the sides of an isosceles triangle; twice the product of the same two quantities is the perpendicular; and twice the difference of their squares is the base" (*Brāhmasphuṭa-siddhānta*, XII, 33). See dialogue note 178.

30. (a) Euclid (VI, 3) gives the case for the internal bisector only. The following proofs are a little simpler than Euclid's and *they apply alike to internal and external division*:

$$\frac{\triangle ABD}{\triangle ACD} = \frac{\frac{1}{2}AB \cdot DQ}{\frac{1}{2}AC \cdot DP} = \frac{AB}{AC},$$

since $DQ = DP$ from congruent triangles ADQ, ADP. Also

$$\frac{\triangle ABD}{\triangle ACD} = \frac{BD}{CD}; \quad \text{hence,} \quad \frac{BD}{CD} = \frac{AB}{AC}.$$

Otherwise: by similar triangles BDX, CDY, $BD : CD = BX : CY$,
and, by similar triangles ABX, ACY, $AB : AC = BX : CY$;
hence, $BD : CD = AB : AC$.

(b) Let the fixed points be B and C and let the moving point be A. If the bisector of $\angle BAC$ intersects BC at D and the bisector of the two exterior angles at A meets BC produced (if $AB > AC$), or CB produced (if $AC >$ AB), at D', then $BD : CD = AB : AC =$ a constant for all allowable positions of A. Also, $BD' : CD' = AB : AC$, the same constant ratio. Hence D and D' are fixed points for all positions of A in its locus. Further $\angle DAD' = \frac{1}{2}$ int. $\angle A + \frac{1}{2}$ ext. $\angle A = 1$ right angle. Hence the locus of A is a circle on diameter DD' (by the *converse* of Euclid III, 31, according to which theorem the angle in a semicircle is a right angle). Proof of this last claim: Let M be the midpoint of the hypotenuse of triangle DAD', right-angled at A. Let MN meet AD perpendicularly at N; then MN, being parallel to $D'A$, bisects DA as well as DD' (equal-intercept theorem). Hence triangles MNA, MND are congruent, and so $MA = MD$ $(= MD')$. Thus for B, C (and hence D, D') fixed points, the plane locus of A is a circle having DD' as diameter.

The locus asked for in part (b), known as the "circle of Apollonius," is of some importance in the history of mathematics and science. A correct theoretical treatment was already known to Aristotle in the fourth century B.C., and he uses it (in *Meteorologica*, III, 5) in connection with his explanation of the rainbow. Over one century later it appeared in a now lost work of Apollonius where (according to Eutocius's *Commentary on the Conics* [of Apollonius]) it was enunciated along the following lines: "Given two points in a plane and a ratio between unequal straight lines, it is possible to describe a circle in a plane such that the straight lines inflected from the given points to the circumference of the circle shall have a ratio the same as the given one"; see Heath's ed. of Euclid, vol. 2, pp. 197–200, notes to Euclid VI, 3. See also Galileo, *Two New Sciences*, First Day, 89–91, where the same result is established by a more complicated argument than that given here, but without recourse to the angle-bisector theorem. If the given constant ratio were 1 : 1, the locus would, of course, be the perpendicular bisector of the line joining the two given fixed points, or, as we say, a circle of infinite radius passing through the midpoint of the join of the given points. But, as Heath remarked in another connection (vol. 2, p. 75 of his edition of Euclid), "Euclid and the Greek geometers generally did not allow themselves to infer the truth of a proposition in a *limiting case* directly from the general case including it, but preferred a separate proof of the limiting case. . . ." And in the present problem, Galileo discusses at some length "this metamorphosis in passing from finite to infinite" whereby the circular locus "changes its being so completely as to lose its existence and its possibility of being [a circle]" (*Two New Sciences*, First Day, 84–85).

31. (a) The approach suggested in this question, applicable to the concave as well as to the convex case, provides a simple alternative to that set out in the dialogue, at the start of Hora Undecima, for proving that any quadrilateral having the sum of one pair of opposite sides equal to the sum of the other pair, is circumscribable about a circle. The proof for the special class where there are pairs of equal adjacent sides is too simple to require discusssion. As indicated in connection with Fig. 68, it is convenient to regard a concave quadrilateral or other polygon as circumscribable provided that all of its sides—*or sides produced*—can be made tangents to the same circle.

(b) (i) Triangles CDE, ABE are similar, $\therefore$ $CD:AB = DE:BE$, from which $CD = 2$ units. Also $ED \cdot EA = EC \cdot EB$, from which EA $(= 4 + DA) = 6$, $\therefore$ $DA = 2$ units.

(ii) Chords CD, DA are equal, therefore their arcs are equal and hence also the angles they subtend at the circumference: thus $\angle CBD = \angle ABD$. Alternatively: the bisector of $\angle ABC$, being the bisector of $\angle ABE$ of $\triangle ABE$, must cut the opposite side AE in the ratio of AB to BE, that is, in the ratio $1:2$. Since $AD:DE = 1:2$, the bisector of $\angle A$ is the line through B, D.

(iii) Since BX bisects $\angle ABC$, $AX:XC = AB:BC = 4:5$, $\therefore$ $\triangle AXD:CXD = 4:5$. . . . Since triangles AXB, CXD are similar, their areas are in the ratio AB^2 to $CD^2 = 4^2:2^2 = 4:1 = 20:5$. . . .

(iv) Since triangles FDA, FBC are similar,

$$FA:FC = DA:BC, \quad \text{i.e., } FA/(FD + 2) = 2/5 \qquad (1)$$
$$\text{and } FD:FB = DA:BC, \quad \text{i.e., } FD/(FA + 4) = 2/5 \qquad (2)$$

From (1) and (2), obtain $FA = 12/7$, $FD = 16/7$ units. Length of tangent, t units say, from F is given by

$$t^2 = FA \cdot FB \ (\text{or } FD \cdot FC) = \frac{12}{7} \times \frac{40}{7} \left(\text{or } \frac{16}{7} \times \frac{30}{7}, \text{ check!}\right)$$

Hence,

$$t = \frac{4}{7}\sqrt{30} \text{ units.}$$

(c) Show that the condition for orthogonal intersection is that $CA \cdot CB$ shall be equal to the square of the radius of the circle centered at C.

32. (a) Perhaps the most obvious of the numerous possible methods is the following: Bisect the angle formed by the bounding radii and construct the tangent to the arc at its midpoint. The required inscribed circle is the incircle of the isosceles right-angled triangle formed by the tangent just constructed and the bounding radii produced.

$$r + r\sqrt{2} = R, \quad \text{giving} \quad r:R = 1:(\sqrt{2} + 1) = (\sqrt{2} - 1):1.$$

Alternative construction methods should be suggested by this calculation and by the relationships indicated in part (b).

(b) (i) Let I be the center of the inscribed circle, then DCI, ECI are isosceles triangles and $\angle DCO = \frac{1}{2}\angle DIO$, . . . etc. Next, join ED and show that triangles AED, BDE are isosceles with exterior angles ODE, OED each equal to half a right angle. . . . Otherwise, use the converse of the theorem about the division of a side of a triangle by a bisector of the opposite angle; that is, show that $OE:EB = AO:AB$ (since each $= 1:\sqrt{2}$), and deduce that AE bisects $\angle OAB$. . . .

(ii) Let NA intersect OC at point X, then triangles OXA, CXN are similar with $OX:CX = OA:CN$ $(= OC:CN) = \sqrt{2}:1$. But $OI:IC = \sqrt{2}:1$ also. Thus the point of intersection X coincides with the incenter I.

(iii) As above, $OI + IC = OC$ yields $r\sqrt{2} + r = R$; also $OI + IC + CF = OF$ gives $r\sqrt{2} + r + CF = R\sqrt{2}$. From these two equations, it follows that $CF = r$, the incircle radius. This suggests what must surely be the simplest construction for determining the incenter of a given quadrant AOB: With centers A, B and radius equal to AO, describe arcs intersecting outside the quadrant at F. Join OF, cutting arc AB at C; with center C and radius CF, describe an arc cutting OC at I, the required incenter.

(c) A graphical solution via a plan and elevation, even if only sketched, focuses the attention on the essential relations almost as well as the construction of a model: First represent the inscribed sphere in "plan" (in Fig. S28, the top view) and draw the tangent lines as indicated forming the top view of two perpendicular bounding planes of the octant, thus determining the distance of their line of intersection, AB, from the center of the insphere. Next project the lower view, an "elevation" (or a cross-section through AB and the center of the insphere). Only then can the outer arc of the circumscribing octant be described, first in the lower view and then in the upper one. Clearly,

$$R = r\sqrt{3} + r; \quad \text{so} \quad R:r = (\sqrt{3} + 1):1.$$

33. (a) The required point P is the point of intersection of the join AB' and XY, for $AP + PB = AP + PB' < AO + OB' = AO + OB$, where O is any point in XY other than P. It is to be noted also that $\angle BPY = \angle B'PY = \angle APX$. This problem, or its equivalent, is known to us from Greek works on "catoptrics," or the theory of reflections. Archimedes wrote a work on the subject that is no longer extant. Euclid's *Optics* has survived, and there is also a *Catoptrics* that was previously ascribed to Euclid but is now regarded as a much later compilation, probably by Theon of Alexandria (late 4th cent. A.D.). This leaves the *Catoptrics* of Hero of Alexandria (second half of 1st cent. A.D., or first half of the

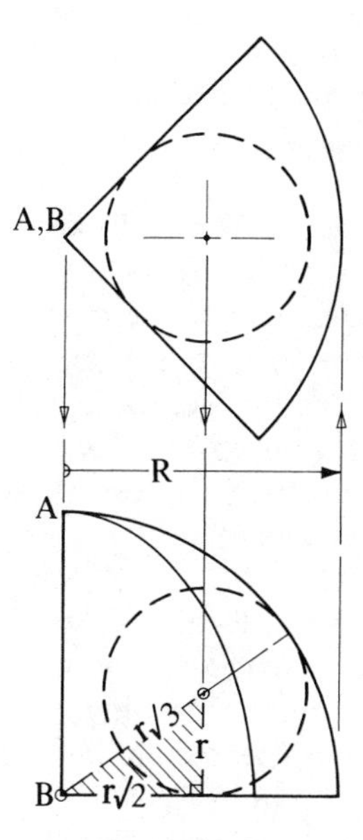

Fig. S28

2nd cent.), as the oldest extant work on reflections. As was usual with the Greeks, Hero assumed that the "visual rays" proceed *from* the eyes *to* the observed object and concludes that these rays move with infinite velocity "for when, after our eyes have been closed, we open them and look up at the sky, no interval of time is required for the visual rays to reach the sky. Indeed, we see the stars as soon as we look up" (cf. Galileo, *Two New Sciences*, First Day, 87–89). Hero proceeds to argue that, because of the infinite velocity, the rays can only travel by the shortest paths, i.e., in straight lines, and in cases of reflections from mirrors (spherical as well as plane), he proves geometrically that the shortest path is in each case that for which the reflected ray leaves the reflecting surface at an angle equal to that at which the incident ray meets it. For an English translation, see, Morris R. Cohen and I. E. Drabkin (eds.) *A Source Book in Greek Science*, New York: McGraw-Hill, 1948, pp. 261–268. Cf. Ivor Thomas, *Selections Illustrating the History of Greek Mathematics*, vol. 2, 1941, pp. 496–503, for a passage from Olympiodorus of Alexandria, 6th cent. A.D., beginning as follows: "For this would be agreed by all, that Nature does nothing in vain nor labours in vain; but if we do not grant that the angles of incidence and reflection are equal, Nature would be labouring in vain by following unequal angles, and instead of the eye apprehending the visible object by the shortest route it would do so by a longer. For straight lines so drawn from the eye to the mirror and thence to the visible object as to make unequal angles will be found to be greater than straight lines so drawn as to make equal angles.'' Then follows a proof, substantially the same as that given by Hero. (In keeping with their respective introductory arguments, both writers assume without comment that the incident and the reflected rays lie in the plane perpendicular to the reflecting surface.) See further, Edward Grant (ed.), *A Source Book in Medieval Science*, Cambridge, Mass.: Harvard Univ. Press, 1974, pp. 403–420, for extracts on visual rays and reflections from Alhazen (fl. A.D. 1000) and the thirteenth century scholars Witelo, Roger Bacon, and John Pecham.

(b) $SP + PQ + QF = S'P + PQ + QF'$, where S', F' are the images of S, F in CD, AB, respectively. Hence the required distance equals the hypotenuse of a right-angled triangle with sides 5, 12 units. $SH + HK + KF = 17$ units, by similar reasoning.

(c) (i) Since Q is on the perpendicular bisector of PP', $PQ = P'Q$; likewise, $RP = RP''$. $\therefore$ $PQ + QR + RP = P'Q + QR + RP''$, which is a minimum when Q, R are collinear with P', P''. Thus, for any arbitrary placement of P in BC, the indicated construction determines the location of the other two vertices of the inscribed triangle of least perimeter. But different positions of P along BC will lead to different values of the least perimeter for triangle PQR. In part (ii) the optimum position of P is sought in order that the corresponding inscribed triangle of minimum perimeter will have less perimeter than the minimum perimeter triangle corresponding to any other position of P on BC.

(ii) Join AP, AP', AP''. Since A is on the perpendicular bisector of PP', $AP' = AP$; likewise $AP'' = AP$. Also $\angle CAP' = \angle CAP$, and $\angle BAP'' = \angle BAP$. Hence, for any position of P along BC, triangle $P'AP''$ is isosceles and has its vertical angle, $P'AP''$, equal to twice angle A of the given triangle ABC. Clearly then, $P'P''$ has its least value when the equal sides AP', AP'' of triangle $P'AP''$ are as short as possible, that is when AP is least.

(iii) It follows from (ii) that the inscribed triangle of absolutely minimum perimeter must have the vertex (P) lying in BC at the foot of the altitude from A. An argument similar to that already given for P, applied to Q in relation to B, and to R in relation to C, could only show that BQ and CR must also be altitudes of triangle ABC in order that PQR be the inscribed triangle of absolutely minimum perimeter. Thus the required inscribed triangle is the pedal or altitude triangle of the given triangle. (Confirmation of this last argument is provided by showing that if D, E, F are given as the feet of the altitudes to sides BC, CA, AB of an acute-angled triangle ABC and FE is produced to X so that $EX = DE$ and EF is produced to Y so that $FY = FD$, then X, Y may readily be shown to be the images of D in CA, AB, respectively. This means that points E and F, defined as the feet of the altitudes from B, C in triangle ABC, are the very same points as Q and R defined by the construction involving the intersection of $P'P''$ (when P is at D) with CA and AB.)

This problem (c) was proposed and a calculus solution provided, in 1775, by G. F. Fagnano dei Toschi (son of the more distinguished mathematician C. C. Fagnano dei Toschi). A geometric solution involving six reflections was given by H. A. Schwarz (1843–1921), while the simpler solution given here was devised by L. Fejér (1880–1959) when still a young student. For a detailed discussion of the problem, see Hans Rademacher and Otto Toeplitz, *The Enjoyment of Mathematics: Selections from Mathematics for the Amateur*, translated by H. Zuckerman, Princeton, N.J.: Princeton Univ. Press, 1957, pp. 28–33 and 197–198.

34. (i) Draw $C'N$ to meet CA perpendicularly at N; then $AA' = NC' = \sqrt{(25^2 - 7^2)} = 24$ (cm).
(ii) Draw $C'N'$ to meet CF produced perpendicularly at N'; then $FF' = N'C' = \sqrt{(25^2 - 15^2)} = 20$ (cm).
(iii) $VF' = VB' = BB' - BV = 24 - BV = 24 - VF$. Also, $VF' = VF + FF' = VF + 20$. $\therefore VF + 20 = 24 - VF$, from which $VF = 2$, $VF' = 22$ (cm).
(iv) $V'F = V'A = AA' - V'A' = 24 - V'A' = 24 - V'F'$; also, $V'F = V'F' + F'F$, etc. . . . $V'F' = 2$, $V'F = 22$ (cm).
(v) $A'V'':AV'' = A'C':AC$, thus $A'V'':(24 + A'V'') = 4:11$, from which $A'V''$ (also $B'V''$) $= 13\frac{5}{7}$(cm).
(vi) Perimeter of $\triangle VV'V'' = 75\frac{3}{7}$ cm. $\therefore$ area $= \frac{1}{2} \times 75\frac{3}{7} \times 4 = 150\frac{6}{7}$ cm².

35. (a) Area ratio $= (\pi - 1):1$.
(b) Let the given line interval be denoted by AB: start by setting out a line BC such that $\angle ABC$ = half a right angle. Describe an arc, center A and with radius a, to cut BC. . . .
(c) In Fig. S29(c), from right triangle ACN, in which $AC = 1 + r$, $(1 + p)^2 + r^2 = (1 + r)^2$.
From right triangle BCN, in which $BC = 2 - r$, $p^2 + r^2 = (2 - r)^2$.
Hence, by subtraction, $(2p + 1)(1) = (3)(2r - 1)$, from which $p = 3r - 2$. Eliminate p, obtaining $r = 8/9$ unit.
(d) In Fig. S29 (d), $CZ = CY = 4$; $AC = 4 - 1 = 3 = AB$. $\therefore$ $\triangle ABC$ is isosceles, and half this triangle, viz., $\triangle ANC$, is similar to $\triangle YMA$. Consequently $MY:NA = AY:CA$, so $\frac{1}{2}XY:\sqrt{8} = 1:3$, from which $XY = (4/3)\sqrt{2}$ units. $PY = (4/3)NA = (8/3)\sqrt{2}$; $PZ = CZ - CN - NP$, where $NP = AM = (1/3)CN$ . . . obtain $PZ = 8/3$. Then, from right triangle PYZ, $YZ = \sqrt{(PY^2 + PZ^2)}$, from which $YZ = (8/3)\sqrt{3}$ units. Finally, $PX = PY - XY = (4/3)\sqrt{2}$, and $XZ = \sqrt{(PX^2 + PZ^2)}$, giving $XZ = (4/3)\sqrt{6}$ units.

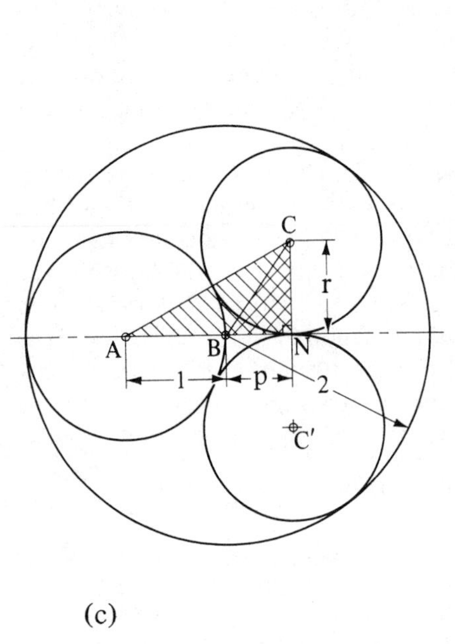

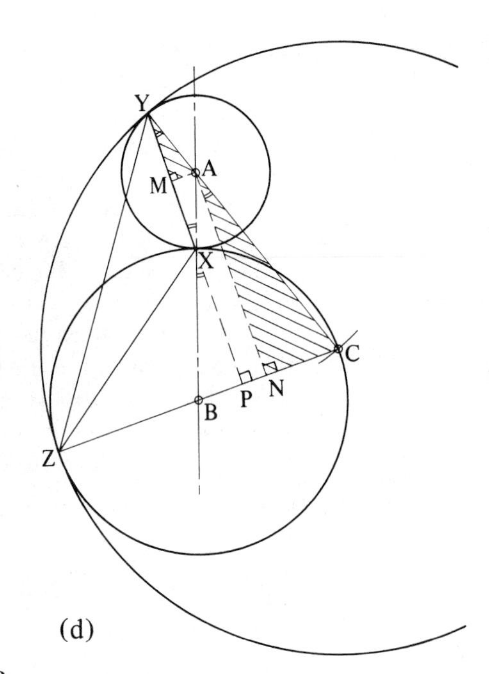

Fig. S29

36. *Note:* For many centuries architects and shipbuilders made use of drawings showing buildings and vessels in plan and elevation, and a number of more or less isolated techniques became associated with the production and interpretation of these drawings. But the rationalization of these techniques awaited the attention of those with more theoretical interests. Albrecht Dürer (1471–1528), Girard Desargues (1591–1661), and Gaspard Monge (1746–1818) are recognized as the great names in the history of orthographic projection, but only with Monge was the time ripe for the general acceptance of a unified subject matter in educational as well as industrial circles. A useful reference is Peter Jeffrey Booker, *A History of Engineering Drawing*, London: Chatto & Windus, 1963. Under the *ancien régime*, Monge had been a professor at the school for military engineers at Mézières, but in that position he had been unable to publish the details of his new "descriptive geometry." As the violence of the revolution subsided somewhat, new schools were opened, and in 1795 Monge gave a course at the short-lived École Normale (at which Lagrange, Laplace, Berthollet, and other distinguished scientists also taught). In the same year some of the same teachers, including Monge, taught at the new École Centrale des Travaux Publics and continued there after 1 September 1795 when that institution was renamed the École Polytechnique. At this most famous of technical schools, mechanics and machine design were described as "applications of descriptive geometry"! Monge's lectures were first published in the *Journal des écoles normales* for 1795, and in book form as *Géométrie descriptive*, 1798. This work was to have great influence through its numerous editions and translations.

In the present example, it should be noted that the true shape of the elliptical face is obtained by taking distances parallel to the major axis of the ellipse from the *elevation*, where these are represented in their true length, while distances parallel to the minor axis are taken from the *plan* (or underneath view), since there they are shown without foreshortening. And it should be clear from this that each of the well-known ellipse constructions shown in Fig. S30 is equivalent to that given by the projection from the oblique face of the truncated cylinder. (For elliptical sections of a cone, see Problems 42 and 43.)

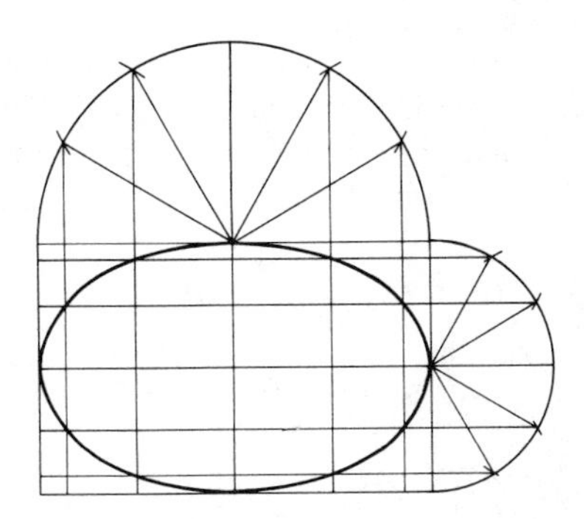

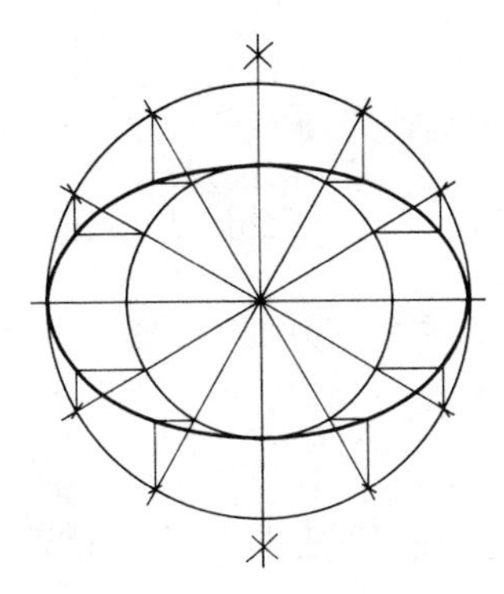

Fig. S30

(i) Major axis of ellipse $= \sqrt{(9^2 + 12^2)} = 15$ units, minor axis $= 9$ units; hence,

$$\text{Area} = \pi \times \frac{15}{2} \times \frac{9}{2} = \frac{135\pi}{4} \text{ square units.}$$

(ii) $x = \frac{1}{2}(2 + 14) = 8; \qquad y = 8 + \frac{1}{2}(14 - 8) = 11;$

$$w = 8 - \frac{1}{2}(8 - 2) = 5;$$

$$z = 8 + \frac{\sqrt{3}}{2}(14 - 8) = 8 + 3\sqrt{3}; \qquad v = 8 - \frac{\sqrt{3}}{2}(8 - 2) = 8 - 3\sqrt{3}.$$

(iii) The curved part of the boundary of the development of the curved surface is easily shown to have the form of a complete cycle of a sine curve, but it is not necessary to make use of this in the required calculations. The simplest approach is to (conceptually) provide the given trun-

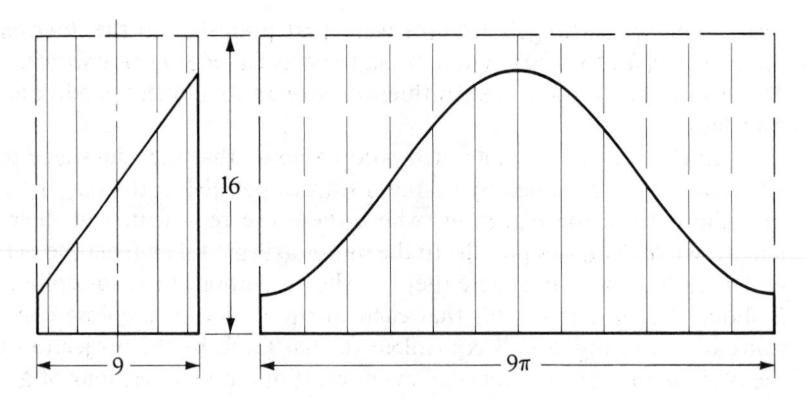

Fig. S31

cated cylinder with a congruent complementary one, as indicated in Fig. S31. This so-called macroreductive or molar approach (dialogue note 43) then allows the curved surface and volume of the truncated cylinder to be found immediately:

$$\text{Area of curved surface} = \frac{1}{2} \times 9\pi \times 16 = 72\pi \text{ square units,}$$

$$\text{Volume} = \frac{1}{2} \times \pi \times \frac{9}{2} \times \frac{9}{2} \times 16 = 162\pi \text{ cubic units.}$$

37. (a) First method: From similar triangles AEK, ADC,

$$AK = \frac{2}{5}AC, \quad \therefore KC = \frac{3}{5}AC \tag{1}$$

$$EK = \frac{2}{5}DC, \quad \therefore EK = \frac{1}{5}BC \tag{2}$$

From similar triangles FEK, FBC,

$$FK : FC = EK : BC \tag{3}$$

From (2) and (3),

$$FK = \frac{1}{5}FC, \quad \therefore KC = \frac{4}{5}FC \tag{4}$$

From (1) and (4),

$$FC = \frac{3}{4}AC, \quad \therefore AF = \frac{1}{4}AC,$$

hence, $AF : FC = 1 : 3$.

Second method: Let S, T denote the respective area measures of triangles CDF, DEF; then triangles DEB, AEF, AEB have area measures $S - T$, $\frac{2}{3}T$, $\frac{2}{3}(S - T)$, respectively. ("Triangles . . . which are under the same height are to one another as their bases"—Euclid, VI, 1.) Hence,

$$\frac{\triangle AFB}{\triangle FCB} = \frac{\frac{2}{3}(S - T) + \frac{2}{3}T}{2S} = \frac{\frac{2}{3}S}{2S} = \frac{1}{3}. \quad \therefore \ AF:FC = 1:3.$$

(b) (i) Draw XA parallel to PZ meeting QR at A; it is easily proved that $\triangle XAY \equiv \triangle ZRY$, from which $XY = YZ$. (Otherwise draw parallels to QR through X and Z. . . .)
(ii) As in part (i), use congruent triangles XAY, ZRY, also similar triangles XQA, PQR; the shortest solution is probably the following:

$$\frac{QY}{YR} + 1 = \frac{QR}{YR} = \frac{2QR}{AR} = \frac{2PQ}{PX}, \quad \text{and} \quad \frac{PZ}{PX} + 1 = \frac{PZ + PX}{PX} = \frac{2PQ}{PX};$$

Hence, $\dfrac{QY}{YR} = \dfrac{PZ}{PX}$ (each being equal to $\dfrac{2PQ}{PX} - 1$).

Because this way of proceeding is so far from being obvious, the following methodological approach will doubtless be preferred:

Let a units denote the length of each of PQ, PR;
Let b " " QX, RZ;
Let c " " AY, YR;
Let d " QA.

We have to prove that QY/YR (i.e., $(c + d)/c) = PX/PZ$ (i.e., $(a + b)/(a - b)$).
Proof: From similar triangles XQA, PQR,

$$\frac{QA}{QR} = \frac{XQ}{PQ}, \quad \text{i.e.,} \quad \frac{d}{d + 2c} = \frac{b}{a}, \text{ from which, } d = \frac{2bc}{a - b}.$$

Substitute this expression for d in the left side of the equation to be established, and simplify to obtain the right side.

(c) Draw AK, HB parallel to EF meeting DR at A and B, then

$$DA = \frac{1}{5}DR, \qquad AB = \frac{3}{5}DR, \qquad AK = \frac{1}{5}RF = \frac{1}{5}ER; \qquad HB = \frac{4}{5}ER.$$

Let X be the point of intersection of HK and DR.
From similar triangles AKX, BHX, $AX:BX = AK:HB = \frac{1}{5}ER : \frac{4}{5}ER = 1:4$.

$$\therefore \ AX = \frac{1}{4}BX = \frac{1}{5}AB = \frac{1}{5} \text{ of } \frac{3}{5} \text{ of } DR = \frac{3}{25}DR;$$

so DX, equal to $DA + AX$, $= \dfrac{1}{5}DR + \dfrac{3}{25}DR = \dfrac{8}{25}DR$;

$$\therefore \ XR = \frac{17}{25}DR.$$

Thus $DX:XR = 8:17$.

Let Y be the point of intersection of HK, FT. Draw KC parallel to FT to meet DE at C. Then, from similar triangles DCK, DTF,

$$DC:DT = DK:DF = 1:5, \quad \text{i.e.,} \quad DC = \frac{1}{5}DT = \frac{1}{10}DE, \text{ so}$$

$$HC = \frac{7}{10}DE; \; HT = \frac{3}{10}DE.$$

From similar triangles HYT, HKC,

$$YT:KC = HT:HC = 3:7, \therefore YT = \frac{3}{7}KC,$$

where $KC = \frac{1}{5}FT$, $\therefore YT = \frac{3}{7} \times \frac{1}{5}FT = \frac{3}{35}FT$, so $FY = \frac{32}{35}FT$.

Thus $FY: YT = 32:3$.

$$\text{Area ratio, } \frac{EFKH}{DEF} = \frac{\Delta DEF - \Delta DHK}{\Delta DEF} = \frac{\frac{1}{2}DE \cdot FP - \frac{1}{2}DH \cdot KQ}{\frac{1}{2}DE \cdot FP},$$

where FP, KQ are altitudes in triangles DEF, DHK. Then, from similar triangles DFP, DKQ,

$$FP:KQ = DF:DK = 5:1; \text{ thus } FP = 5KQ.$$

Hence, required area ratio

$$= \frac{DE \cdot (5KQ) - DH \cdot KQ}{DE \cdot (5KQ)} = \frac{5DE - \frac{4}{5}DE}{5DE} = \frac{4\frac{1}{5}}{5} = \frac{21}{25}.$$

38. First case: Obtain the distance between parallel edges be, cd, and hence find area of trapezium $bcde = \frac{1}{2}(16 + 4) \cdot 5\sqrt{13} = 50\sqrt{13}$ square units. Next, working in the plane through A, M, M', where M, M' are the midpoints of be, CD, respectively, and letting p, p' units, denote the respective perpendicular distances from A to the plane of $bcde$, and from M to the face $CDdc$, obtain $p' = 400/AM'$ (from $\frac{1}{2} \times \frac{3}{5}AM' \times p' = \frac{1}{2} \times 16 \times 15$), and further, by similar triangles, $p:p' = \frac{1}{5}AM'/5\sqrt{13}$. Hence, $p = 16/\sqrt{13}$, and volume of removed pyramid $Abcde = \frac{1}{3} \times 50\sqrt{13} \times 16/\sqrt{13} = 266\frac{2}{3}$ cubic units.

Second case: Obtain $bd = 5\sqrt{17}$, $ce = (32/5)\sqrt{2}$; area of $bcde = 16\sqrt{34}$ square units. Then find the perpendicular distance from b to edge Dd ($= 400\sqrt{2}/AD$), and, by similar triangles, perpendicular from A to plane of $bcde = (16/17)\sqrt{34}$ units; hence, volume of the removed pyramid $Abcde = 170\frac{2}{3}$ cubic units.

Note: The lack of symmetry about a point in the case of the pyramidal sections, $bcde$, does *not* carry over to the analogous "oval" sections of a cone; they are in fact true ellipses with central symmetry. See Problem 42.

39. For the ellipse, $FF' = 8$ units, directrices are 18 units apart; $EF = \frac{2}{3}\ell$, $FE' = \ell \sim 5$, $E'F' = 13 \sim \ell$, $EF' = \sqrt{\{(EF^2 - E'F^2) + E'F'^2\}}$, which comes to $\frac{2}{3}(18 - \ell)$. (*Note:* $\ell \sim 5$ denotes $\ell - 5$ for $\ell \nless 5$ and denotes $5 - \ell$ for $\ell \ngtr 5$. $18 - \ell$ is certainly positive since $3 \leq \ell \leq 15$.) $EF + EF' = \frac{2}{3}\ell + \frac{2}{3}(18 - \ell) = 12$ units = length of the major axis = $\frac{2}{3}$ distance between directrices. Minor axis = $2\sqrt{(6^2 - 4^2)} = 4\sqrt{5}$ units. See again note 69.

For the hyperbola, $FF'' = 18$ units, directrices are 8 units apart; $HF'' \sim HF = \sqrt{\{(13 + \ell)^2 + (9/4)\ell^2 - (\ell - 5)^2\}} - (3/2)\ell$, which simplifies to 12 units, and this is equal to the transverse axis (the join of the vertices) or to 3/2 the distance between the directrices. The asymptotes are the two straight lines, intersecting at the midpoint of FF'', so positioned that the distances from H on the curve to one or other of these lines approaches zero as the distances HF, HF'' increase without limit. Clearly, then, the asymptotes will have to have gradients such that the ratio of any interval along an asymptote to the projection of this interval onto the axis through FF'' is equal to the constant ratio, HF to HH''. In the present case this ratio is 3 : 2, and consequently the gradients of the asymptotes are $\pm\frac{1}{2}\sqrt{5}$. If the asymptotes were at right angles, the eccentricity would be $\sqrt{2}$.

For an outline of the ancient (but surprisingly sophisticated) work on conic sections, see Heath's *History of Greek Mathematics*, vol. 2, chap. 14.

> It is remarkable that the directrix does not appear at all in Apollonius's great treatise on conics. The focal properties of the central conics are given by Apollonius, but the foci are obtained in a different way without any reference to the directrix; the focus of the parabola does not appear at all. We may perhaps conclude that neither did Euclid's *Conics* contain the focus-directrix property, for, according to Pappus, Apollonius based his first four books on Euclid's four books, while filling them out and adding to them. Yet Pappus gives the proposition as a lemma to Euclid's *Surface-Loci*, from which we cannot but infer that it was assumed in that treatise without proof. If, then, Euclid did not take it from his own *Conics*, what more likely than that it was contained in Aristaeus's *Solid Loci*?

(P. 119.) The works of Euclid mentioned in this passage, and all the writings of Aristaeus (fl. 3rd quarter of the 4th cent. B.C.), are lost. Pappus wrote in the first half of the 4th century A.D.: see his *Collection*, VII, 235–238; ed. of P. Ver Eecke, vol. 2, pp. 792–802. For Apollonius (fl. late 3rd cent. B.C.), see Apollonius of Perga, *Treatise on Conic Sections,* translated and edited by Thomas L. Heath, 1896. See also J. H. Weaver, "On Foci of Conics," *Bulletin of the American Mathematical Society 23* (1917):357–365, and the articles on Menaechmus (who is credited with the discovery of the three kinds of conic section, mid 4th cent. B.C.), Aristaeus, Euclid, Apollonius, and Pappus in the *Dictionary of Scientific Biography*.

40. (a) $$\frac{A'P'^2}{AP^2} = \frac{U'A' \cdot A'W'}{UA \cdot AW} = \frac{A'W'}{AW} = \frac{OA'}{OA}.$$

($UAA'U'$ is a parallelogram; triangles $OA'W$, OAW are similar.)

(b) $$\frac{OB^2}{BP} = \frac{AP^2}{OA} = \frac{FP^2 - FA^2}{OF + FA} = \frac{NP^2 - FA^2}{OF + FA}$$
$$= \frac{(NP + FA)(NP - FA)}{OF + FA} = \frac{(2 \cdot OF + 2 \cdot FA)(2 \cdot OF)}{OF + FA}$$

$\therefore$ $OB^2/BP = 4 \cdot OF$, which is constant. In modern symbolism (introduced by Descartes, 1637), $y^2/x = 4a$.

41. (a) First consider point T anywhere on the first-quadrant branch of the curve (see Fig. S32). Then, since

$$FT : NT = \sqrt{2} : 1, \quad \{(x_T - x_F)^2 + (y_T - y_F)^2\} : NT^2 = 2 : 1,$$

or simply

$$(x - k)^2 + (y - k)^2 = 2NT^2 = QT^2 = (QS + ST)^2,$$
where $QS = y_S - y_Q = x_T - x_F = x - k$,
so $(QS + ST)^2 = (x - k + y)^2$.

Hence,

$$x^2 - 2kx + k^2 + y^2 - 2ky + k^2 = x^2 + k^2 + y^2 - 2kx - 2ky + 2xy,$$

from which, $xy = \frac{1}{2}k^2$.

It is shown likewise that $F'T : N'T = \sqrt{2} : 1$ leads to exactly the same equation, $xy = \frac{1}{2}k^2$. And, of course, the tracing point may be on either branch of the hyperbola; for suppose we work with a point T' on the curve in the third quadrant, then

$$FT' : N''T' = \sqrt{2} : 1 \Rightarrow \{(x_F - x_{T'})^2 + (y_F - y_{T'})^2\} : N''T'^2 = 2 : 1,$$

or simply

$$(k - x)^2 + (k - y)^2 = 2N''T'^2 = T'Q''^2 = (k - x - y)^2.$$

Hence,

$$k^2 - 2kx + x^2 + k^2 - 2ky + y^2 = k^2 + x^2 + y^2 - 2kx + 2xy - 2ky,$$

from which, $xy = \frac{1}{2}k^2$, as before.

(b) For any point T on the curve in the first quadrant:

$$F'T - FT = \sqrt{\{(x_T - x_{F'})^2 + (y_T - y_{F'})^2\}}$$
$$- \sqrt{\{(x_T - x_F)^2 + (y_T + y_F)^2\}},$$

or simply,

$$\sqrt{\{(x + k)^2 + (y + k)^2\}} - \sqrt{\{(x - k)^2 + (y - k)^2\}}$$
$$= \sqrt{\{x^2 + 2kx + k^2 + y^2 + 2ky + k^2\}}$$
$$- \sqrt{\{x^2 - 2kx + k^2 + y^2 - 2ky + k^2\}},$$

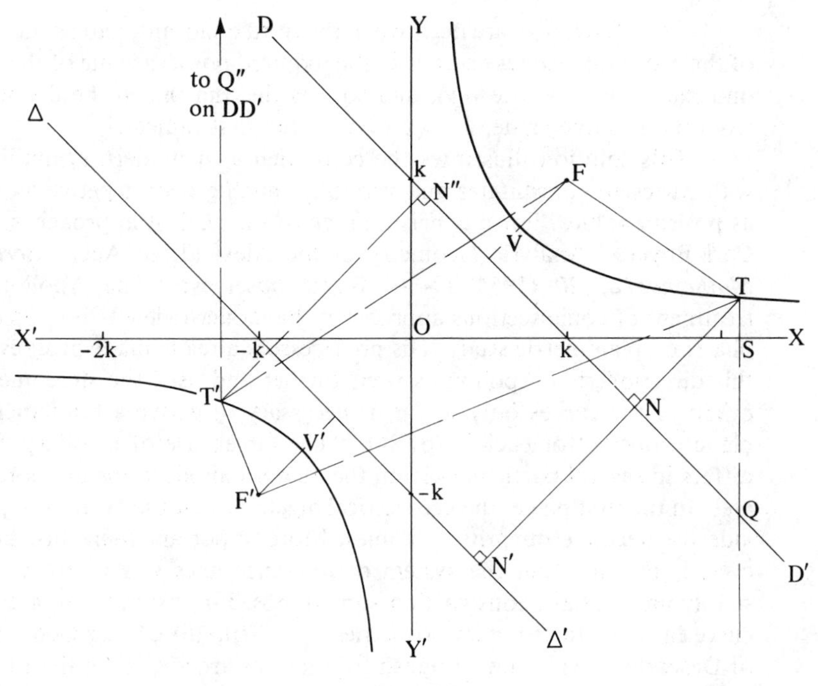

Fig. S32

where, from part (a), $k^2 = 2xy$. Substitute $2xy$ for *one* k^2 in each radical:

$$\begin{aligned} F'T - FT &= \sqrt{\{x^2 + 2xy + y^2 + 2k(x + y) + k^2\}} \\ &\quad - \sqrt{\{x^2 + 2xy + y^2 - 2k(x + y) + k^2\}} \\ &= (x + y + k) - (x + y - k) \\ &= 2k. \end{aligned}$$

With any point T' on the third-quadrant branch of the hyperbola:

$$\begin{aligned} FT' - F'T' &= \sqrt{\{(x_F - x_{T'})^2 + (y_F - y_{T'})^2\}} \\ &\quad - \sqrt{\{(x_{F'} - x_{T'})^2 + (y_{F'} - y_{T'})^2\}}, \end{aligned}$$

or simply,

$$\begin{aligned} &\sqrt{\{(k - x)^2 + (k - y)^2\}} - \sqrt{\{(-k - x)^2 + (-k - y)^2\}} \\ &= \sqrt{\{k^2 - 2kx + x^2 + k^2 - 2ky + y^2\}} \\ &\quad - \sqrt{\{k^2 + 2kx + x^2 + k^2 + 2ky + y^2\}} \\ &= \sqrt{\{k^2 - 2k(x + y) + (x^2 + 2xy + y^2)\}} \\ &\quad - \sqrt{\{k^2 + 2k(x + y) + (x^2 + 2xy + y^2)\}} \\ &= k - (x + y) - [-(k + x + y)] \\ &= 2k, \quad \text{as before.} \end{aligned}$$

Note: Since x, y are negative in the third quadrant, and as the sum of their absolute values exceeds k, the required *positive* value of the second radical is $-(k + x + y)$, and so it is this that has to be deducted from the positive value, $k - (x + y)$, of the first radical.

This solution illustrates the convenience of modern symbolism, with cartesian coordinates and variables ranging over negative as well as positive values. For a concise survey of the ancient approaches, see Carl Boyer, "Analytic Geometry in the Alexandrian Age," *Scripta Mathematica, 20* (1954):30–36. Boyer observes: "The Apollonian treatment of conic sections approaches the modern view . . . in its emphasis on planimetric study. His predecessors already made progress in this direction, but Apollonius went further and used the stereometric origin of the curves only so far as necessary to derive a fundamental plane property for each." (p. 34). "The Greek use of auxiliary lines differs in several particulars from the modern applications of coordinates. In the first place, the geometrical algebra of antiquity did not provide for negative quantities or lines. More important than this, however, is the fact that the system of reference lines was in every case simply an auxiliary construction superimposed *a posteriori* on a given curve in order to study its properties. . . ." (p. 36) Cf. *La Géométrie* of Descartes (1637). Two English translations are available; the first is *The Geometry of René Descartes,* a facsimile of the original with an English translation and notes by D. E. Smith and M. L. Latham, Chicago: Open Court: 1925, and New York: Dover reprint, 1954; the second by P. J. Olscamp (with the *Discourse on Method,* etc.), is titled *Discourse on Method, Optics, Geometry, and Meteorology*, Indianapolis: Bobbs-Merril, 1965. Descartes himself used only positive coordinates. Negative as well as positive coordinates became well known only following their use by Newton, though "he had been anticipated to some extent by others, notably Wallis and Lahire" (Carl Boyer, *History of Analytic Geometry*, New York: Scripta Mathematica, Yeshiva University, 1956, p. 139). See further, three articles in *Historia Mathematica*: A. G. Molland, "Shifting the Foundations: Descartes's Transformation of Ancient Geometry," *Historia Mathematica 3* (1976):21–49; Eric G. Forbes, "Descartes and the Birth of Analytic Geometry," ibid. *4* (1977):141–151; Timothy Lenoir, "Descartes and the Geometrization of Thought: The Methodological Background of Descartes' *Géométrie*," ibid. *6* (1979):355–379. The last of these articles incidentally throws some light on the first of quoted passages used to introduce Hora Prima of the dialogue. Cf. George Polya, *How to Solve It, A New Aspect of Mathematical Method*, 2nd ed., Garden City, N.Y.: Doubleday Anchor 1957, pp. 112, 134, 141ff., and see Solution 7(d), supra.

42. (i) $PF = PQ$ (equal tangents), likewise $PF' = PQ'$; $\therefore$ $PF + PF' = QQ' = BB' = AA'$.

(ii) $PF + PF' = 24$ cm. Consider P moved to end of minor axis; then $PF = PF' = 12$ cm. PF is the hypotenuse of a right triangle having base equal to $\frac{1}{2}FF' = 10$ cm, and altitude = semiminor axis = $\sqrt{(12^2 - 10^2)}$, from which, minor axis = $4\sqrt{11}$ cm.

(iii) By similar triangles—or by equal alternative expressions for the area of triangle CAV'', viz., $\frac{1}{2}(\frac{1}{2}AB) \cdot CV''$ and $\frac{1}{2}CA \cdot AV''$—obtain $AB = 2CA \cdot AV''/CV''$, where CV'' is given by CV'': $(CV'' - 25) = 11:4$. Thus find $CV'' = 39\frac{2}{7}$ cm, $AB = 21\frac{3}{25}$ cm (radius, $10\frac{14}{25}$ cm). $A'B'$ may be found by the same method, or, more simply, from $A'B':AB = 4:11$. $A'B' = 7\frac{7}{17}$ cm (radius, $3\frac{21}{25}$ cm).

(iv) (Proof by congruent triangles, may be shortened by the equal-intercept theorem.)

$$\text{Diameter at level of } NM = \frac{1}{2}\left(7\frac{17}{25} + 21\frac{3}{25}\right) = 14\frac{2}{5} \text{ cm.;}$$

$$\text{radius} = 7\frac{1}{5} \text{ cm.}$$

(v) $$\text{Radius at level of } V = 10\frac{14}{25} - \frac{2}{24}\left(10\frac{14}{25} - 3\frac{21}{25}\right),$$

which comes to 10 cm exactly.

$$\text{Radius at level of } V' = 3\frac{21}{25} + \frac{2}{24}\left(10\frac{14}{25} - 3\frac{21}{25}\right) = 4\frac{2}{5} \text{ cm.}$$

Check that $\frac{1}{2}(10 + 4\frac{2}{5}) = 7\frac{1}{5}$ cm = radius already calculated for level of NM. NM is found from the equation,

$$\text{Radius at } V' + NM = \text{radius at } V - NM,$$

from which $NM = \frac{1}{2}(10 - 4\frac{2}{5}) = 2\frac{4}{5}$ cm.

(vi) The minor axis of the elliptical section coincides with the chord drawn through M, perpendicular to NM, in the circle of radius $7\frac{1}{5}$ cm, and so has length $2\sqrt{\{(7\frac{1}{5})^2 - (2\frac{4}{5})^2\}} = 2\sqrt{\{10 \times 4\frac{2}{5}\}} = 4\sqrt{11}$ cm.

Note: The result that $PF + PF'$ = a constant, obtained in part (i) of this example, indicates by its form that the "oval" section will *not* be egg-shaped, i.e., it will not be narrower toward the vertex of the cone but will be symmetrical about the perpendicular bisector of FF' as well as about the line through FF'. That the definition of an ellipse as a conic section whose plane is inclined to the axis of the cone at a greater angle than a generator and the alternative definition as the plane locus of a point P for which $FP:FN$ = a constant less than unity really do define the same curve is evident from the fact that each of these definitions implies the *common further definition* of the ellipse as the plane locus of P for which $PF + PF' =$ constant (cf. Problem 39).

An interesting case of erroneous geometrical intuition is evidenced by the famous woodcut reproduced in Fig. S33 from Albrecht Dürer's *Unterweysung der Messung mit dem Zirckel und Richtscheyt*, 1525. This diagram illustrates a theoretically correct construction method for obtaining the true shape of an elliptical section of a cone, *yet this section is nevertheless distorted to appear egg-shaped in the drawing*, apparently in accordance with the preconceived idea that the

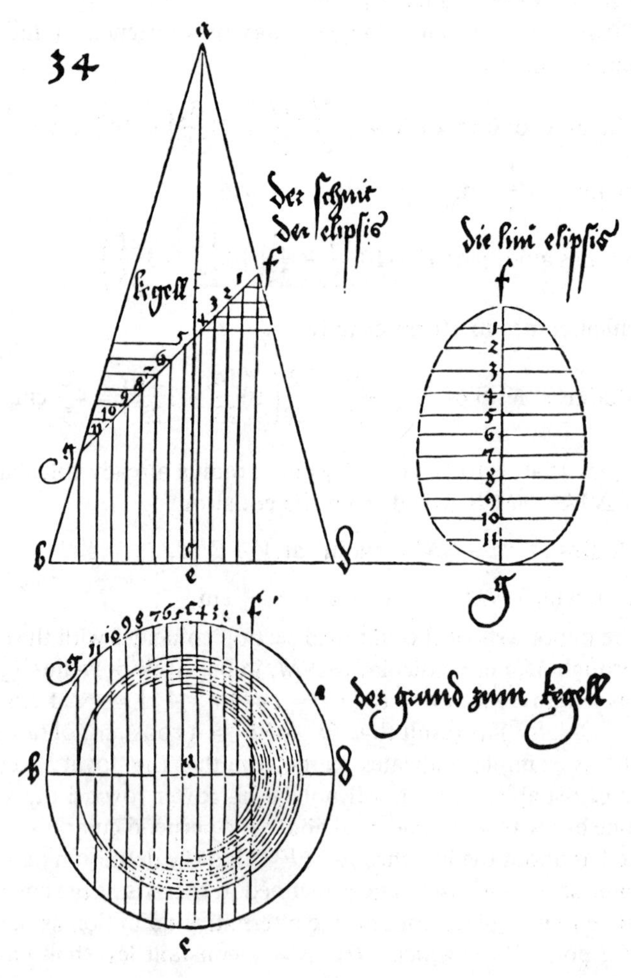

Fig. S33

oval possessed only one axis of symmetry. Dürer actually chose the term *Eierlinie* ("egg line") to describe the section. In a communication to the *Australian Mathematics Teacher 30* (1974): 222–226, Daniel Pedoe tried to argue that Dürer's egg-shaped ellipse arose from a woodcut error rather than from a conceptual mistake, but Pedoe does not make a very convincing case.

The spheres used in Problem 42 are known as Dandelin spheres, after G. P. Dandelin (1794–1847), a Frenchman who became professor of mechanics at the University of Liège. According to the Theorem of Dandelin, if a sphere be inscribed in a cone so as to touch any plane, its point of contact with that plane is a focus and the intersection with that plane by the plane of the circle of contact of the sphere and cone is a directrix of the section of the cone made by the first plane. A somewhat different version of this result had already been established by Hugh Hamilton in his *Treatise of Conic Sections*, 1758, as was pointed out by V. Kommerell in his contribution to the fourth volume of Moritz Cantor's *Vorlesungen über Geschichte der Mathematik*, Leipzig: Teubner, 1908; refer pp. 461–464.

43. (a) $MM' = 6\frac{1}{2}$ cm, $ON = 1\frac{1}{4}$ cm. Triangles ABX and BCM are similar, $\therefore$ $AB : BC = BX : CM$, where $CM = \frac{1}{2}CD = \frac{1}{2}$ of $(5/9)CA = (5/18)AB$, from which $AB = (27/5)\sqrt{5}$cm.

(b) Minor axis = chord of circle of diameter $M'M$, the chord passing through O perpendicular to $M'M$. $\therefore$ minor axis has length $2\sqrt{(M'O \cdot OM)} = 6$ cm.

(c) Eccentricity, ϵ, i.e., the distance between foci divided by the major axis, $= FF'/BD$, where $FF' = 2\sqrt{\{(4\frac{1}{2})^2 - 3^2\}}$, from which, $\epsilon = \frac{1}{3}\sqrt{5}$.

(d) The major axis of the elliptical section is equal to the hypotenuse of a right triangle with sides 39, 52 mm. Therefore the major axis has length $13\sqrt{(3^2 + 4^2)} = 65$ mm. The minor axis of ellipse is equal to a chord of a circle of radius $19\frac{1}{2}$ mm, distant $7\frac{1}{2}$ mm from the center. Therefore, the minor axis has length $2\sqrt{\{(19\frac{1}{2})^2 - (7\frac{1}{2})^2\}} = 2\sqrt{(27 \times 12)} = 36$ mm.

For the inclined section of the cylinder, refer to Fig. S34. The general treatment illustrated in Problem 42 is applicable here to show that the section $V'EV$ is an ellipse ($EC + EC' = ET + ET' = OO' =$ constant).

Alternatively:

(i) Let $VV' = 2a$ (= 65 mm), diameter of cylinder = $2b$ (= 36 mm), then inscribed spheres have radii b (= 18 mm). By similar triangles OMC, $V'VN$, $MC : p = b : 2b \Rightarrow MC = \frac{1}{2}p$, $\therefore CC' = p = \sqrt{\{(2a)^2 - (2b)^2\}}$ (= $\sqrt{2292}$ mm), and this is precisely the distance between the foci of an ellipse with semiaxes a, b [= $2\sqrt{(a^2 - b^2)}$].

(ii) Again by similar triangles OMC, $V'VN$, $OM : 2a = b : 2b$, giving $OM = a$, $OO' = 2a$.

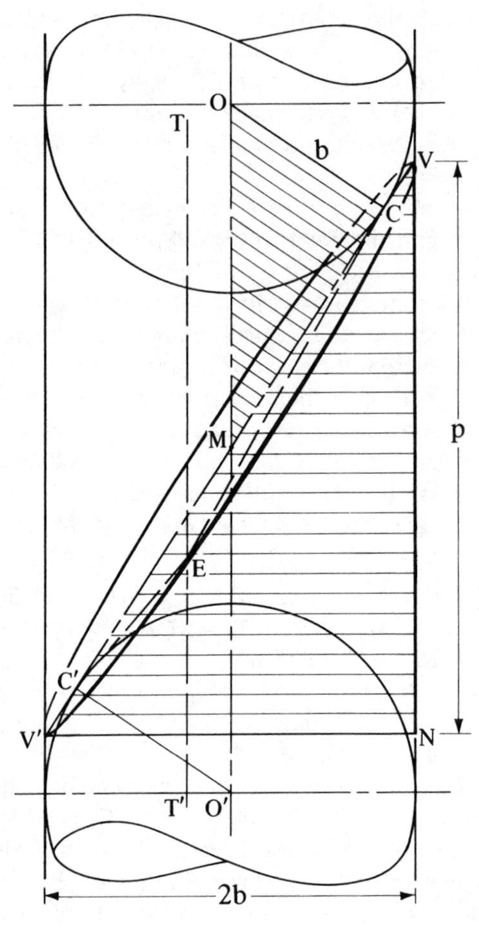

Fig. S34

44. It has to be shown that $V:U = r\sqrt{r}:R\sqrt{R}$; for example, if $r:R = 4:9$, then $V:U = 8:27 \Rightarrow V:(V+U) = 8:35$. The complementary cone has volume $C = \frac{1}{3}\pi r^2 h'$, where $h':h = r:(R-r)$; hence

$$C = \frac{\pi r^3 h}{3(R-r)} \tag{1}$$

$V + C = \frac{1}{3}\pi abp$, where $p:(R+r) = (h'+h''):2a$, and $h'':h = r:(R+r)$. Simplifying, $p = Rrh/a(R-r)$ is obtained, so $V + C = \frac{1}{3}\pi b \cdot Rrh/(R-r)$, where a has been eliminated, and

$$b = \sqrt{\left[\left\{\frac{1}{2}(R+r)\right\}^2 - \left\{\frac{1}{2}(R-r)\right\}^2\right]},$$

which comes to $b = \sqrt{(Rr)}$, so

$$V + C = \frac{\pi h \cdot Rr\sqrt{(Rr)}}{3(R - r)} \tag{2}$$

$U + V + C = \frac{1}{3}\pi R^2(h + h')$, where $(h + h') : h' = R : r$, and $h' = rh/(R - r)$ from above. $\therefore\ h + h' = Rh/(R - r)$, and so,

$$U + V + C = \frac{\pi R^3 h}{3(R - r)} \tag{3}$$

(Check, from (1) and (3), $(U + V + C) : C = R^3 : r^3$, as it must.) From (3) minus (1):

$$U + V = \frac{\pi h}{3(R - r)} \{R^3 - r^3\},$$

and from (2) minus (1):

$$V = \frac{\pi h}{3(R - r)} \{Rr\sqrt{(Rr)} - r^3\}.$$

Subtracting:

$$U = \frac{\pi h}{3(R - r)} \{R^3 - Rr\sqrt{(Rr)}\}.$$

$$\therefore \frac{V}{U} = \frac{r\sqrt{r} \cdot (R\sqrt{R} - r\sqrt{r})}{R\sqrt{R} \cdot (R\sqrt{R} - r\sqrt{r})} = \frac{r\sqrt{r}}{R\sqrt{R}}.$$

45. For a discussion of Cavalieri's theorem or principle, see dialogue note 187. Cavalieri enunciated this as follows: "If between the same parallels any two plane figures are constructed, and if in them, any straight lines being drawn equidistant from the parallels, the included portions of any one of these lines are equal, the plane figures are also equal to one another; and, if between the same parallel planes any solid figures are constructed, and if in them, any planes being drawn equidistant from the parallel planes, the included plane figures out of any one of the planes so drawn are equal, the solid figures are likewise equal to one another. The figures so compared let us call analogues, the solid as well as the plane. . . ." (P. 448 of the translation by G. W. Evans, *American Mathematical Monthly 24* (1917):447–451.)

(a) Note that the tetrahedron is not a regular one; for convenience of calculation of its volume, the top edge is taken perpendicular to the triangular face of area $\frac{1}{2}\ell \cdot 2R$. Its volume is then given by $V = \frac{1}{3}(\ell R)\ell$. The value of ℓ is to be chosen that the corresponding cross-sections of the sphere and the tetrahedron shall be equal to each other at all levels, i.e., for any value of z. By similar triangles, $x/\ell = z/2R$, and $y/\ell = (2R - z)/2R$; hence,

$$\text{Area of cross-section at height } z = xy = (z\ell/2R) \cdot \ell(2R - z)/2R$$
$$= \ell^2 \cdot z(2R - z)/4R^2.$$

In the sphere, the corresponding cross-section has area

$$\pi r^2 = \pi\{R^2 - (z - R)^2\} = \pi z(2R - z).$$

For the two solids to be analogues of each other,

$$xy = \pi r^2, \quad \text{i.e.,} \quad \ell^2 z(2R - z)/4R^2 = \pi z(2R - z);$$

hence ℓ is to be made equal to $2R\sqrt{\pi}$. Then,

$$\text{Volume of sphere} = \text{volume of tetrahedron}$$
$$= \frac{1}{3} \text{ area of face} \times \text{perpendicular edge length}$$
$$= \frac{1}{3}(\ell R)\ell = \frac{1}{3}R(4R^2\pi) = \frac{4}{3}\pi R^3.$$

For an alternative method see Fig. S35.

$$\text{Sphere: } A_1 = \pi(R^2 - z^2).$$
$$\text{Double cone: } A_2 = \pi z^2$$
$$\text{Cylinder: } A_3 = \pi R^2$$
$$\therefore A_1 + A_2 = A_3 \text{ for all } z \text{ (from } z = -R \text{ to } z = +R);$$
$$\therefore V_1 + V_2 = V_3; \text{ thus}$$
$$V_{\text{sph}} + 2\left(\frac{1}{3}\pi R^2 \cdot R\right) = \pi R^2(2R) \Rightarrow V_{\text{sph}} = \frac{4}{3}\pi R^3.$$

George Bruce Halsted claimed the first method as his own ("The Criterion for the Two-Term Prismoidal Formulae," *Transactions of the Texas Academy of Science 1*, no. 5 (1896):19–32, on p. 23). In his *Mensuration,* he proved by substantially this method that "Any sphere is equal in volume to a tetrahedron whose midsection is equivalent to a

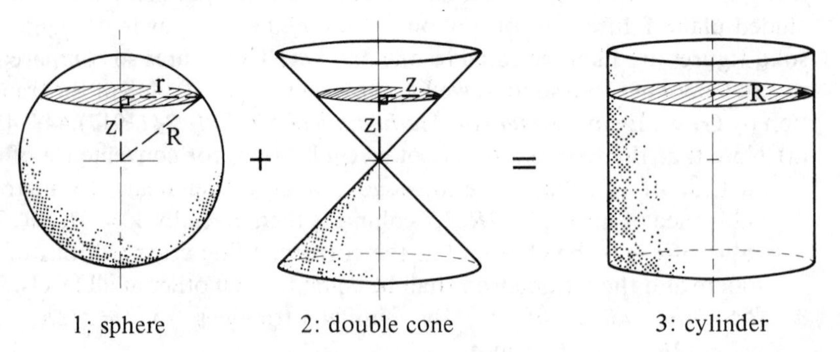

Fig. S35

great circle of the sphere, and whose altitude equals a diameter'' (*Metrical Geometry: An Elementary Treatise on Mensuration*, 2nd ed., Boston: Ginn-Heath, 1881, pp. 110–111). Actually, the determination of the volume of a double "ungula" by Gregory of St. Vincent (probably before 1630, published 1647) contains the essential insight. Refer, Margaret Baron, *The Origins of the Infinitesimal Calculus*, Oxford: Pergamon Press, 1969, pp. 141–142. For the geometrical relations involved in the second method, see dialogue discusssion relating to Fig. 83, note 167, and end of note 187.

(b) That the prism shown in Fig. S36 is an analogue (Cavalierian transformation) of the paraboloid of revolution is established by showing that $kz \cdot R = \pi R^2$. In the paraboloid, $r^2/R^2 = z/h$; therefore area of circular section at depth z is given by $\pi z R^2/h$. In the prism, the rectangular section at depth z has area kzR, where $k = \pi R/h$, that is, area $= \pi z R^2/h$, precisely as for the paraboloid. Thus, for all z, the corresponding sections of the two solids are equal, so, by Cavalieri's theorem,

$$\text{Volume of paraboloid} = \text{volume of prism} = \frac{1}{2}(khR) \cdot h = \frac{1}{2}\pi R^2 h.$$

For other ways of establishing this result, see Archimedes, "On Conoids and Spheroids," Propositions 21, 22, and "The Method," Proposition 4 (Heath (ed.) *The Works of Archimedes*, pp. 131–133, and Heath's Supplement, pp. 24–25). The methods of al-Haytham, *c*. A.D. 1000, and the sixteenth- and seventeenth-century European mathematicians are indicated in Baron, *Origins of the Infinitesimal Calculus*, pp. 66–68 and chap. 3.

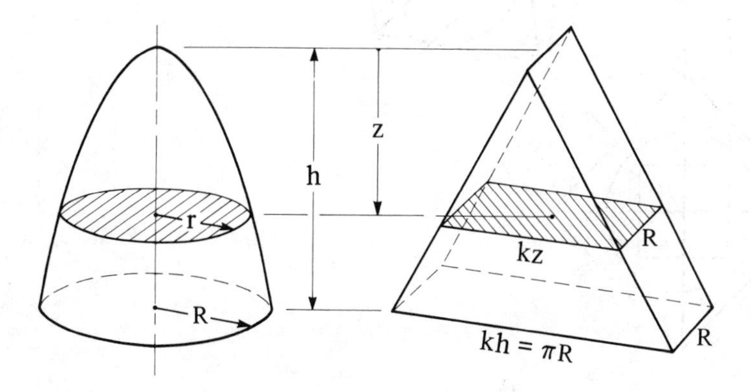

Fig. S36

46. In examples such as (a) and (d), one's intuitions can be brought into focus, and the appropriate calculations planned, by reference to carefully considered diagrams, which may either be perspective sketches or plans and elevations projected from each other. The various views of a regular octahedron shown in Fig. S37 illustrate different properties and suggest alternative calculation methods, which might otherwise be overlooked.

(a) Start by showing that the "height" of the cuboid is one-third of the distance between opposite vertices of the octahedron, and that the "top" (or "bottom") face diagonal is two-thirds of the edge length of the octahedron.

(i) Edge length ratio, $e_c : e_o = \sqrt{2} : 3$; (ii) surface area ratio, cube to octahedron, $= 2\sqrt{3} : 9$; (iii) volume ratio $= 2 : 9$. (iv) Radius of sphere $= e_o\sqrt{6}/6 = e_c\sqrt{3}/2$. (*Check:* One-third surface of octahedron times radius of insphere = volume of octahedron.)

(b) $a : r = 13(\sqrt{2} - 1) : 18$.

(c) $V_s : V_p = A_s : A_p = 10\pi : 81$.

(d) Radius of insphere $= e_t\sqrt{6}/12 = e_o\sqrt{6}/6$.

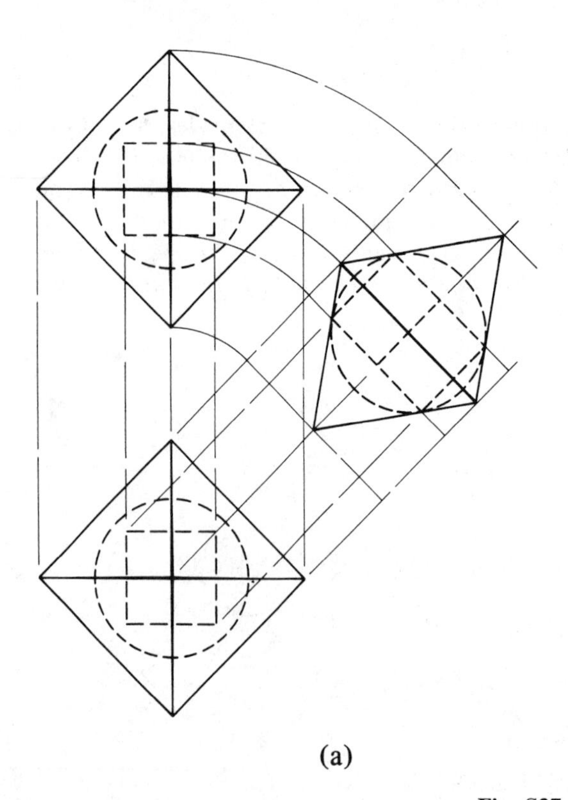

(a)

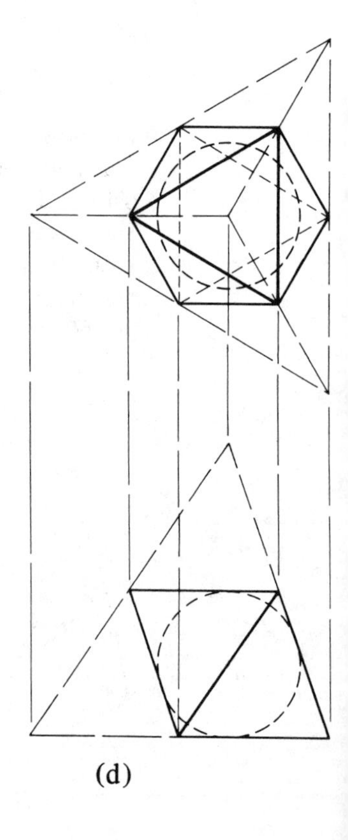

(d)

Fig. S37

47. (a) Let E, F, G, H be the centers of the squares $ABST$, $BCUV$, $CDWX$, $DAYZ$, respectively. (See Fig. S38.) The congruency of triangles AHE, BFE, CFG, DHG is easily established. Hence, $HE = EF = FG = GH$, and $\angle HEF = 90 - \epsilon_1 + \epsilon_2$ degrees. Thus $EFGH$ is a right-angled equilateral quadrilateral; that is, it is a square.

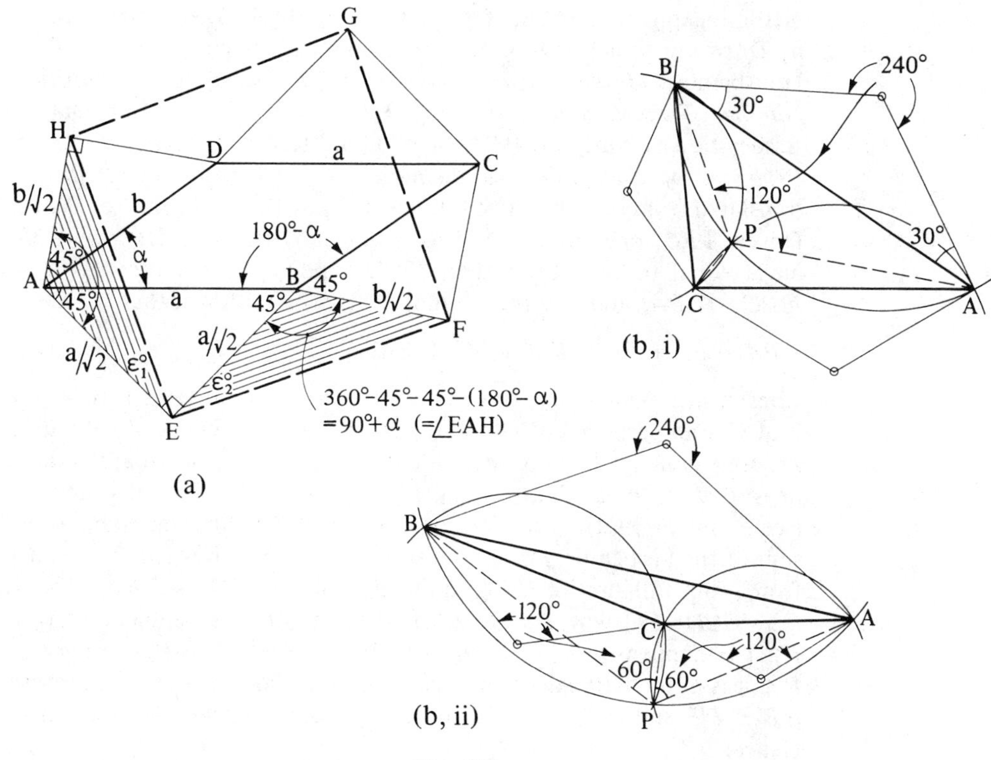

Fig. S38

(b) As long as the largest angle of the triangle is less than 120°, the required point P lies within the triangle and is readily located by the construction indicated in Fig. S38(b, i). If the same construction is applied to a triangle having an angle exceeding 120°, the point P so determined lies outside the triangle and only the longest side subtends an angle of 120° at P; the other two sides subtend angles of 60°, as shown in S38(b, ii).

Note: This construction is of interest because of its connection with a problem proposed by Fermat: to determine the position of a point in the plane of any given triangle in order that the sum of the distances from the point to the vertices has the least possible value. Cavalieri, Torricelli, Viviani, and Roberval all concerned themselves with this problem and several successful solutions were devised at the time.

Richard Courant and Herbert Robbins discuss this problem in chapter 7 of their *What Is Mathematics?* (London: Oxford Univ. Press, 1941) under the heading "Steiner's Problem"! But the essentials of the treatment they outline are to be found in Torricelli (*Opere, di Evangelista Torricelli,* edited by G. Loria and G. Vassura, vol. 1, part 2, Faenza: Montenari, 1919, pp. 90–97, and vol. 3, pp. 426–431, and in Cavalieri, *Exercitationes geometricae sex,* 1647, pp. 504–510. Yet another method, explicitly attributed to Torricelli by Honsberger, and implicitly by Dörrie, is worth giving here on account of its elegance. (Cf. Ross Honsberger, *Mathematical Gems, from Elementary Combinatorics, Number Theory, and Geometry,* Washington, D.C.: Mathematical Association of America, 1973, chap. 3, and Heinrich Dörrie, *100 Great Problems of Elementary Mathematics, Their History and Solution,* translated from the German edition of 1948 by David Antin, New York: Dover, 1965, pp. 361–363.) We consider any triangle ABC, with no angle equal to or greater than 120°, a point P such that $\angle APB = \angle BPC = \angle CPA$, and any other point P'. It will be shown that

$$PA + PB + PC < P'A + P'B + P'C.$$

Construct an auxiliary triangle with sides through the vertices of $\triangle ABC$ and perpendicular to PA, PB, PC, respectively. Since PA, PB, PC are equally inclined to each other, so are the perpendiculars to these lines; that is, the auxiliary triangle is equilateral. By Viviani's theorem (see Problem 19(d)), for any point within the equilateral triangle the sum of the perpendiculars to the sides is constant.Thus for P', distinct from P but still interior to the equilateral triangle, $P'A' + P'B' + P'C' = PA + PB + PC$, where A', B', C' are the feet of the perpendiculars from P'. If P' happens to be collinear with P, A, then $P'A' = P'A$; otherwise $P'A'A$ is a right triangle with hypotenuse $P'A$, so $P'A' < P'A$. Likewise $P'B' \leq P'B$ and $P'C' \leq P'C$, with the equal sign applying at most once. Hence:

$$PA + PB + PC\ (= P'A' + P'B' + P'C') < P'A + P'B + P'C.$$

Clearly, the effect of moving P' outside the auxiliary triangle (or away from the plane of A, B, C) is only to increase the right side of this inequality. Thus P, as obtained by the construction of Fig. S38 (b, i), is the required "Fermat point" for which the sum of the distances to the vertices is minimized. For a triangle having an angle equal to or greater than 120°, it can be shown that the Fermat point coincides with the vertex of this largest angle.

(c) One method is to express the altitude from X in $\triangle AXB$ in terms of the length of the side of the square, by first establishing that in a right triangle with an acute angle of 15°, the ratio of the sides containing the right angle is $(2 + \sqrt{3}) : 1$. Then, by subtraction, establish that in

$\triangle CXD$, the altitude from X is $\sqrt{3}/2$ times CX, from which XC, XD each equal CD. Otherwise, make use of a triangle YBC congruent to $\triangle XAB$, with Y inside $ABCD$, and establish that $\triangle YXC$ is also congruent to $\triangle XAB$. . . .

(d) Let A, B, C denote the respective centers of the circles of radii 8 cm, 2 cm, and r cm, where r is the variable radius of the circle touching the first two circles. Then, if the variable circle does not enclose the 2-cm circle, $AC = 8 - r$, and $BC = 2 + r$, $\therefore AC + BC = 10$, and so the locus is an ellipse with A, B as foci, major axis 10 cm, minor axis $5\sqrt{3}$ cm. If, on the other hand, the variable circle encloses the 2-cm circle so as to touch it, then $AC = 8 - r$, as before, but $BC = r - 2$, so $AB + BC = 6$, and the locus is the ellipse with A, B as foci, major axis 6 cm, minor axis $\sqrt{11}$ cm.

(e) In each case make use of one branch of a hyperbola traced by a point P such that $PF : PN = 2 : 1$, with focus F at an end of one axis of the ellipse and with the directrix along the other axis.

48. (a) Inspection of the triangle with sides a, a, $b - a$ (Fig. S39) shows that

$$2\alpha \gtreqless 180° - 4\alpha \text{ according as } a \gtreqless b - a,$$
$$\text{or } 6\alpha \gtreqless 180° \quad '' \quad '' \quad 2a \gtreqless b.$$

Hence, arc $\gtreqless$ semicircle according as $b \lesseqgtr 2a$.

(b) The designated segments are similar iff $\beta_1 = \alpha_1$. Now $\alpha_1 = \alpha_2 = \alpha_3 = \alpha_4$, and $\beta_1 = \beta_2$. Hence the segments are similar iff $\beta_2 = \alpha_4$, that is, iff AB lies along AD. It is presumed that problems (a) and (b) were solved by Hippocrates of Chios since they are subsidiary to his famous investigations into lunes (already mentioned in Solutions 1 and 21).

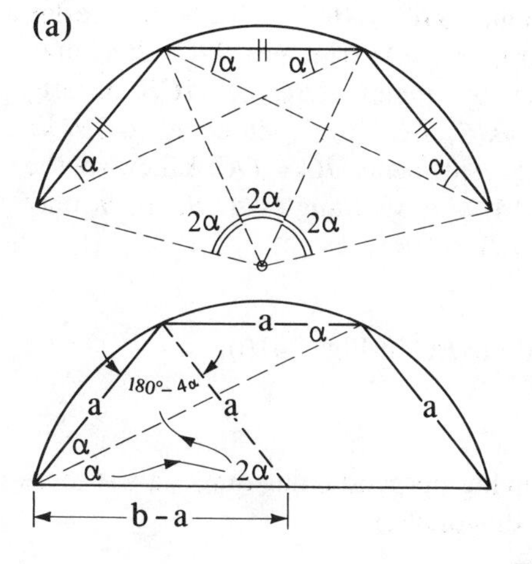

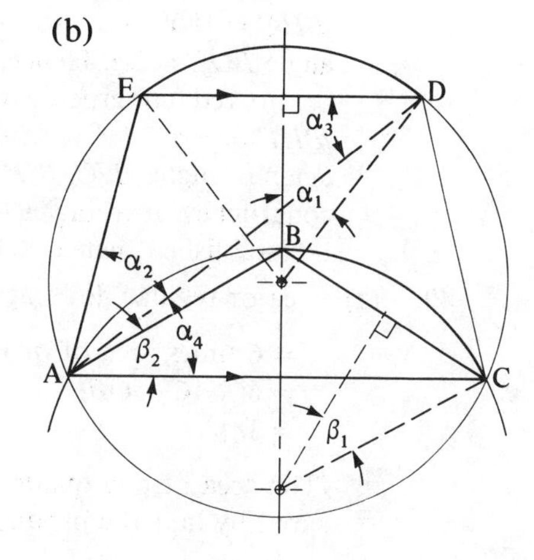

Fig. S39

(c) One method is to join *BC, AE,* complete equilateral triangle *ACD,* and to show that $\angle CAE = \angle CBE = \frac{1}{2}\angle CAD = 30°$. Hence, CE = radius of $\odot ABC$. Further $\triangle DAE \equiv \triangle CAE$, from which $DE = CE$.

Another method is indicated in Fig. S40: show that isosceles triangles *COA, CED* are congruent. . . . (This problem and the simple construction corresponding to it, for finding the radius and hence the center of a given circle, appears to have been first devised by J. H. Swale in 1830 or a little earlier. It was reported by T. S. Davies, *Philosophical Magazine,* 4th ser., *1* (1851):536–544.)

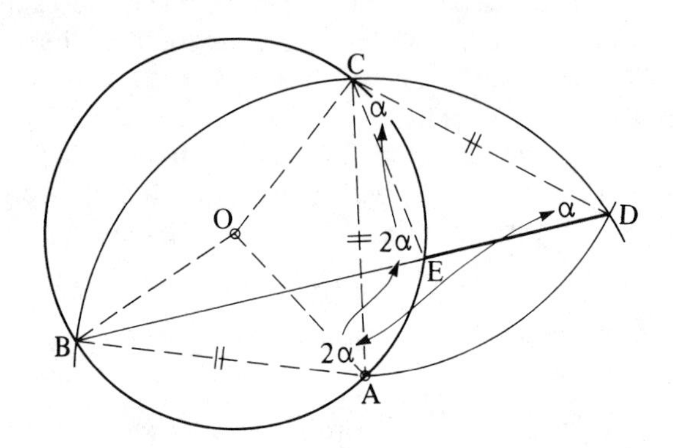

Fig. S40

(d) Let *O, J, K, L* denote the circumcenters of triangles *ABC, BHC, CHA, AHB,* respectively. For the acute-angled case, if $\angle BAC = \alpha$, then $\angle BHC = 180° - \alpha$, and reflex angle $BJC = 360° - 2\alpha$. Hence prove isosceles triangles *BJC, BOC* congruent. For $\angle BAC$ $(= \alpha)$ obtuse, $\angle BHC = 180° - \alpha$, so obtuse angle $BJC = 360° - 2\alpha$, while reflex angle $BOC = 2\alpha$. Hence, as before, isosceles triangles *BJC, BOC* may be proved congruent. For either of angles *ABC* or *ACB* obtuse, $\angle BHF = \angle BAC$ $(= \alpha)$ while angles *BJC, BOC* are each equal to 2α, so again triangles *BJC, BOC* are congruent; hence $JC = OC$. Likewise, the congruency of triangles *CKA, COA,* and of triangles *ALB, AOB,* may be established; hence $KA = OA$; $LB = OB$. . . .

49. (a) Area of regular dodecagon

$$= 6 \text{ times area of quadrilateral } OABC \text{ in Fig. S41(i),}$$
$$= 6(\tfrac{1}{2}AC \cdot OB)$$
$$= 3r^2.$$

(The area of any quadrilateral having perpendicular diagonals is measured by half the product of the diagonals.)

For an alternative method see Fig. S41(ii), suggested by an illustration from an old Chinese mathematical text reproduced in Joseph Needham, *Science and Civilization in China,* vol. 3 (with the collaboration of Wang Ling), Cambridge, England: Cambridge Univ. Press, 1959, p. 29. The dissection shows that the dodecagon = $\frac{3}{4}$ area of outer square = $\frac{3}{4}(4r^2) = 3r^2$.

Area in terms of the side length, a units say: From Fig. S41(iii),

$$b = a\sqrt{3}/2, \qquad c = a/2. \qquad \therefore a + 2b + 2c = 2a + a\sqrt{3}.$$

Hence, from the right triangle shown, $a^2(2 + \sqrt{3})^2 + a^2 = (2r)^2$, from which $r^2 = (2 + \sqrt{3})a^2$.

$$\therefore \text{Area } (= 3r^2) = 3(2 + \sqrt{3})a^2.$$

Otherwise (independently of the result, area = $3r^2$), see Fig. S41(iv):

$$\text{Area} = (a + 2b)^2 + 4ac + 4bc, \quad \text{where} \quad b = a\sqrt{3}/2, \ c = a/2,$$

as before. Hence,

$$\text{Area} = 3(2 + \sqrt{3})a^2.$$

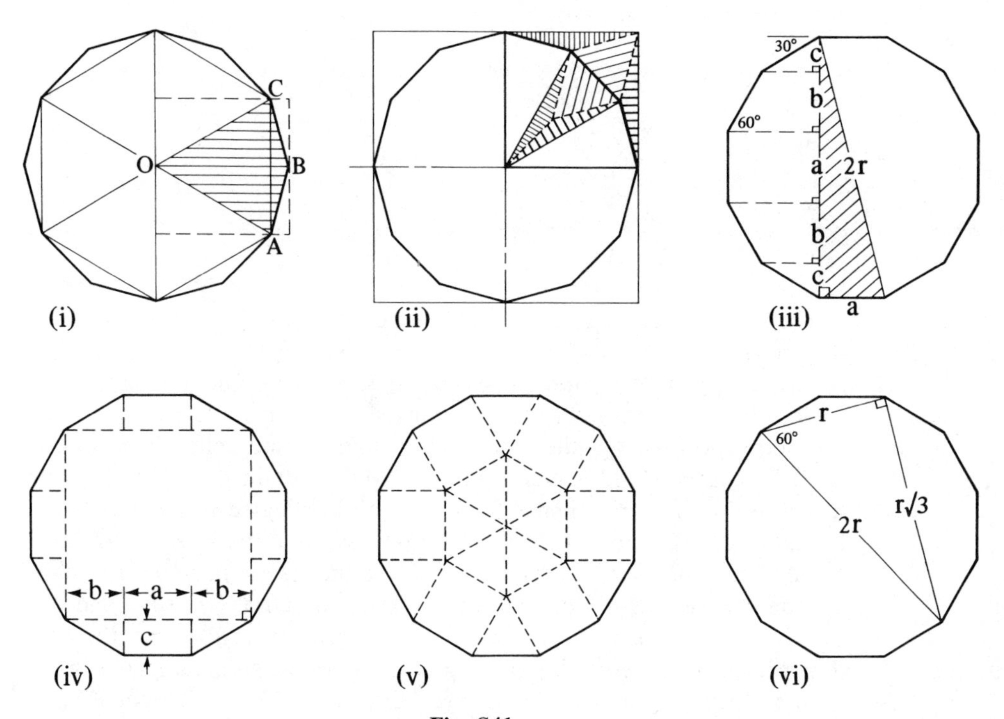

Fig. S41

Another method, suggested by an old tiling pattern, a Romanesque example of which is illustrated in Owen Jones, *The Grammar of Ornament,* London: Quaritch, 1910 ed., Pl. 30, fig. 19, is indicated in Fig. S41(v).

$$\begin{aligned}\text{Dodecagon} &= 6 \text{ squares and } 12 \text{ equilateral triangles, all of side } a\\ &= 6a^2 + 12(a^2\sqrt{3}/4)\\ &= 3(2+\sqrt{3})a^2\end{aligned}$$

Area in terms of a diagonal subtending four sides of the dodecagon: as shown in Fig. S41(vi), such a diagonal has length $r\sqrt{3}$.

$$\therefore \text{ Area } (=3r^2) = (\text{diagonal})^2.$$

The ingenious six-piece dissections of regular dodecagons and the rearrangements to form squares of equal area, shown in Fig. S42, are redrawn from Harry Lindgren, *Geometric Dissections,* Princeton, N.J.: Van Nostrand, 1964 (revised and enlarged by G. Frederickson as *Recreational Problems in Geometric Dissections,* New York: Dover, 1972; see p. 41 of either edition.)

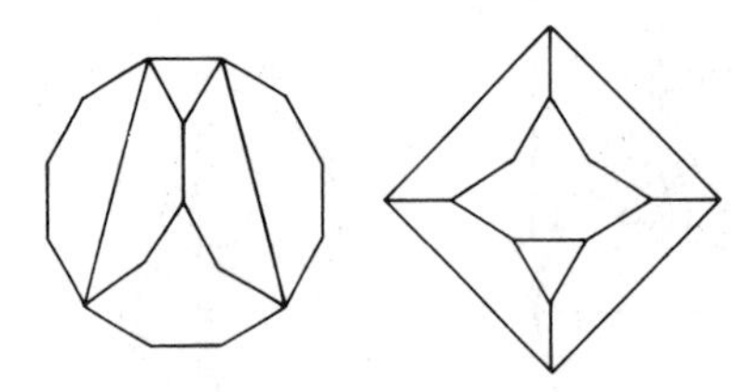
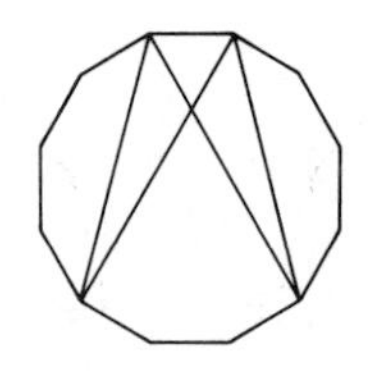
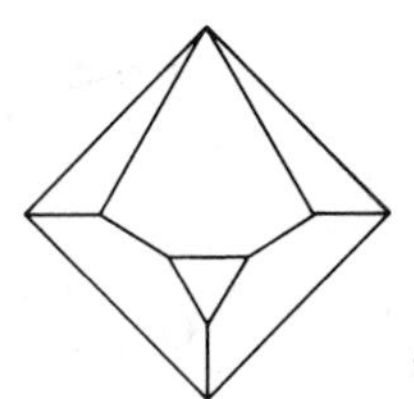

Fig. S42

(b) The best method to apply is that analogous to the first method given in part (a).

(c) The term *golden section* came into vogue only in modern times. Luca Pacioli (1509) referred to the same division of a line interval as the "divine proportion," while the Greek geometers had called it a division into "extreme and mean ratio." If AB is divided at P such that $AB:AP = AP:PB$, then AB was said to be divided into extreme and mean ratio (Euclid, VI, Def. 3); we now say that the interval AB is divided in golden section. It will be noticed that AP ($= p$, say) is the mean proportional, or geometric mean, between AB ($= a$, say) and PB ($= a - p$), since $a:p = p:(a-p)$. In general, if $a:p = p:c$, the construction for determining p when a and c are given is indicated in the first of the diagrams in Fig. S43 (cf. Euclid VI, 13, or dialogue Fig. 7). But in the special case of extreme and mean proportion, where a is given

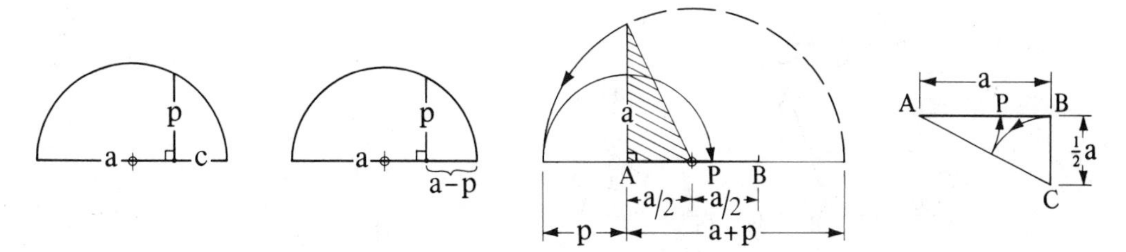

Fig. S43

and p is to be determined, $a - p$ is initially unknown and the corresponding diagram cannot be constructed directly. But since

$$p^2 = a(a - p) \Rightarrow p^2 + ap = a^2, \quad \text{or} \quad p(a + p) = a^2,$$

the construction indicated in the third diagram suggests itself and this does enable p to be determined, given only a. (Cf. Problem 21a, diagram (i) of Fig. P12.) In the present case, the sides of the shaded right triangle being $a/2$, a, and $a\sqrt{5}/2$, the value of p can be read off the diagram as $a(\sqrt{5} - 1)/2$. (Cf. Euclid, II, 11, and VI, 30.) In modern books, it is customary first to obtain $p = a(\sqrt{5} - 1)/2$ algebraically and then to give the more compact construction shown in the fourth diagram.

50. (a) Let $ABCDE$ be a regular pentagon (see Fig. S44, first diagram); its exterior angles, each equal to 72°, are also exterior angles of isosceles triangles ABC, BCD, etc. If diagonals AC, BD intersect at X, the triangles ABX, CDX are easily shown to have angles of 36°, 72°, 72°, and consequently $AX = AB = BC = CD = DX$, $= a$ say. And from congruent triangles ABC, BCD, $AC = BD$, $= d$, say; then $CX = BX = d - a$. From similar triangles ABC, BXC, $AC:BC = BC:XC$, that is $d:a = a:(d - a)$. Thus the diagonals cut each other in golden section.

(b) From part (a) of this problem and part (c) of the previous one, we have, $a/d = (\sqrt{5} - 1)/2$, or inverting and rationalizing the denominator, $d/a = (\sqrt{5} + 1)/2$. A simple construction for obtaining diagonal d $(= a/2 + (a\sqrt{5})/2)$, given side a, is indicated in the third diagram of Fig. S44. The fourth diagram shows an equivalent construction for obtaining AP $(= d)$, and hence vertex D at the point of intersection of the arc, center A and radius AP, with the perpendicular bisector of AB $(= a)$. The remaining vertices, C and E, of a regular pentagon having AB as a side, may then be located by the intersection of arcs of radius a, centered at D, A, B.

(c) Let $AB = a$, $OA = R$, $ON = r$, then BX, equal to half a diagonal, $= a(\sqrt{5} + 1)/4$. Since triangles OAN and BAX are similar, $ON:OA = BX:BA$, giving $r:R = (\sqrt{5} + 1):4$.

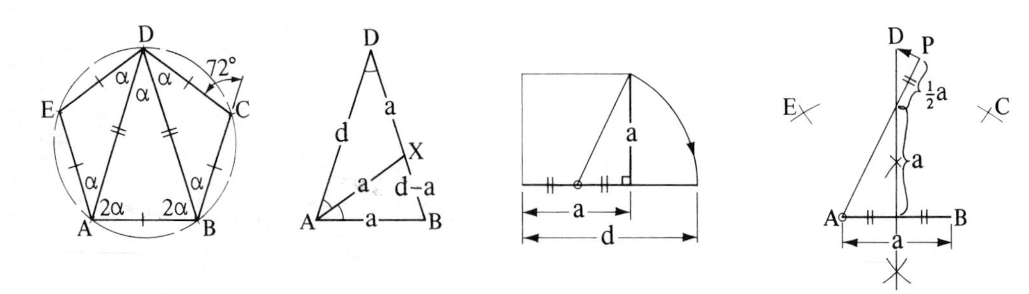

Fig. S44

(e) Side of regular pentagon, a_5 say, and of regular decagon, a_{10} say, are required in terms of the circumradius R. From the similar triangles BAX, OAN, used in part (c), $BA : OA = AX : AN$, where $BA = a_5$, $OA = R$, $AN = \frac{1}{2}a_5$, and $AX = \sqrt{(BA^2 - BX^2)}$, reducing to $AX = a_5\sqrt{(10 - 2\sqrt{5})/4}$. Hence obtain $a_5 = R\sqrt{(10 - 2\sqrt{5})/2}$. To find a_{10}, make use of $r = \frac{1}{2}(a_{10} + R)$, proved in (d), and $r = R(\sqrt{5} + 1)/4$, from (c). Hence obtain $a_{10} = R(\sqrt{5} - 1)/2$. It is easily shown that the given construction produces these values.

This construction was discussed by Ptolemy in his *Syntaxis* (ἡ μαθηματικὴ σύνταξις, "the mathematical [here astronomical] compilation"), *c.* A.D. 150, better known as the *Almagest*, from *almagestum* or *almagesti*, from *al-majisti* from ἡ μεγίστη σύνταξις or ἡ μεγάλη σύνταξις (the greatest, or great, compilation). English translation by R. C. Taliaferro in *Great Books of the Western World*, vol. 16, pp. 1–478. For the regular pentagon and decagon, see pp. 14–15.

51. (a) Referring to the first diagram of Fig. S45, we have from the previous examples, $d : a = a : (d - a)$, from which $d = \frac{1}{2}a(\sqrt{5} + 1)$. Further, $d = 2(d - a) + b$, from which $b = \frac{1}{2}a(3 - \sqrt{5})$. Area ratio, innermost region : complete figure, $= b^2 : a^2$, which reduces to $(7 - 3\sqrt{5}) : 2$.

(b) $\triangle AOD \equiv \triangle AED$, $\therefore \angle AOD = \angle AED$, that is, the angle between two nonadjacent slant edges = the interior angle of a regular pentagon = 108°. Also, altitudes ON, EN of the congruent triangles AOD, AED are equal; $\therefore \triangle ONE$ (shown in its true shape in the cross-section in the second diagram of Fig. S45) is isosceles. Hence, exterior angle $ONM = 2\angle OEN$; that is, the inclination to the base of the plane through OA, OD is twice the inclination of a slant edge.

(c) From the similar triangles OAN, BAX, of Problem 50(c), $ON : BX = AN : AX$. where $ON = p$, $BX = \frac{1}{2}$ diagonal $= a(\sqrt{5} + 1)/4$, $AN = a/2$, $AX = \sqrt{(BA^2 - BX^2)}$ reducing to $\frac{1}{4}a\sqrt{(10 - 2\sqrt{5})}$, as in Problem 50(e). Hence obtain $p = \frac{1}{10}a\sqrt{\{5(5 + 2\sqrt{5})\}}$. Volume of pyramid, $V = \frac{1}{3}(2\frac{1}{2}ap)z$, where perpendicular height $z = \sqrt{(s^2 - p^2)}$, slant height $s = \frac{1}{2}a\sqrt{3}$. Expressing V in terms of a alone, and simplifying, obtain $V = \frac{1}{24}a^3 \cdot \sqrt{\{10(3 + \sqrt{5})\}}$.

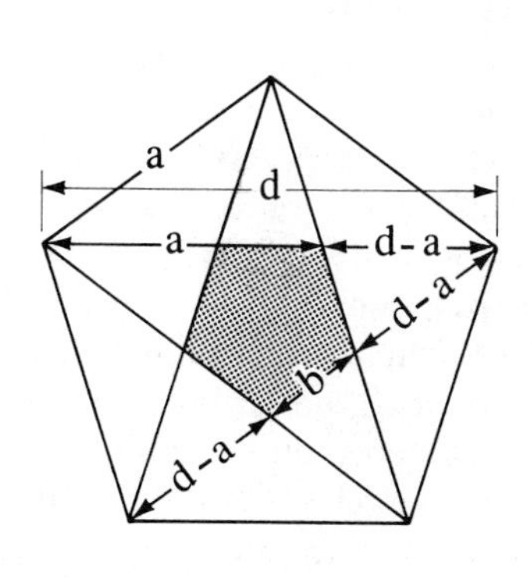

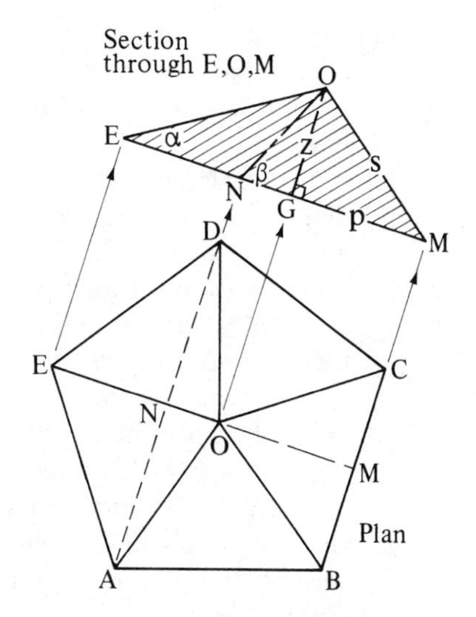

Fig. S45

52. (a) Area of regular pentagon, $P = 5(\frac{1}{2}ar)$, where side a and perpendicular to side (= incircle radius r) are each to be expressed in terms of the circumcircle radius ρ. From the similar triangles shown in diagram (i) of Fig. P28 given for Problem 50, obtain $a = \rho\sqrt{(10 - 2\sqrt{5})}/2$, and $r = \rho(\sqrt{5} + 1)/4$. Substituting for a and r in $P = 2\frac{1}{2}\,ar$, and simplifying, obtain $P = \frac{5}{8}\{\sqrt{(10 + 2\sqrt{5})}\}\rho^2$. It is easy to show that the side length and perpendicular height of an equilateral triangle inscribed in a circle of radius ρ are, respectively, $\rho\sqrt{3}$ and $1\frac{1}{2}\,\rho$, giving area $T = \frac{3}{4}\sqrt{3} \cdot \rho^2$. Then $12P/20T$ simplifies to $\sqrt{\{6(5 + \sqrt{5})\}}/6 \approx 1.098$, showing that $12P$ exceeds $20T$ by almost 10 percent.

(b) (i) R_d = half-diagonal of cube of edge e_c,

$$\therefore\ R_d = e_c\sqrt{3}/2 \qquad (1)$$

ρ_d = circumradius of regular pentagon with diagonal equal to edge e_c.

$$r_d^2 : R_d^2 = (R_d^2 - \rho_d^2) : R_d^2 = 1 - \rho_d^2 : R_d^2 \qquad (2)$$

From part (a), $a = \rho\sqrt{(10 - 2\sqrt{5})}/2$, $\therefore\ \rho = 2a/\sqrt{(10 - 2\sqrt{5})}$, where side length $= \frac{1}{2}(\sqrt{5} - 1)$ times diagonal, that is, $a = \frac{1}{2}(\sqrt{5} - 1) \cdot e_c$; from which,

$$\rho\ (\text{or, here, } \rho_d) = e_c(\sqrt{5} - 1)/\sqrt{(10 - 2\sqrt{5})} \qquad (3)$$

Substituting from (1) and (3) into (2) and simplifying, obtain

$$r_d^2 : R_d^2 = (5 + 2\sqrt{5})/15.$$

(ii) R_i = half-diagonal of rectangle of width e_i and length $\frac{1}{2}(\sqrt{5} + 1)e_i$. Hence obtain $R_i = \frac{1}{4}e_i\sqrt{(10 + 2\sqrt{5})}$. ρ_i = two-thirds of median of equilateral triangle of side e_i. $\therefore\ \rho_i = e_i/\sqrt{3}$. $r_i^2 : R_i^2$, equal to $1 - \rho_i^2/R_i^2$, can now be readily simplified to $(5 + 2\sqrt{5})/15$.

It is thus shown that $r_d^2 : R_d^2$ and $r_i^2 : R_i^2$ have a common value, hence (positive values only being relevant), $r_d : R_d = r_i : R_i$.

(c) It follows from part (b) that the right-angled triangles shown shaded in diagrams (i) and (ii) of Fig. P29 are *similar*. Hence, we need only to be given any *one* of the following equalities, $R_i = R_d$, $r_i = r_d$, $\rho_i = \rho_d$, in order to establish that these triangles are congruent, and so to establish that the two remaining equalities hold also. Thus, if a regular dodecahedron and a regular icosahedron have equal circumscribing spheres, then not only do they have equal inscribed spheres, but in addition the faces of each solid are inscribable within equal circles—and so the findings of part (a) are directly applicable to the surface areas of the two solids. Further, conceiving two such polyhedra as composed of sets of right pyramids having their bases coinciding with the faces of the respective figures and their apices meeting at the centers, we have, for the volume ratio,

$$V_d : V_i = 12(\tfrac{1}{3}Pr) : 20(\tfrac{1}{3}Tr) = 12P : 20T = \text{surface ratio, } A_d : A_i.$$

Consequently, since $12P/20T \approx 1.098$, for a regular dodecahedron and a regular icosahedron circumscribed about equal spheres, or inscribed within equal spheres, the volume and surface of the dodecahedron exceed the volume and surface of the icosahedron by almost 10 percent in each case.

(d) Pappus (first half of the 4th cent. A.D.) proved in the concluding propositions of Book V of his *Collection* that for any pair of the five regular solids for which the surfaces are equal, that which has the more faces is the greater in volume. For the particular case of the icosahedron and the dodecahedron, this relation may be shown on the basis of the result established in part (c). For if the icosahedron in that example be enlarged (by almost 5 percent lineally) so that its surface becomes equal to that of the dodecahedron, then the volume is increased (by about 15 percent); consequently the volume of the icosahedron will then exceed that of the dodecahedron (by about 5 percent).

In more detail: let A_i, A_d, V_i, V_d denote the respective areas and volumes of a regular icosahedron and a regular dodecahedron inscribed in the same sphere; then, as has been shown, $A_d = kA_i$, and $V_d = kV_i$, where $k = \sqrt{\{6(5 + \sqrt{5})\}}/6 \approx 1.098$. Further, let A_I and V_I denote the area and volume of a regular icosahedron such that $A_I = A_d$ ($\approx 1.098\ A_i$), then, in terms of the edge lengths,

$$V_I : V_i = e_I^3 : e_i^3, \quad \text{where} \quad e_I^2 : e_i^2 = A_I : A_i.$$

Hence,

$$V_I : V_i = A_I^{3/2} : A_i^{3/2} = k^{3/2} : 1.$$

Then, $V_I : V_d$, the ratio sought, may be found from

$$V_I/V_d = (V_I/V_i)(V_i/V_d) = (k^{3/2}/1)(1/k) = k^{1/2} : 1;$$
$$\text{so } V_I : V_d \approx \sqrt{1.098} : 1 \approx 1.048 : 1;$$

that is, the volume of the icosahedron exceeds that of the dodecahedron by nearly 5 percent when the surface areas are equal.

Historical Note

Diagram (i) Fig. P29 for Problem 52, in which the twelve edges of a cube are made to coincide with twelve face-diagonals of a regular dodecahedron, illustrates the relation between these two solid figures (both inscribable within the same sphere) utilized by Euclid in XIII, 17 of his *Elements.* Diagram (ii), relating a regular icosahedron with a certain enclosing cube, depicts an association not to be found in Euclid. Heath, in his commentary following Euclid XIII, 16, attributes this construction to the Cambridge mathematician H. M. Taylor (1842–1927). However, the icosahedron and cube are connected in the very same way in Book 3 of Luca Pacioli's *Divina proportione,* 1509, from which the first of the diagrams in Fig. S46 is reproduced. This final book of the *Divina* is usually understood to consist of an unacknowledged translation (from Latin into Italian) of a large part of Piero della Francesca's *De quinque corporibus regularibus, c.* 1480, a transcript of which is preserved in the Vatican Library. For a spirited defence of Pacioli against the charge of plagiarism, see Robert Emmet Taylor, *No Royal Road: Luca Pacioli and His Times,* Chapel Hill: Univ. of North Carolina Press, 1942, pp. 285–286, 342–355. Cf., however, Margaret Daly Davis,

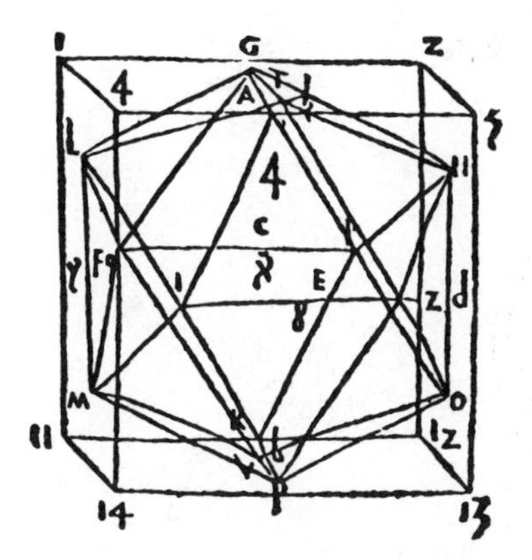

Fig. S46

Piero della Francesca's Mathematical Treatises, Ravenna: Longo, 1977, appendices I and II.

In Euclid's classic treatment of the regular icosahedron (Euclid, XIII, 16), the solid is regarded as composed of two pentagonal pyramids separated by a polygonal "drum" or "antiprism." This is shown in the right-hand diagram of Fig. S46, reproduced from Kepler's *Harmonice Mundi,* 1619 (*Johannes Kepler Gesammelte Werke,* Munich: Beck, vol. 6, p. 79). From the same work is reproduced Fig. S47, Kepler's diagram of the geometrical model that he devised for the Copernican solar system (ibid., p. 298). In this scheme, the five regular solids were arranged so as to nest between—and determine the relative sizes of—six concentric spheres, one sphere for each of the known planets, including the earth. An

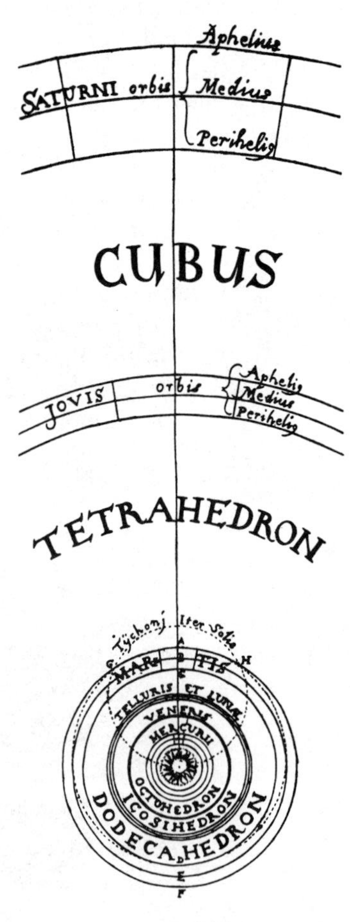

Fig. S47

earlier version of this model had appeared in the *Mysterium cosmographicum* of 1596, and Kepler returned to it also in Book IV of the *Epitome astronomiae Copernicanae,* 1620. The three works mentioned show Kepler's inspired endeavor to reveal theological significances and final causes. He was, of course, very well aware that the ratio of the radii of the circumscribed and inscribed spheres is the same for the dodecahedron as for the icosahedron (and also that it is the same for the octahedron as for the cube). The radius of the sphere inscribed in each of the regular solids was given relative to a radius of 100,000 assumed for the circumscribing sphere. These values Kepler tabulated correct to five significant figures: for the cube, 57,735; for the tetrahedron, 33,333; for the dodecahedron 79,465; for the icosahedron, 79,465; for the octahedron, 57,735 (*Epitome,* IV, 1, iii; *Gesammelte Werke,* vol. 7, p. 273). English translations of Books IV and V of the *Epitome* and of Book V of *Harmonice Mundi* appear in vol. 16 of *Great Books of the Western World.*

The metaphysical role of mathematics displayed in Kepler's works is clearly in the Pythagorean tradition as known to us, and of course to Kepler, in the doctrine of Plato's *Timaeus.* In this tradition, not merely are astronomical measures to be found to save the appearances, they must further be comprehended by some unifying interrelation. As Kepler tells us, the astronomer may rightly hope to contemplate the true form of the edifice of the world and to understand something of the archetypal principles of Creation. But with the Greeks, outside the Pythagorean Order, mathematics was already separated from philosophy, and in the century of Theatetus, Aristaeus, and Euclid, the theory of the regular polyhedra was worked out independently of extraneous significances that mystical minds might wish to discover for these fascinating figures. (There is now doubt as to whether the Aristaeus who wrote a lost work entitled *Concerning the Comparison of the Five Regular Solids* was the same Aristaeus who flourished a generation before Euclid and who played a major part in the development of conic section theory; it might have been a younger Aristaeus who perhaps followed Euclid in his investigations into the regular solids.)

A so-called Book XIV of the *Elements,* written not by Euclid, but by Hypsicles in the first half of the second century B.C., was presented as a commentary or addendum to a lost work on the comparison of the dodecahedron and the icosahedron written by Apollonius (fl. c. 200 B.C.). An English translation of this Book XIV is included in Heath's edition of Euclid's *Elements* as it forms an interesting supplement to Euclid's Book XIII. Hypsicles' introduction to the second of his eight propositions is worth quoting:

> [Prop. 2.] The same circle circumscribes both the pentagon of the dodecahedron and the triangle of the icosahedron inscribed in the same sphere.
>
> This is proved by Aristaeus in his work entitled *Comparison of the Five Regular Figures.* But Apollonius proves in the second edition of his *Comparison of the Dodecahedron with the Icosahedron* that, as the surface of the dodecahedron is to the surface of the icosahedron, so also is the [volume of the] dodecahedron itself to

> the [volume of the] icosahedron, because the perpendicular from the centre of the sphere to the pentagon of the dodecahedron and to the triangle of the icosahedron is the same.

It remains to be said that the Greek geometers were handicapped in their investigations of the polyhedra, and in the composition of their demonstrations, by the lack of any notation allowing for the convenient manipulation of surd expressions. In reading some of the more extended parts of Euclid (not to mention the even more demanding *Conics* of Apollonius!), modern students will find themselves confronted from time to time with *tours de force* which are not easily assimilated. But persistence will be rewarded by a rich harvest of geometrical insights.

XVIII

Learning without thought is labour lost; thought without learning is perilous.

—Confucius

A particle of tuition conveys science to a comprehensive mind; and, having reached it, expands of its own impulse. As oil poured upon water, as a secret entrusted to the vile, as alms bestowed upon the worthy, however little, so does science infused into a wise mind spread by intrinsic force.

—Bhāskara

I hope that posterity will judge me kindly, not only as to the things which I have explained, but also as to those which I have intentionally omitted so as to leave to others the pleasure of discovery.

—Descartes

Man's infancy is intrinsically frustrating, and it is this fact which makes him restlessly creative, searching for 'ideal' solutions, whether these be in the shape of scientific world-views, philosophies, religions, or the integrative patterns of art.

—Anthony Storr

The mathematician's patterns, like the painter's or the poet's, must be beautiful. *The ideas, like the colours or the words, must fit together in a harmonious way. Beauty is the first test: there is no permanent place in the world for ugly mathematics.*

—G. H. Hardy

Art is not an end in itself, but a means of addressing humanity.

—M. P. Mussorgsky

An artist never really finishes his work; he merely abandons it.

—Paul Valéry

BIBLIOGRAPHIA SELECTA

The books and articles included here, amounting to approximately one-third of those referred to in the notes, are the ones most likely to be found important by readers wishing to study further the topics introduced in the foregoing pages. Most of the monographs, reference works, and journals in this list should be found in any major university library.

ALHAZEN. See Ibn al-HAYTHAM.

APOLLONIUS of Perga. *Treatise on Conic Sections.* Translated and edited in modern notation with introductions by Thomas L. Heath. Cambridge, England: Cambridge Univ. Press, 1896.

ARCHIMEDES. *The Works of Archimedes,* 1897. Supplement to this, titled *The Method of Archimedes,* 1912. Both publications translated, edited and introduced by Thomas L. Heath, Cambridge, England: Cambridge Univ. Press; reprinted New York: Dover, n.d. (See also E. J. DIJKSTERHUIS.)

ARISTARCHUS. See Thomas L. HEATH, 1913.

ARISTOTLE. *The Works of Aristotle.* Translated under the editorship of J. A. Smith and W. D. Ross. 12 vols. Oxford: Clarendon Press, 1908–52. One-volume abridgement, *The Basic Works of Aristotle,* edited and introduced by Richard McKeon, New York: Random House, 1941.
See also John CORCORAN, Thomas L. HEATH, 1949, and references given in dialogue note 146.

ĀRYABHAṬA. *The Āryabhaṭīya of Āryabhaṭa, An Ancient Indian Work on Mathematics and Astronomy.* Translated, with notes, by Walter Eugene Clark. Chicago: Univ. of Chicago Press, 1930.

Francis BACON. *The Works of Francis Bacon.* Edited by James Spedding, R. L. Ellis, and D. D. Heath. 14 vols. London, 1857–74. (See also *Great Books of the Western World,* vol. 30, for a translation of *Novum Organum.*)

Margaret E. BARON. *The Origins of the Infinitesimal Calculus.* Oxford: Pergamon Press, 1969.

Evert W. BETH and Jean PIAGET. *Mathematical Epistemology and Psychology.* Translated by W. Mays. Dordrecht: Reidel, 1966.

BHĀSKARA. *Līlāvatī* [a work on arithmetic and mensuration dedicated to Bhāskara's daughter, Līlāvatī], and *Bijagaṇita* [or *Vija-gaṇita,* = "Calculation with elements," or "Algebra"], both about A.D. 1150. See Henry Thomas COLEBROOKE for English translations.

David BLOOR. "Polyhedra and the Abominations of Leviticus." *British Journal for the History of Science 11* (1978):245–272. (See also John WORRALL.)

H. J. M. BOS. "Differentials, Higher-Order Differentials and the Derivative in the Leibnizian Calculus." *Archive for History of Exact Sciences 14* (1974):1–90.

Carl B. BOYER. *The History of the Calculus and Its Conceptual Development.* First published in 1939 as *The Concepts of the Calculus.* Corrected 1949. Reprinted New York: Dover, 1959.

______. *A History of Mathematics.* New York: Wiley, 1968.

______. "Cavalieri, Limits and Discarded Infinitesimals." *Scripta Mathematica 8* (1941):79–91.

______. "Early Rectifications of Curves." *Mélanges Alexandre Koyré.* vol. 1. Paris: Hermann, 1964, pp. 30–39.

BRAHMAGUPTA. *Brāhma-sphuṭa-siddhānta* ["The Revised System of Brahma"], *c.* A.D. 628. See Henry Thomas COLEBROOKE for English translation of chap. 12 (Arithmetic) and chap. 18 (Algebra).

Buffalo Symposium on Modernist Interpretations of Ancient Logic. See John CORCORAN (ed.).

Walter BURKERT. *Lore and Science in Ancient Pythagoreanism.* Translated by E. L. Minar. Cambridge, Mass.: Harvard Univ. Press, 1972.

Robert E. BUTTS and Joseph C. PITT (eds.). *New Perspectives on Galileo, Papers Deriving from and Related to a Workshop on Galileo Held at Virginia Polytechnic Institute and State University, 1975.* Dordrecht: Reidel, 1978.

Florian CAJORI. *A History of the Conceptions of Limits and Fluxions in Great Britain from Newton to Woodhouse.* Chicago: Open Court, 1919.

______. *A History of Mathematical Notations.* 2 vols. Chicago: Open Court, 1928–29.

CAVALIERI [P. Bonaventura Cavalerio]. *Geometria indivisibilibus continuorum nova quadam ratione promota.* Bologna, 1st ed. 1635, 2nd ed. 1653.

______. *Exercitationes geometricae sex.* Bologna, 1647.

The Century Dictionary, An Encyclopedic Lexicon of the English Language. Editor-in-chief, William D. Whitney [1st ed., 1889–91]. Rev. ed., 8 vols., London: The Times, and New York: Century, 1903.
(A work of remarkable comprehensiveness: "In the mathematical work the aim has been to define all the older English terms, and all the modern ones that can be considered to be in general use . . ."—Preface, p. xiv.)

Morris R. COHEN and I. E. DRABKIN (eds.). *A Source Book in Greek Science.* New York: McGraw-Hill, 1948.

Henry Thomas COLEBROOKE. *Algebra with Arithmetic and Mensuration from the Sanscrit of Brahmegupta and Bhāscara.* Translated by Colebrooke. London: Murray, 1817.

Julian Lowell COOLIDGE. *The Mathematics of Great Amateurs.* Oxford: Clarendon Press, 1949. Reprinted, New York: Dover, 1963.

John CORCORAN (ed.). *Ancient Logic and Its Modern Interpretations.* Proceedings of the Buffalo Symposium on Modernist Interpretations of Ancient Logic, 1972. Synthese Historical Library, vol. 9. Dordrecht: Reidel, 1974.

F. M. CORNFORD. See PLATO.

Richard COURANT and Herbert ROBBINS. *What Is Mathematics?* London: Oxford Univ. Press, 1941.

H. S. M. COXETER. *Introduction to Geometry.* New York: Wiley, 1st ed. 1961, 2nd ed. 1969.

Michael J. CROWE. "Ten 'Laws' Concerning Patterns of Change in the History of Mathematics." *Historia Mathematica 2* (1975):161–166. (See Herbert MEHRTENS for a related article.)

DESCARTES. *The Geometry of René Descartes.* Facsimile of the original French text of 1637 with English translation and notes by D. E. Smith and M. L. Latham. Chicago: Open Court, 1925. Reprinted, New York: Dover, 1954.
(An alternative translation is contained in the following book.)

______. *Discourse on Method, Optics, Geometry, and Meteorology,* Translated and introduced by P. J. Olscamp. Indianapolis: Bobbs-Merrill, 1965.

(Translations of the *Discourse on Method* and *The Geometry* are included in vol. 31 of *Great Books of the Western World.*) (See also references given at the end of Solution 41.)

Dictionary of Scientific Biography. Edited by C. C. Gillispie. 16 vols. New York: Scribners, 1970–1980.

E. J. DIJKSTERHUIS. *Archimedes.* Copenhagen: Munksgaard, 1956.

Heinrich DÖRRIE. *100 Great Problems of Elementary Mathematics, Their History and Solution.* Translated from the German edition of 1948 by David Antin. New York: Dover, 1965.

Stillman DRAKE. *Galileo Studies: Personality, Tradition and Revolution.* Ann Arbor: Univ. of Michigan Press, 1970.

______. *Galileo at Work: His Scientific Biography.* Chicago: Univ. of Chicago Press, 1978.

______. *Galileo.* Past Masters Series. Oxford: Oxford Univ. Press, 1980.

Albrecht DÜRER. *Underweysung der Messung mit dem Zirckel und Richtscheyt.* 1525. Facsimile, Portland, Ore.: Collegium Graphicum, 1972.

______. *Vier Bücher von menschlicher Proportion.* 1528. Facsimile, Zürich: Stocker, 1969.

EUCLID. *The Thirteen Books of Euclid's Elements.* Translated, with introduction and commentary, by Thomas L. Heath. 3 vols. Cambridge, England: Cambridge Univ. Press, 1st ed. 1908, 2nd ed., 1926. Facsimile of 2nd ed., New York: Dover, 1956.

______. "The Optics of Euclid." English translation by H. E. Burton. *Journal of the Optical Society of America 35* (1945):357–372.

Howard EVES. *An Introduction to the History of Mathematics* [1953]. New York: Holt, Rinehart and Winston, 2nd ed. 1964; 3rd ed. 1969; 4th ed. 1976.

Leonardo [Pisano] FIBONACCI. *Scritti di Leonardo Pisano,* vol. 1: *Liber Abbaci;* vol. 2: *Practica Geometriae.* Edited by Baldassarre Boncompagni. Rome: 1857–62.

Hans FREUDENTHAL. *Mathematics as an Educational Task.* Dordrecht: Reidel, 1973.

GALILEO. [*Dialogo di Galileo Linceo . . . Doue ne i congressi de giornate si discorre sopra i due Massimi Sistemi del Mondo,* 1632. English.] *Dialogue Concerning the Two*

Chief World Systems—Ptolemaic and Copernican. Translated by Stillman Drake. Foreword by Albert Einstein. Berkeley: Univ. of California Press, 1953.

______. [*Discorsi e Dimostrazioni Matematiche intorno à Due Nuove Scienze,* 1638. English.] *Two New Sciences Including Centers of Gravity and Force of Percussion.* Translated, with introduction and notes, by Stillman Drake. Madison: Univ. of Wisconsin Press, 1974.

______. *Discoveries and Opinions of Galileo.* Translated, with introduction and notes, by Stillman Drake. Garden City, N.Y.: Doubleday, 1957. (Contains: "The Starry Messenger," "Letters on Sunspots," "Letter to the Grand Duchess Christina," and excerpts from "The Assayer.")

______. *Le Opere di Galileo Galilei.* Edizione nazionale, edited by Antonio Favaro. 20 vols. in 21. Florence: G. Barbèra, 1890–1909. Reprinted with slight revision, 1929–1939, and again 1964–1966.

______. See also R. E. BUTTS and J. C. PITT, S. DRAKE, L. GEYMONAT, and E. McMULLIN.

Solomon GANDZ. "The Mishnat ha Middot, The First Hebrew Geometry of about 150 C.E., and the Geometry of Muhammad ibn Musa al-Khowarizmi, The First Arabic Geometry (*c.* 850 C.E.) . . . ," Introduction, translation, and notes by S. Gandz. *Quellen und Studien zur Geschichte der Mathematik, Astronomie und Physik A, 2* (1932):1–96 and plates. Reprinted in Gandz's *Studies in Hebrew Astronomy and Mathematics.* New York: Ktav, 1970.

Ludovico GEYMONAT. *Galileo Galilei: A Biography and Inquiry into His Philosophy of Science.* Translated with notes, by Stillman Drake. New York: McGraw-Hill, 1965.

Herman H. GOLDSTINE. *A History of Numerical Analysis from the 16th through the 19th Century.* New York: Springer, 1977.

Edward GRANT (ed.). *A Source Book in Medieval Science.* Cambridge, Mass.: Harvard Univ. Press, 1974.

Ivor GRATTAN-GUINNESS. *The Development of the Foundations of Mathematical Analysis from Euler to Riemann.* Cambridge, Mass.: MIT Press, 1970.

______. (ed.). *From Calculus to Set Theory, 1630–1910.* London: Duckworth, 1980.

Great Books of the Western World. Edited by R. M. Hutchins. 54 vols. Chicago: William Benton and Encyclopaedia Britannica, 1952.

GREGORY. *James Gregory Tercentenary Memorial Volume.* Containing his correspondence with John Collins and his hitherto unpublished mathematical manuscripts. Edited by H. W. Turnbull. London: Bell, 1939.

W. K. C. GUTHRIE. *A History of Greek Philosophy.* 6 vols. Cambridge, England: Cambridge Univ. Press, 1962–1981.

Ian HACKING. *The Emergence of Probability: A Philosophical Study of Early Ideas about Probability, Induction and Statistical Inference.* London: Cambridge Univ. Press, 1975. (For review, see Colin HOWSON.)

Jacques HADAMARD. *An Essay on the Psychology of Invention in the Mathematical Field.* Princeton, N.J.: Princeton Univ. Press, 1945. Reprinted, New York: Dover, 1954.

G. H. HARDY. *A Mathematician's Apology.* Cambridge, England: Cambridge Univ. Press, 1st ed. 1940; 2nd ed. (with a foreword by C. P. Snow), 1967.

Ibn al-HAYTHAM. "Die Abhandlung über die Ausmessung des Paraboloids von el Ḥasan b. el-Ḥasan b. el-Haitham." Translated, with commentary by Heinrich Suter. *Bibliotheca Mathematica,* 3rd ser., *12* (1911-12):289-332.

Thomas L. HEATH. *Aristarchus of Samos, The Ancient Copernicus: A History of Greek Astronomy to Aristarchus together with Aristarchus's Treatise on the Sizes and Distances of the Sun and Moon.* Oxford: Clarendon Press, 1913. Reprinted 1959, 1966.

______. *A History of Greek Mathematics.* 2 vols. Oxford: Clarendon Press, 1921. Reprinted 1960, 1965.

______. *A Manual of Greek Mathematics.* Oxford: Clarendon Press, 1931. Reprinted, New York: Dover, 1963.

______. *Greek Astronomy.* London: Dent, 1932.

______. *Mathematics in Aristotle.* Oxford: Clarendon Press, 1949.

______. For works edited by HEATH, see APOLLONIUS, ARCHIMEDES, and EUCLID.

A History of Technology. Edited by Charles Singer, E. J. Holmyard, A. R. Hall and Trevor Williams. 5 vols. Oxford: Clarendon Press, 1954-1958.

Joseph E. HOFMANN. *Leibniz in Paris 1672-1676, His Growth to Mathematical Maturity.* Translated by A. Prag and D. T. Whiteside. London: Cambridge Univ. Press, 1974.

Ross HONSBERGER. *Mathematical Gems, from Elementary Combinatorics, Number Theory, and Geometry.* Washington, D.C.: Mathematical Association of America, 1973.

Colin HOWSON. "The Prehistory of Chance [Review of Hacking's *Emergence of Probability*]." *British Journal for the Philosophy of Science 29* (1978):274-280.

Charles HUTTON. *Mathematical and Philosophical Dictionary.* Facsimile of 1st ed. (1795-96). Hildesheim [Germany] and New York: Olms, 1973.

Christiaan HUYGENS. *Oeuvres complètes.* 22 vols. The Hague: Nijhoff, for the Société hollandaise des sciences, 1888-1950.

Glenn JAMES and Robert C. JAMES (eds.). *Mathematics Dictionary.* New York: Van Nostrand 1st ed. 1949; 2nd ed. 1959; 3rd ed. 1968; 4th ed. 1976.

KEPLER. *Johannes Kepler Gesammelte Werke.* Edited by W. von Dyck, M. Caspar, and F. Hammer, 19 of approximately 22 planned vols. published to date, Munich: Beck, 1937-. (For English translations of the *Epitome astronomiae Copernicanae,* IV & V, and *Harmonice Mundi,* V, see *Great Books of the Western World,* vol. 16.)

Felix KLEIN. *Elementary Mathematics from an Advanced Standpoint,* vol. 1: *Arithmetic, Algebra, Analysis;* vol. 2: *Geometry.* Translated from the 3rd German ed. (1924) by E. R. Hedrick and C. A. Noble. New York: Dover, n.d.

______. *Famous Problems of Elementary Geometry.* Translated by W. W. Beman and D. E. Smith, 1897. 2nd ed. with notes by R. C. Archibald, 1930. Reprinted, New York: Dover, 1956. Reprinted under the title *Famous Problems and Other Monographs,* by Felix Klein and Others, New York: Chelsea, 1962.

Morris KLINE. *Mathematical Thought from Ancient to Modern Times.* New York: Oxford Univ. Press, 1972.

______. *Why the Professor Can't Teach: Mathematics and the Dilemma of University Education.* New York: St Martin's Press, 1977.

Thomas S. KUHN. *The Structure of Scientific Revolutions.* (International Encyclopedia of Unified Science, vol. 2, no. 2; with 1969 Postscript.) 1st ed. 1962; 2nd ed. (enlarged), Chicago: Univ. of Chicago Press, 1970.
(See also L. LAUDAN, and H. MEHRTENS.)

Imre LAKATOS. *Proofs and Refutations: The Logic of Mathematical Discovery.* Edited by John Worrall and Elie Zahar. Cambridge: Cambridge Univ. Press, 1976. (Expanded version of the four-part dialogue in vol. 14 of the *British Journal for the Philosophy of Science,* 1963–64.)
(See also D. BLOOR, L. LAUDAN, and J. WORRALL for discussions of Lakatos's work.)

Larry LAUDAN. *Progress and Its Problems: Towards a Theory of Scientific Growth.* Berkeley: Univ. of California Press, 1977.

LEIBNIZ. See Joseph E. HOFMANN.

LEONARDO PISANO. See FIBONACCI.

Ernan McMULLIN (ed.), *Galileo, Man of Science.* New York: Basic Books, 1967.

MAHĀVĪRA. *The Ganita-sāra-saṅgraha of Mahāvīrācārya,* with English translation and notes by M. Raṅgācārya. Introduction by D. E. Smith. Madras: Government Press, 1912.

Mathematics Dictionary. See Glenn JAMES and Robert C. JAMES (eds.).

Kenneth O. MAY. *Bibliography and Research Manual of the History of Mathematics.* Toronto: Univ. of Toronto Press, 1973.

Herbert MEHRTENS. "T. S. Kuhn's Theories and Mathematics: A Discussion Paper on the 'New Historiography' of Mathematics." *Historia Mathematica 3* (1976):297–320.

Charles MUGLER. *Dictionnaire historique de la terminologie géométrique des Grecs.* Issued in 2 parts. Paris: Klincksieck, 1958–59.

Joseph NEEDHAM. *Science and Civilization in China.* (5 vols. in 10 to 1982. Cambridge, England: Cambridge Univ. Press, 1954–, especially vol. 3, in collaboration with Wang Ling: *Mathematics and the Sciences of the Heavens and the Earth,* 1959.

James R. NEWMAN (ed.). *The World of Mathematics, A Small Library of the Literature of Mathematics from A'h-mosé the Scribe to Albert Einstein.* With commentaries and notes. 4 vols. New York: Simon and Schuster, 1956.

Isaac NEWTON. *Philosophiae Naturalis Principia Mathematica.* 1st ed. 1687; 2nd ed. 1713; 3rd ed. 1726 translated into English by Andrew Motte, 1729; this translation revised and supplied with a historical and explanatory appendix by Floran Cajori, Cambridge, England: Cambridge Univ. Press, and Berkeley: Univ. of California Press, 1934. The revised translation, without Cajori's valuable appendix, is reproduced in *Great Books of the Western World,* vol. 34. (Also, *The Third Edition with Variant Readings* [but without English translation] assembled and edited by Alexandre Koyré and I. Bernard Cohen. 2 vols. Cambridge Univ. Press, 1972.)

______. *The Method of Fluxions and Infinite Series, with its Application to the Geometry of Curve-lines, by the Inventor Sir Isaac Newton . . .* "translated from the Author's Latin Original not yet made publick . . ." by John Colson, London, 1736.
(See also B. ROBINS and F. CAJORI.)

______. *The Mathematical Papers of Isaac Newton.* Edited by D. T. Whiteside. 8 vols. Cambridge, England: Cambridge Univ. Press, 1967–1981.

Thomas NICKLES (ed.). *Scientific Discovery, Logic and Rationality.* Boston Studies in the Philosophy of Science, vol. 56. Dordrecht: Reidel, 1980.

PAPPUS of Alexandria. [*Collectiones mathematicae.* French.] *La collection mathématique.* Translated, with introduction and notes, by P. Ver Eecke. 2 vols. Paris: De Brouwer/Blanchard, 1933.

Olaf PEDERSEN. *A Survey of the Almagest.* Odense, Norway: Odense Univ. Press, 1974.

Daniel PEDOE. *A Course of Geometry for Colleges and Universities.* Cambridge, England: Cambridge Univ. Press, 1970.

PLATO. *The Collected Dialogues of Plato including the Letters.* [Various translators], edited by E. Hamilton and H. Cairns. Bollingen Series, 71. Princeton, N.J.: Princeton Univ. Press, 1961. Corrected reprint 1963.

______. *Plato's Cosmology, The Timaeus of Plato.* Translated, with a running commentary, by Francis Macdonald Cornford. London: Kegan Paul, Trench, Trubner, 1937.

Jules Henri POINCARÉ [''Le raisonnement mathématique,'' 1908. English.] Translated by Halsted (as ''Mathematical Creation'') and by Maitland (as ''Mathematical Discovery''). See pp. 383–294 and pp. 46–63, respectively, in the following English editions.

______. *The Foundations of Science;* [consisting of] *Science and Hypothesis, The Value of Science, Science and Method.* Translated by G. B. Halsted. New York: Science Press, 1913; reprinted 1921. Reprinted again Lancaster, Pa.: Science Press, 1946.

______. *Science and Method.* Translated by Francis Maitland. New York: Scribners, 1914. Reprinted New York: Dover, n.d. (Maitland's translation of the lecture ''Mathematical Discovery'' is reproduced in Newman's *World of Mathematics,* vol. 4, pp. 2041–2050.)

George POLYA. *How to Solve It, A New Aspect of Mathematical Method,* 1st ed. Princeton, N.J.: Princeton Univ. Press, 1945. 2nd ed., Garden City, N.Y.: Doubleday Anchor, 1957.

______. *Mathematics and Plausible Reasoning,* vol. 1: *Induction and Analogy in Mathematics;* vol. 2: *Patterns of Plausible Inference.* Princeton, N.J.: Princeton Univ. Press, 1954.

______. *Mathematical Discovery: On Understanding, Learning, and Teaching Problem Solving.* New York: Wiley, vol. 1, 1962, and vol. 2, 1965.

Karl R. POPPER. *Conjectures and Refutations, The Growth of Scientific Knowledge.* London: Routledge and Kegan Paul, 1963.

______. *The Philosophy of Karl Popper.* Edited by Paul Arthur Schilpp. 2 vols. Library of Living Philosophers. La Salle, Ill.: Open Court, 1974.

John POTTAGE. ''The Mensuration of Quadrilaterals and the Generation of Pythagorean Triads: A Mathematical, Heuristical and Historical Study with Special Reference to Brahmagupta's Rules.'' *Archive for History of Exact Sciences 12* (1974):299–354.

PTOLEMY [Claudius Ptolemaeus, *Syntaxis.* English.] ''The Almagest.'' Translated by R. C. Taliaferro. In *Great Books of the Western World,* vol. 16.
(For a commentary, see Olaf PEDERSON.)

Hans RADEMACHER and Otto TOEPLITZ. *The Enjoyment of Mathematics, Selections from Mathematics for the Amateur.* Translated by H. Zuckerman. Princeton, N.J.: Princeton Univ. Press, 1957.

ROBERVAL. See under Evelyn WALKER.

Benjamin ROBINS. *A Discourse Concerning the Nature and Certainty of Sir Isaac Newton's Methods of Fluxions and of Prime and Ultimate Ratios.* 1st ed. 1735. Reprinted with miscellaneous other pieces in vol. 2 of *Mathematical Tracts of the Late Benjamin Robins,* edited by James Wilson, 2 vols, London, 1761.

J. F. SCOTT. *The Mathematical Work of John Wallis, D.D., F.R.S. (1616–1703).* London: Taylor and Francis, 1938.

David Eugene SMITH (ed.). *A Source Book in Mathematics.* New York: McGraw-Hill, 1929. Reprinted in 2 vol., New York: Dover, 1959.

David Eugene SMITH and Yoshio MIKAMI. *A History of Japanese Mathematics.* Chicago: Open Court, 1914.

SOCIOLOGY OF MATHEMATICS. Theme of special issue of *Social Studies of Science* part 1 of vol. 8 (February 1978).

ŚRĪDHARA. *The Pāṭīgaṇita of Śrīdharācārya with an ancient Sanskrit Commentary.* Edited with introduction, English translation, and notes, by K. S. Shukla. Lucknow: Dept. of Mathematics and Astronomy, Lucknow Univ., 1959.

STEVIN. *The Principal Works of Simon Stevin.* Edited by E. J. Dijksterhuis, D. J. Struik, and others. 5 vols in 6. Amsterdam: Swets and Zeitlinger, 1955–1966.

Dirk J. STRUIK (ed.). *A Source Book in Mathematics, 1200–1800.* Cambridge, Mass.: Harvard Univ. Press, 1969.

Árpád SZABÓ. *The Beginnings of Greek Mathematics.* Translated from the German edition of 1969 by A. M. Ungar. Synthese Historical Library, vol. 17. Dordrecht: Reidel, 1978.

Ivor THOMAS. *Selections Illustrating the History of Greek Mathematics.* Greek with English translation by I. Thomas. 2 vols. Loeb Classical Library. London: Heinemann, 1939–1941.

D'Arcy Wentworth THOMPSON. *On Growth and Form.* 1st ed. 1917. 2nd ed., 2 vols., Cambridge, England: Cambridge Univ. Press, 1942. Single volume, abridged, Cambridge Univ. Press, 1961.

TORRICELLI. *Opere di Evangelista Torricelli.* Edited by G. Loria and G. Vassura, Vols. 1–3, Faenza: Montanari, 1919. Vol. 4, Faenza: Lega, 1944.

Bartel L. van der WAERDEN. *Science Awakening I.* Translated from the Dutch edition of 1950 by A. Dresden, 1954. 2nd English ed., Groningen: Noordhoff, 1961.

______. *Science Awakening II, The Birth of Astronomy.* With contributions by Peter Huber. 1st ed. in German, 1965. Completely revised English version, Leyden: Noordhoff, and New York: Oxford Univ. Press, 1974.

Evelyn WALKER. *A Study of the Traité des Indivisibles of Gilles Persone de Roberval. . . .* New York: Teachers College, Columbia University, 1932.

John WALLIS. See J. F. SCOTT.

D. T. WHITESIDE. "Mathematical Thought in the Later 17th Century." *Archive for History of Exact Sciences 1* (1961):179–388.

Ludwig WITTGENSTEIN. *Bemerkungen über die Grundlagen der Mathematik/Remarks on the Foundations of Mathematics.* German text with translation by G. E. M. Anscombe, edited by G. H. von Wright, R. Rhees, and G. E. M. Anscombe. Oxford: Blackwell, 1956.

John WORRALL. "A Reply to David Bloor [with a rejoinder by Bloor]." *British Journal for the History of Science 12* (1979):71–81.

INDEX

INDEX

Unprefixed numbers refer to pages, n denotes a note, P a problem, S a solution.